Springer Series in Materials Science

Volume 270

The Springer Series in Materials Science covers the complete spectrum of materials physics, including fundamental principles, physical properties, materials theory and design. Recognizing the increasing importance of materials science in future device technologies, the book titles in this series reflect the state-of-the-art in understanding and controlling the structure and properties of all important classes of materials.

More information about this series at http://www.springer.com/series/856

Cor Claeys · Eddy Simoen

Metal Impurities in Silicon- and Germanium-Based Technologies

Origin, Characterization, Control, and Device Impact

Springer

Cor Claeys
KU Leuven
Leuven, Belgium

Eddy Simoen
imec
Leuven, Belgium

ISSN 0933-033X ISSN 2196-2812 (electronic)
Springer Series in Materials Science
ISBN 978-3-030-06747-2 ISBN 978-3-319-93925-4 (eBook)
https://doi.org/10.1007/978-3-319-93925-4

Printed on acid-free paper

This Springer imprint is published by the registered company Springer International Publishing AG part of Springer Nature
The registered company address is: Gewerbestrasse 11, 6330 Cham, Switzerland

Prof. Jan Vanhellemont, an outstanding material scientist who stimulated many young students and researchers and who tried to find joy in every facet of his professional and family life. He would surely have been enthusiastic to contribute to this book.

Preface

The control of impurities in semiconductor materials has been a critical issue since the invention of the transistor due to their severe impact on the electrical device performance. Impurities can be introduced during either the crystal growing process and/or the device fabrication of the material itself. Since the 1960's much attention has been given to transition metals and over the years the number of publications has increased exponentially. Whereas originally both germanium and silicon were studied, the breakthrough of the planar technology shifted the focus more to silicon. However, during the last two decades germanium has known a renewed interest for deep submicron technologies, and good performance data is reported for e.g. FinFETs, tunnelFETs and nanowires. Therefore, this book will discuss metallic impurities in both Si- and Ge-based technologies.

An excellent work on metal impurities in silicon-device fabrication is the book by K. Graff, with a first edition in 1994 and an updated version in 1999. The present book is based on that work and will refer to some of the given information, without repeating all the different references. An important consideration is that the number of metals used in device manufacturing (e.g. high-k dielectrics, via filling, metallization and interconnects, silicidation/germanidation, ferroelectric memories, 3D integration, etc.) has strongly increased in recent years. This was one of the main motivations for writing this book. In parallel, there is a continuous push to reduce thermal budgets in processing by either using lower temperatures (< 500°C for 3D integration) or shorter times (ns and ms laser annealing). This allows less options for gettering treatments. Finally, scaling the device dimensions to the nano-era makes them sensitive to a smaller number of defects, so that specifications and contamination control become tighter and tighter, challenging today's metal contamination metrology. Ultimately, the device itself becomes the most sensitive tool to detect single metal atoms. All these factors have led to new demands and insights, so that there is definitively a need for a critical assessment on the recent literature on metal behavior in group IV semiconductors.

The main intention is 1) to extend the information by including some additional and new insights reported during the last twenty years and 2) to include recent information on the origin, characterization and electrical impact of the metal impurities. Especially references that appears after 2000 are critically discussed. Over the years the sensitivity of the detection techniques has been increased and new non-destructive techniques that can be used in-line during wafer processing have been developed. The International Roadmap for Semiconductor Technologies (ITRS) has for advanced technology nodes put forward more stringent specifications for the allowable metals. Special attention is also given to the behavior of metals in photovoltaic devices. Improved and/or new models for metal gettering are addressed. In addition, a chapter on ab-initio calculations and density functional theory (DFT) has been included.

The book is designed for a wide audience, including students, PhD students, researchers, development engineers, process engineers and application engineers working at universities, research centers or in industry. Information is given on the underlying physics involved and the state of the art in the field is addressed. The extensive reference list after each chapter is making it an indispensable reference book for more in depth studies of the discussed topics. It is the first time that Si, SiGe and Ge technologies are simultaneously addressed.

Leuven, Belgium
August 2018

Prof. Dr. Cor Claeys
Prof. Dr. Eddy Simoen

Contents

Abbreviations

A	Acceptor impurity
AAS	Atomic Absorption Spectroscopy
AB	Anti-bonding
ac	Alternating Current
AES	Auger Electron Spectroscopy
AFM	Atomic Force Microscope
AG	As-Grown-Generation
AHT	As-Grown Hole Trap
ALD	Atomic Layer Deposition
AlG	Aluminium Gettering
APM	Ammonium Peroxide Mixture
APT	Atom Probe Tomography
BB	Band-to-Band
BC	Bond-Centered
BD	Breakdown
BDG	Boron Diffusion Gettering
BMD	Bulk Micro Defect
BEOL	Back-End of Line
BESOI	Bonded and Etched-back Silicon-On-Insulator
BOX	Buried Oxide
BPC	Back-side Photo-Current
BTBT	Band-to-Band-Tunneling
BTI	Bias Temperature Instability
BTS	Bias Temperature Stress
C33	Trichloroethane
CB	Conduction Band
CC	Configuration Coordinate
CCD	Charge-Coupled Device
CDI	Carrier Density Imaging
CDTA	1,2-Cyclohexane Diamino Tetra Acetic acid

CESL	Contact Etch Stop Layer
CET	Capacitance Equivalent Thickness
CF	Conductive Filament
CIS	CMOS Image Sensor
CL	Cathodoluminescence
CMOS	Complementary Metal–Oxide–Semiconductor
CMP	Chemical Mechanical Polishing
COP	Crystal-Originated Particle
CP	Charge Pumping
CPD	Contact Potential Difference
CRE	Chemical Rate Equations
CTE	Constant of Thermal Expansion
CVD	Chemical Vapor Deposition
Cz	Czochralski
C-AFM	Conductive Atomic Force Microscopy
D	Donor impurity
dc	Direct Current
DB	Dangling Bond
DCM	Defect-Centric Model
DCR	Dark Count Rate
DCS	Dark Current Spectroscopy
DFT	Density Functional Theory
DHF	Diluted Hydrogen Fluoride
DIW	De-ionized Water
DL	Detection Limit
DLTS	Deep-Level Transient Spectroscopy
DP	Dry Polish
DRAM	Dynamic Random-Access Memory
DRM	Depletion Region Modulation
DS	Directional solidification
DTPA	Diethylenetriaminopentaacetic acid
DTPMP	Diethylene Triamine Penta Methylene Phosphonic acid
DUV	Deep Ultra-Violet
DZ	Denuded (or Defect-lean) Zone
EBIC	Electron-Beam-Induced Current
ECD	Electrochemical Deposition
EDD	Energy Dispersive Detector
EDS	Energy Dispersive Spectroscopy
EDTA	Ethylene Diamine Tetra Acetic acid
EDX	Energy Dispersive X-ray
EFG	Edge-Defined Film-fed Ribbon Growth
ehp	Electron–Hole Pairs
ELYMAT	Electrolytic Metal Tracer
ELYMOS	Electrolytic Metal Tracer Metal–Oxide–Semiconductor
EOR	End-Of-Range

EOT	Equivalent Oxide Thickness
EPR	Electron Paramagnetic Resonance
ESR	Electron Spin Resonance
ET	Extremely Thin
EVB	Electron Valence Band
EXAFS	Extended X-ray Absorption Fine Structure
FA	Furnace Anneal
FD	Fully Depleted
FDSOI	Fully Depleted Silicon-on-Insulator
FEOL	Front-End-Of-Line
FeRAM	Ferroelectric Random-Access Memory
FESEM	Field Emission Secondary Electron Microscope
FET	Field Effect Transistor
FF	Fill Factor
FG	Floating Gate
FGA	Forming Gas Annealing
FIB	Focused Ion Beam
FM	Ferromagnetic
FN	Fowler-Nordheim
FOUP	Front Opening Unified Pod
FPC	Front-side Photo-Current
FT	Fourier-Transform
FUSI	Fully Silicided
FWHM	Full Width at Half Maximum
FZ	Float Zone
GAA	Gate-All-Around
GB	Grain Boundary
GD	Generated Defect
GGA	Generalized-Gradient Approximation
GIDL	Gate-Induced Drain Leakage
GOI	Gate Oxide Integrity
GOX	Gate Oxide
GR	Generation-Recombination
HBD	Hard Breakdown
HC	Hot Carrier
HCI	Hot Carrier Injection
hf	Hyperfine
HF	Hartree–Fock
HI	High Injection
HKMG	High-k Metal Gate
HPGe	High-Purity Germanium
HPM	Hydrogen Peroxide Mixture
HRTEM	High-Resolution Transmission Electron Microscopy
HTFGA	High-Temperature Forming Gas Anneal
I	Interstitial

IBSC	Intermediate Band Solar Cell
IC	Integrated Circuit
ICP-AES	Inductively-Coupled Plasma Atomic Emission Spectroscopy
ICP-MS	Inductively-Coupled Plasma Mass Spectroscopy
IDLS	Injection-dependent Lifetime Spectroscopy
IG	Internal Gettering
IL	Interfacial Layer
ILD	Interlayer Dielectric
IoT	Internet of Things
IQE	Internal Quantum Efficiency
IR	Infrared
ITR	Incremental Trap Response
ITRS	International Technology Roadmap for Semiconductors
JL	Junctionless
KOZ	Keep Out Zone
LBIC	Laser-Beam-Induced Current
LCAO	Linear Combinations of Atomic Orbitals
LDA	Local Density Approximation
LDD	Lowly Doped Drain
LER	Line Edge Roughness
LF	Low Frequency
LFN	Low-Frequency Noise
LID	Light-Induced Degradation
LLS	Local Light Scatterers
LMTO	Linear Muffin-Tin-Orbital
LOCOS	LOCal Oxidation of Silicon
LP	Liquid Phase
LRS	Low Resistivity State
M	Metal
mc	Multi-crystalline
MD	Molecular Dynamics
MDD	Medium-Doped Drain
MDZ	Magic Denuded Zone
MEMS	Microelectromechanical System
MG	Metallurgical Grade
MIM	Metal–Insulator–Metal
MOS	Metal–Oxide–Semiconductor
MOSFET	Metal–Oxide–Semiconductor Field Effect Transistor
MRE	Metal Removal Efficiency
MW	Microwave
MWIR	Medium Wavelength Infrared
MW-PC	Microwave Photoconductive
MW-PCD	Microwave Photoconductive Decay
NAA	Neutron Activation Analysis
NCz	Nitrogen-Doped Czochralski Silicon

NN	Nearest Neighbor
NVM	Non-volatile Memory
NW	Nanowire
NWFET	Nanowire FET
OBIC	Optical Beam-Induced Current
OISF	Oxidation-Induced Stacking Fault
OSF	Oxidation-Induced Stacking Fault
P	Precipitate
PAI	Pre-amorphization Implantation
PBG	Polysilicon Backside Gettering
PBTI	Positive Bias Temperature Instability
PC	PhotoConductivity
PCD	Photoconductive Decay
PCM	Phase Change Memory
PC1D	Personal Computer One Dimensional
PD	Point Defect
PDC	Precipitate-Dislocation Complex
PDG	Phosphorus Diffusion Gettering
PECVD	Plasma-Enhanced Chemical Vapor Deposition
PF	Poole–Frenkel
PID	Potential-Induced Degradation
PiP	Package in Package
PL	Photoluminescence
PM	Plasma Mimic
PMA	Post Metal-deposition Anneal
POGO	Pre-oxidation Gettering OSFs
PoP	Package on Package
PP	Percolation Path
ppb	Part Per Billion
ppm	part per million atoms
PSD	Power Spectral Density
PSG	PhosphoSilicate Glass
PSII	Pre-silicidation Ion Implant
PT	Post Treatment
PV	Photovoltaic
PVD	Physical Vapor Deposition
PZT	$Pb(Zr,Ti)O_3$
P/E	Program/Erase
QSS-PC	Quasi Steady-State Photoconductance
RBS	Rutherford Back-Scattering
RCA	Radio Corporation of America
ReRAM	Resistive Random-Access Memory
RIE	Reactive Ion Etching
RF	Radio Frequency
RS	Raman Spectroscopy

RT	Room Temperature
RTA	Rapid Thermal Anneal
RTN	Random Telegraph Noise
RTP	Rapid Thermal Processing
RTS	Random Telegraph Signal
RW	Random Walk
R/W	Read/Write
S	Substitutional
SAGB	Small-Angle Grain Boundary
SB	Schottky Barrier
SBT	$SrBi_2TaO_9$
sc	semiconductor
SC-1	Standard Clean 1
SC-2	Standard Clean 2
scr	Space Charge Region
S/D	Source/Drain
SEM	Scanning Electron Microscope
SF	Stacking Fault
SILC	Stress-Induced Leakage Current
SIMOX	Separation by Implanted Oxygen
SIMS	Secondary Ion Mass Spectrometry
SIRM	Scanning Infra-Red Microscopy
SMT	Stress Memorization Technique
SNWT	Silicon NanoWire Transistor
SoC	System-on-Chip
SoG	Solar-Grade
SOG	Spin-on-Glass
SOI	Silicon-on-Insulator
SON	Silicon-on-Nothing
SPER	Solid-Phase Epitaxial Regrowth
s-pits	Saucer pits
SPICE	Simulation Program with Integrated Circuit Emphasis
SPM	Sulphur Peroxide Mixture
SPV	Surface Photovoltage
SRAM	Static Random-Access Memory
SRH	Shockley–Read–Hall
SR-TXRF	Synchrotron-Radiation TXRF
SS	Stainless Steel
STI	Shallow Trench Isolation
STM	Scanning Tunneling Microscope
S/D	Source/Drain
T	Tetrahedral
TAT	Trap-Assisted Tunneling
TCAD	Technology Computer-Aided Design
TCE	Trichloroethylene

TDDB	Time-Dependent Dielectric Breakdown
TDLS	Temperature-Dependent Lifetime Spectroscopy
TDPAC	Time-Dependent Perturbed Angular Correlation
TEA	Tri Ethanol Amine
TEM	Transmission Electron Microscope
TID	Transient Ion Drift
TM	Transition Metal
TMAH	Tetra Methyl Ammonium Hydroxide
TOF-SIMS	Time-of-Flight SIMS
TSV	Through Silicon Vias
TTDB	Time-To Dielectric Breakdown
TTF	Time to Failure
TTHA	Triethylene tetramine hexaacetate
TXRF	Total Reflection X-ray Fluorescence
ULSI	Ultra-Large Scale Integration
UPG	Ultra-PoliGrind
UPS	Ultraviolet Photoelectron Spectroscopy
UPW	Ultra-Pure Water
UTBB	Ultra-Thin Body and Box
UV	Ultraviolet
V	Vacancy
VB	Valence Band
VPD	Vapor Phase Decomposition
VPD-AAS	Vapor Phase Decomposition Atomic Absorption Spectroscopy
VPD-DC	Vapor Phase Decomposition Droplet Collection
WF	Work Function
WID	Within Die
WKB	Wentzel, Kramers, and Brillouin
wt	weight
XANES	X-ray Absorption Near-Edge Structure
XAS	X-ray Absorption Spectroscopy
XBIC	X-ray Beam-Induced Current
XPS	X-ray Photoelectron Spectroscopy
XRF	X-ray Fluorescence
YAG	Yttrium Aluminium Garnet
2DEG	Two-dimensional Electron Gas
3D	Three Dimensional
μ-XRF	Micro X-ray Fluorescence

Symbols

A	Area (m^2)
a	Nearest neighbor distance (m)
a_0	Lattice parameter (m)
A_G	Gate area (m^2)
A_R	Richardson constant ($Am^{-2}K^{-2}$)
A_S	Area of the lateral depletion region (m^2)
b	Capture radius (m)
b_{Cu}	Jump distance of a Cu atom (m)
B	Strength of a magnetic field (G)
B_0	Bulk modulus (m^2/N)
b	Burgers vector (m)
Br	Magnetic field at resonance (G)
C^{act}	Activation energy constant in the diffusivity of atoms (eV/Nm)
$C(M^+)$	Metal donor concentration (m^{-3})
$C(M^+X^-)$	Concentration of the M^+X pair (m^{-3})
$C(X^-)$	Concentration of the acceptor dopant X^- (m^{-3})
C*	Actual concentration (m^{-3})
C′	Calibration factor in the determination of the iron concentration from lifetime measurements ($m^{-3}s$)
C_0	Concentration in the melt at the start of solidification (m^{-3})
C(t)	Capacitance at time t (F)
C_e	Contrast of the EBIC image
C_{eq}	Concentration at equilibrium (m^{-3})
C_f	Steady-state capacitance (F)
C_G	Gate capacitance (F/m^2)
C_I	Concentration of interstitials (m^{-3})
C_I^*	Equilibrium concentration of interstitials (m^{-3})
C_L	Concentration in the melt (m^{-3})
C_M	Metal concentration (m^{-3})
C_M^*	Metal concentration (m^{-3})

$C_{M\text{-}diss}$	Dissolved metal concentration (m^{-3})
C_{ox}	Oxide capacitance (F)
C_{τ}	Lifetime calibration factor ($m^{-3}s$)
c_n	Electron capture rate ($m^{-3}s^{-1}$)
c_p	Hole capture rate ($m^{-3}s^{-1}$)
C_S	Concentration in the solid (m^{-3})
C_V	Concentration of vacancies (m^{-3})
C_V^*	Equilibrium concentration of vacancies (m^{-3})
D	Diffusion coefficient (m^2s^{-1})
D	Layer thickness (m)
D_{Cu}	Diffusion coefficient of Cu (m^2s^{-1})
D_{Cu-ox}	Diffusion coefficient of Cu on SiO_2 (m^2s^{-1})
d_0	Jump distance during a diffusion step (m)
D_0	Pre-exponential diffusion factor (m^2s^{-1})
D_{As}	As implantation dose (m^{-2})
D_{BMD}	Bulk micro defect density (m^{-3})
D_{eff}	Effective diffusion coefficient (m^2s^{-1})
D_{int}	Intrinsic diffusion coefficient (m^2s^{-1})
D_{it}	Density of interface traps ($m^{-2}eV^{-1}$)
D_{liq}	Diffusion coefficient in the liquid (m^2s^{-1})
D_n	Minority carrier diffusion coefficient (for electrons) (m^2s^{-1})
D_M	Metal diffusivity (m^2s^{-1})
D_{min}	Minority carrier diffusion coefficient (m^2/s)
d_n	Dissolution rate of a precipitate (s^{-1})
d_{pr}	Size precipitate (m)
D_p	Minority carrier diffusion coefficient (for holes) (m^2s^{-1})
D_{resist}	Diffusivity in the resist (m^2s^{-1})
D_{sub}	Diffusivity in the substrate (m^2s^{-1})
D_T	Trap density in the depletion layer (m^{-2})
D_{void}	Void density (m^{-3})
D_{wafer}	Wafer thickness (m)
E	Energy per implanted ion (eV)
E_a	Activation energy (eV)
E_b	Binding energy of a defect pair (eV)
E_B	Energy barrier for carrier capture (eV)
E_c	Critical energy for amorphization (eV)
E_C	Conduction band minimum (eV)
E_C^L	Bottom of the conduction band in the L-valley
E_C^X	Bottom of the conduction band in the X-valley
E_{cap}	Activation energy for capturing (eV)
E_{Coul}	Coulomb energy (eV)
E_{Cor}	Correction to the activation energy (eV)
E_d	Diffusion barrier (eV)
E_{d0}	Calculated position of deep level in DFT (eV)

E_{diss}	Dissociation energy (eV)
E_{em}	Activation energy for emission (eV)
E_f	Formation energy (eV)
E_{Fn}	Quasi Fermi level for electrons (eV)
E_{Fp}	Quasi Fermi level for holes (eV)
E_F	Surface Fermi level (eV)
E_g	Band gap energy (eV)
$E_{g\text{-}Si}$	Band gap energy for silicon (eV)
$E_{g\text{-}Ge}$	Band gap energy for germanium (eV)
E_{GR}	Activation energy to create a hole at the interface in silicon (eV)
E_i	Intrinsic energy level (eV)
E_{el}^{H}	Elastic energy in the H site (eV)
E_{el}^{T}	Elastic energy in the T site (eV)
E_{Id}	Ionization energy of a defect (eV)
E_{Is}	Ionization energy of a standard defect (eV)
E_{i0}	Interaction energy between a dislocation and a metal atom (eV)
E_M	Deep level associated with a metal impurity (eV)
E_m	Migration energy (eV)
E_p	Upper energy of the band (eV)
E_p	Hole emission rate (s^{-1})
E_{relax}	Energy related to the displacement of lattice atoms (eV)
E_s	Donor level of a standard or marker defect in DFT (eV)
E_T	Energy level in the band gap (eV)
E_{tot}	Total energy of a supercell in DFT (eV)
E_V	Valence band maximum (eV)
E_Z	Zeeman energy (eV)
E_0	Electronic part of the energy due to a trapping event (eV)
F	Electric field (V/m)
f	Frequency (Hz)
f_{chem}	Chemical driving force for precipitation (eV)
f_d	Diffusion correlation factor (−)
f_E	Energy fraction for lattice disorder (–)
F_{ox}	Oxide electric field (V/m)
F_p	Total occupation of a precipitate band (–)
$F_p^{(N)}$	Fraction of occupied states when the precipitate is neutral (–)
F_s	Surface electric field (V/m)
f_s	Fraction solidified (–)
f_T	Trap occupancy by an electron (–)
F_{max}^{loc}	Maximum local breakdown field (V/m)
g	Trap degeneracy factor (–)
G	Conductance (Ω^{-1})
G_o	Optical generation rate ($m^{-3}s^{-1}$)
G_D	Drain conductance (–)
G_g	Temperature gradient at the melt–solid interface (K/m)

g_{free}	g_l value of a free electron (−)
G_{GB}	Conductance across a grain boundary (Ω^{-1})
g_l	Landé factor (–)
g_m	Transconductance (Ω^{-1})
g_n	Growth rate of a precipitate with n metal atoms (s^{-1})
G_{min}	Minority carrier generation rate ($s^{-1}cm^{-3}$)
G_{th}	Thermal generation rate of minority carriers ($m^{-3}s^{-1}$)
H	Activation enthalpy (J)
h	Planck's constant (Js)
H_m	Migration enthalpy (J)
H_S	Solution enthalpy (J)
I	Angular spin momentum (–)
I_B	Substrate current (A)
I_{Bpc}	Backside photocurrent (A)
I_{cp}	Contact potential current (A)
$I_{dir\text{-}tunn}$	Direct tunneling current (A)
I_{EDT}	Gate edge tunnel current (A)
I_d	Displacement current (A)
I_{dark}	Dark current (A)
I_{diff}	Diffusion current (A)
I_D	Drain current (A)
I_{defect}	Generation current per defect (A)
I_G	Gate current (A)
I_{gen}	Generation current (A)
I_{GIDL}	Gate-induced drain leakage current (A)
I_N	Average leakage current in a CIS pixel containing N metal atoms (A)
I_R	Reverse current (at reverse bias V_R) (A)
I_{sat}	Saturation current (A)
I_{SC}	Short-circuit current of a solar cell (A)
I_{SILC}	Stress-induced leakage current (A)
I_{TE}	Current top electrode (A)
J	Flux (m^{-2})
J_e	Electron flux (Nm/C)
J_G	Gate current density (A/m^2)
J_{SC}	Short-circuit current density of a solar cell (A/m^2)
J_{0e}	Emitter saturation current density (A/m^2)
k	Ratio of electron and hole capture cross section (–)
K	Central force constant per atom in Si (eV/m^2)
k_M	Segregation coefficient of a metal (−)
k_0	Distribution coefficient (–)
k*	Distribution coefficient in equilibrium (–)
k_B	Boltzmann constant (J/K)
k_{eff}	Effective distribution coefficient (–)
k_{Fe}	Iron segregation coefficient (–)
k_{int}	Segregation coefficient at the interface (–)

K_T	Pre-factor of DLTS Arrhenius plot (sK^2)
L	Gate length (m)
L_{diff}	Minority carrier diffusion length (m)
L_{eff}	Effective gate length (m)
L_n	Diffusion length for electrons (m)
L_p	Diffusion length for holes (m)
L_t	Trap length of a square-shaped trap area (m)
L_{to}	Trap length of a charge square-shaped trap area (m)
M_{AlSi}	Metal concentration in AlSi liquid (m^{-3})
M_{Si}	Metal concentration in silicon (m^{-3})
m_0	Free electron mass (kg)
m_e^*	Electron effective mass (kg)
m_h^*	Hole effective mass (kg)
m_l	Longitudinal effective mass (kg)
m_{ox}	Electron effective mass in SiO_2 (kg)
n	Carrier concentration (m^{-3})
n_0	Equilibrium electron density (m^{-3})
n_1	Equilibrium electron density when $E_F = E_T$ (m^{-3})
N_a	Dopant acceptor concentration (m^{-3})
N_{at}	Atomic density in a host material (m^{-3})
N_{av}	Average number of atoms per pixel (m^{-3})
N_C	Density of states in the conduction band (m^{-3})
N_{crit}	Critical trap density (m^{-3})
N_d	Total density of active core states (m^{-2})
N_{dop}	Doping density (m^{-3})
N_e	Number of d electrons of a TM (−)
N_i	Number of interstitial sites (−)
n_i	Intrinsic carrier concentrations (m^{-3})
N_{ip}	Total density of electrically interface-active states (m^{-3})
N_{ox}	Density of oxygen precipitates (m^{-3})
N_p	Density of states of a precipitate in the energy band ($m^{-3}eV^{-1}$)
N_{prec}	Density of precipitates (m^{-3})
N_{Si}	Number of silicon lattice sites (−)
n_s	Surface carrier concentration (m^{-2})
N_T	Density of trapping levels (m^{-3})
N_{Teff}	Effective density of traps (m^{-3})
n_t	Number of traps (–)
N_V	Density of states in the valence band (m^{-3})
p	Free hole concentration (m^{-3})
p_0	Equilibrium hole density (m^{-3})
p_1	Equilibrium hole density when $E_F = E_T$ (m^{-3})
P_C	Capture probability (–)
P_{etun}	Capture probability by tunneling (–)
P_g	Parameter Cz crystal pulling (m^2/Ks)
P_{gcrit}	Critical value of the Cz crystal pulling parameter (m^2/Ks)

P_{htun}	Hole tunneling probability (–)
P_N	Probability for a pixel containing N metal atoms (–)
q	Electron charge (C)
Q	Charge (C)
Q_B	Bulk charge density (C/m^2)
Q_{BD}	Charge to breakdown (C)
Q_{Cu}	Charge associated with Cu diffusion in oxide (C)
Q_G	Image charge (C)
Q_{inv}	Inversion charge density (C/m^2)
Q_n	Channel charge density (C/m^2)
Q_{tot}	Total oxide charge (C)
r	Radius (m)
R	Reflectivity (–)
r_{ass}	Association rate (s^{-1})
r_c	Critical radius (m)
R_c	Radius for V=K_BT (m)
R_C	Capture rate for electrons (s^{-1})
R_D	Channel resistance (C/m^2)
r_{diss}	Dissociation rate (s^{-1})
R_e	Emission rate of an electron (s^{-1})
r_{ext}	Covalent radius external inpurities (m)
R_{gen}	Carrier generation rate (s^{-1})
R_h	Capture rate of a hole (s^{-1})
r_H	Radius for hydrogen capture (m)
r_{Hse}	Radius of hexagonal interstitial site in cubic lattice (m)
r_{ox}	Radius of an oxygen precipitate (m)
r_{prec}	Precipitate radius (m)
R_p	Projected ion implantation range (m)
R_Q	Radial charge position (m)
R_S	Channel resistance (Ω)
r_{SD}	Radius where the lattice distortion of a dislocation acts (m)
r_{si}	Silicon radius (m)
r_T	Radius of the opening of the tetragonal interstitial site (m)
R_{thr}	Rate of thermal excitation (s^{-1})
r_0	Average radius of an as-grown precipitate (m)
s	Recombination velocity (m/s)
S	Solubility (m^{-3})
S	Electron Spin (–)
S_f	formation energy (J/K)
S_0	Pre-exponential solubility factor (–)
s_c	Back-surface generation velocity (m/s)
s_g	Surface generation velocity (m/s)
s_r	Surface recombination velocity (m/s)
SS	Subthreshold swing (mV/decade)
S_S	Solution entropy (J/K)

S_{tot}^{eq}	Total solubility at equilibrium (m^3)
t	Time (s)
t_a	Annealing time (s)
$t_{1/2}$	Half-life of a radioactive isotope (s)
T	Temperature (K)
T_a	Annealing temperature (K)
T_c	Critical growth temperature (K)
T_d	Diffusion temperature (K)
t_b	Barrier thickness (m)
T_{bake}	Bake temperature (K)
t_d	Diffusion time (s)
t_{dp}	Thickness of polysilicon depletion layer (m)
t_{dp}	Polysilicon depletion (m)
T_e	Electron temperature (K)
t_{film}	Film thickness (m)
t_h	Time in the RTS high state (s)
t_{HK}	High-κ dielectric layer thickness (m)
t_i	Incubation time for precipitation (s)
t_{IL}	Interfacial layer thickness (m)
t_{inv}	Inversion layer thickness (m)
t_l	Time in the RTS low state (s)
T_M	Melting temperature (K)
t_{ox}	Oxide thickness (m)
T_p	Peak temperature in DLTS (K)
t_p	Pulse duration (s)
t_{relax}	Relaxation time (s)
t_{sc}	Thickness of a solar cell (m)
t_w	Sampling period (s)
V	Volume (m^3)
V(r)	Potential at distance r (V)
V_{att}	Attractive Coulomb potential for pair formation (V)
V_B	Potential drop in the substrate (V)
V_{BS}	Substrate voltage (V)
V_{BD}	Breakdown voltage (V)
V_c	Voltage difference between the quasi Fermi levels (V)
V_{cp}	Contact potential difference (V)
V_{DS}	Drain voltage (V)
V_{DSFmax}	Drain voltage for maximum τ_c/τ_e ratio in forward operation (V)
V_{DSRmax}	Drain voltage for maximum τ_c/τ_e ratio in reverse operation (V)
V_{eff}	Effective volume of a metal atom in a precipitate (m^3)
V_{FB}	Flat-band voltage (V)
v_g	Crystal growth rate (m/s)
V_G	Gate voltage (V)
V_{GS}	Gate-source voltage (V)

VG_{stress}	Stresss voltage (V)
v_n	Carrier velocity (m/s)
V_{oc}	Open-circuit voltage of a solar cell (V)
V_P	Probe voltage (V)
V_R	Reverse bias (V)
V_{rep}	Repulsive potential in pair formation (V)
V_s	Surface photovoltage (V)
V_T	Threshold voltage (V)
V_{TE}	Voltage top electrode (V)
$V_{term\text{-}e}$	Thermionic emission potential (V)
v_{th}	Thermal velocity (m/s)
v_{thn}	Thermal velocity for electrons (m/s)
v_{thp}	Thermal velocity for holes (m/s)
w	Width depletion layer (m)
W	Gate width (m)
W_{eff}	Effective gate width (m)
W_{Fin}	Fin width (m)
W_t	Width of a trap (m)
X_j	Junction depth (m)
X_T	Trap depth from the interface (m)
x_T^{CL}	Trap depth from the interface (classical calculation) (m)
x_T^{QM}	Trap depth from the interface (quantum mechanical calculation) (m)
y_T	Lateral trap position in the channel (m)
Z	Number of possible defect pair orientations (–)
Z_a	Atomic number (–)
Z_{eff}	Effective trap position in the dielectric in the HRS (m)
z_H^{NN}	Number of next neighbors at the H site (–)
z_T^{NN}	Number of next neighbors at the H site (–)

Greek Symbols

α	Concentration ratio trapped over mobile metal ions (–)
α_o	Optical absorption coefficient (m^{-1})
α_{sc}	Scattering coefficient (Vs)
α_t	Tunneling parameter (m)
β	Absorbed vacancies per precipitated metal atom (–)
β_c	Lattice contraction coefficient (–)
γ	Emitted interstitials per precipitated metal atom (–)
γ_c	Tunnel rate for capturing (s^{-1})
γ_e	Tunnel rate for emission (s^{-1})
Γ_i	Internal equilibration rate in a band of defect states (s^{-1})
δ	Linear misfit (–)
δ_b	Boundary layer at the melt (m)
ΔC	Capacitance transient amplitude in DLTS (F)
ΔE	Energy difference (eV)
ΔE_B	Activation energy of the capture cross section (eV)
ΔE_{el}	Change in elastic energy of an interstitial in the H minus the T site (eV)
ΔE_p	Width of an energy band associated with a precipitate (eV)
ΔE_Z	Zeeman splitting (eV)
Δf_{chem}	Chemical driving force (eV)
ΔG_T	Gibbs free energy (kJ/mol)
ΔH_T	Formation enthalpy (kJ/mol)
Δn	Excess electron density (m^{-3})
Δp	Excess hole density (m^{-3})
ΔS	Migration entropy (J/K)
ΔT	Degree of undercooling (K)
ε	Constrained strain (–)
ε_0	Permittivity of vacuum (F/cm)
ε_{eff}	Effective permittivity (F/m)
ε_{HK}	Dielectric constant of high-κ dielectric (–)

ε_{IL}	Dielectric constant of interfacial layer (–)
ε_{ox}	Dielectric constant of SiO_2 (–)
ε_s	Dielectric constant of the semiconductor (–)
ε_{Si}	Dielectric constant of silicon (–)
Φ	Photon flux ($m^{-2}s^{-1}$)
ϕ_B	Angle between the magnetic field and the reference axis (–)
ϕ_c	Critical dose for amorphization (m^{-2})
η	Injection level (–)
η_{sc}	Solar cell efficiency (–)
κ	Dielectric constant (–)
λ	Tunneling parameter (m)
λ_o	Optical wavelength (m)
μ	Mobility ($m^2V^{-1}s^{-1}$)
μ_0	Low field mobility ($m^2V^{-1}s^{-1}$)
μ_B	Bohr magneton (G/s)
μ_M	Chemical potential of the metal (eV)
μ_{SC}	Chemical potential of the semiconductor (eV)
μ_s	Shear modulus (–)
ν	Attempt frequency (s^{-1})
ξ	Ratio of the volume difference between a precipitate and the host (–)
ρ_d	Surface density of dislocations (m^{-2})
σ	Interface energy silicide/host (eV)
σ_c	Effective capture cross section (m^2)
σ_n	Capture cross section for electrons (m^2)
σ_i	Intrinsic contribution to the variation of the bandwidth (−)
σ_N	Bandwidth of a peak with N metal atoms (−)
$\sigma_{n,po}$	Capture cross section (electrons/holes) extrapolated to infinite temperature (m^2)
σ_{min}	Minority carrier capture cross section (m^2)
σ_p	Capture cross section for holes (m^2)
σ_T	Trap defined contribution to the bandwidth (−)
σ_{var}	Standard deviation of a variable (A)
σ_0	Temperature-independent cross-section pre-factor (m^2)
τ_{ass}	Association time constant (s)
τ_{Auger}	Auger recombination lifetime (s)
τ_c	Capture time constant (s)
τ_d	Time constant for ion drift (s)
τ_{diss}	Dissociation time constant (s)
τ_e	Emission time constant (s)
τ_{emit}	Recombination lifetime in the emitter of a solar cell (s)
τ_{ep}	Emission time constant for holes (s)
τ_{eff}	Effective lifetime (s)
τ_g	Generation lifetime (s)
$\tau_{g,eff}$	Effective generation lifetime (s)
τ_{other}	Recombination lifetime due to background defects (s)

τ_n	Minority carrier lifetime (for electrons) (s)
τ_{n0}	Capture time constant for electrons (s)
τ_p	Minority carrier lifetime (for holes) (s)
τ_P	Time constant for precipitation (s)
τ_{n0}	Capture time constant for holes (s)
τ_{prec}	Carrier lifetime in the presence of a precipitate (s)
τ_r	Recombination lifetime (s)
τ_s	Surface lifetime (s)
τ_{sg}	Lateral surface generation lifetime (s)
τ_{SRH}	SRH lifetime (s)
τ_1	Average time constant (s)
φ_b	Schottky barrier height for electrons (V)
ϕ_e	Electron affinity difference between Si and SiO_2 (V)
Ψ_s	Band bending at the semiconductor surface (V)
$\delta\Psi_s^T$	Surface potential shift near a trap (V)
ϖ	Radial frequency (s^{-1})
$\theta(F)$	Electric field enhancement factor (–)
Ω	Capturing volume around an ion (m^3)
Ω_{vol}	Average atomic volume (m^3)
Ω_0	Volume of a sphere with radius R_C (m^3)
Ω_P	Molecular volume silicide phase (m^3/mole)
$\Omega_{M_x Si_y}$	Volume of a M_xSi_y precipitate (m^3)
Ω_{Si}	Volume of silicon (m^3)

Chapter 1
Introduction

Since the invention of the point-contact amplifier or transistor in 1947, the purity of the semiconductor material was a key concern. In those days, germanium was the first choice as semiconductor material due to the difficulty to grow high quality and defect-free silicon crystals. The discovery a decade later of the integrated circuit based on the planar technology, triggered the switch-over to silicon technology, which is up to today dominating the exploded and still growing semiconductor market. Even the Ge-based technologies and/or the heterogeneous co-integration of Ge and III–V devices are fabricated on a Si-platform.

The electrical impact on germanium [1, 2] and silicon [3] devices of crystal defects such as dislocations has since the beginning been studied and resulted in improved crystal growth techniques. However, the observation in 1960 by Goetzberger and Shockley [4], that also the presence of metals such as Cu, Fe, Mn, Au, Zn and Ni degraded the electrical performance of $n^{+}p$ Si diodes, triggered the research on the control and elimination of metals both during crystal growth and device processing. Over the years the number of publications in the field increased exponentially and a new research field, called defect engineering and gettering, was initiated.

Metal contamination on the wafer surface can during subsequent thermal processing, depending on the type of metal and the anneal conditions, evaporate from the surface, diffuse in the bulk of the semiconductor or become incorporated in e.g. the growing film on the surface (e.g. oxide, nitride, epitaxial layer etc.). Figure 1.1 illustrates the in-diffusion in SiO_2 of different metals on the silicon surface before a thermal anneal step at 950 °C for 30 min in O_2 and N_2, respectively. Some of these metals have a better affinity with oxide and form complexes like FeO, Fe_2O_3, etc. during the anneal in 100% O_2 [5]. In the case of Mo and Mn most of the metal diffuses in the native oxide during an anneal in nitrogen. If most of the metal concentration is in the oxide then it can be stated there is only limited diffusion in the semiconductor. From a practical device viewpoint it means that metals may besides impacting the excess leakage current of p-n junctions also degrade the gate oxide integrity.

C. Claeys and E. Simoen, *Metal Impurities in Silicon- and Germanium-Based Technologies*, Springer Series in Materials Science 270,
https://doi.org/10.1007/978-3-319-93925-4_1

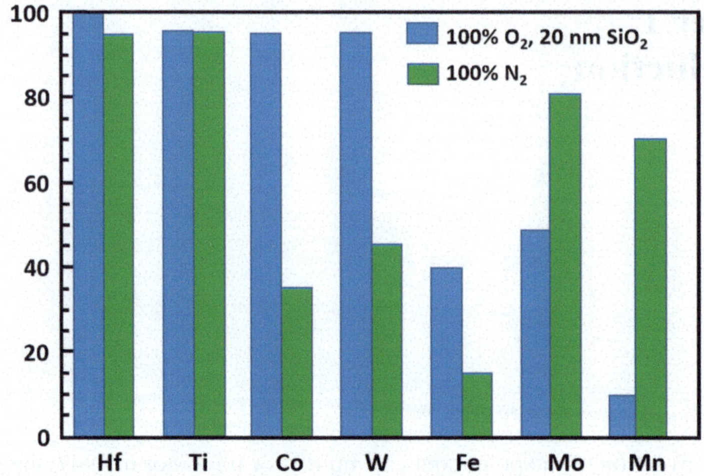

Fig. 1.1 Influence of the anneal ambient on the ratio of metal contaminants (%) into SiO_2 during a treatment in O_2 or N_2 at 950 °C for 30 min of metal contaminated Si wafers. Reproduced from Borde et al. [5] by permission of ECS-The Electrochemical Society

As it became soon clear that it is hard to eliminate all possible metallic contamination during device manufacturing, processing techniques have been developed in order to remove the metals to 'harmless' regions in the wafers where they don't impact the electrical performance of the devices in the active regions. For both Ge [6] and Si [4], gettering based on the use of a phosphorus (P_2O_5) or boron doped glass layer (B_2O_3) on the backside of the wafer turned out to be effective. Advanced process technologies impose stringent demands on the metal specifications so that a variety of different gettering techniques have been developed and evaluated.

When studying metallic contamination in semiconductor devices it is important to take the following issues into consideration: (1) which are the metals one is interested in and under which configuration will they be incorporated in the semiconductor lattice (interstitially, substitutionally, complexes, cluster, etc.), (2) what is the electrical activity of the metals (trapping levels in the band gap, capture cross sections), (3) what is the metal impact on the electrical device performance and how will the device/circuit yield be influenced, (4) how can the presence of metals be revealed and identified, (5) what is the origin of the metals, and (6) how can the contamination be avoided and/or removed or transformed into an electrically non-active state.

Although there is a very tight control of the metals during crystal growth of integrated circuit (IC) quality wafers, this is not the case for some of the feedstock used for metallurgical-grade solar cells, where cost considerations are the driving factor. For these applications, the allowable contamination level will be determined by the solar cell specification. Figure 1.2 shows typical concentrations of metals in metallurgical-grade (MG-Si), solar-grade (SoG-Si) and multicrystalline (mc-Si) Si solar

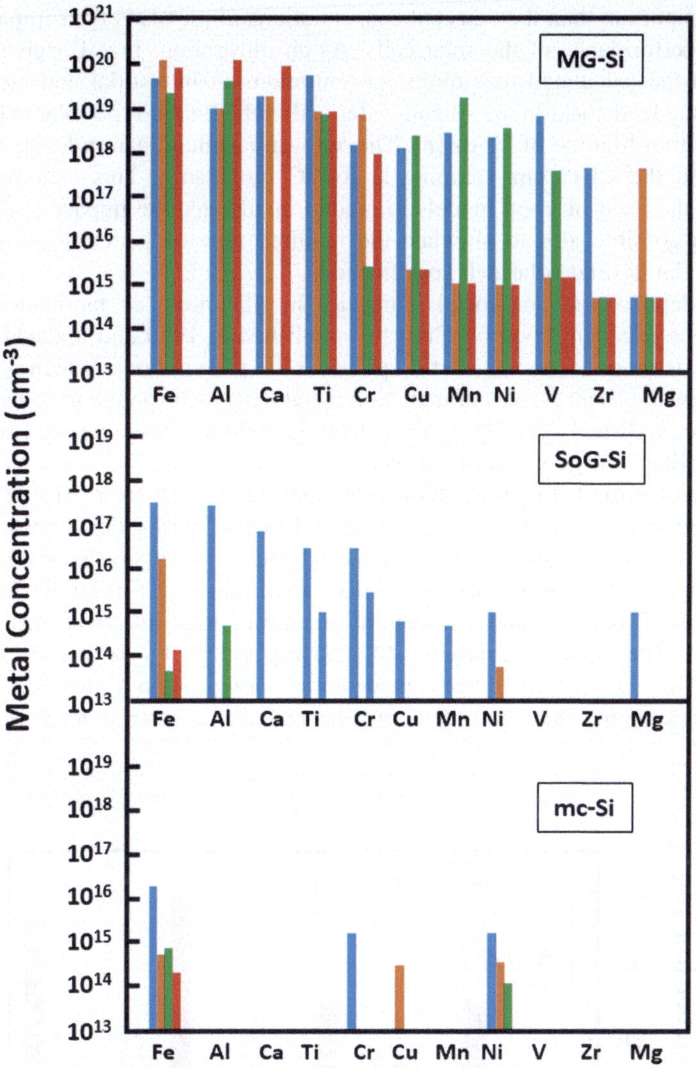

Fig. 1.2 Typical concentrations of metals in metallurgical-grade Si (MG-Si), solar-grade Si (SoG-Si) and multicrystalline (mc-Si) Si solar cells as reported in different publications. Reproduced from Istratov et al. [7] with permission of Elsevier

cells as reported in different publications and summarized by Istratov et al. [7]. It is important to remark that in this figure no bar doesn't mean that the metal is not present in concentrations above the detection limit as for some of the experiments the specific metal was not investigated. The absolute values strongly depend on the fabrication techniques used, but the chart is clearly indicating the zoo of metals that can be present in these materials.

More important than the detected concentrations of metals is their impact on the electrical performance of the solar cells. As an illustration, Fig. 1.3 gives a comparison of the calculated maximum concentrations of interstitial and precipitated metals in the feedstock, in the silicon wafer and in the final p-type solar cell, aiming for an electron lifetime of 20 μs [8]. The allowable contamination levels are much higher than the $<10^{10}$ cm^{-3} aiming at for IC processing. This example clearly illustrates the need of good models to predict in advance the impact of the metals and when possible also to simulate the required type of gettering approach for achieving the desired solar cell specifications.

An in-depth review on metal impurities in silicon-device fabrication can be found in the excellent book by Graff [9], published in 1990 and updated in 2000. That work is more or less the starting point for the present book, which is intended to (i) review the basic aspects of transition metals, (ii) focus on the new insights and knowledge published after 2000, (iii) extend the field by including information om metals in SiGe and germanium materials.

Whereas for the 0.13 μm CMOS technology the switch from Al to Cu wiring was a huge step, nowadays the use of several metals during device processing is becoming common practice. Some of the advanced processing schemes such as silicidation/germanidation processes, wafer metallization, 3D wafer stacking based on through-silicon-vias, and ferroelectric memories may introduce metallic contamination. The book is reviewing different aspects of 3*d*, 4*d* and 5*d* transition metals (TMs), whereby reference is made to the configuration of the outer electrons of the neutral atom. The metals that are investigated are shown in the periodic table in Fig. 1.4.

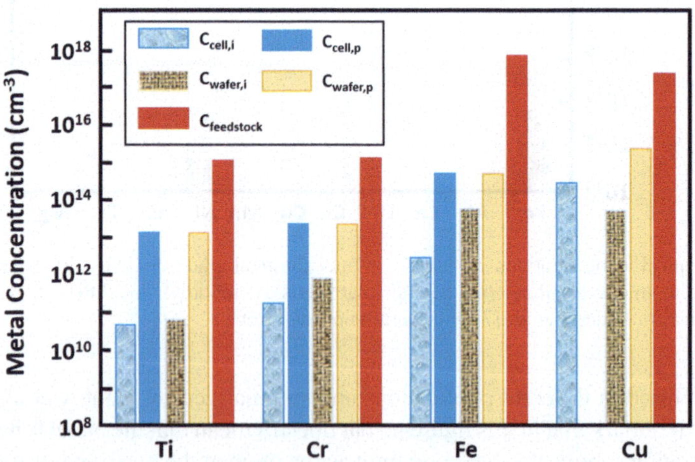

Fig. 1.3 Acceptable metal concentration of interstitial and precipitated Ti, Cr, Fe and Cu, respectively, in a p-type solar cell ($C_{cell,i/p}$) with an electron lifetime of 20 μs, in the silicon wafer ($C_{wafer,i/p}$) and in the silicon feedstock ($C_{feedstock}$). Reproduced from Hofstetter et al. [8] with permission of Elsevier

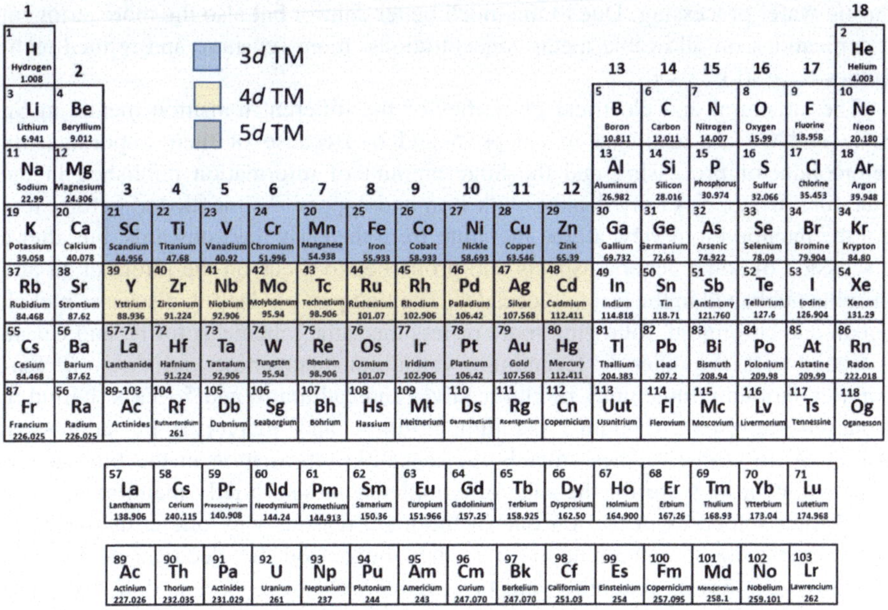

Fig. 1.4 Periodic table of the chemical elements indicating the 3*d*, 4*d* and 5*d* transition metals. For each element the symbol, the atomic number and the atomic weight is given

The book is divided in 9 Chapters. Chapter 2 first gives a brief description of the physical background of some general properties of transition metals such as their solubility, diffusivity, segregation and precipitation. The degrading performance by TMs is related to the associated defect states in the band gap, the donor or acceptor nature of the traps, their activation energy and their capture cross sections. This chapter forms the basis for a good understanding of the other chapters in the book.

Chapter 3 is devoted to the origin of the metallic contamination during crystal growth and device processing. Although there is a very tight control of the metals during crystal growth of integrated circuit (IC) quality wafers, this is not the case for some of the feedstock used for metallurgical grade solar cells, where cost considerations are the driving factor. As will be outlined, each step of the device processing from wafer cleaning to metal wiring, can introduce TMs and as demonstrated for advanced processing special attention has to be given to ion implantation steps. Therefore, statistical control-charts have to be generated on a regular basis. Some of the advanced processing schemes are using more exotic metals which can lead to metallic contamination.

The used characterization techniques for detecting, monitoring and identification of metallic contamination are reviewed in Chap. 4. Different chemical analysis techniques, structural characterization techniques to reveal metal-related defects and electrical measurement techniques are discussed, indicating their potential and their limitations. Very important are non-destructive techniques that can be used in line during wafer processing. Due to the much better control but also the more stringent specifications on allowable metal concentrations, more sensitive and refined techniques have to be used.

The structural and electrical properties of the different transition metals in Si, SiGe and Ge are analyzed in Chaps. 5 and 6. Because of their importance in semiconductor processing and the huge amount of information published in the literature, Chap. 5 is only dealing with iron and copper. As mentioned before, it is not the intention to review all the literature from the early days onwards, but after a discussion of some general aspects and properties to focus on the information and insights published in the last two decades. The first part of Chap. 6 discusses nickel, cobalt and chromium, following more or less the same scheme as for Fe and Cu. In a second part, the other 3*d* (scandium, vanadium, manganese and zinc), *4d* (zirconium, niobium, ruthenium, rhodium, silver and cadmium) and 5*d* TMs (hafnium, tantalum, tungsten, rhenium, osmium, iridium and mercury) are addressed. Although, for some of these metals the available information in the literature is rather limited, they are for the sake of completeness also briefly mentioned.

Chapter 7 focuses in fact on one of the main reasons why one is interested in controlling metallic contamination, i.e., their impact on the electrical device performance, the device reliability and the overall yield of devices and circuits. The impact on MOS capacitors, p-n junctions and Schottky barriers is reviewed, before outlining the impact on transistors and circuits, including Charge-Coupled Devices (CCDs) and CMOS Image Sensors (CIS). In addition, it will be shown that some devices are sensitive even to single metal atoms, enabling the development of novel detection methods with ultimate sensitivity.

As already mentioned it is not always possible to avoid all metallic contamination so that it is necessary to implement gettering techniques. After outlining some general aspects of gettering strategies, different gettering approaches are discussed in Chap. 8. The aim is to give a type of synopsis and to focus on the understanding and the new insights that have been obtained since the beginning of the century. For advanced scaled down process schemes one has to take into account the strongly reduced thermal budget so that proximity gettering near the active region is gaining in importance. A dedicated approach is required for gettering in photovoltaic processes as in that case the gettering step is often an inherent part of the cell processing.

Much insight into the structure and the electrical activity of metal defects and complexes is obtained by numerical calculations. This is the topic of Chap. 9, outlining first the evolution of ab initio calculations to determine from an energy viewpoint the most stable defect configuration (interstitial or substitutional) for isolated metals, metal-impurity (dopant atoms, hydrogen) complexes and metal clusters. The Density Functional Theory (DFT) calculations are nowadays commonly used by researchers studying theoretical defect properties which can be used as a basis to explain the experimental observations. In addition, also physics-based analytical models will be reviewed, providing a deeper insight in the trends of important transition metal parameters, like, the diffusivity or pairing with dopant atoms.

Each of the chapters is ending with a reference list. It was surely not the aim to given a complete list of all the published papers, but to limit the list to the most relevant papers supporting the different statements and to include important review papers which can be used as a reference for further reading.

References

1. A.D. Kurtz, S.A. Kulin, B.L. Averbach, Effect of dislocations on the minority carrier lifetime in semiconductors. Phys. Rev. **101**, 1285–1291 (1956). https://doi.org/10.1103/PhysRev.101.1285
2. J. Okada, Recombination of excess carriers in semiconductors. J. Phys. Soc. Jpn. **12**, 1338–1344 (1957). https://doi.org/10.1143/JPSJ.12.1338
3. A.G. Chynoweth, G.L. Pearson, Effect of dislocations on breakdown in silicon p-n junctions. J. Appl. Phys. **29**, 1103–1110 (1958). https://doi.org/10.1063/1.1723368
4. A. Goetzberger, W. Shockley, Metal precipitates in silicon p-n junctions. J. Appl. Phys. **31**, 1821–1824 (1960). https://doi.org/10.1063/1.1735455
5. Y. Borde, A. Maurel, A. Danel, A. Roche, M. Veillerot, Impact of metallic contamination on Si: short loop issues. ECS Trans. **11**(2), 151–158 (2007).https://doi.org/10.1149/1.2779374
6. M. Kikuchi, S. Iizima, A simpler method for removing copper from germanium. J. Phys. Soc. Jpn. **12**, 982 (1957). https://doi.org/10.1143/JPSJ.12.982
7. A.A. Istratov, T. Buonassisi, M.D. Pickett, M. Heuer, E.R. Weber, Control of metal impurities in “dirty” multicrystalline silicon for solar cells. Mater. Sci. Eng. B **134**, 282–286 (2006). https://doi.org/10.1016/j.mseb.2006.06.023
8. J. Hofstetter, J.F. Lelièvre, C. del Cañizo, A. Luque, Acceptable contamination levels in solar grade silicon: from feedstock to solar cell. Mater. Sci. Eng. B **159–160**, 299–304 (2009). https://doi.org/10.1016/j.mseb.2008.05.021
9. D. Graff, *Metal Impurities in Silicon-Device Fabrication* (Springer, Heidelberg, Germany). https://doi.org/10.1007/978-3-642-57121-3

Chapter 2
Basic Properties of Transition Metals in Semiconductors

Before addressing the source and the characterization/detection of transition metals in silicon- and germanium-based technologies, this chapter discusses some common properties of metals in semiconductors and is aimed to provide a general physical background. This should facilitate the discussions on the electrical properties of specific transition metals covered in Chaps. 5 and 6.

Attention is given to parameters like maximum solid solubility, diffusivity, segregation and precipitation. As will be outlined, there exists an interaction between these parameters which is influenced by external factors such as the experimental conditions (temperature, cooling rate …), the doping of the material, the presence of lattice defects, the growth conditions of the crystal etc.

Several electrical properties will be directly influenced by the presence of metallic contamination. The dissolved impurities lead to defect states in the band gap with donor or acceptor activity, thereby influencing the free carrier concentration and the Fermi-level. The determination of the activation energy of the centers facilitates their identification. The degrading impact of impurity centers strongly depends on their capture cross sections. It is e.g. well known that Fe compared to Cu is very harmful in p-type Si, while the opposite is the case for n-type Si. In many cases the standard Shockley-Read-Hall (SRH) theory can be applied to extract information on minority carrier lifetime and generation-recombination processes.

Since the early days it was observed that metallic impurities influence diode leakage and gate oxide breakdown. An in-depth analysis of the different leakage current components can generate useful information to determine the origin and the identification of the metallic contamination. This topic is addressed in detail in Chap. 7.

C. Claeys and E. Simoen, *Metal Impurities in Silicon- and Germanium-Based Technologies*, Springer Series in Materials Science 270,
https://doi.org/10.1007/978-3-319-93925-4_2

2.1 Solid Solubility

The solid solubility of an impurity is defined as the maximum impurity concentration that at thermal equilibrium and a given temperature can be dissolved in the host lattice. This implies an equilibrium between the impurity in- and out-diffusion from the surface, where the infinite source is located. The solubility S depends exponentially on the temperature and is given by an Arrhenius law

$$S = S_0 N_{at} \exp\left(S_S - \frac{H_S}{k_B T}\right) \tag{2.1}$$

with S_0 a pre-exponential factor, N_{at} the atomic density of the host material (5×10^{22} cm^{-3} for Si and 4.42×10^{22} cm^{-3} for Ge at 300 K), k_B the Boltzmann constant, T the absolute temperature and S_S and H_S the solution entropy and enthalpy, respectively. The expression is only valid in a temperature range below the eutectic temperature and well below the melting temperature of the host material. For metals dissolving in substitutional lattice positions (e.g. Au and Pt) S_0 depends on the host material and the vacancy formation energy. Thermal equilibrium is influenced by the in- and out-diffusion of the metal at the surface. Quenching contaminated samples from a high temperature causes supersaturation above the solubility.

Solubility data for some 3*d* transition metals (TMs) in intrinsic silicon as function of temperature is given in Table 2.1 [1]. The solubility S(1100 °C) values are calculated using (2.1). For Ni and Cu the given numbers are obtained by extrapolating the low temperature data and may therefore be overestimated. With the exception of Zn, the solution enthalpy H_s decreases with atomic number. It should be remarked that for the 3*d* TMs in Table 2.1, Zn is the only metal where at thermal equilibrium the substitutional defect concentration is larger than the interstitial one [2]. Good information on the ratio of interstitial to substitutional solubility is lacking for most of the metals. For some of the metals both interstitial and substitutional configurations have been observed. This will further be elaborated in Chaps. 5 and 6.

Table 2.1 Solubility data for 3*d* transition metals in intrinsic silicon. S is calculated using (2.1)

Metal	S_S(J/K)	H_S (eV)	T-region (°C)	S (cm^{-3}), 1000 °C	References
Ti	4.22	3.05	950–1200	2.1×10^{13}	[3]
Cr	4.7	2.79	900–1300	3.1×10^{14}	[4]
Mn	7.11	2.80	900–1200	3.2×10^{15}	[4, 5]
Fe	8.2	2.94	900–1200	2.9×10^{15}	[4]
Co	7.6	2.83	700–1200	4.0×10^{15}	[4]
Ni	3.2	1.68	500–950	5.0×10^{17}	[4]
Cu	2.4	1.49	500–800	8.0×10^{17}	[4]
Zn	7.26	2.49	840–1200	5.2×10^{16}	[6]

The temperature dependence of the solubility, based on (2.1) and the data given in Table 2.1, is graphically represented in Fig. 2.1. The solubility strongly decreases for decreasing temperatures. This is important for what is happening during the cooling cycle at the end of a thermal treatment, i.e., there can exist an impurity super-saturation in the lattice which afterwards can lead to precipitation into a metal silicide as will be discussed in Sect. 2.3.

The solubility is not only influenced by the interstitial/substitutional defect formation but also depends on the doping concentration and type (P, B, In, Ga, As …) of the crystal. The solubility increases for higher doping concentrations due to the metal segregation forming doping-metal pairs (see Sect. 2.2.1). For Mn, Fe and Co a solubility enhancement has been observed for P or B concentrations above 5×10^{18} cm^{-3} in a temperature range between 700 and 850 °C and <900 °C, respectively [7]. Also the oxygen concentration in Czochralski grown Si may have an impact [7].

The solubility of metals in Ge has been studied for more than half a century and was first reviewed in 1960 by Trumbore [8]. The typical retrograde behavior, as illustrated in Fig. 2.2, is observed with an increasing solubility when lowering the temperature from the melting point, until a maximum is reached at an intermediate

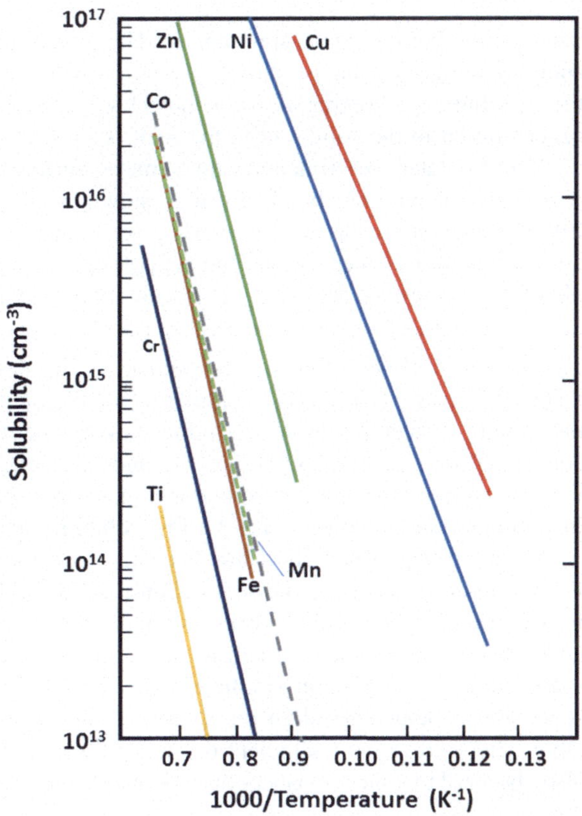

Fig. 2.1 Solubility of 3*d* transition metals in silicon as a function of temperature, calculated using (2.1) and the data given in Table 2.1. Reproduced from Graff [1] by permission of Springer

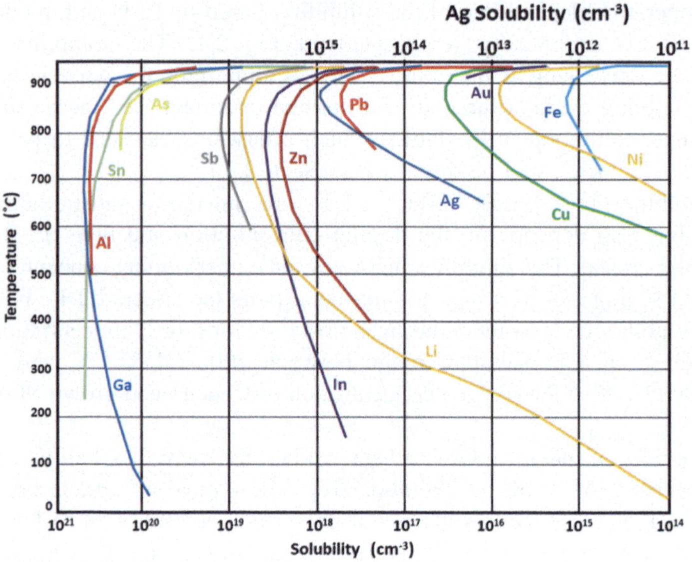

Fig. 2.2 Solubility in germanium as a function of the temperature. After Trumbore [8]

temperature before decreasing again. The given data have been obtained by studying impurity behavior during crystal growth and can be directly correlated to the distribution or segregation coefficient k_0 defined as the ratio of the impurity concentration in the solid and in the melt ($k_0 = C_S/C_L$).

Also for later experiments with a metal surface layer as impurity source, a retrograde behavior can be observed as shown in Fig. 2.3 for Au, Ag and Cu in Ge [9]. The analysis is based on spreading resistance measurements whereby for the conversion into concentrations the deep levels associated with the substitutional dissolved impurities are used. The solid lines are based on (2.1) with S_0 and $S_S - \frac{H_S}{k_B T}$ equal to 7.8 and −1.56 eV/k_BT for Cu_s, 2.8 and =1.85 eV/k_BT for Ag_s and 12 and −1.46 eV/k_BT for Au_s, respectively. The dashed line is for Cu_i with as values 0.19 and −1.46 eV/k_BT. Similar as for silicon the solubility also depends on the doping concentration which influences the Fermi level and the free carrier concentration, i.e., holes in p-type Ge and electrons n-type Ge compared to the intrinsic carrier concentration in undoped material [10]. Most of the 3*d* TMs in Ge, incorporated on lattice sites, have a low solubility and a high diffusivity.

An interesting observation has been reported for the case when the impurity incorporation is occurring during the liquid-solid solidification process such as e.g. in the case of crystal pulling from the melt. The thermal equilibrium between the concentration in the melt C_L and the concentration in the solid C_S is given by the segregation or distribution coefficient $k_0 = C_S/C_L$ as illustrated in Fig. 2.4a. The segregation coefficients align on a curve according to their group in the periodic table. The segregation coefficient decreases for increasing atom radius [13]. It was also observed that there exists both in Ge and Si a correlation between the maximum

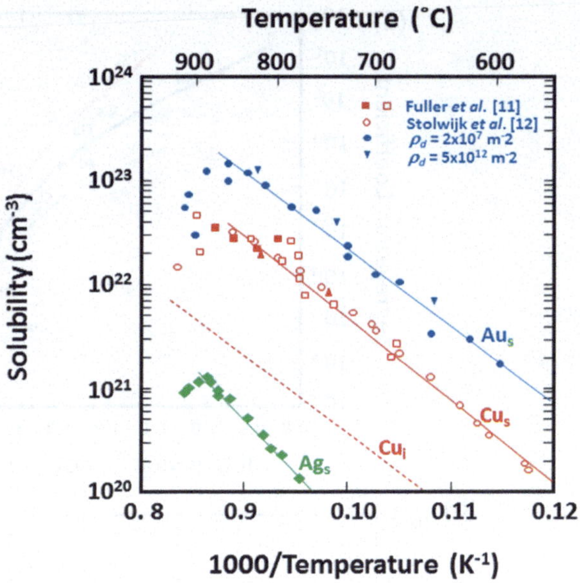

Fig. 2.3 Solubility of Au, Ag and Cu in germanium as a function of the temperature for 2 different dislocation densities ρ_d(m^{-2}). Data from Fuller et al. [11] and Stolwijk et al. [12] is also shown. Reprinted from Bracht et al. [9]. Copyright (1991) by The American Physical Society

solid solubility and the segregation coefficient, as shown in Fig. 2.4b [14]. The solid line for Ge corresponds with $S = 4.4 \times 10^{21}\ k_0$. Compared to Ge, Fe in Si is behaving somewhat anomalous.

Although during Ge processing the maximum temperature is kept below 600 °C, the solubility of metals is higher than in Si at the same temperature. Especially in highly doped n^+ and/or p^+ regions such as source and drain of a Metal-Oxide-Semiconductor (MOS) transistor this may trigger the gettering of metals as discussed in Chap. 8.

2.2 Diffusivity

The diffusivity of a metal depends on the lattice position it occupies, i.e., interstitial or substitutional, and on the diffusion mechanism (interstitial, vacancy, kick-out, Frank-Turnbull or dissociative). A detailed discussion of the diffusion mechanisms can be found in reference books such as e.g. [15]. Interstitial diffusion is faster than diffusion processes involving vacancies and/or self-interstitials. For impurities in a diamond lattice like Si and Ge, most 3*d* TMs will occupy predominantly tetrahedral interstitial sites, i.e., the center of the tetrahedron formed by four lattice atoms (*T*-sites). The 4*d*- and 5*d* TMs are predominantly dissolved on substitutional sites. Ab initio calculations (Density Functional Theory—DFT) of the formation energy are used to determine the most stable configuration. This topic is addressed in Chap. 9. The diffusivity is higher for impurities on interstitial sites compared to those dissolved at substitutional sites at the same temperature.

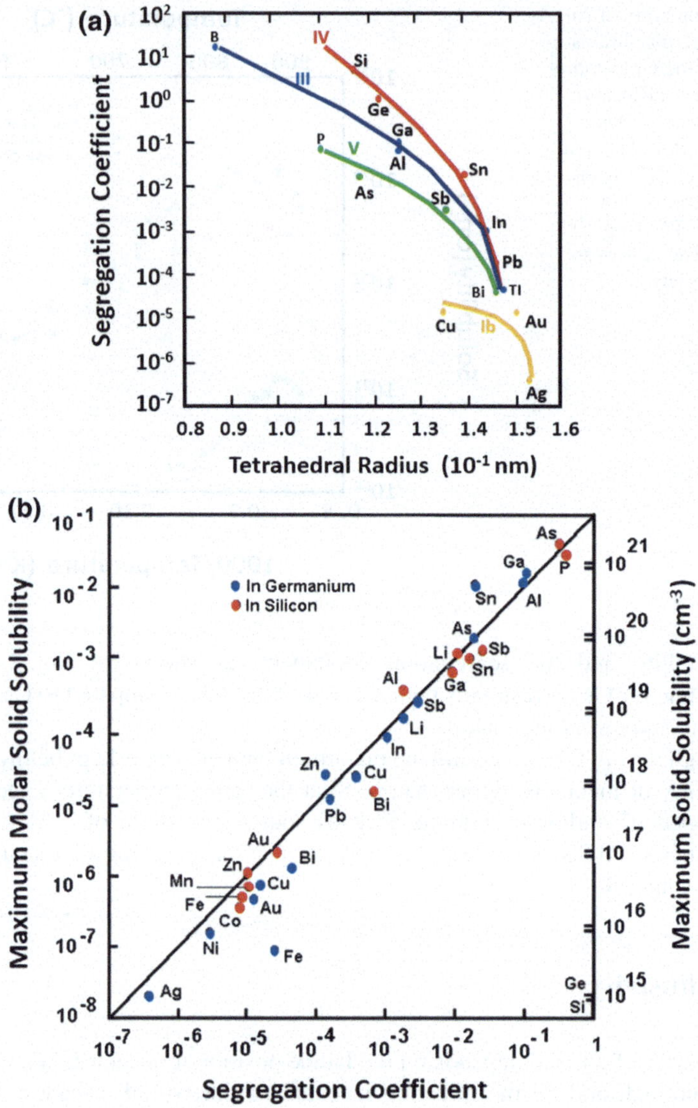

Fig. 2.4 **a** Segregation coefficients of impurities at the melting point of germanium as a function of the tetrahedral radii. After Trumbore [8]. **b** maximum molar solid solubility and maximum solid solubility versus segregation coefficient at the melting point for impurities in Ge and Si. Reproduced from Fischler [14]. Copyright 1991 AIP Publishing

A summary of the diffusivity of impurities in Si is given in Fig. 2.5a, showing very high diffusivity for 3d transition metals like Cu, Fe, Co and Ni [16]. Beside the absolute value of the diffusivity its behavior with temperature is very important. The diffusivity of metals and some common dopants in Ge is given in Fig. 2.5b [17].

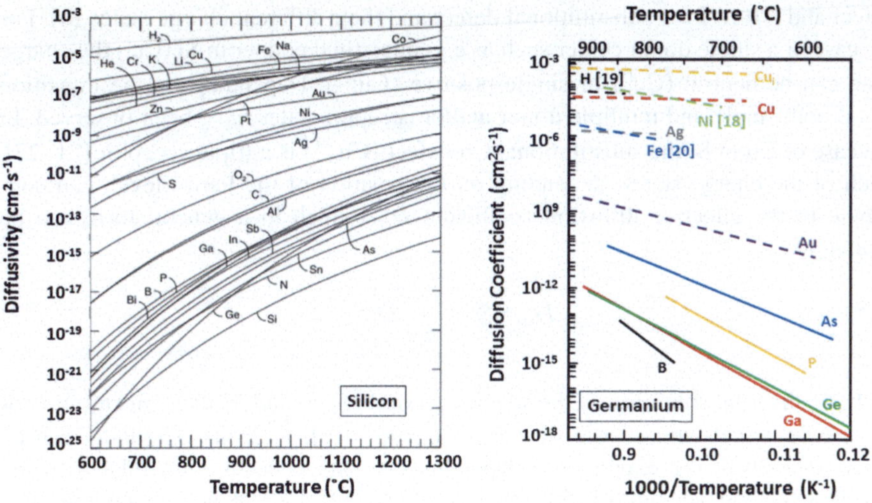

Fig. 2.5 Diffusivity in silicon (after Jones [16]) and germanium (Reproduced from Bracht [17] by permission of Elsevier) as a function of the temperature. The diffusion data in Ge for Ni, H and Fe are from Van der Maesen and Brenkman [18], Van Wieringen and Warmholtz [19] and Boltaks [20], respectively

Also in Ge the metals have a much higher diffusivity compared to self-diffusion and the dopant elements. Comparison of Fig. 2.5a, b clearly indicates that the metals have a higher diffusivity in Ge than in Si at the same temperature.

The diffusivity can be expressed by an empirical relation

$$D = D_0 \exp\left(-\frac{H}{k_B T}\right) \tag{2.2}$$

with D_0 a pre-exponential factor and H the activation enthalpy. Since most of the $3d$ TMs in Si have an interstitial position in the lattice, the activation energy will be determined by the migration enthalpy H_m which decreases for increasing atomic number (going from Cr to Fe), resulting in a higher diffusivity. Cr_i, Mn_i and Fe_i can easily be kept as isolated species after moderate quenching. In contrast to the solubility which is extremely low at room temperature, the same metals remain rather mobile enabling ion pairing reactions with doping atoms. The low solubility implies that during the cooling cycle the supersatured impurities will out-diffuse to the wafer surface, especially for low cooling rates. For high cooling rates the metal will precipitate in the bulk, forming a new phase such as metal-silicide. The slow diffusing metals can upon cooling remain dissolved in the lattice, yielding metastable defects at room temperature. These defects are at the basis of the observed deep levels.

One has to take into account that the effective diffusion coefficient is composed of a summation over all the impurity species (x) whereby $x = \mathrm{I}$ for an interstitial defect and x = V for a substitutional defect and their different charge states (n). For the case of a single donor center such as e.g. interstitial copper in Si (Cu_i) the charge state can be neutral (Cu_i^0) or single positive (Cu_i^+). Depending on the transition metal both single and multiple donor and/or acceptor states have been observed. In the case of Cu in Si the substitutional Cu defect (Cu_s^{-3}) is a triple acceptor [21, 22]. Each of the charge states, depending on the position of the Fermi level, can contribute to the effective diffusion coefficient D_{eff} which in a general form can be written as

$$D_{eff} S_{tot}^{eq} = \sum_{x,n} D_x^n S_x^n \tag{2.3}$$

with S_{tot}^{eq} the total solubility of the species at equilibrium and S_x^n the concentration of the x type defect with charge state n (−m, −2, −1, 0, 1, 2, m). The latter can be determined from the exponential dependence of the concentration related to the Fermi level and accounting for the ionization energy. The neutral charge state concentration is independent of the Fermi level.

The mobile defects (donor- or acceptor-type) can interact with doping related defects of the other type by Coulomb interaction, a process called ion pairing and which is explained next. Obviously that this has a strong impact on the diffusivities calculated from the experimental data.

2.2.1 Ion Pairing and Doping Effects

It was already reported in the fifties of the previous century that mobile metal donors or acceptors can be trapped by immobile substitutional doping atoms such as B, P, As etc. [23] or by the charges around a dislocation promoting precipitation [24]. This implies that compared to intrinsic Si or Ge the TM diffusivity will be reduced in doped and/or deformed crystals.

Dopants like B, Ga, Al and In are forming shallow acceptor states on substitutional sites. TMs like e.g. Cu, Fe, Cr, Mn are present in Si or Ge as mobile interstitial donors. The donor (acceptor) defect is either neutral or has a positive (negative) charge. This will lead to one of the following pairing reactions between the metal M and the dopant X

$$[M^+] + [X^-] \leftrightarrow [M^+X^-] \quad \text{or} \quad [M^-] + [X^+] \leftrightarrow [M^-X^+] \tag{2.4}$$

This equilibrium reaction will take place until metal precipitates are formed and/or pair dissociation occurs. At room temperature a donor defect like Cu_i^+ will be attracted by an immobile negatively charged substitutional boron ion (B_s^-) in p-type material if the potential energy of the donor-acceptor interaction $V(r)$ between the ions (separated at a distance r from each other) exceeds k_BT. The original acceptor-donor pairing treatment by Reiss et al. [23] is based on the fact that a mobile defect becomes immobilized as soon as it comes within the so-called capturing volume around the dopant ion given by

$$\Omega = 4\pi \int_a^b r^2 \exp\left[\frac{V(r)}{k_B T}\right] dr \tag{2.5}$$

with a the nearest-neighbor equilibrium distance, b the capture radius of the potential $V(r)$. Ω_0 is the volume of a sphere with capture radius R_C given by $V(R_C) = k_B T$, i.e., the case where there is no attractive potential. The capturing volume is called the pairing constant. When pairing effectively occurs, the distance between the ions becomes much smaller. In the case of Cu in B-doped Si the capture radius R_C is about 7 nm, while the Cu_i^+ to B_s^- distance in the CuB pair is only 0.234 nm [25]. For the FeB pair the distance is about the same, $r = 0.235$ nm [26].

In (2.5) the pairing constant is calculated based on the electrostatic or Coulomb interaction given by

$$V(r) = \frac{1}{4\pi\varepsilon_{Si}\varepsilon_0}\left[\frac{q^2}{r}\right] \tag{2.6}$$

with ε_{Si} the permittivity of silicon and ε_0 the dielectric constant of vacuum, respectively, and q the elementary charge. This enables to calculate R_C. For doping concentrations higher than 10^{17} cm^{-3} an exponential correction factor has to be including in the equation, resulting in [26]

$$k_B T = \frac{q^2}{4\pi\varepsilon_{Si}\varepsilon_0 R_C} \exp\left[-\frac{R_C}{\left(\frac{\varepsilon_{Si}\varepsilon_0 k_B T}{q^2 p}\right)^{1/2}}\right] \tag{2.7}$$

with p the free hole concentration.

A systematic study based on infrared absorption spectroscopy and capacitance measurements of copper-acceptor interactions in Si doped with different p-type dopants pointed out that the dissociation energy depends on the acceptor type (0.61 ± 0.02 eV for CuB, 0.69 ± 0.02 eV for CuIn, 0.70 ± 0.02 eV for CuAl, 0.71 ± 0.02 eV for CuGa and 0.85 ± 0.05 eV for CuPt) [25]. This observation which is in agreement with ab initio calculations [27] indicates the presence of a covalent component so that the pairing constant is underestimated [28]. Another additional effect that can lead to an underestimated pairing coefficient determination is the presence of a strain component. This is taken into account in the DFT calculations enabling for pairs to determine the different components of the pair formation energy as has in detail been studied for e.g. co-doping effects in silicon [29]. The formed metal-dopant pairs are electrically active and form a defect state in the band gap, as will be detailed in Chaps. 5 and 6.

The stability of the MX pair depends on the dissociation energy E_{diss}, which in a first approximation is the sum of the binding energy E_b and the diffusion barrier E_d. In the case of FeB and CuB pairs in Si the binding energy (≈0.52 eV) is nearly the same but the diffusion barrier for Fe_i^+ is 0.68 eV compared to 0.18 V for Cu_i^+ so that the dissociation energy is much larger for FeB compared to CuB [30]. In the literature, there is a large discrepancy between the reported values for the binding energy and for the diffusion barrier of Fe. However, based on the temperature behavior of the dissociation process, a dissociation energy of 1.17–2.0 eV is found in agreement with the value given above [31]. Dissociation of pairs can be promoted by increasing the temperature, by light illumination, by ultrasound and by minority carrier injection [26]. CuB pairs are already unstable at room temperature, while FeB pairs dissociate around 200 °C. The latter can be exploited to identify Fe contamination in p-type silicon by lifetime measurements (Chaps. 4 and 5).

As ion pairing has an impact on the diffusion coefficient, pairing should be suppressed in order to determine the intrinsic diffusion coefficient. For copper in Si this has been studied using transient ion drift (TID) experiments on quenched Cu contaminated, weak boron-doped samples enabling to determine (2.8a) [24], while for Fe (2.8b) is obtained

$$D_{int}(Cu_i) = (3.0 \pm 0.3) \times 10^{-4} \exp\left(-\frac{0.18 \pm 0.01\,\text{eV}}{k_B T}\right)\left(\frac{\text{cm}^2}{\text{s}}\right) \tag{2.8a}$$

$$D_{int}(Fe_i) = (1.0^{+0.8}_{-0.4}) \times 10^{-3} \exp\left(-\frac{0.67\,\text{eV}}{k_B T}\right)\left(\frac{\text{cm}^2}{\text{s}}\right) \tag{2.8b}$$

The intrinsic diffusion equation has been modified to account for ion pairing, resulting in the following expression for the effective diffusion coefficient [23]

$$D_{eff} = \frac{D_{int}}{1 + \alpha} \tag{2.9}$$

α is the concentration ratio of trapped over mobile metals. In the case of Cu it is the ratio of the immobile Cu_i^+ to mobile Cu_i^+. The immobile Cu_i^+ concentration is higher than the CuB pairs since Cu_i^+ is trapped as soon as it comes within the capture radius R_C as explained above. Only a fraction of the B^- acceptor traps will be transformed into CuB complexes. The correlation between the concentrations in equilibrium and without precipitation effects of the different species involved is given by

$$\frac{C(M^+X^-)}{C(M^+)C(X^-)} = \frac{Z}{N_i} \exp\left(\frac{E_b}{k_B T}\right) \tag{2.10}$$

with $C(M^+X^-)$ the pair concentration, $C(M^+)$ the metal donor concentration, $C(X^-)$ the dopant acceptor concentration (also N_a), Z the number of possible orientations of the pair with the same symmetry around the acceptor atom

(4 for tetrahedral symmetry) and N_i the number of interstitial sites in the lattice. In (2.9) α can be replaced by the product of the pairing coefficient and the number of acceptor traps ΩN_a, whereby Ω can be calculated from (2.5).

The pairing effect is influenced by the doping concentration, i.e., in the example of Cu diffusion in Si by the acceptor concentration given by the boron doping. Experimental TID studies pointed out that for Cu in Si the following expression can be derived for the effective diffusion coefficient for boron concentrations $\leq 10^{17}$ cm^{-3} [28].

$$D_{eff} = \frac{3 \times 10^{-4} \exp\left(-\frac{2090}{T}\right)}{1 + 2.584 \times 10^{-20} \exp\left(\frac{4990}{T}\right)\left(\frac{N_a}{T}\right)} \left(\frac{\mathrm{cm}^2}{\mathrm{s}}\right) \tag{2.11}$$

This expression clearly indicates that the Cu diffusivity decreases for increasing boron doping levels and for lower temperatures. At room temperature Cu can diffuse through a 300 mm diameter intrinsic Si wafer (thickness about 750–800 μm) in less than 4 h. Also the formation of CuB pairs is a fast process at room temperature, similar as the case for the FeB formation. The same expression can be used to calculated the amount of Cu out-diffusion during either a temperature treatment, a cooling cycle or storage at room temperature [32]. The constants in (2.11) have to be adjusted for other doping types than boron. Similar expressions can be derived for other TMs in Si or Ge as will be shown in Chap. 5.

The impact of the boron doping concentration on the diffusivity is illustrated in Fig. 2.6 for Cu in boron-doped Si [28]. The experimental diffusion data obtained by TID investigations are in agreement with the calculated curves using (2.11). The earlier data of Hall and Racette [10] for a high B-doping (N_a = 2 × 10^{20} cm^{-3}) can

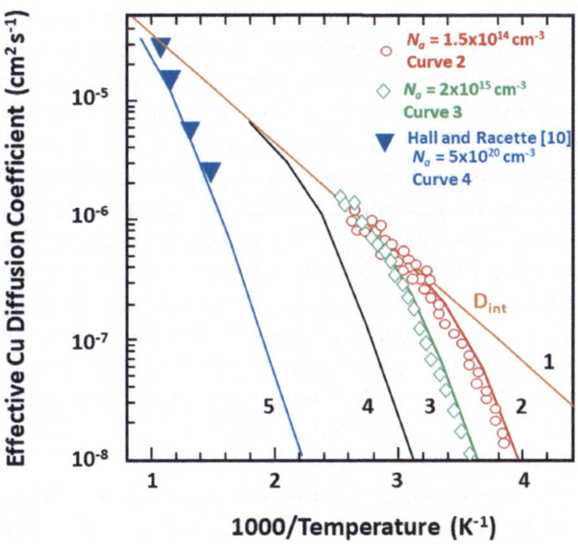

Fig. 2.6 Effective diffusion coefficient for Cu in Si for two different B-doping concentrations. The data of Hall and Racette [10] are also indicated. Curve 1 is the intrinsic diffusion coefficient (2.8a). Curves 2–5 are calculated according to (2.11). Curve 4 is for N_a = 1 × 10^{17} cm^{-3}. Reproduced from Istratov et al. [28]. Copyright (1998) by the American Physical Society

also reasonably be fitted. In contrast to what was mentioned in the original paper, the obtained diffusion coefficients are not intrinsic but effective values.

The maximum interstitial Cu concentration that can be measured is in agreement with the maximum Cu solubility concentration if the copper concentration is less than a B-doping concentration of about 1×10^{16} cm^{-3} [33]. This phenomenon is related to Cu precipitation and will be addressed in the next section. Cu is a shallow donor and will compensate the boron acceptors thereby shifting the position of the Fermi level further away from the valence band. For high Cu concentrations exceeding the doping concentration type-inversion can occur. The Cu precipitates create defect band states in the upper half of the band gap. Depending on the position of the Fermi level the precipitates can have different charge states. If they are positively charged they will repel the positive Cu_i^+ ions. This indicates an important interaction between Cu precipitation and Cu out-diffusion. This is also the case for some of the other transition metals in Si or Ge.

The solubility and diffusion equations for Cu given above can be used for medium doping levels but should be modified for very high doping concentrations. Especially for highly doped n-type Si there is a larger tendency to form Cu_s and thereby reducing the Cu diffusivity [21]. This is extremely important as highly-doped layers of phosphorus or boron are used for gettering purposes. In this case the doping concentration is in the range 10^{20}–10^{21} cm^{-3}, which leads to a reduction of the metal diffusion coefficient and the formation of ternary compounds [21]. This topic is in detail treated in Chap. 8.

2.3 Segregation and Precipitation

If at a certain temperature Si or Ge is contaminated with a metal, the metal atoms will diffuse into the bulk. During the cooling cycle of such a metal-contaminated sample different processes may occur. The impurities can remain dissolved in the bulk resulting in a super-saturation above the solid solubility. The concentration will be higher for a metal with a low diffusion coefficient and a high cooling rate. Fast diffusing metals such as Cu, Ni and Co are not so easy to quench-in [1]. Fast diffusers form preferentially defects on interstitial sites, while slow diffusing metals prefer substitutional positions. As mentioned earlier for several metals it is not straightforward to determine the ratio of the interstitial over substitutional defect concentrations. These isolated impurities are electrically active centers (donors or acceptor type) with one or more deep levels in the band gap and different charge states and therefore impacting the electrical device properties such as minority carrier lifetime and leakage current. Depending on the type of metal, more than one stable defect can be formed. Interactions with doping atoms lead to ion pairing as discussed in Sect. 2.2.1.

The metals in Si can also pair with other metals (e.g. Au with Fe, Ni, Cu, Mn, Cr and V) (see e.g. [34, 35]). The paring of Fe with Au, Zn, Pd, Pt and Ag has been studied in detail (see e.g. Istratov et al. [26]). It is also possible to form pairs of interstitial and substitutional defects of the same metal such as e.g. Cu_sCu_i, first postulated based on photoluminescence studies in Si [36]. The model has been adjusted and the defect responsible for the photoluminescence spectrum most likely consists of a Cu_s coupled to 3 Cu_i, i.e., a Cu_s–$3Cu_i$ complex [37]. Theoretical DFT calculations [38] and Deep-Level Transient Spectroscopy (DLTS) studies [39] support this hypothesis. In Ge the Cu_sCu_i pair has been identified by DLTS experiments as a hole trap with an acceptor level at E_V+ 0.09 eV [40]. Metals can also interact with hydrogen present in the material, leading to complex formation and passivation effects (e.g. [32–42]).

For moderate cooling rates metals with a sufficient diffusivity may precipitate in the bulk of the material. This can occur either homogeneously or heterogeneously via lattice defects or other impurity precipitates present in the material. This requires during the cooling process a diffusion of the metal towards the nuclei. These precipitates are electrically active and influence the electrical device parameters. The size of the precipitates is an important parameter and often inversely proportional to the density. Precipitates residing at the wafer surface will have a degrading effect on the quality of the gate dielectric of a Metal-Oxide-Semiconductor (MOS) capacitor. The surface precipitation is at the basis of the well-known haze test [1], which has been studied and applied for different metals such as Fe, Co, Cu, Ni and Pd [43]. After rapid cooling the defects are delineated by defect etching and investigated by optical microscopy. The haze is associated with the etch pit density, called S-pits or Saucer pits. The sensitivity is about 10^{11} cm^{-2} for Ni, 10^{12} cm^{-2} for Cu, and 10^{13} cm^{-2} for Fe [44]. More details will be shown in Chap. 4.

The precipitates are amphoteric in nature so that their charge state depends on the position of the Fermi level. For precipitation to occur at a certain temperature the metal should be super-saturated in the lattice, i.e., a concentration C^* higher than the equilibrium concentration C_{eq}, and the diffusion coefficient must be high enough. The chemical driving force f_{chem} is given by

$$f_{chem} = k_B T \cdot \ln \frac{C^*}{C_{eq}} \tag{2.12}$$

As shown in Fig. 2.7 the temperature has an opposite effect on the driving force and the diffusivity [45]. The driving force decreases for increasing temperature while the diffusivity increases. Although Cu has a lower solution enthalpy than Fe it is diffusing much faster and will therefore easier precipitate. Cu already precipitates during the cooling cycle itself. In defect-free material the precipitation process starts from the surface and further extends into the bulk of the material. The depth of the precipitation zone under the surface depends on the cooling rate and can for a slow cooling (e.g. $T < 5$ K/s) be hundreds of microns [45]. The near surface precipitation gives the haze.

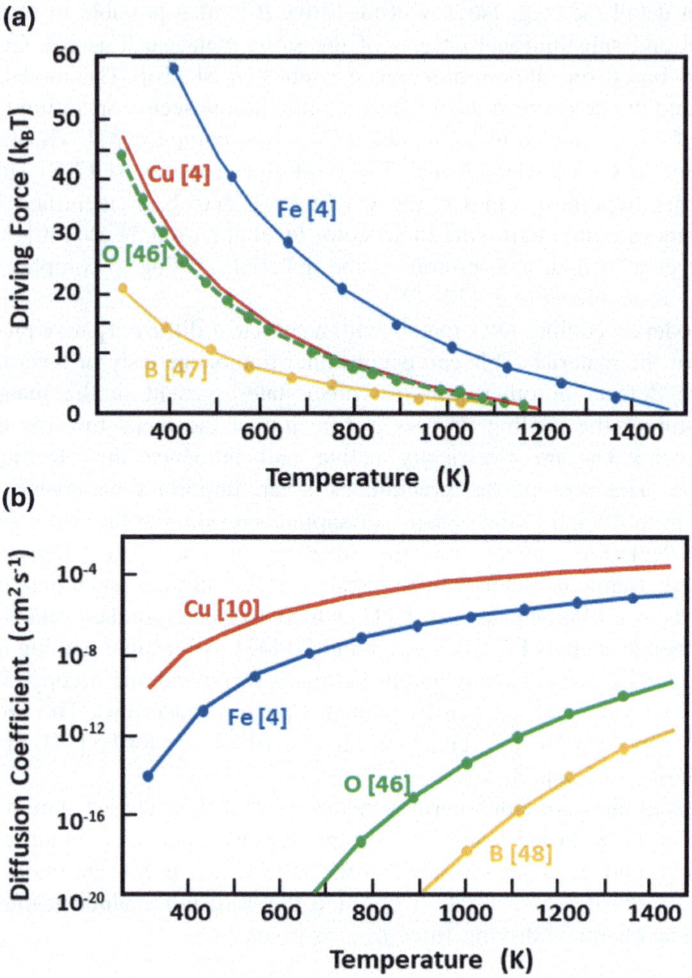

Fig. 2.7 Chemical driving force for precipitation **a** and diffusivity **b** of impurities in Si as a function of temperature. Data is included from Hall and Racette [10], Weber [4], Mikkelsen et al. [46], Nobili [47] and Brown et al. [48]. Reproduced from Seibt [45] by permission of ECS-The Electrochemical Society

It has been shown that interstitially dissolved metal impurities M_i such as Fe, Co and Ni precipitate in the disilicide form (MSi_2), while Cu and Pd form metal-rich silicides (M_xSi, $x \geq 2$). In general, the quasi-chemical reaction can be written as

$$M_i + P_n \leftrightarrow P_{n+1} + \gamma I + \beta V \tag{2.13}$$

with P_n a precipitate containing n metal atoms. During the reaction γ self- interstitials (I) are emitted and/or β vacancies (V) are absorbed per precipitated metal atom in order to compensate for the strain associated with the volume change of the precipitate. Equation (2.13) indicates that point defects like self-interstitials and vacancies play an important role in the precipitation process. The critical radius of a spherical metal precipitate is given by [49, 50]

$$r_c = \frac{2\sigma\Omega_P}{Ek_BT\ \ln\left(\frac{C_M}{C_M^*}\right)\left(\frac{C_I}{C_I^*}\right)^{\gamma}\left(\frac{C_V}{C_V^*}\right)^{\beta} - 6\mu_s\delta\varepsilon} \tag{2.14}$$

with σ the silicide/host interface energy per unit area, Ω_P the molecular volume of the silicide phase, μ_s the shear modulus, δ the linear misfit, ε the 'constrained strain', and $E = (1+\delta-\varepsilon)^3$. The strain energy associated with a volume change dV is $6\mu\delta\varepsilon dV$. Formula (2.14) is also valid for the co-precipitation of different metals M_{zi} simultaneously. In that case $\left(\frac{C_M}{C_M^*}\right)$ has to be replaced by $\sum_{i-1}^{k}\left(\frac{C_M}{C_M^*}\right)^{z_i}$ [51]. The volume difference between a precipitate M_xSi_y is given by $\xi = \Omega_{M_xSi_y}/y\Omega_{Si}$. Table 2.2 summarizes for some reported metal silicides the ξ-factor and the number of emitted self-interstitials per precipitated metal atom. If the ξ-factor is <1 the precipitation will preferentially occur in tensile stressed areas (e.g. a SiO_2 precipitate in Si, dislocated area) so that according to (2.14) the radius will increase. Considerations on the ξ-factor are important for metal gettering. For some metal silicides the volume expansion is so large that secondary defects are formed as e.g. for Cu_3Si where the about 150% volume increase may lead to Cu colonies bounded by dislocations as illustrated in Fig. 2.8 [52]. In the case of Ni or Co the formation of punched-out dislocations may only occur for large precipitates and as a result of cooling due to the different thermal expansion coefficients of the disilicide and the silicon. More discussion on the composition of some of these silicides will be given in Chaps. 5 and 6 dealing with an elaborated review of the different transition metals in silicon and germanium.

Table 2.2 ξ-factor and number of self-interstitials emitted per precipitated metal atom for metal silicides. Negative values correspond with self-interstitial absorption or vacancy emission

Metal	Silicide	ξ-factor	Emitted Si γ	Secondary defects
Fe	α-$FeSi_2$	0.94	−0.11 (RT)	
	Fe_3Si	2.26	0.42 (RT)	
	FeSi	1.13	0.13 (RT)	
Co	$CoSi_2$	0.96	−0.08 (RT)	Punched out dislocations?
		0.99	−0.03 (1000 °C)	
Ni	Ni_3Si	2.16	0.39 (RT)	Punched out dislocations?
	Ni_2Si	1.64	0.32 (RT)	
	NiSi	1.22	0.22 (RT)	
	$NiSi_2$	0.99	−0.025 (RT)	
			0 (≈400 °C)	
			−0.036 (1000 °C)	
Cu	Cu_3Si	2.5	≈0.5 (RT)	Metastable stacking faults
	CuSi	2.01	1.01 (RT)	Frank type SFs
				Edge dislocations

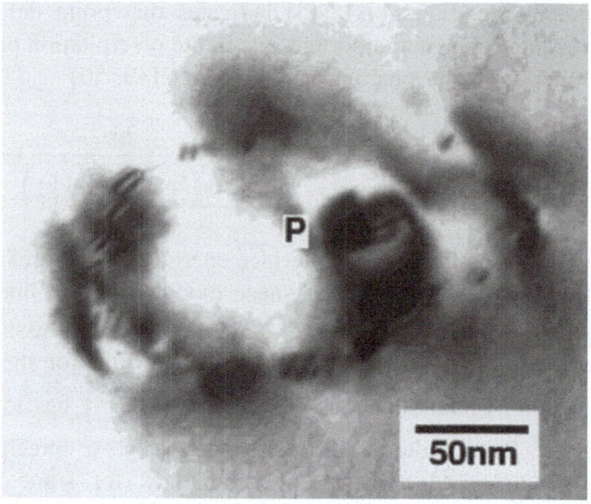

Fig. 2.8 TEM micrograph of a Cu_3Si precipitate. The central precipitate is surrounded by spherical particles attached to a dislocation loop. Reproduced from Seibt et al. [52]. Copyright 1999 WILEY-VCH Verlag GmbH & Co

Another way of getting general insight in the precipitation process is by looking at the lattice parameter of the precipitate compared to the lattice constant of Si as a function of temperature. This is illustrated in Fig. 2.9 [1]. The $CoSi_2$ lattice constant is always smaller than Si implying that there will be tensile stress in the material. For $NiSi_2$ the lattice constant is equal to Si at 500 °C, so that in this case there is no stress implying that no emission/absorption of interstitials/vacancies will occur and consequently no generation of secondary defects. The stress around the Ni-precipitate is compressive at higher temperatures and tensile at lower temperatures.

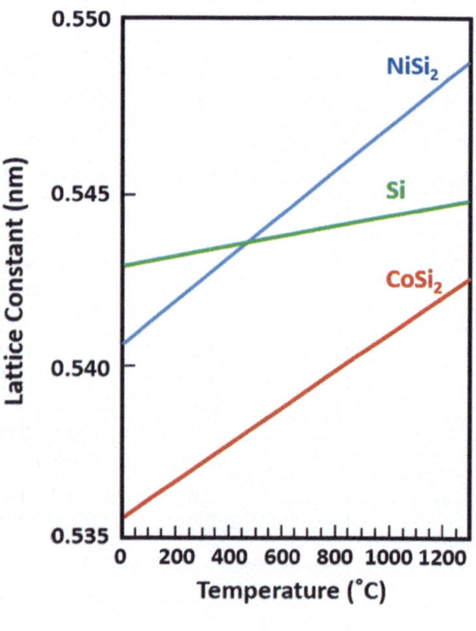

Fig. 2.9 Lattice constants of Si, $NiSi_2$ and $CoSi_2$ as function of temperature. Reproduced from Graff [1] by permission of Springer

The precipitation of Ni into $NiSi_2$ after quenching from a high temperature proceeds via the formation of coherent {111} platelets with a radius which can go up to 1 μm but a relative small thickness (<1 nm) [53]. The platelets are associated with a stacking fault and bounded by a dislocation. In the early precipitation phase only 2 Ni layers are involved. Most of the Ni in these metastable precipitates has a sevenfold coordination and is located at the interface. Further evolution occurs via Ostwald ripening whereby smaller precipitates dissolve and larger ones further grow, depending on the anneal temperature. For temperatures above 500 °C the density of particles will decrease while the size increases, as illustrated in Fig. 2.10. The driving force for Ostwald ripening is the change in interface energy [53]. The process will be different in case that defects present in the material promote heterogeneous precipitation to occur.

The metals can also segregate at extended defects present in the materials such as e.g. dislocations or oxygen precipitates. The interaction between the precipitate and the dislocation can be caused by (i) the strain field around the dislocation, (ii) chemical binding to the dislocation core, or (iii) accumulation close to the dislocations because of the intrinsic point defects present [54]. The latter effect will occur for impurities mainly dissolved on vacancy sites such as Au and Pt in Si. The strain field around a dislocation is well known as being a very efficient getter area for metal.

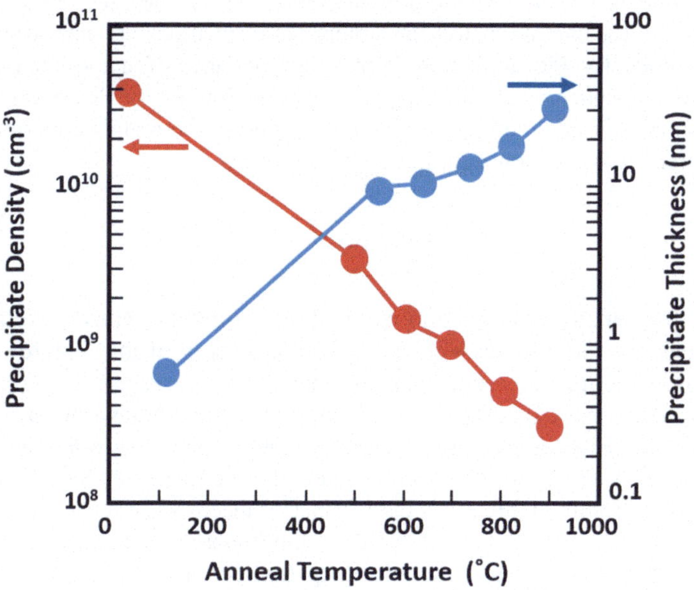

Fig. 2.10 Precipitate density and size after 20 min anneal of Ni contaminated samples quenched-in from 1050 °C, illustrating Ostwald ripening

2.4 Electrical Properties of Transition Metals

As mentioned before transition metals form after quenching from a high temperature isolated interstitial and/or substitutional defects, electrically characterized by energy levels in the band gap. Already sixty years ago it was observed that Au contamination creates donor and acceptor levels in both Ge ($E_V + 0.05$ (D), $E_V + 0.15$ (A), $E_C + 0.20$ (A) and $E_C - 0.04$ (A)) and Si ($E_V + 0.35$ (D), $E_V + 0.62$ (A)) [55]. The presence of four Au levels in Ge was later confirmed by DLTS investigations and they were assigned to a substitutional Au defect [56]. Zeeman spectroscopy enabled to identify that the two Au-levels in Si also belong to the same isolated substitutional Au defect [57]. After diffusion the metals may form pairs and/or can precipitate into metal silicides and/or segregate at other extended defects present in the material. Many of these defects have extensively been studied and are electrically active via either deep or shallow levels. The deep levels act as traps for carriers and therefore impact the electrical device properties (e.g. minority carrier lifetime), depending on the location of the trap level and their capture cross sections. Although this will be extensively discussed in following chapters, here some general properties are briefly addressed.

The electrical activity of the TMs has been studied by a large variety of techniques such as e.g. resistivity measurements, electron paramagnetic resonance (EPR), Hall measurements, DLTS, electron-beam induced current measurement (EBIC), Mössbauer spectroscopy, photoluminescence (PL) etc. Interpretation of the published results is not always straightforward as besides the used analytical technique also the sample history and preparation methods are important. Rather accurate values for the activation energy are obtained if determined from the temperature dependence of the capture and emission rates. However, when an Arrhenius plot is used a correction factor ΔE_B, due to the temperature dependence of the capture cross section $\boldsymbol{\sigma_{n,p}}$, has to be included

$$\sigma_{n,p} = \sigma_{n,p0} \exp\left(-\frac{\Delta E_B}{k_B T}\right) \tag{2.14}$$

with $\sigma_{n,p0}$ the capture cross section extrapolated to infinite temperature [58]. The correction factor corresponds with the activation energy of the capture cross section. Within experimental error, for many metals almost similar results have been reported so that these data can be considered as reliable. Metals can form single or multiple charged donors (D) and/or acceptors (A). This is for some dissolved 3*d*, 4*d* and 5*d* metals in silicon illustrated in Table 2.3, published in 2000 by Graff [1]. The first column in the table gives the number of electrons in the *s*- and *d*-shell of the free atom, as Lemke [59] noted that this is an important parameter for observing trends in both activation energies and segregation coefficients.

A more elaborated discussion of the defect levels is given in later chapters. It is, however, interesting to remark that for most metals the energy levels are not so different between silicon and germanium, as shown in Table 2.4. The main

Table 2.3 Activation energies of 3*d*, 4*d* and 5*d* TMs in silicon. *N* is the sum of *s* + *d* electrons. ΔE_B is a correction factor to be added if the activation energy is demined from an Arrhenius plot. After Graff [1]

N	Metal	I/V	DDD	DD	ΔE_B(eV)	D	ΔE_B(eV)	A	AA
3	Sc	I	+0.20	−0.50		−0.21	(0.004)		
	Y	I							
	La	–							
4	Ti	I		+0.28	(0.038)	−0.27		−0.08	
	Zr	I		+0.32		−0.42		−0.13	
	Hf	I		+0.32		−0.40		−0.10	
5	V	I		+0.32	(0.14)	−0.45		−0.18	
	Nb	I		+0.18		+0.62		−0.28	
	Ta	I?		+0.19		+0.58		−0.22	
6	Cr	I				−0.22			+0.00
	Mo	I				+0.28			
	W	I?				+0.40			
7	Mn	I		+0.27	(0.07)	−0.42		−0.12	
	Tc	I?				−0.35		−0.07	
	Re								
8	Fe	I				+0.39	(0.043)	−0.14	
	Ru	I				+0.26		−0.22	
	Os	I?				+0.30			
9	Co	S				+0.41		−0.41	
	Rh	S				−0.58		−0.32	
	Ir	S				−0.62		−0.24	
10	Ni	S		+0.12		+0.17		−0.41	−0.07
	Pd	S		+0.08		+0.30		−0.21	
	Pt	S				+0.32		−0.23	
11	Cu	S				+0.22		+0.46	−0.16
	Ag	S				+0.38		−0.55	
	Au	S				+0.34		−0.55	
12	Zn	S						+0.32	−0.53
	Cd	S						+0.5	−0.45
	Hg								

Table 2.4 Energy levels (in eV) of Mn, Fe, Co, Ni and Cu in silicon and germanium [60]

	Silicon [59]			Germanium [61, 62]		
Mn	$E_V + 0.27$	$E_C - 0.27$	$E_C - 0.11$	$E_V + 0.17$	$E_C - 0.37$	
Fe	$E_V + 0.39$			$E_V + 0.34$	$E_C - 0.27$	
Co	$E_V + 0.40$	$E_C - 0.41$		$E_V + 0.09$	$E_V + 0.25$	$E_C - 0.31$
Ni	$E_V + 0.17$	$E_C - 0.38$	$E_C - 0.06$	$E_V + 0.22$	$E_C - 0.30$	
Cu	$E_V + 0.23$	$E_C - 0.46$	$E_C - 0.16$	$E_V + 0.04$	$E_V + 0.33$	$E_C - 0.26$

difference between TMs in Si and Ge is their incorporation in the lattice, i.e., most TMs in Si are on interstitial sites, while in Ge they are on lattice sites. This explains the multiple acceptor nature of the levels in Ge. The different nature of the same metal in Si or Ge (e.g. Fe) is at the basis of the different electrical impact as discussed in the next section.

Similar as dissolved metals, metal precipitates are also electrically active and their activity is characterized by their energy level in the band gap and their capture cross sections. The latter are a key parameter for lifetime measurement analysis, device modelling and leakage current interpretation. The electrical activity is influenced by the carrier drift, carrier diffusion, the electrical field, the charge state of the precipitate and the recombination activity at the precipitate/substrate interface. A metal precipitate is a silicide forming a Schottky barrier with the Si (or Ge) host and has an effective charge and a space charge region around it, attracting minority carriers and repulsing majority carriers. To calculate the dominant recombination mechanism, one has to take into account both the carrier recombination at interface states, characterized by the surface recombination velocity, and the recombination inside the precipitate, depending on the charge transport across the Schottky barrier and controlled by thermionic emission. A quantitative model has been elaborated for spherical precipitates in both n- and p-type Si, leading to the conclusion that thermionic emission is the dominant mechanism [63]. The thermionic emission potential for electrons $V_{term\text{-}e}$ is given by

$$V_{term-e} = \frac{k_B T}{q} \ln\left[1 - \frac{J_e m_0}{A_R T^2 m_e^*} \exp\left(\frac{\varphi_b}{k_B T}\right)\right] \tag{2.15}$$

with J_e the electron flux, m_0 the free electron mass, m_e^* the effective electron mass in Si, A_R the Richardson constant and φ_b the Schottky barrier height for electrons. A symmetric equation (with negative sign) holds for the hole potential. This expression can be used for numerical simulations.

Besides the metal energy level in the band gap E_T, the hole and electron capture cross section have a dominant impact on key parameters such as carrier lifetime and leakage current. The generation τ_g and recombination lifetime τ_r can be expressed as [64]

$$\tau_g = 2\tau_r \cosh\left[\frac{E_T - E_i}{k_B T}\right] \tag{2.16a}$$

$$\tau_r = 2\tau_n = \frac{1}{N_T \sigma_n \upsilon_{thn}} = \frac{1}{N_T c_n} \tag{2.16b}$$

with τ_r the recombination lifetime (given in (2.16b) for p-type material), which for low carrier injection is also called the minority carrier lifetime (i.e. τ_n for p-type and τ_p for n-type), υ_{thn} the thermal velocity of electrons, and c_n the electron capture rate. At room temperature the thermal velocity for electrons (holes) is 2.3×10^7 cm/s (1.65×10^7 cm/s) and 3.1×10^7 cm/s (1.9×10^7 cm/s) in Si and

Ge, respectively [60]. $E_i \approx E_g/2$ is the intrinsic level with E_g the band gap of the host material given by the temperature dependent expressions

$$E_{g-Si} = 1.17 - \frac{4.73 \times 10^{-4} T^2}{T + 6.36} (\text{eV}) \tag{2.17a}$$

$$E_{g-Ge} = 7.42 - \frac{4.8 \times 10^{-4} T^2}{T + 6.235} (\text{eV}) \tag{2.17b}$$

for respectively silicon and germanium. The lifetimes given by (2.16) can e.g. be determined from diffusion current measurements on diodes [65]. The importance of the metal capture cross section (σ_n, σ_p) is shown in Fig. 2.11 for Ge, giving the capture rate ($c_{n,p} = \sigma_{n,p} \upsilon_{thn,p}$) for different metals [66]. In p-type Ge, Fe, Co and Ni are effective lifetime killers, while because of the smaller electron capture cross

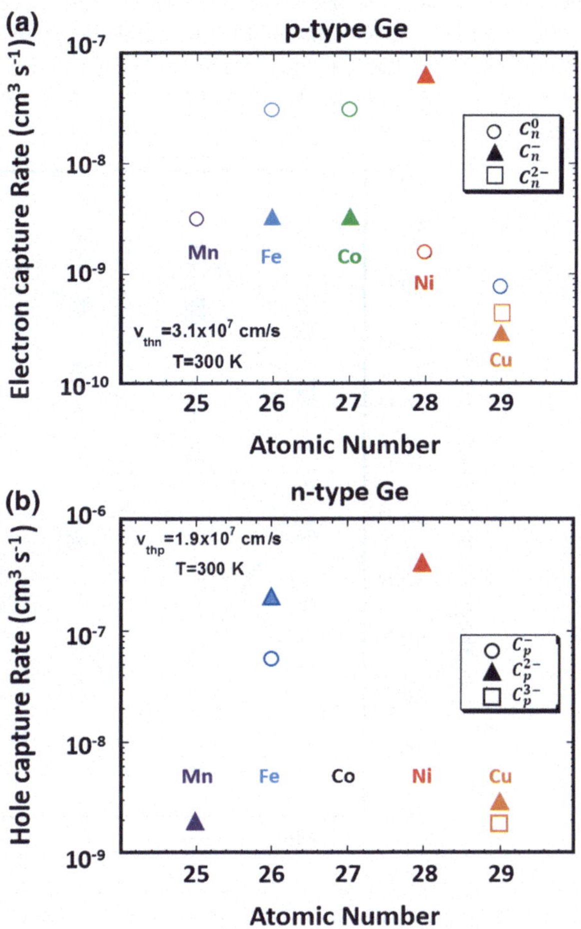

Fig. 2.11 **a** Electron capture rate at 300 K for TMs in p-type Ge. The charge state of the empty substitutional metal centers is either neutral (O), single (▲) or double negative (□), **b** hole capture rate at 300 K for TMs in n-type Ge. The charge state of the empty substitutional metal centers is either single (O), double (▲) or triple negative (□). Reproduced from Simoen et al. [66] by permission of Springer

section Cu is less harmful. The same trend holds for n-type Ge. In general the capture rates in n-type Ge are about one decade higher than in p-type Ge, what is the opposite for Si. This implies that for the same metal with the same concentration the recombination lifetime is higher in p-type Ge than in n-type Si. As mentioned before, this is related to the different nature of the metal defects in Si and Ge.

One can also calculate the generation current per defect center, given by

$$I_{defect} = \frac{qn_i\sigma_{n,p}\upsilon_{thn,p}}{2\cosh\left(\frac{E_T-E_i}{k_BT}\right)} \tag{2.18}$$

This expression is graphically represented in Fig. 2.12 for Mn, Fe, Co, Ni and Cu in Ge and Si. The metals with an energy level close to mid gap (like Fe and Ni) and having small minority capture cross sections are the most effective leakage current generators. If one takes into account that the intrinsic carrier concentration depends exponentially on the band gap, then one can calculate a normalized 'room' temperature for Ge, i.e., the same intrinsic carrier concentration as for Si at 300 K, which equals 222 K. This enables to point out from Fig. 2.12, that metals have a larger impact on the leakage current in Ge than in Si. A similar type of analysis can be done for SiGe alloys. It has to be remarked that this will not lead to a parallel

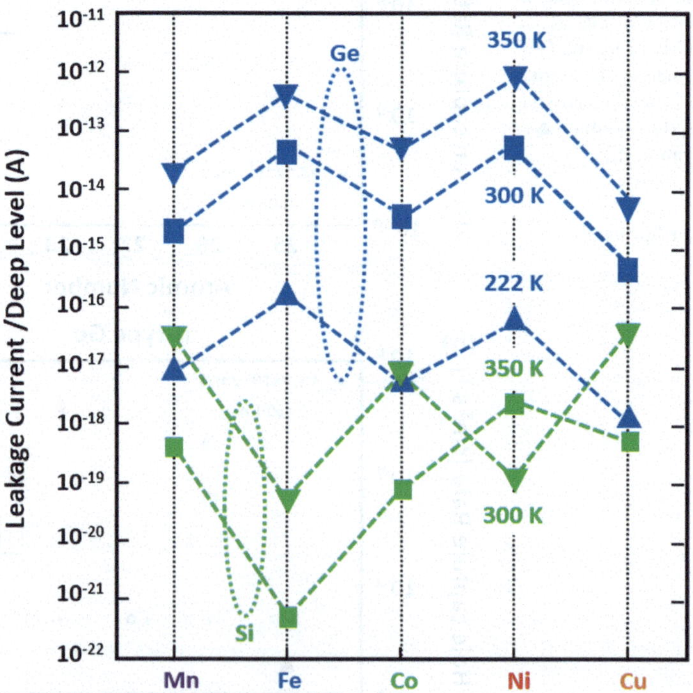

Fig. 2.12 Leakage current per deep level in Ge and Si. Reproduced from Vanhellemont and Simoen [60] by permission of ECS-The Electrochemical Society

shift of the curves as the final result depends on both the band gap energy of the material and the electrical activity of the metal. The correlation between the solubility and the maximum observed leakage current is not straightforward as for supersaturation of metals, phenomena such as ion pairing and precipitation play an important role.

The electrical activity of the metals depends on the way they are incorporated in the lattice, i.e., as isolated interstitial or substitutional atoms, as complexes due to pairing or as precipitates, and also geometrical dimensions of the metal inclusion. In the case of metal precipitates, it has been suggested to use the following equation to determine their impact on the carrier lifetime [67]

$$\tau_{prec} = \frac{1}{4\pi N_{prec} D_n r_{prec}} + \frac{1}{4\pi N_{prec} D_n s r_{prec}^2} \tag{2.19}$$

with τ_{prec} the carrier lifetime in the presence of precipitates, N_{prec} the precipitate density, r_{prec} the radius of the precipitate and s the surface recombination velocity. For charged precipitates the radius has to be replaced by the radius of the space charge region and the recombination velocity by the effective surface recombination velocity at the edge of the space charge region. Reasonable estimations of the carrier lifetime have been obtained by using parametrization techniques taking into account recombination effects at the precipitates associated with the internal Schottky contact [68]. As an illustration of this approach, Fig. 2.13 gives the calculated carrier-dependent carrier lifetime for an Fe concentration of 1×10^{14} cm^{-3} in p-type silicon. Calculations are done for different cases, i.e., all Fe in isolated

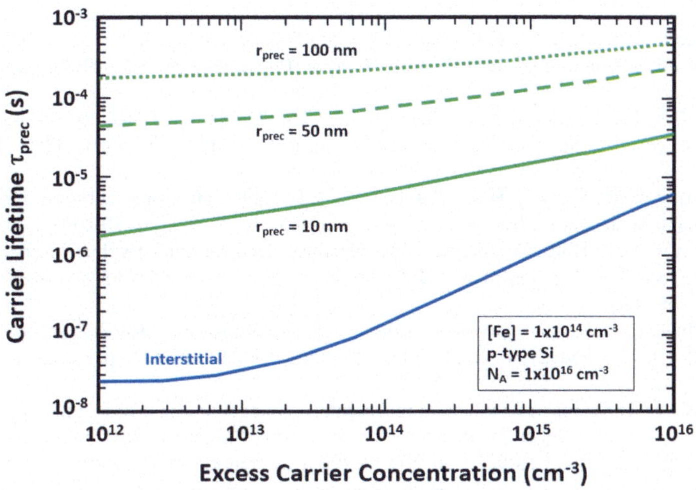

Fig. 2.13 Calculated excess carrier-dependent carrier lifetime for Fe (1×10^{14} cm^{-3}) in p-type Si, for isolated and 100% precipitated positions, respectively. Reproduced from Kwapil et al. [68]. Copyright 2015 IEEE

interstitial atoms and 100% precipitation. It can clearly be noticed that the lifetime is dependent on the radius of the precipitate and increases for larger precipitates. The lowest carrier lifetime is obtained when all Fe is present in interstitial form.

References

1. D. Graff, Metal impurities in silicon-device fabrication (2000) (Springer, Heidelberg, Germany). https://doi.org/10.1007/978-3-642-57121-3
2. H. Bracht, H. Overhof, Kinetics of interstitial-substitutional exchange of Zn, Pt, and Au in Si: experimental results and theoretical calculations. Phys. Status Solidi A **158**, 47–55 (1996). https://doi.org/10.1002/pssa.2211580107
3. S. Hocine, D. Mathiot, Diffusion and solubility of titanium in silicon. Mater. Sci. Forum **38–42**, 725–728 (1989). https://doi.org/10.4028/www.scientific.net/MSF
4. E.R. Weber, Transition metals in silicon. Appl. Phys. A **30**, 1–22 (1983). https://doi.org/10.1007/BF00617708
5. D. Gilles, W. Bergholz, W. Schröter, Diffusion of manganese in silicon studied by deep-level transient spectroscopy and tracer measurements. J. Appl. Phys. **59**, 3590–3593 (1986). https://doi.org/10.1063/1.337042
6. D. Grünebaum, Th. Czekalla, N.A. Stolwijk, H. Mehrer, I. Yonenaga, K. Sumino, Diffusion and solubility of zinc in dislocation-free and plastically deformed silicon-crystals. Appl. Phys. A **53**, 65–74 (1991). https://doi.org/10.1007/BF00323437
7. D. Gilles, W. Schröter, W. Bergholz, Impact of the electronic structure on the solubility and diffusion of 3*d* transition elements in silicon. Phys. Rev. B **41**, 5770–5782 (1990). https://doi.org/10.1103/PhysRevB.41.5770
8. F.A. Trumbore, Solid solubility of impurity elements in germanium and silicon. Bell Syst. Techn. J. **39**, 205–233 (1960). https://doi.org/10.1002/j.1538-7305.1960.tb03928.x
9. H. Bracht, N.A. Stolwijk, H. Mehrer, Diffusion and solubility of copper, silver and gold in germanium. Phys. Rev. B **43**, 14465–14477 (1991). https://doi.org/10.1103/PhysRevB.43.14465
10. R.N. Hall, J.H. Racette, Diffusion and solubility of copper in extrinsic and intrinsic germanium, silicon and gallium arsenide. J. Appl. Phys. **35**, 379–397 (1964). https://doi.org/10.1063/1.1713322
11. C.S. Fuller, J.D. Struthers, J.A. Ditzenberger, K.B. Wolfstirn, Diffusivity and solubility of copper in germanium. Phys. Rev. **93**, 1182–1189 (1954). https://doi.org/10.1103/physrev.93.1182
12. N.A. Stolwijk, W. Frank, J. Hölzl, S.J. Pearton, E.E. Haller, Diffusion and solubility of copper in germanium. J. Appl. Phys. **57**, 5211–5219 (1985). https://doi.org/10.1063/1.335259
13. J.A. Burton, E.D. Kolb, W. Slichter, J.D. Struthers, Distribution of solute in crystals grown from the melt Part II. Experimental. J. Chem. Phys. **21**, 1991–1996 (1953). https://doi.org/10.1063/1.1698729
14. S. Fischler, Correlation between maximum solid solubility and distribution coefficient for impurities in Ge and Si. J. Appl. Phys. **33**, 1615–1622 (1962). https://doi.org/10.1063/1.1728792
15. H. Mehrer, Diffusion in solids—fundamentals, methods, materials, diffusion-controlled processes (2016) (Springer, Heidelberg, Germany) https://doi.org/10.1007/978-3-642-57121-3
16. S.W. Jones, Diffusion in silicon. IC Knowledge LLC report 1–68 (2000)
17. H. Bracht, Copper related diffusion phenomena in germanium. Mater. Sci. Eng. **7**, 113–124 (2004). https://doi.org/10.1016/j.mssp.2004.06.001
18. F. Van der Maesen, J.A. Brenkman, The solid solubility and the diffusion of nickel in germanium. Philips Res. Rep. **9**, 225–230 (1954)

19. A. Van Wieringen, N. Warmholtz, On the permeation of hydrogen and helium in single crystal silicon and germanium at elevated temperatures. Physica **22**, 849–865 (1956). https://doi.org/10.1016/S0031-8914(56)90039-8
20. B.I. Boltaks, *Diffusion in semiconductors* (Infosearch Ltd., London, 1963)
21. R.L. Meek, T.E. Seidel, Enhanced solubility and ion pairing of Cu and Au in heavily doped silicon at high temperatures. J. Phys. Chem. Solids **36**, 731–740 (1975). https://doi.org/10.1016/0022-3697(75)90096-7
22. A. Fazzio, M.J. Caldas, A. Zunger, Electronic structure of copper, silver and gold impurities in silicon. Phys. Rev. B **32**, 934–954 (1985). https://doi.org/10.1103/PhysRevB.32.934
23. H. Reiss, C.S. Fuller, F.J. Morin, Chemical interaction among defects in germanium and silicon. Bell Syst. Techn. J. **3**, 535–636 (1956). https://doi.org/10.1002/j.1538-7305.1956.tb02393
24. F.C. Frank, D. Turnbull, Mechanism of diffusion of copper in germanium. Phys. Rev. **104**, 617–618 (1956). doi/https://doi.org/10.1103/physrev.104.617
25. P. Wagner, H. Hage, H. Prigge, T.H. Prescha, J. Weber, Properties of copper-induced complexes in silicon. Semicond. Silicon **90**(7), 675–686 (1990)
26. A.A. Istratov, H. Hieslmair, E.R. Weber, Iron and its complexes in silicon. Appl. Phys. A **69**, 13–44 (1999). https://doi.org/10.1007/s003399900059
27. S.K. Estreicher, Copper, lithium and hydrogen in boron c-Si. Phys. Rev. B **41**, 5447–5450 (1990). https://doi.org/10.1103/PhysRevB.41.5447
28. A.A. Istratov, C. Flink, H. Hieslmair, E.R. Weber, Intrinsic diffusion coefficient of interstitial copper in silicon. Phys. Rev. Lett. **81**, 1243–1246 (1998). https://doi.org/10.1103/PhysRevLett.81.1243
29. C. Ahn, S.T. Dunham, Predicative models for the co-doping effects between combinations of donors (P/As/Sb) and acceptors (B/Ga/In). In Proceedings of 2006 International Conference on Simulation of Semiconductor Processes and Devices SISPAD 2006 83–86 (2006). https://doi.org/10.1109/sispad.2006.282844
30. A.A. Istratov, Weber ER (2000) The importance of understanding of the physics of copper in silicon for development and characterization of copper diffusion barrier. ECS Proc. **27**, 90–111 (2000)
31. H. Feichtinger, Electrical and kinetic properties of thermally activated iron in silicon. Acta Phys. Austr. **51**, 161–184 (1979)
32. M.B. Shabani, S. Okuuchi, Y. Shamanuki, Kinetics of low-temperature out-diffusion of copper from silicon wafers. ECS Proc. **99–16**, 510–525 (1999)
33. A.A. Istratov, X. Flink, H. Hieslmair, S.A. McHugo, E.R. Weber, Diffusion, solubility and gettering of copper in silicon. Mater. Sci. Eng. B **72**, 99–104 (2000). https://doi.org/10.1016/S0921-5107(99)00514-0
34. S.D. Brotherton, P. Bradley, A. Gill, E.R. Weber, Electrical observation of the Au–Fe complex in silicon. J. Appl. Phys. **55**, 952–956 (1984). https://doi.org/10.1063/1.333149
35. R. Czaputa, Transition metal impurities in silicon: new defect reactions. Appl. Phys. A **49**, 431–436 (1989). https://doi.org/10.1007/BF00615028
36. J. Weber, H. Bauch, R. Sauer, Optical properties of copper in silicon: Excitons bound to isotopic copper pairs. Phys. Rev. B **25**, 7688–7699 (1982). https://doi.org/10.1103/physrevb.25.7688
37. K. Shirai, H. Yamaguchi, A. Yanase, H. Katayama-Yosida, A new photoluminescence complex in Si and its photoluminescence. J. Phys. Condens. Mater. **21**, 064209 (2009). https://doi.org/10.1088/0953-8984/21/6/064249
38. A. Carvalho, D.J. Backlund, S.K. Estreicher, Four-copper complexes in Si and Cu-photoluminescence defect: a first principle study. Phys. Rev. B **84**, 155322 (2011). https://doi.org/10.1103/PhysRevB.84.155322
39. N.A. Yarykin, J. Weber, Identification of copper-copper and copper-hydrogen complexes in silicon. Semiconductors **47**, 275–278 (2013). https://doi.org/10.1134/S1063782613020231

40. Y. Kamiura, F. Hashimoto, Copper-related deep levels and their annealing kinetics in germanium. Jpn. J. Appl. Phys. **28**, 763–769 (1989). iopscience.iop.org/article/https://doi.org/10.1143/jjap.28.763
41. J.U. Sachse, E.Ö. Sveinbjörnsson, N. Yarykin, J. Weber, Similarities in the electrical properties of transition metal-hydrogen complexes in silicon. Mater. Sci. Eng., B **58**, 134–140 (1999). https://doi.org/10.1016/S0921-5107(98)00282-7
42. T.M. Gibbons, S.J. Backlund, S.K. Estreicher, Cobalt-related defects in silicon. J. Appl. Phys. **121**, 045704 (2017). https://doi.org/10.1063/1.4975034
43. P. Alpern, W. Bergholz, R. Kakoschke, Detection of fast diffusing metal impurities in silicon by haze test and by modulated optical reflectance: a comparison. J. Electrochem. Soc. **136**, 3841–3848 (1989). jes.ecsdl.org/content/136/12/3841
44. M. Hourai, T. Murakami, T. Shigematsu, N. Fujino, T. Shiraiwa, Behaviour of defects induced by metallic impurities on Si(100) surfaces. Jpn. J. Appl. Phys. **28**, 2413–2420 (1989). https://doi.org/10.1143/JJAP.28.2413
45. M. Seibt, Metal impurity precipitation in silicon. Proc. Electrochem Soc. **97**(22), 243–258 (1997)
46. J.C. Mikkelsen Jr., The diffusivity and solubility of oxygen in silicon. Mater. Res. Soc. **59**, 19–30 (1986). https://doi.org/10.1557/proc-59-19
47. D. Nobili, Properties of silicon. EMIS Datareviews Ser **4**, 384 (Inspec London 1988)
48. A.A. Brown, P.B. Moymagh, P.J. Rosser, Properties of silicon. EMIS Datareviews Ser **4**, 327 (Inspec London 1988)
49. J. Vanhellemont, C. Claeys, A theoretical study of the critical radius of precipitates and its application to silicon oxide in silicon. J. Appl. Phys. **62**, 3960–3967 (1987). https://doi.org/10.1063/1.339194
50. J. Vanhellemont, C. Claeys, Erratum: a theoretical study of the critical radius of precipitates and its application to silicon oxide in silicon. J. Appl. Phys. **71**, 1073 (1992). https://doi.org/10.1063/1.351408
51. J. Vanhellemont, C. Claeys, On the role of point defects in gettering processes. Mater. Sci. Forum **38–41**, 171–176 (1989). https://doi.org/10.4028/www.scientific.net/MSF.38-41.171
52. M. Seibt, H. Hedemann, A.A. Istratov, F. Riedel, A. Sattler, W. Schröter, Structural and electrical properties of metal silicide precipitated in silicon. Phys. Status Solidi A **171**, 301–310 (1999). https://doi.org/10.1002/(SICI)1521-396X(199901)171:13.3.CO;2-G
53. M. Seibt, W. Schröter, Precipitation behaviour of nickel in silicon. Philos. Mag. **A59**, 337–352 (1989)
54. M. Seibt, R. Khalil, V. Kveder, W. Schröter, Electronic states at dislocations and metal silicide precipitates in crystalline silicon and their role in solar cell materials. Appl. Phys. A **96**, 235–253 (2009). https://doi.org/10.1007/s00339-008-5027-8
55. C.B. Collins, R.O. Carlson, C.J. Gallagher, Properties of gold-doped silicon. Phys. Rev. **105**, 1168–1173 (1957). https://doi.org/10.1103/PhysRev.105.1168
56. E. Simoen, P. Clauws, G. Huylebroeck, J. Vennik, DLTS of gold impurities in germanium. Semicond. Sci. Technol. **2**, 507–512 (1987). https://doi.org/10.1088/0268-1242/2/8/006
57. G.D. Watkins, M. Kleverman, A. Thilderkvist, H.G. Grimmeiss, Structure of gold in silicon. Phys. Rev. Lett. **67**, 1149–1152 (1991). https://doi.org/10.1103/PhysRevLett.67.1149
58. J.-W. Chen, A.G. Milnes, A. Rohatgi, Titanium in silicon: a deep level impurity. Solid-State Electron. **22**, 801–808 (1979). https://doi.org/10.1016/0038-1101(79)90130-8
59. M. Lemke, Characterisation of transition-metal-doped silicon crystals prepared by the float zone technique. Proc. Electrochem. Soc. **94**(10), 695–710 (1994)
60. J. Vanhellemont, E. Simoen, Brother silicon, sister germanium. J. Electrochem. Soc. **154**, H572–H583 (2007). https://doi.org/10.1149/1.272221
61. P. Clauws, E. Simoen, Metals in germanium. Mater. Sci. Semicond. Process. **9**, 546–553 (2006). https://doi.org/10.1016/j.mssp.2006.08.047

62. J. Vanhellemont, O. De Gryse, S. Hens, P. Vanmeerbeek, D. Poelman, P. Clauws, E. Simoen, C. Claeys, I. Romandic, A. Theuwis, G. Raskin, H. Vercammen, P. Mylemans, Grown-in defects and diffusion in Czochralski-grown germanium. Defect Diff. Forum **230–232**, 149–176 (2004). https://www.scientific.net/DDF.230-232.14
63. P.S. Plekhanov, T.Y. Tan, Schottky effect model of electrical activity of metal precipitates in silicon. Appl. Phys. Lett. **76**, 3777–3779 (2000). https://doi.org/10.1063/1.126778
64. E. Simoen, C. Claeys, On the impact of the capture rates on the generation/recombination lifetime ratio of a single deep level. IEEE Trans. Electron. Dev. **46**, 1487–1488 (1999). https://doi.org/10.1109/16.772495
65. E. Simoen, C. Claeys, J. Vanhellemont, Defect analysis in semiconductor materials based on pn junction diode characteristics. Defect Diff. Forum **261–262**, 1–24 (2007)
66. E. Simoen, C. Claeys, S. Sioncke, J. Van Steenbergen, M. Meuris, S. Forment, J. Vanhellemont, P. Clauws, A. Theuwis, Lifetime and leakage current considerations in metal-doped germanium. J. Mater. Sci.: Mater. Electron. **18**, 799–804 (2007). https://doi.org/10.1007/s10854-006-9110-7
67. J.M. Hwang, D.K. Schroder, Recombination properties of oxygen-precipitated silicon. J. Appl. Phys. **59**, 2476–2486 (1986). https://doi.org/10.1063/1.336993
68. W. Kwapil, J. Schön, W. Warta, M.C. Schubert, Carrier recombination at metallic precipitates in p- and n-type silicon. IEEE J. Photovol. **5**, 1285–1292 (2015). https://doi.org/10.1109/JPHOTOV.2015.2438634

Chapter 3
Source of Metals in Si and Ge Crystal Growth and Processing

This chapter gives an overview of the possible sources leading to metallic contamination in Si or Ge wafers. Several of these sources are briefly discussed and illustrated, without going into detail about the properties of the metals and their impact on the device performance as these issues are addressed in other chapters in the book.

First of all the silicon wafer fabrication process going from crystal pulling to wafering is addressed. Whereas stringent requirements are put forward for microelectronic grade silicon employed for manufacturing integrated circuits (ICs), this is not the case for the starting material used for the fabrication of solar cells. As overall cost reduction is here a key element, metallurgical grade silicon or solar-grade silicon with a higher level of metallic contamination is allowed. Quite often gettering techniques are implemented in the process flow to reduce or even eliminate the metal concentration.

About two decades ago, extensive research started worldwide on the correlation between the quality or purity of the chemical solutions used in wafer cleaning processes and metallic contamination of the wafers. The in-depth studies resulted in a clear understanding of the underlying physical and chemical mechanisms and even in the use of optimized cost-effective and environment-friendly cleaning recipes.

As for advanced scaled-down technologies the number of masking steps strongly increases with every new generation, it is obvious that photoresist as a possible source of contamination must be tightly controlled. Associated with that are the different wet and/or dry processes used for resist stripping. In addition, the wafer handling in general (tweezers, chucks, susceptors, wafer boxes etc.) has been recognized as an important source of metallic contamination and has therefore been investigated and strongly improved.

A very important source of metallic contamination is the ion implantation process. Not only the loading of the wafers into the implantation tool, but also metal sputtering processes and the possible cross-contamination have to be taken into account. In addition, mass interference effects are observed. Therefore, implantation

C. Claeys and E. Simoen, *Metal Impurities in Silicon- and Germanium-Based Technologies*, Springer Series in Materials Science 270,
https://doi.org/10.1007/978-3-319-93925-4_3

processes are standardly monitored by so-called statistical control-charts tracing the metallic contamination behavior as a function of time.

Thermal processing may also lead to metallic contamination. Special attention has to be given to rapid thermal processing as this may involve high temperatures and fast cooling cycles, thereby promoting the quenching of metal impurities with concentrations above the solid solubility.

Although metallic contamination should be avoided, during device fabrication, metallic layers are often intentionally deposited on the wafers. Typical examples are silicidation/germanidation processes, wafer metallization, 3D wafer stacking based on through-silicon-vias, and ferroelectric memories. The potential risk of contamination has therefore to be studied and well controlled.

Overall it is essential to avoid at any time metallic contamination and when this is not possible to keep the concentration levels low enough not to impact the electrical device properties. Implementation of gettering techniques is a common strategy to follow and will be addressed in Chap. 8.

3.1 Crystal Growth

The manufacturing of single crystalline Si is mainly based on crystal pulling from the melt via either the Czochralsi (Cz) (>80% of the IC market) or Float Zone (FZ) technique, whereas for multi-crystalline Si (mc-Si) used for photovoltaics (PV) applications the directional solidification (DS) process is commonly used. The crystal growth process on itself can already be an important source for transition metals in the semiconductor material. The contamination level, however, depends not only on the quality of the feedstock but is also influenced by the contact between the hot zone of the melt and the environment (crucibles, susceptor, puller, atmosphere etc.). The possible sources in a Cz puller are schematically shown in Fig. 3.1. The dopants can be introduced either via the melt or via the ambient gas.

The metal impurities in such a crystal pulling system have been investigated by Harada et al. [1]. They observed that Al and Ni in the molten Si were mainly introduced from the crucible and the polysilicon, respectively, while both sources can result in the incorporation of Fe. For this experiment the calculated concentrations were for Fe and Ni about 10^8–10^9 cm^{-3} and for Al 10^{11}–10^{12} cm^{-3}.

For PV applications the main driving force is a cost reduction to maintain competitiveness. The production processes, aiming at a higher throughput and reduction in energy consumption during the fabrication process, are therefore using alternative growth processes such as directional solidification, casting, sheet formation, ribbon growth etc. The impurity control of the feedstock is also reduced so that higher concentration levels of transition metals are obtained than normally allowed for IC-grade material. Although much progress has been achieved over the last decade, Table 3.1 [2] gives an illustration of possible metal concentration levels that can be observed. Astropower is mc-Si sheet material, Baysix is cast material and EFG is edge-defined film-fed ribbon growth mc-Si. It should be remarked that the measured

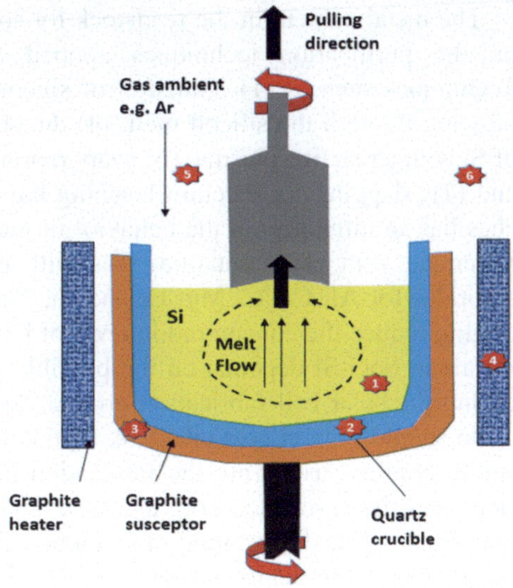

Fig. 3.1 Schematics of a Cz crystal puller and possible sources of contamination: (1) feedstock (Si + dopants), (2) quartz crucible, (3) graphite susceptor, (4) liners, (5) gas ambient and (6) the puller system

levels are often near the detection limit so that optimized analytical techniques and procedures must be used. This topic is elaborated in Chap. 4. Since the early days [3] neutron activation analysis (NAA) has been commonly used and resulted in reliable data. This technique was also used to generate the data of Table 3.1. The table indicates that rather high levels of metal contamination are observed for Fe, Ni and Cr.

Table 3.1 Average metal concentration measured by NAA for three different types of mc-Si solar cell material

Element	Astropower (cm^{-3})	BaySix (cm^{-3})	EFG (cm^{-3})
Fe	1.50×10^{16}	4.00×10^{14}	6.00×10^{14}
Ni	1.80×10^{15}	less than DL (3×10^{14})	less than DL (1.1×10^{14})
Co	9.70×10^{13}	2.10×10^{13}	1.70×10^{12}
Cu	less than DL (7.8×10^{12})	$<2.4 \times 10^{14}$	$<1.3 \times 10^{14}$
Cr	1.80×10^{15}	1.00×10^{13}	1.70×10^{12}
Hf	less than DL (1.8×10^{12})	7.80×10^{12}	less than DL (6.8×10^{11})
Mo	4.60×10^{13}	1.50×10^{13}	6.40×10^{12}
W	2.00×10^{13}	less than DL (2.2×10^{11})	less than DL (8×10^{10})
Au	4.80×10^{11}	6.50×10^{10}	2.00×10^{10}
As	4.70×10^{13}	3.40×10^{12}	less than DL (2.2×10^{10})
Sb	2.40×10^{14}	1.70×10^{12}	1.20×10^{11}
Ga	2.50×10^{13}	less than DL	9.00×10^{12}

DL stands for below detection limit (after Istratov et al. [2])

The metal content in the feedstock for solar-grade Si (SoG-Si) strongly depends on the purification techniques applied to the metallurgical-grade (MG) Si. Techniques such as (1) reduction of silicon by carbon, (2) acid leaching, (3) gas blowing through the silicon melt, (4) direct solidification, (5) melting and refining of Si with a reactive plasma, (6) evaporation of phosphorus from the Si melt surface and (7) 'slagging' or calcium leaching have been reported. Each of these approaches has an influence on the behavior of the transition metals [4]. The gas blowing technique with HCl-containing gas will lead to the formation of volatile metal chlorides for Al, Ca, C, Mg, Fe and Ti. The calcium leaching can successfully be used to reduce the concentration level of Fe and Ti. The allowable concentration of metals in SoG-Si depends on the possible implementation of gettering techniques during the solar cell fabrication process, keeping in mind that the overall process cycle should remain cost effective. Optimal understanding of defect engineering is surely required as during the fabrication process the different thermal processing steps and the associated cooling cycles may impact the metal behavior which f.i. may result in the dissociation of metal precipitates or metal complexes rendering the metals more electrically active.

An important parameter, especially for direct solidification and zone refining, is the in Chap. 2 mentioned distribution or segregation coefficient k_o giving the ratio of the impurity concentration in the solid over the concentration in the melt. As shown in Fig. 2.4, this coefficient is for most of the impurities smaller than 1, leading to a higher concentration in the melt as the solidification process proceeds. This process can be described by Scheil's equation, valid for no back diffusion into the solid and complete diffusion in the liquid, given by [5, 6]

$$C_S = k_0 C_0 (1 - f_s)^{(k_0 - 1)} \tag{3.1}$$

with C_O the concentration in the melt at the start of the solidification and f_s the fraction solidified. Due to the incomplete mixing, there exists a boundary layer δ_b at the melt in front of the solidifying interface causing an increased impurity concentration in that region. To take into account the influence of the crystal growth on the distribution coefficient, one can define an effective distribution coefficient k_{eff} given by

$$k_{eff} = \frac{1}{k^* + (1 - k^*) e^{-v_g \delta_b / D_{liq}}} \tag{3.2}$$

with k^* the distribution coefficient at equilibrium, v_g the growth rate and D_{liq} the impurity diffusion coefficient in the liquid. Fitting experimental data with Scheil's expression resulted for Fe in a $k_{eff} < 0.05$ [6]. However, by performing dedicated experiments by growing intentionally Fe contaminated mc-Si ingots a much lower value of $k_{eff} = 2 \times 10^{-5}$ is obtained [7].

For the fabrication of PV Si high throughput low-cost sawing techniques, such as wire sawing, are used for cutting the ingot into wafers. To remove the saw damage and contamination related with the cutting slurry, including heavy metallic

Table 3.2 As-cut surface contamination determined by TXRF analysis of the liquid surface etching solution (after Rip et al. [9])

Surface concentration (10^{10} cm^{-2})	Cr	Fe	Ni	Cu	Zn
Supplier A	10–12	14–70	11–16	200–1700	45–120
Supplier B	8–15	3300–4600	33–37	1300–1900	280–390

contamination, appropriate cleaning techniques are used [8]. Analysis based on vapor phase decomposition combined with Total Reflection X-Ray Fluorescence (TXRF) analysis of as cut Si PV wafers from two different suppliers pointed out trace metal levels of Cr, Fe, Ni and Cu with surface concentrations that can exceed 10^{12} cm^{-2}, as shown in Table 3.2 [9]. This can be related to the wire sawing which is in most cases a brass-coated (Cu/Zn alloy) steel wire with an abrasive, either suspended (e.g. SiC) or deposited on it.

3.2 Wet Wafer Cleaning Processes

3.2.1 Contamination in Si Cleaning Technology

Since the early days the RCA process, developed in 1970 by Werner Kern and based on hot alkaline and acidic mixtures of hydrogen peroxide, has commonly been used for cleaning silicon wafers [10]. The first step, referred to as standard solution '1' or SC-1 or APM, is a mixture of ammonium hydroxide, hydrogen peroxide and water (NH_4OH:H_2O_2:H_2O (1:1:5) at 70 °C) intended to remove organic surface residues, several metals and inorganic particles. The particles are removed by slight etching of the surface and lift off of the contamination from the surface by electrostatic repulsive forces between the particle and the surface. This is followed by a second cleaning step (SC-2 or HPM) with a mixture of dilute hydrochloric acid and hydrogen peroxide (HCl:H_2O_2:H_2O (1:1:5) at 70 °C) designed to remove alkali ions and cations such as Al^{+3}, Fe^{+3} and Mg^{+3} forming insoluble hydroxides in an alkaline solution. To remove heavy organic contamination from the wafer surface the two-step cycle is often preceded by a SPM step with hot sulphuric acid and hydrogen peroxide (H_2SO_4:H_2O_2 (4:1) at 90 °C). To improve the efficiency of SC-2 a dip in diluted HF (DHF) is included after SC-1 for removing the oxide layer at the surface which may have trapped metallic contamination during the APM step.

The beneficial effect of the cleaning cycle can be hampered or even reversed as the chemicals in the cleaning solutions may be a source of metallic contamination. The metal deposition on the surface is caused by the oxidation reduction reaction between the metal ion and the Si as the metal can take electrons from the Si [11]. The Si surface under the metal particle is oxidized and etched away by the OH^- in the solution. The etching rate depends on both the NHO_4 and the H_2O_2

concentrations. NHO_4 increases the etching rate, while it is decreased by the H_2O_2. Therefore, as will be explained further, the H_2O_2 decomposition rate has an important impact on the metal behavior and on the silicon surface roughness. The shape and the chemical composition of the metal contaminants adhering on the surface have been studied, such as e.g. for Cu [12]. The Cu ion has a higher redox potential than Si, extracting electrons from the Si and resulting into metallic Cu on the surface. The Cu nucleus is negatively charged and will attract other Cu ions coming in the vicinity, resulting in a growth of the nucleus. In general, the underlying physics of the different aspects of the cleaning process is well understood and several process optimizations have been proposed (see e.g. [11–17]).

The metal concentration level resulting from different cleaning steps is illustrated in Fig. 3.2, giving after a typical SPM-APM-DHF-HPM cycle the Fe, Cu and Zn concentrations measured at the surface by TXRF [18]. The indicated measurement error is the 1 sigma value for the averages of the measured concentrations. The Fe and Cu concentrations are in the 10^{11} cm^{-2} range. After APM a high Fe concentration is measured. Changing the ratio of the mixture has only a marginal impact. The figure clearly illustrates that metals deposited from the SC-1 solutions are partially removed during the SC-2 cleaning step.

The chemical supplier is also impacting the in-situ metal concentration. Therefore, Fig. 3.3 illustrates the impact of the chemical supplier in the case of an APM clean [18]. It can be noticed that the Fe concentration is more difficult to

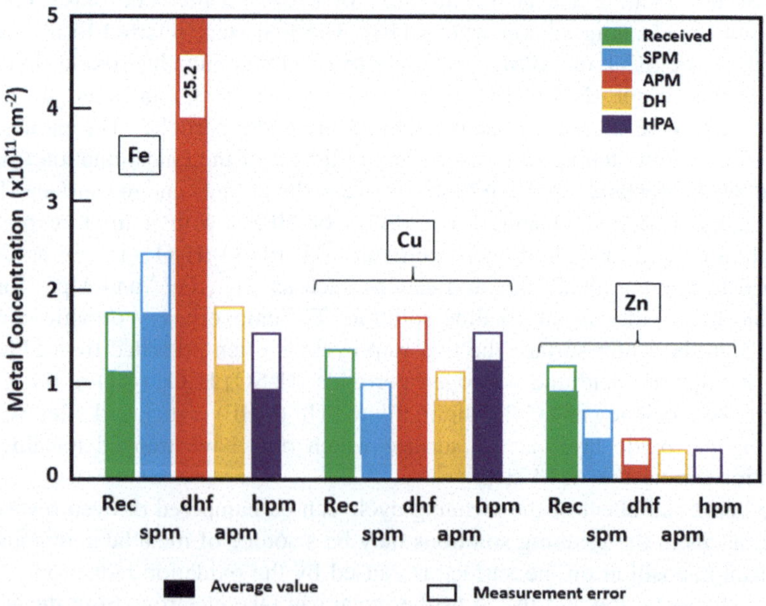

Fig. 3.2 Metal concentrations after a SPM-APM-DHF-HPM cleaning cycle. Reproduced from Anttila et al. [18] by permission of ECS-The Electrochemical Society

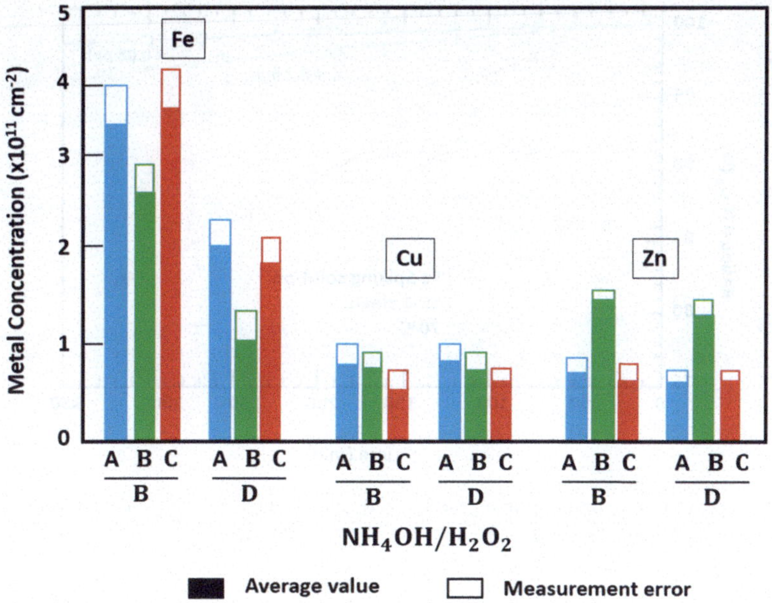

Fig. 3.3 Metal concentrations after a SPM-APM-DHF-HPM cleaning cycle using chemicals from different suppliers. Vendors B and D are the best for low Fe contaminations in the H_2O_2. Reproduced from Anttila et al. [18] by permission of ECS-The Electrochemical Society

control, while Cu and Zn is in the 1×10^{11} cm^{-2} range. Similar investigations for the HPM step indicated that the chemical supplier is less important and mostly results in final metal concentrations in the order of 1×10^{11} cm^{-2} [18].

To obtain a complete picture of the metal behavior during wet cleaning, one has also to take into account the time of the cleaning step and the age of the cleaning solution. In automated cleaning benches the chemicals have to be continuously refreshed as the decomposition of H_2O_2 leads to the evaporation of the NH_3 and H_2O according to the equation

$$2H_2O_2(aq) \leftrightarrow O_2(g) + 2\,H_2O \tag{3.3}$$

The decomposition rate depends on both the metallic and non-metallic contaminants, the *pH* of the solution and the temperature of the bath [14, 19]. The half-life of the solution decreases with increasing Fe concentration, related to the oxidation change between Fe^{3+} and Fe^{2+}, while Cu has a lower catalytic action. When both metals are in the solution they enhance each other. The impact of Fe on the decomposition rate is shown in Fig. 3.4 [14], based on Fe spiking experiments of the SC-1 cleaning solution. The impact of Fe and Cu during the SC-2 step is less pronounced. The decomposition can be stabilized by using complexing agents such as triethanolamine (TEA), ethylene diamine tetra acetic acid (EDTA), 1,2-cyclohexanediaminotetraacetic acid (CDTA), diethylenetriaminopentaacetic

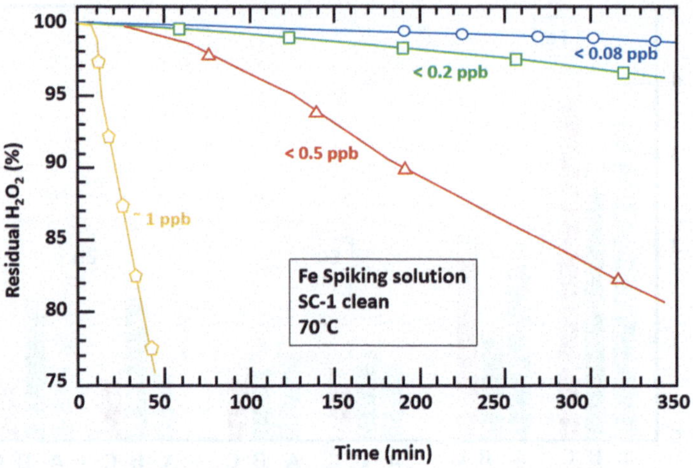

Fig. 3.4 Impact of Fe concentration on the H_2O_2 decomposition rate. Reproduced from Schmidt et al. [14]. Copyright 1995 The Japan Society of Applied Physics

acid (DTPA) and diethylenetriaminepentamethylenephosphonic acid (DTPMP) [20, 21]. The chelating agent is reacting with the metals most likely to form soluble metal hydroxides, avoiding them to be deposited on the wafer surface. The need for less refreshment cycles of the bath reduces the metallic contamination buildup in the solution and lowers the cost of the cleaning cycle. In the extreme case the complexing agents can be so effective during the SC-1 step that there is no need any more to use a SC-2 cleaning step for removing the metals.

Because of the lower levels of alkali metals such as Li, Na and K, the low volatility and cost considerations, research has been performed to add to the diluted ($NHO_4 + H_2O$) in the SC-1 clean the surfactant tetramethylammoniumhydroxide (TMAH) and/or the chelating agent EDTA [22]. Only TMAH is effective in particle removal but doesn't suppress the metal contamination. It has even been observed that the Ni concentration increases for higher TMAH concentrations and/or for longer exposure times [23]. The best results, from a viewpoint of particles and metals, are achieved by spiking TMAH with EDTA (100 ppb). This one-step clean recipe has been electrically tested for both pre-gate [22] and polysilicon post-CMP cleaning [24].

The rinse in ultra-pure water (UPW) between the cleaning steps can also be a source of metallic contamination. As pure water has a $pH = 7$ some metal oxides (e.g. Fe oxide) are according to the Pourbaix diagram thermodynamically stable [25]. The amount of metallic contamination on the silicon surface can be lowered by using an ion exchange filter [26, 27]. Elimination of the metals can be achieved by modifying the pH of the UPW as shown in Fig. 3.5, illustrating the effectiveness for removing metals like Fe, Al and Zn [28]. The addition of 0.01% HCl removes most of the metals from the spiked water solution. Good Fe and Cu removal was also achieved by UPW/O_3 [29]. In this case additional HCl spiking has no significant impact.

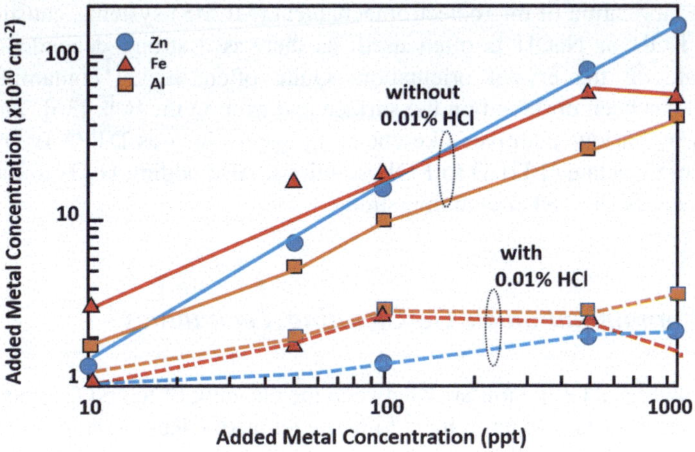

Fig. 3.5 Metal removal by 0.01% HCl ($pH = 2.6$) added to intentionally metal contaminated DI water containing less than 10 ppt contaminants. Reproduced from Tardif et al. [28] by permission of ECS-The Electrochemical Society

The majority of the metal contamination data discussed above has been obtained by investigating blanket wafers. The metal removal process, however, is more different in the case of deep trenches as then the wetting of the surface by the chemicals is more complex because of the high aspect ratio. By studying the metal removal efficiency (MRE) of intentionally contaminated samples for a blanket wafer and a wafer with 4 μm deep trenches [30], it was observed that for elements like Cr, Mo, Nb, Ta, Ti, W and Zr the metal removal efficiency is rather poor due to the poor chemical wetting in the deep trenches. Adding a chelating element and a surfactant has no strong impact on the MRE. Best results are reported for a HF-SC1-HCl sequence [30].

As a final remark it should be mentioned that several Si cleaning recipes have been proposed by different groups based on the use of diluted chemicals at room temperature [28, 31–33]. Generally, use is made of ozonated water rinsing and dilute HF treatments, whereby the ozone step removes organics and noble metals and oxidizes the wafer surface. The subsequent HF treatment removes the oxide as well as particles and metals. As a last step another ozone treatment may be used to make the surface hydrophilic. Besides effective particle removal and low metallic concentrations, these recipes are more environment friendly, require a shorter processing time and are very cost effective.

Even when using a supplier of high purity chemicals, there may be metallic contamination originating from the cleaning tank itself. For most of the fluoropolymers the reported metal contamination levels for Fe, Cu, Al, and Zn, due to bleaching out of the polymer by the wet chemicals, are in the ppb to ppm range [34]. The final concentration depends on parameters such as the pre-cleaning of the tank, the type of chemical used, the operating temperature, the lifetime of the tank etc. [35].

For the fabrication of microelectromechanical (MEMS) systems, caustic etching based on KOH or NaOH is often used, as there is a strong dependence of the etching rate on the crystal orientation. Quite often, nickel contamination by adsorption has been observed on the surface and even in the bulk [36]. This can be controlled by various additives like chelating agents such as DTPA or triethylene tetramine hexaacetate (TTHA) to KOH solutions. Also adding H_2O_2 to the etching solution reduces the nickel contamination.

3.2.2 Contamination in Ge Cleaning Technology

Although there is a lot of similarity between the cleaning of Si and Ge wafers, some special aspects for Ge cleaning have to be mentioned. Cleaning steps are based on oxidizing the surface and then removal of the oxide to reduce the particle count and the metallic impurities. The etching behavior of Ge surfaces in $HF/H_2O_2/H_2O$ has been studied in the fifties of the previous century [37, 38] and later extended for $x/H_2O_2/H_2O$ mixtures with x being acetic, tartaric, citric, oxalic acid etc. [39, 40]. Due to the renewed interest in Ge processing for both PV and IC applications, a large variety of wet cleaning chemistries have been systematically investigated for both n- and p-type Ge wafers [41]. In comparison to Si, the etching rate of the Ge surface is much faster. While for Si the formation of a thin SiO_2 passivating film inhibits the Si surface etching in non HF-containing solutions, for Ge the formed GeO_2 is water soluble so that simultaneous oxidation and etching of the oxide leads to a net Ge surface etching [42]. The Ge surface will be etched for most of the common cleaning recipes used in Si processing.

The standard SC-1 and SC-2 clean used for Si to remove particles and metals cannot be used for Ge as their etching rate for Ge is too high even if highly diluted H_2O_2 is used [41]. Therefore, alternative cleaning solutions have been proposed such as HCl/H_2O and HF/H_2O [43]. As the cleaning chemicals used can also be a source of contamination, Fig. 3.6a shows the impact of a contaminated solution on the metal absorption on the Ge surface as a results of a 1 min contact between the contaminated solution and the Ge surface. The used solution was $NH_4OH/H_2O_2/H_2O$ (1/10/x) with different H_2O_2 concentrations, whereby the NH_4OH was spiked with Ca, Ti, Fe, Cr, Cu, K and Ni. The metal concentration was measured using VPD-DC-TXRF (Vapor Phase Decomposition-Droplet Collection TXRF). It can clearly be seen that for a more concentrated SC-1 solution less contamination is found on the surface. This suggests that a higher etch rate prevents the metal deposition on the surface [43]. In contrast to Si, the metal deposition on a Ge surface is not reduced by lowering the *pH* of the solution.

Figure 3.6b illustrates the influence of the HF concentration on the metal removal rate of intentionally contaminated (spin coating, surface concentration $> 1 \times 10^{12}$ cm^{-2}) Ge wafers [44]. Although HF is efficient to remove the metallic contamination and is not etching the Ge, a concentration of 5% HF is needed to decrease the investigated contamination concentration below the 10^{10} cm^{-2} level.

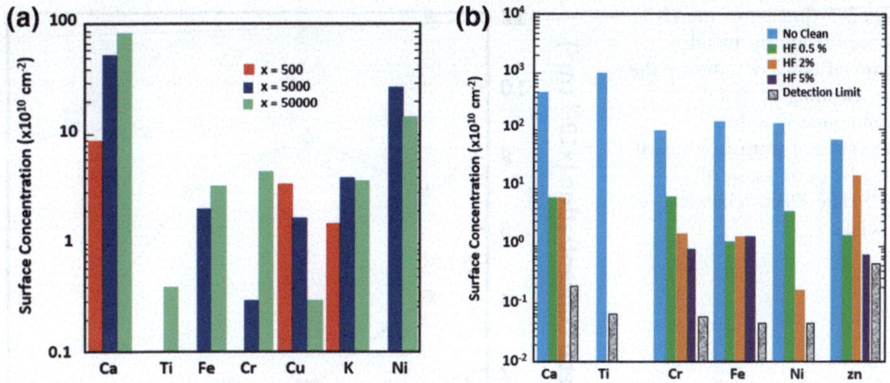

Fig. 3.6 **a** Absorption of metals on the Ge surface after a 1 min contact with an intentionally metal contaminated SC-1 cleaning solution $NH_4OH/H_2O_2/H_2O$ (1/10/x) with different concentrations of H_2O. No bar indicates that the concentration is below the detection limit (after Sioncke et al. [43]). **b** Influence of the HF concentration on the metal removal rate of intentionally contaminated (spin coating, surface concentration > 1×10^{12} at/cm^2) Ge wafers. The surface concentration is measured with VPD-TXRF. Reproduced from Hellin et al. [44] by permission of ECS-The Electrochemical Society

3.3 Dry Vapor Phase Wafer Cleaning

Dry vapor phase cleaning approaches can be used to avoid metallic contamination by the chemicals used in the cleaning solutions. Also here the offer is quite large so that this section will be restricted to outline some general aspects. It has been suggested to use a HCl-based recipe by heating the wafer to 650 °C in an HCl, 2% O_2 and N_2 ambient for 30 min [45]. During the heat cycle the metals react with O_2 and HCl to form volatile reaction products, metal chlorides and/or oxychlorides which are eliminated by evaporation. The metal removal efficiency was tested by using intentionally contaminated wafers by metal spiking the SC-1 solution. The results are shown in Fig. 3.7 and clearly indicate the sensitivity to the HCl concentration. Total metal removal is achieved for 3% HCl. It is important that the metals can evaporate before they in-diffuse into the wafer. In case of NO/HCl mixtures, whereby nitrosyl compounds are formed which are also highly volatile, the use of high pressure (100 Torr) gives also good results at 200 °C [46]. Gas phase cleaning is very benign from a material consumption viewpoint.

3.4 Photoresist Deposition and Stripping

The scaling trend aiming at smaller geometrical dimensions also resulted in the development of new photoresists to cope with i-line, immersion and deep ultra-violet (DUV) lithography, respectively. Although high-purity (10–30 ppb) resists have

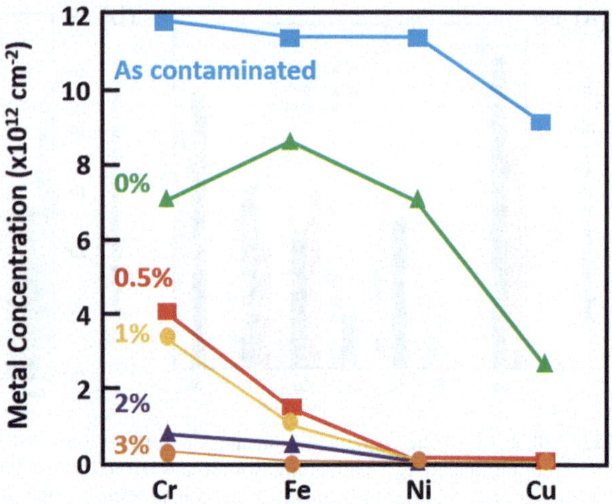

Fig. 3.7 Impact of the HCl concentration on metal removal efficiency during the dry cleaning of metal contaminated wafer. Reproduced from Limb et al. [45] by permission of ECS-The Electrochemical Society

been developed and most of the focus went to particle control and removal, there are reports on metallic contamination and its behavior during the different bake cycles. A variety of analytical techniques are available and used for this purpose [47]. Based on radioactive trace experiments it was observed that Mn and Zn are present in the DUV photoresist and their concentration in the silicon depends on the resist bake temperature [48]. As the concentration is higher in the resist than in the Si, most of the metals will remain in a near surface region. For Zn the concentration in the substrate decreases for higher baking temperatures (going from 80 °C to 120 °C), whereas for Mn it goes through an optimum around 100 °C. The model to explain this is based on two opposite driving forces, i.e., the migration of the metal ion and its solvation with the residual H_2O or photoresist before evaporation [48]. The behavior will be different depending on the underlying substrate (bare Si, SiO_2, Si_3N_4 etc.). In the case of Si the electronegativity of Mn or Zn is lower than Si so that no electrons can be picked up from the Si. The enthalpy of oxide formation will determine the morphology of the precipitates and the surface adsorption depends on the electrochemical potential and the *pH* of the solution. SC-2, SPM and DHF are even at room temperature efficient cleans to remove the Zn and Mn contamination.

The qualitative model given above is based on the definition of a *K*-factor related to the diffusivity of the impurity in the photoresist D_{resist} and in the substrate D_{sub} and the segregation coefficient at the interface k_{int}, given by

$$K^{-1} = \frac{1}{\sqrt{D_{resist}}} + \frac{1}{k_{int}\sqrt{D_{sub}}} \tag{3.4}$$

As the metal diffusivity in the resist is higher than in the solid substrate, the *K*-factor at typical bake temperatures will be $K = k_{int}\sqrt{D_{sub}}$.

To take explicitly the bake temperature in account, a simple model for the metal impurity out-diffusion from the resist to the underlying film such as Si, oxide or nitride has been proposed [49]. The model is based on the Fick's diffusion law in combination with an Arrhenius law relating the diffusivities (D_x, with x = resist or x = sub) and interfacial segregation coefficient of the metal impurity to the baking temperature T_{bake}

$$D_x = D_{x,0}\exp\left(-\frac{E_x}{k_B T_{bake}}\right) \tag{3.5a}$$

$$k_{int} = k_{int,0}\exp\left(-\frac{E_k}{k_B T_{bake}}\right) \tag{3.5b}$$

resulting in

$$K^{-1} = \frac{1}{\sqrt{D_{resist}}} + \frac{1}{k_{int}\sqrt{D_{sub}}} = \alpha_{resist}\exp\left(\frac{\lambda_{resist}}{T_{bake}}\right) + \alpha_{sub}\exp\left(\frac{\lambda_{sub}}{T_{bake}}\right) \tag{3.6}$$

with $\alpha_{resist} = (D_{resist,0})^{-1/2}$, $\lambda_{resist} = E_{resist}/2k_B$, $\alpha_{sub} = (D_{int,O})^{-1}(D_{sub,O})^{-1/2}$, $\lambda_{Sub} = (E_{sub} - 2E_k)/2k_B$. Equation (3.6) enables to explain the experimentally determined K-variation as function of the experimental parameters.

The dry etching of photoresist may result in heavy metal contamination [50]. To exclude the etching equipment itself as the main cause for metal contamination, wafers with an without photoresist have been compared as shown in Fig. 3.8. The Secondary Ion Mass Spectroscopy (SIMS) study indicates that the Fe and Pb are coming from the resist. Metallic contamination in the resist can thus be transferred onto the silicon wafer. The oxide layer underneath can depending on its thickness reduce but not eliminate the contamination, although all parameters have to be taken into account such as contamination level of the resist, energy of the plasma, design of the etching reactor, etching process, temperature etc. Some of the metals will also be embedded in the oxide layer, making it more difficult to remove them by a wet cleaning step. For modern device processing most of the resist and

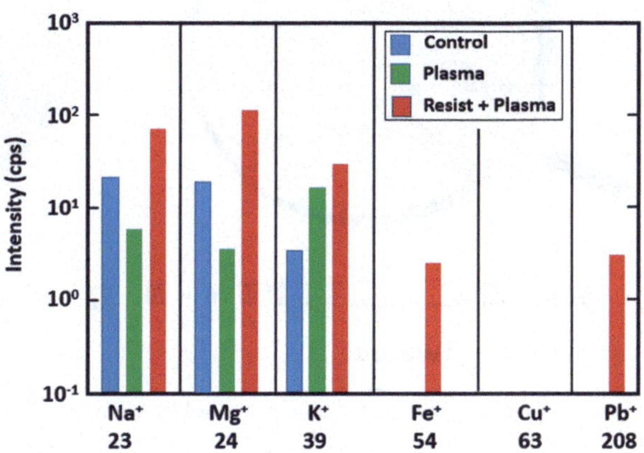

Fig. 3.8 SIMS evaluation of the Si wafer surface after resist stripping. Plasma treatment without resist performed for 1 h. The resist stripping and coating were repeated seven times. The mass of the detected elements is also indicated. Reproduced from Fujimura and Yano [50] by permission of ECS—The Electrochemical Society

stripping problems have been adequately solved. Metallic contamination is also a problem when photoresists are implanted as discussed in Sect. 3.6.

Laser cleaning enables to reduces particulates from the wafer surface without the use of water or chemicals and doesn't produce hazardous reaction products. It also offers the possibility of cleaning small defects with less than half the size of the minimum defect sizes that are removed by conventional wet cleaning. Different lasers can be used (KrF or excimer) depending on the manufacturer. The used laser power should be low enough to avoid any laser-induced surface damage. The removal efficiency is higher for Au and Cu than for W particles [51]. Typical reported metal concentration levels are 3.3×10^9 at/cm^2 and 4.6×10^{10} at/cm^2 for Fe and Al, respectively.

3.5 Wafer Handling

For small diameter wafers it was common to handle the wafers with tweezers, whereby the physical contact with the Si was an important source of metallic contamination as the tweezers were fabricated in stainless steel. The use of vacuum tweezer with a fluorocarbon coating was an improvement. However, for several decades now the use of tweezers has been abandoned, also as a result of the use of large diameter wafers. Another potential source of metal contamination are the wafer chucks. Figure 3.9 illustrates the Fe contamination, measured by surface photoconductive decay (PCD), caused by a tweezer and a wafer chuck, respectively. In the first case there is an imprint at the edge, while the chuck leads to a central spot with a lower lifetime.

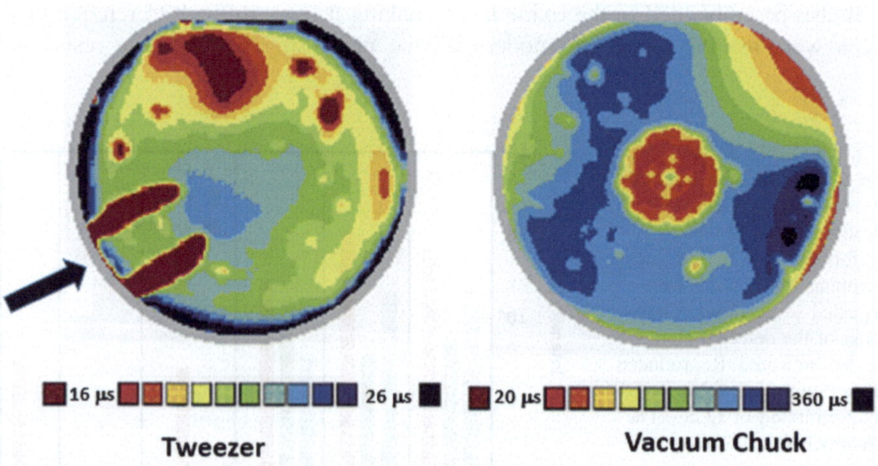

Fig. 3.9 Impact of Fe contamination, caused by wafer handling with a tweezer or vacuum chuck, on the carrier lifetime, measured by photoconductive decay. *Courtesy* Semilab

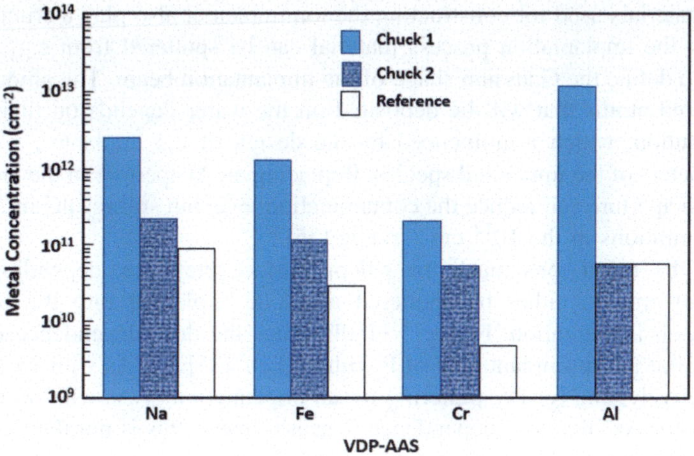

Fig. 3.10 VPD-AAS analysis of metal contamination caused by two different vacuum chucks. Reproduced from Kroninger et al. [52] with permission of Elsevier

Investigations by atomic absorption spectroscopy (AAS) combined with vapor phase decomposition (VPD), by placing the polished wafer side on the chuck and measuring at the backside, pointed out Fe, Cr and Al concentrations in the 10^{11}–10^{13} cm^{-2} range as illustrated in Fig. 3.10 for two different chucks [35, 52]. The origin of the very high Al contamination for chuck 1 was related to the chipping of the Al_2O_3 chuck coating. Overall both the number of particles and the metallic contamination should be controlled. Essential is the appropriate coating of the chucks and the control of the friction during handling.

For 300 mm wafer transfer the use of FOUPs (Front Opening Unified Pods) with a well-controlled inside ambient has become standard. Care has to be taken to avoid possible cross contamination originating from a metal contaminated wafer placed in a FOUP that afterwards is used for transferring contamination-free Si wafers [53].

3.6 Ion Implantation

Ion implantation can in different ways be the source to metallic contamination:

(1) The loading of the wafers into the ion implanter. Depending on the design of the implanter, both the clamping of the wafers and the wafer transport system can lead to metallic contamination [54]. The materials in physical contact with the wafer are either from stainless steel or a type of polymer. The shape of the clamping system in the end-station has been observed in the pattern of the lifetime mapping and can be traced back to Fe [35, 55]. Most of the aluminum alloys used contain about 1% Fe.

(2) The materials used for constructing the ion implanter also play a crucial role as during the implantation process material can be sputtered from e.g. apertures used to define the Gaussian shape of the implantation beam. The amount of the sputtered atoms that will be deposited on the wafer depends on their angular distribution, which is influenced by the design of the implanter, and on the parameters of the implanted species. Replacing an Al aperture in the end-station by a Si aperture can reduce the contamination level but still results in Al and Fe concentrations in the 10^{12} cm^{-2} range [35].

Once the metal ions reach the silicon surface they can, depending on the implant energy, either be sputtered away or implanted into the Si by the knock-on implantation. Figure 3.11 illustrates the dose dependence of the Al sputtering for an implantation of P with 45 keV [35] and As with 25 keV [54], respectively, and for Fe sputtering for an implantation of As with 60 keV [55]. While for As^+ the dose dependence is nearly linear, this is not the case for the P^+ implantation.

Figures like Fig. 3.11 are mainly intended as an illustration only, as for a detailed analysis of the contamination related to the sputter process one has to take into account the metal (Fe, Al, Mo …) that is sputtered away, the implanted species (As, P, B …), the dose and the used energy. For the same dose and energy there exists a correlation between the sputter yield (number of sputtered atoms/number of incident ions) and the metallic concentration as illustrated in Fig. 3.12 for Fe in Si [57].The Fe concentration increases with the sputter yield (or ion mass).

In the case of As^+-implantation, it has also been demonstrated that by using a Si-based behind sputtering protection board, avoiding sputtering from behind

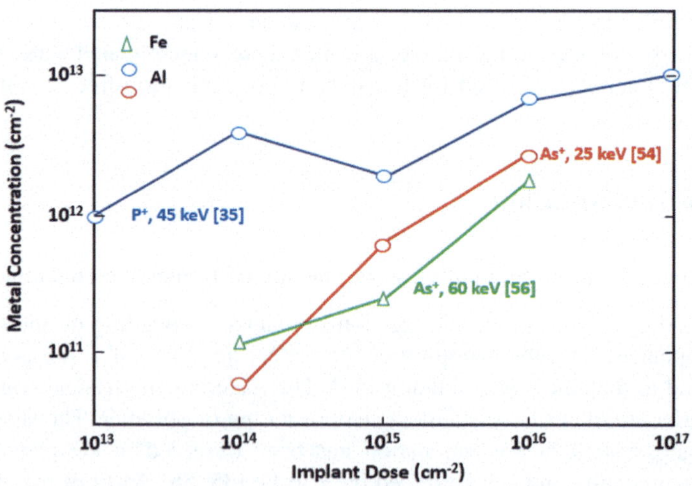

Fig. 3.11 Al concentration due to sputtering during an ion implantation (P^+, 45 keV and As^+, 25 keV), and Fe concentration (As^+, 60 keV). The data is collected from the indicated references

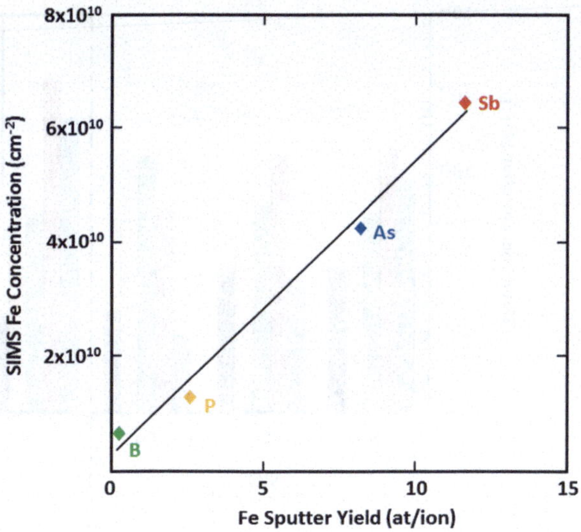

Fig. 3.12 Fe concentration measured by SIMS versus the Fe sputter yield as a function of ion mass for 10^{15} cm^{-2} implants at 80 keV. Reproduced from Todorov et al. [57]. Copyright 1999 IEEE

the wafer to reach the Si wafer by rebound, the Al concentration can be reduced by one order of magnitude [54]. For more recent generations of ion implanters it has been shown that the As diffusion length L_{As} decreases with increasing implant dose D_{As} according to [56]

$$L_{As} \approx D_{\mathrm{As}}^{-1/2} \tag{3.7}$$

When the support is coated with silicon the Fe concentration becomes independent of the implant dose and can be fully associated with the Fe contamination of the wafer backside, caused by the mechanical contact between wafer and support [56].

The penetration of the sputtered metals in the Si wafer can, similar as for dry etching, be reduced by using an implant oxide. The penetration of metals with a low diffusivity in SiO_2 is controlled by the thickness of the oxide layer, but this is not the case for fast diffusing metals such as Fe and Cu. The Fe penetration depth in Si decreases exponentially with increasing oxide thickness and the decrease is slower for high energy implants [57]. There are differences between using SiO_2 or Si_3N_4 as a screening layer, as illustrated in Fig. 3.13 for Fe, Cr and Ni [58]. The surface was before implantation (As, 10^{14} cm^{-2}, 1 MeV) by chemical spiking experiments contaminated with about 10^{13} cm^{-2} metal. After annealing at 950 °C for 10 min, it was observed that for both screening layers, most of the Fe is on the surface and in the layer itself. Cr and Cu, however, diffuse through the film into the silicon, an effect which is more pronounced for Si_3N_4 compared to SiO_2. This indicates that the film density is not a dominant factor as SiO_2 has a lower density. Ab initio calculations, using cristobalite SiO_2 and β-Si_3N_4 point out that Cr has a lower diffusion barrier in SiO_2 [58].

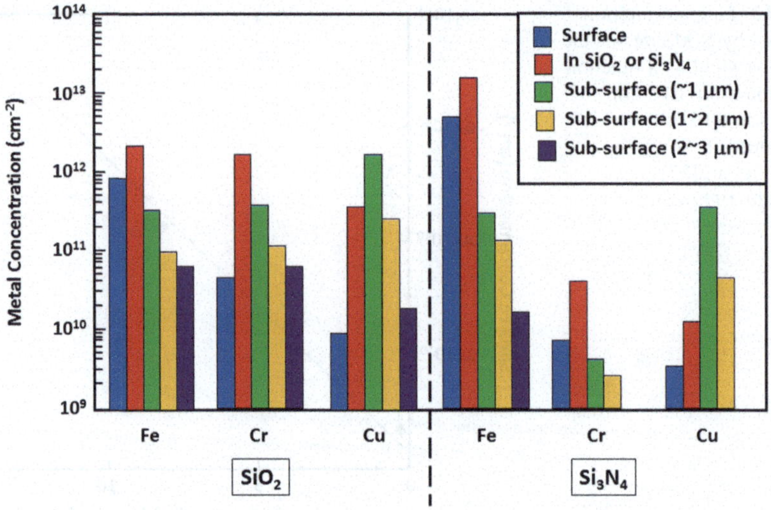

Fig. 3.13 Depth distribution of the metal concentration in Si wafers implanted with As ions at 1 MeV through a 20 nm screening oxide or nitride. The wafer received an anneal at 950 °C for 10 min. Reproduced from Saga et al. [58] by permission of ECS-The Electrochemical Society

(3) Contamination due to sputtering can also occur if an implantation is performed in an implanter which previously was used for implanting wafers with an exposed metal layer [59]. This has been demonstrated by performing an As implantation on clean wafers in an ion implanter previously used to implant wafers with a W layer. Whereas implantation in a clean implanter gives about 1.5×10^{10} cm^{-2} W, measured by TXRF, this increased after the implant in the contaminated system to 8.15×10^{10} cm^{-2} [59]. A lower content is achieved be performing several decontamination As implants on dummy wafers before the final implantation. Full W removal is only obtained for a dose as high as 6.5×10^{16} cm^{-2} As. For this type of experiments one has to use very sensitive analytical techniques such as e.g. DLTS (see Chap. 4).

(4) Contamination can also take place due to mass interference. A well-known example is the interaction with molybdenum during BF_2 implantations, whereby in the silicon the Mo level can be as high as 2×10^{18} cm^{-3} [50]. The Mo is coming from the Mo arc source and during implantation the double ionized Mo^{2+} is not filtered by the mass analyzer magnet since it has the same charge-to-mass ratio than $^{11}B^{19}F_2^{+}$. The co-implanted Mo during BF_2 implants increases exponentially with increasing ion source current [60]. For typical used BF_2 implantations in advanced devices the Mo surface concentration is in the 10^{11} cm^{-2} range [59]. A solution to this problem is to use another material for the source such as e.g. Ta. The impact of Mo on the device performance is studied in detail in Chap. 7.

Monitoring the cleanliness of the ion implantation is an issue that has to be performed on a regular basis. It is therefore possible to monitor with contamination control charts the implanter condition not only after periodic maintenance (e.g. beam line control) but also on a weekly or even daily basis by using surface photovoltage (SPV) measurement to determine the diffusion length [61].

The risk for contamination is higher for high-current implanters. Therefore, special attention has been given to the fabrication process for Silicon-on Insulator (SOI) substrates based on separation by implanted oxygen (SIMOX). The process consists of a high dose oxygen implantation followed by a high temperature anneal, typically 1×10^{18} cm^{-2} and 1250–1320 °C anneal [62]. High implantation currents are required to reduce the total process time. After implantation Fe, Ni and Cr concentrations in the 10^{17}–10^{18} cm^{-3} range have been reported and these metals have a degrading impact on the back-channel leakage current after irradiation, which can increase 300× compared to the pre-irradiation values [63]. The concentration of some metals reduces after the anneal step to about 1×10^{10} cm^{-2} [64]. Another approach to achieve metallic concentrations in the low 10^9 cm^{-2} range is by using a low dose oxygen implantation (4×10^{17} cm^{-2}) followed by a high temperature anneal (1350 °C) [65].

3.7 Thermal Processing

In the early days of silicon processing, metallic contamination was a serious problem, resulting in metal concentrations in the 10^{13}–10^{14} cm^{-3} range [3, 66]. Oxidation in a chlorine containing ambient has a strong positive impact on lowering the contamination level by the formation of volatile metal compounds [67]. HCl gas gettering is still used for metal removal during the fabrication of solar cells as will be discussed in more detail in Chap. 8. Nowadays the metallic contamination during thermal processing is limited, although at higher temperatures metals are volatile and can in the gas phase be transported from the tube or the boat to the wafer surface. This can be minimized by using high quality SiC [68].

In advanced fabrication technologies Rapid Thermal Processing (RTP) or Rapid Thermal Anneal (RTA) techniques are commonly used. Their low thermal budget enables the formation of shallow junctions, but the higher temperatures and faster cooling cycles used may result in the quenching of the metal impurities present in the material and yielding concentration levels well above the solute solubility. In addition, low-temperature RTP may dissolve Cu- and Ni-rich precipitates, while $FeSi_2$ precipitates remain unaltered [69]. High-temperature RTP will partially dissolve all three types of precipitates. It is therefore important to keep in mind that the observed behavior of metallic contamination at the end of a full process cycle will strongly depend on the number, type and sequence of the RTP steps used. This is also essential for the performance of some solar cell technologies as clearly illustrated in Fig. 3.14, giving the impact of thermal processing on the size and spatial distribution of Cu defects, thereby impacting the diffusion length [70]. This type of

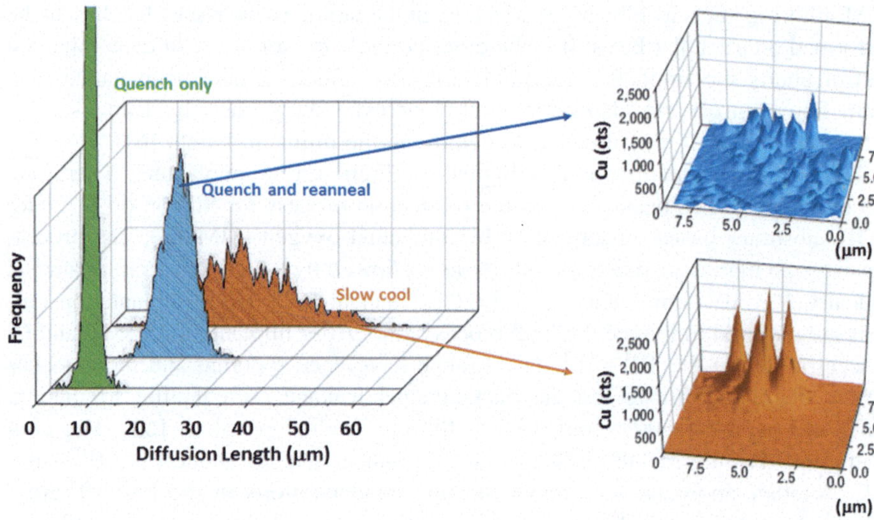

Fig. 3.14 Impact of RTP processing on the size and spatial distribution of Cu defects in Si, via lifetime measurements, showing the influence of an RTP step (quench 200 °C/s), RTP + anneal (655 °C) and RTP with slow cooling (7–8 °C/s). Reproduced from Buonassisi et al. [70] by permission of Springer

defect engineering should be optimized in view of the further used gettering techniques with as final goal the use of low cost SoG-Si for good performing solar cells.

There exists some discussion on the impact of RTA on the metal diffusivity. Experiments using wafers with Cu, Ni and Pd contamination on the backside receiving a RTA treatment show the formation of haze on the silicon front surface due to the metal flux through the wafer, which could indicate that RTA leads to an enhanced diffusion of the metal atoms [71]. Direct information on the metal diffusivity has been obtained by sputtering a metal layer and giving the wafers an RTA treatment [72]. For these experiments wafers with either a metal layer or a metal silicide (600 °C furnace pre-treatment) have been studied. As shown in Fig. 3.15 for Ti, the diffusivity data for the $TiSi_2$/Si sample is in agreement with the equilibrium value, while a lower diffusivity is observed for the Ti/Si sample due to the increase of the diffusion activation energy from 1.79 eV to about 2.9 eV. The lower diffusivity for Ti/Si may be caused by the injection of vacancies during the silicidation reaction.

The Ti solubility after RTA, however, is for both conditions strongly enhanced compared to the equilibrium value. The solubility activation energy for the Ti/Si case (2 eV) is decreased from the equilibrium value (3 eV), while it is increased for the $TiSi_2$/Si structure (5 eV). Similar experiments for Cr contamination, which is a fast diffuser compared to Ti, points out that for $CrSi_2$/Si the solubility is only slightly enhanced without a change in activation energy [72].

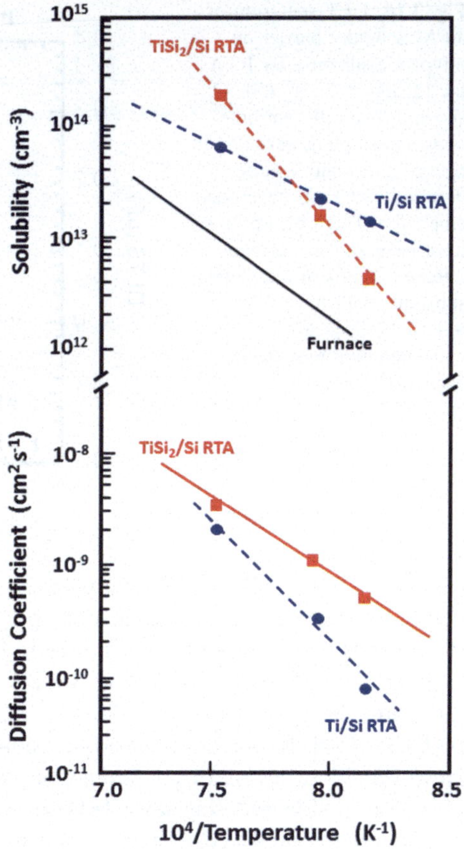

Fig. 3.15 Diffusion coefficient and solubility of Ti in Si after a RTA step of $TiSi_2$/Si and Ti/Si structures, respectively. The solid lines corresponds with the equilibrium (furnace anneal) condition. Reproduced from Mathiot et al. [72]. Copyright 1991 AIP Publishing

DLTS investigation of p-type wafers after RTA (750 °C, 15 min, N_2) reveals the presence of hole traps, as illustrated in Fig. 3.16. In the given case two hole traps are found at 160 and 210 K and the peak height increases towards the surface as shown by changing the pulse height. These acceptor level around E_V+ 0.3 eV and E_V+ 0.4 eV have in the past been assigned to Fe_sB_s complexes [73]. Scanning more in detail the literature indicates that several researchers have found hole traps for similar type of treatments, with in most cases a dominant trap at E_V+ 0.30, which is attributed to the presence of 3*d* transition metals such as Cr, Ni, Ti or Fe [74–77]. The trap can be inhibited by the growth of an oxide layer or a silicidation reaction during the RTP step, indicating that the generation is not dominated by a specific intrinsic point defect such as vacancies or interstitials, and is lower in FZ than in Cz or epitaxial wafers [77].

Heavy metals can also be introduced during the growth of epitaxial layers leading to a non-uniform distribution of the measured diffusion lengths over the wafer [78]. Copper contamination has also been reported for low-temperature epitaxial Si deposition using ion-assisted deposition. For temperatures in the range

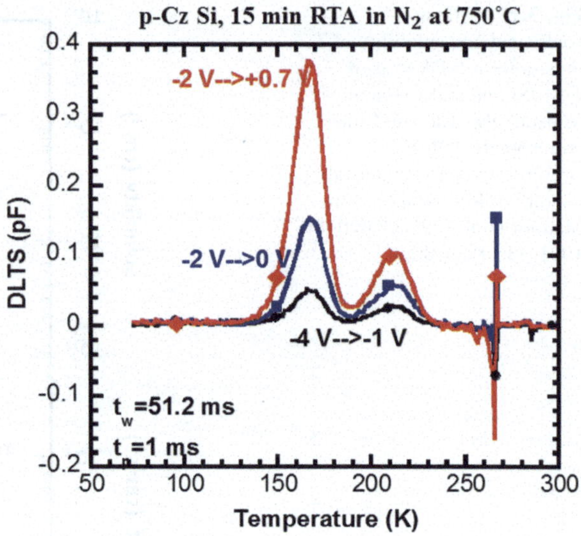

Fig. 3.16 DLT spectrum of an Al Schottky barrier on p-type Cz silicon after RTA for 15 min in N_2 at 750 °C, showing two different hole traps. A sampling period t_w = 51.2 ms and a pulse duration t_p = 1 ms have been employed. Three bias pulses have been applied, probing different depths in the silicon substrate, with the −2 V–> +0.7 V spectrum corresponding with the near-surface region

between 460 and 510 °C the relatively short recombination lifetime of the layers is determined by the incorporation of deep-level recombination centers, which, from PL analysis were shown to be Cu-related [79].

Even in commercial epitaxial wafers, Ti contamination in the range 10^9–10^{11} cm^{-3} has been observed, directly impacting the charge transfer efficiency in pn-CCDs [80]. In this case the origin could have been the SiC covered susceptor. However, these levels of contamination can be removed by a phosphorus doped poly-Si backside gettering layer [81].

If not well controlled, every thermal step can introduce metallic contamination. This has also been demonstrated for atomic layer deposition (ALD) of high-quality SiO_2 layers [82]. Using electron spin resonance (ESR) signals have been detected which could be correlated with interstitial Cr_i (5 × 10^{11} cm^{-2}) and interstitial Fe_i (10 ppt) located in a near interfacial layer of a few microns.

3.8 Metal Layers in Device Fabrication

Although metallic contamination has to be eliminated, for some fabrication processes there is a need to deposit a metal layer on the wafer surface, such as in the case of silicidation, metallization and some approaches for 3D wafer stacking by e.g. through silicon vias (TSVs). This is also the case for the metal electrodes of ferroelectric random access memories. Obvious that special pre-cautions have to be taken to minimize potential metal in-diffusion in the bulk silicon. Without going too much into details, some of the potential risks are briefly discussed and illustrated. Further chapters are dealing more in detail with the properties of the metals and their electrical impact.

3.8.1 Silicidation and Germanidation

Silicidation, based on the deposition of a metal layer on silicon followed by a low-temperature anneal step to form a metal alloy (e.g. Ti on Si to form $TiSi_2$), is commonly used to lower the contact resistivity and to replace the polysilicon interconnects in scaled down technologies, see e.g. [83–86]. Although a large variety of different metal-silicides have been studied during the last two decades in view of their suitable work function [87], Ti, Co and Ni silicides are the most frequently used alloys in Si processing. Besides a low contact resistivity, process optimization is required to control the morphology of the silicide phase, to obtain a good thermal stability, to lower the defect formation, to suppress lateral encroachment, to enable silicidation of small feature size structures and to take into account the impact of the dopant level in the silicon. Although $NiPt_{(1-x)}Si$ can be used for 22 nm technologies, attention is also given to the use of refractory metals like Er, Yb and Ir [88]. PtSi has also been considered for a long time, and successfully implemented in medium wavelength infrared (MWIR) detectors for thermal imaging [89], but the Pt radioactive decay can be a source of alpha particles and can cause significant soft error problems in memory devices.

As strain engineering is required to improve the drive current of the devices, like e.g. by the use of $Si_{1-x}Ge_x$ raised source/drain regions, the strain effect on the silicidation process has to be taken into account. For Ge-based devices a germanidation process using NiGe is extensively studied for both p- and n-type contacts [90–93]. Besides the stability of the contacts, attention has to be given to the enhanced doping effects, segregation of dopants near the interface and control of the Fermi-level pinning effect. The latter can be overcome by increasing the doping concentration and/or the use of barrier layers [92].

Special attention has to be given to possible metal contamination in the silicon resulting in an increased leakage current. This is illustrated in Fig. 3.17 for $TiSi_2$ n^+p and p^+n Si junctions for different Ar anneal temperatures [94]. It can be noticed that the n^+p leakage current increases after 300 °C anneal and reaches a maximum at 460 °C which is about 6 orders of magnitude higher than for the as-deposited sample. The p^+n junction has an opposite behavior with a minimum at 460 °C which is two orders of magnitude lower than for the as-deposited diode. Activation energy studies pointed out that for the p^+n junction the measured activation energy is about 0.9 eV which can be associated with the band gap energy, pointing towards a diffusion current as the origin of the leakage. For the n^+p junction the impact of a generation-recombination level at 0.16 eV is observed. The in-diffusion of Ti in the n^+-region causes an additional tunnel current. It is assumed that the Ti difussion occurs at low temperature before the start of the silicidation process, which is known to be dominated by Si diffusion. Systematic analysis of the different leakage current components (area, perimeter and corner) of p^+n junctions points out that a Ti-silicidation mainly influences the corner component [95]. Also for n^+p junctions the corner component plays a dominant role [96].

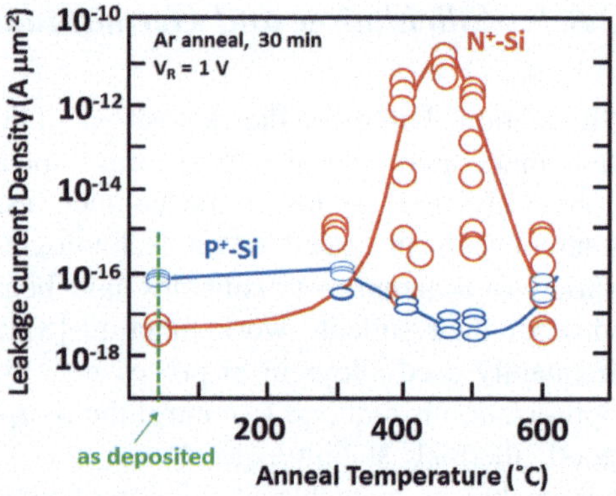

Fig. 3.17 Leakage current density at a reverse voltage $V_R = 1$ V of Ti-silicided n^+- and p^+-Si junctions as a function of the anneal temperature. Reproduced from Sakata et al. [92]. Copyright 1997 The Japan Society of Applied Physics

During Co-silicidation Co is the diffusing species instead of Si. Also in this case an increased leakage current has been observed for n^+p diodes, while for the p^+n diodes there is only an impact on the forward current [97]. A direct correlation has been observed between the in diffusion of Co and the leakage current profile as illustrated in Fig. 3.18 [98]. The connection between the two profiles enables to calculate the effective leakage per Co atom. The calculated Co capture cross section is 2.4×10^{-13} cm^2. Implementing additional RTA steps and/or using a TiN capping only slightly impacts the leakage current depth profile so that the profile is more or less set during the initial phase of the $CoSi_2$ formation process.

It is important to remark that an increased leakage current can also be caused by defect generation during the silicidation process, as depending on the type of silicide, silicon interstitials and/or vacancies are generated. As mentioned before also stress present in the layers can have an influence. This has been confirmed as the leakage current increases with trench depth for both $TiSi_2$ and $CoSi_2$ [98]. An associated level at 0.6 eV has been observed to influence both p^+n and n^+p junctions and its concentration can be influenced by a pre-amorphization implantation

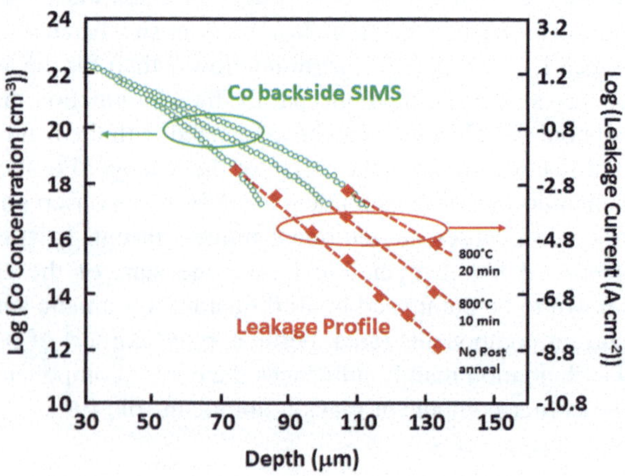

Fig. 3.18 Comparison between the Co SIMS profile and the leakage-depth profile after different anneal steps. The 20 nm stretch of the depletion layer is taken into account. The connection of the profiles allows a rough estimate of the effective leakage current per Co atom. Reproduced from Tsuchiaki et al. [98]. Copyright 2002 The Japan Society of Applied Physics

(PAI) of Ge ions before silicidation [99, 100]. This implant releases stress at the interface and has a beneficial impact on the interface roughness. Additional factors impacting the leakage current level, but not discussed within the scope of this Chapter, are the morphology of the silicide, defect formation inside the silicide layer and silicide spiking. The morphology and defect formation can be controlled by using a Si capping layer on the source/drain junctions.

The metals cannot only diffusive in the silicon substrate but also in the dielectric layers (isolation layers, gate dielectrics, spacers), resulting in shorts between, e.g., source/drain and the gate or between devices. In the case of rare earths residual contamination levels of 1.5–87 × 10^{12} at/cm^2 have been observed and for Yb the concentration cannot be lowered by more aggressive cleaning steps [101]. On the other hand Er can be completely removed.

For achieving low contact resistivity on Ge, fundamental investigations have been performed on a large variety of different metals for the germanidation process [90]. X-ray diffraction pointed out that NiGe and PdGe enables not only the lowest sheet resistivity but also remain stable over a wide temperature range. Therefore, Ni, Pd and Pt have been further studied from a process integration viewpoint for n-type Ge [102]. For Ni only deep levels, above the DLTS detection limit of $\sim 10^{10}$ cm^{-3} are found starting around 400 °C as shown in Fig. 3.19 [103]. However, the maximum Ni concentration found at 450 °C is about 3.9×10^{11} cm^{-3}, which is too low to significantly impact the device performance in a circuit as at room temperature a generation current of $\sim 8 \times 10^{-15}$ A is created per substitutional Ni atom.

In the case of Pd or Pt deep levels at $E_C - 0.36$ eV and $E_C - 0.285$ eV are observed related to defects most likely associated with vacancies as excess

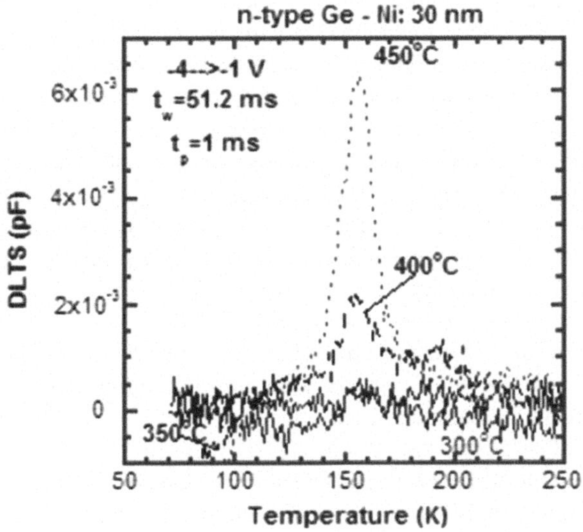

Fig. 3.19 Deep-Level Transient spectra (t_w = 51.2 ms, t_p = 1 ms and a pulse from −4 to −1 V) for a 4 mm diameter NiGe Schottky barrier on n-Ge for different RTA temperatures: 300, 350, 400 and 450 °C

vacancies are generated due to the Ge diffusion during the germanidation process [102]. These levels anneal out at temperatures around 400 °C. In contrast to Ni there is clearly no in-diffusion of metals.

3.8.2 *Metallization*

Where for several decades Al and its alloys have been used as standard metallization technique, more advanced technologies switched in the late nineties over to Cu metallization to improve the electrical conductivity and electromigration resistance [104]. A number of copper deposition processes such as electrochemical deposition (ECD), physical vapor deposition (PVD), and chemical vapor deposition (CVD) are pursued for damascene structure. Metal nitride barrier layers can be implemented to prevent migration of Cu within the wafer. The deposition processes can result in process-induced metal contamination via an uncovered barrier layer. In addition, in the exposed areas there can be poor adhesion near the wafer edge of the wafer so that material can flake off during subsequent wafer handling, including chemical-mechanical polishing (CMP) and/or metrology steps for determining layer properties, linewidth control, uniformity etc. This will lead to metal contamination during further processing. Metallic contamination at the backside of the wafer will not only result in metal in-diffusion but can also be the cause for cross-contamination, thereby contaminating the used equipment, wafers processed in the same batch and/or wafers processed later in the same tool. It is therefore common practice to use dedicated tools in the fabrication area for processing, metrology inspection and handling wafers with a Cu layer [105]. Control charts, based on TXRF measurement of control samples, can be used to monitor the statistical variation of the Cu level in the Cu processing area [106].

Backside contamination early in the process results in Cu in-diffusion, thereby impacting the device junction leakage, breakdown voltage and reliability if the concentration is high enough [107]. Cu backside contamination at the back-end stage of the processing, however, is less a problem as even after 10 h anneal at 400 °C no impact on the performance of the devices on the front-side of the wafer have been noticed [108], although the risk for cross contamination remains. In order to prevent Cu in-diffusion in the dielectric oxides between different metal levels, diffusion barriers such as single or double layers of e.g. TiN, TaN, SiO_2, Si_3N_4 have to be implemented. A poor barrier layer quality may result in Cu contamination in the polysilicon and/or source/drain regions, resulting in spiking through the junction and/or near extension regions and possibly also through the gate dielectric, degrading the front-end yield and reliability [109]. Figure 3.20 is a Focused Ion Beam—Scanning Electron Microscope (FIB-SEM) illustration of Cu-silicide defects. At the drain the defect protrudes upwards in the gate with a gate-to-drain short as a result. The Cu-silicide is extending deeper in the p^+-substrate contact than in the n^+ source/drain region.

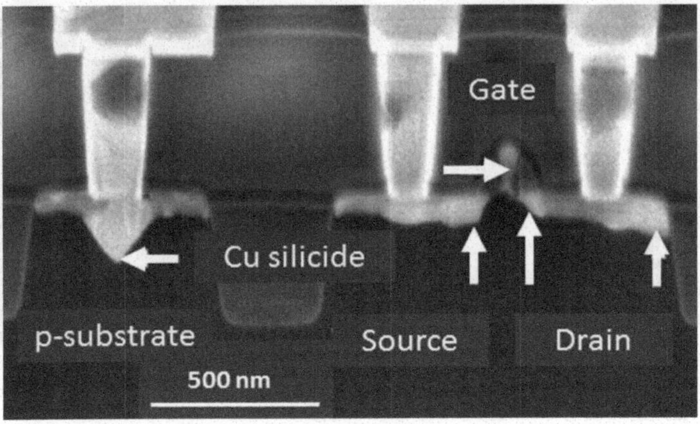

Fig. 3.20 FIB/SEM photograph of an nMOS transistor with excessive gate-to-drain leakage. The arrows indicate Cu-silicide defects protruding into the gate with a gate-to-drain short as a result. The Cu-silicide extends deeper into the p^+-substrate than in the n^+ source/drain regions (after Van den bosch et al. [109])

In case of multiple metallization layers the connection between the layers (i.e., vias) is done by metal plugs, based on either W or Cu. Reliability studies pointed out that there is no difference between using W or Cu plugs [110].

CMP is used to planarize shallow trench isolation, oxide inter-level dielectrics, W contacts or vias, and copper interconnects in damascene architectures [e.g. 111–113]. Depending on the used slurry for the etching, this may lead to metallic contamination. A typical example is the alkali-based $Cu(NO_3)_2$ + NH_4F etch, relying on the reaction [114]:

$$Si + 2Cu^{2+} + 6F^- \rightarrow SiF_6^{2-} + 2Cu^\circ \tag{3.8}$$

to remove Cu from the surface. One of the major effects by Cu in-diffusion is the passivation of shallow acceptors (B, Al, In, Ga) observed in electrical measurements [114–116]. The silica slurry can also be the source of Ni contamination [116]. Experiments with intentional Cu or Ni contaminated slurry can be used to estimate the metal diffusivity at low temperatures [117]. The obtained room temperature Ni diffusivity is $>10^{-9}$ cm^2s^{-1}, which is about two orders of magnitude smaller than the value obtained by extrapolating data measured at higher temperatures [118].

3.8.3 *3D Integration—Through Silicon Vias (TSV)*

To increase the functional packing density much attention is given to 3D integration approaches going beyond System on Chip (SoC) and leading to stacked-die with

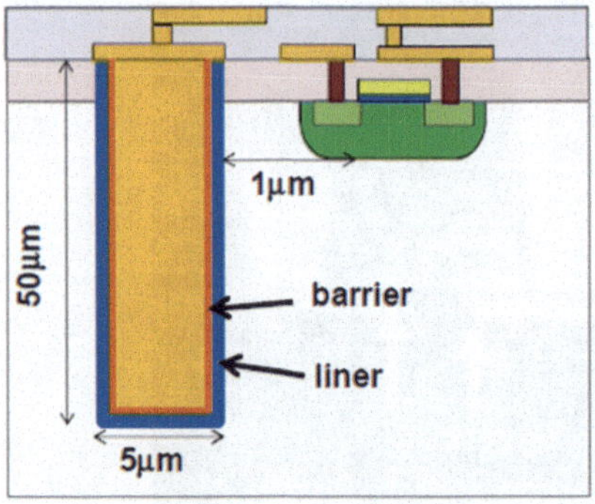

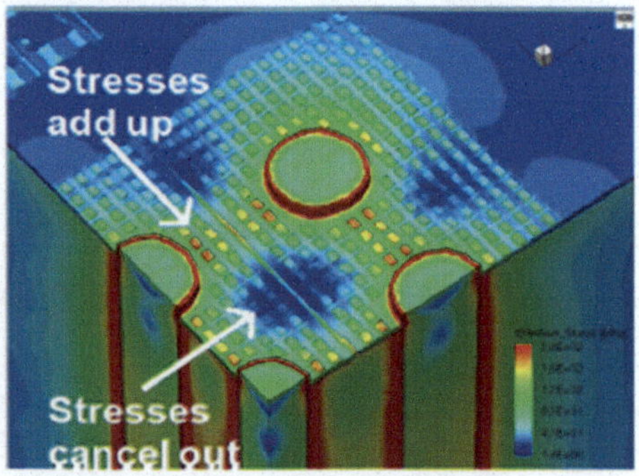

Fig. 3.21 Schematic illustration of the TSV process (**a**) and top-view showing mechanical stress around the vias (**b**)

wirebond, stacked-die via through silicon vias, package on package (PoP) and package in package (PiP). The advantage of stacking dies on each other is besides the higher packing density also a much higher interconnect density and a reduction of the packaging cost, although the design complexity and the testing for good die increase [119]. A schematic illustration of the TSV process is given in Fig. 3.21. After etching the silicon, a liner (e.g. TEOS/O_3 CVD) is deposited before depositing a barrier layer (e.g. Ta PVD) followed by the via filling with a metal. At the end of the process CMP is used. Standardly Cu is used as a metal to fill the via. However, Cu has a high constant of thermal expansion (CTE) (2.55 ppm/°C at 25 °C) in relation to the silicon substrate (16.6 ppm/°C at 25 °C). This mismatch results in a large accumulation of thermomechanical stress when TSVs are exposed to high temperatures and/or temperature cycles, potentially resulting in device failure. During thermal cycling this may results in cracks.

The stress around vias could influence the nearby circuits and Cu diffusion. The region around the contacts where no devices should be placed is called the Keep Out Zone. Leakage current measurements and lifetime analysis indicated that there is no degradation due to possible Cu contamination [120]. The measured Cu concentration in the Si region was around 5×10^{10} cm^{-3} [121]. The use of TSV has also no impact on the transistor degradation, liner breakdown or Cu diffusion when the Keep Out Zone is respected [122, 123].

To minimize the mismatch in CTE, an Invar alloy has been proposed as via metal fill [124]. Invar, a low-cost ferromagnetic material commercially available in wire form and enabling very high aspect-ratio TSVs with inherently void-free metal cores, is a Ni-Fe alloy containing 36% of nickel featuring an extremely low CTE that is well matched to the CTE of silicon. No metal contamination has been reported.

3.8.4 Ferroelectric Memories

Ferroelectric random access memories (FeRAM), based on perovskite $Pb(Zr,Ti)O_3$ (PZT), $SrBi_2TaO_9$ (SBT) or $BiFeO_3$ as capacitor dielectric, are using an oxidizing ambient to crystallize the layer so that noble metals like Pt, Ir or Ru or conductive oxides like IrO_2 or RuO_2 are commonly used as an electrode [125]. A schematic drawing of a ferroelectric memory is shown in Fig. 3.22.

Pt is a fast diffusing element and can degrade the device performance. Experiments on controlled contaminated Pt devices indicated that even for a Pt contamination of 4×10^{14} cm^{-2} the increase in leakage current is only 3 nA/cm^2 [126]. Gate oxide integrity measurements of capacitors with the gate oxide grown on intentionally contaminated samples (spiked solutions) indicated that Pt has no degrading impact [127]. This is also observed for several other metals as illustrated in Fig. 3.23 giving the oxide defect density of 4.5 nm gate oxides at 11 MV/cm [128]. For memory applications Pt, Ba, Sr and Ru are important. The most detrimental are the column III elements and V and Co.

It should be remarked that metal intentionally deposited on the wafer surface behaves differently in an oxidizing or inert ambient as illustrated in Fig. 3.24. The metal contaminated wafers are either oxidized at 800 °C (4.5 nm SiO_2) or 950 °C (20 nm SiO_2) or treated for 30 min in an inert ambient at 950 °C [129]. Without making an in-depth analysis it can be seen that (i) some metals like Ca, Co, W and Fe have a larger affinity with oxide most likely due to the formation of oxide complexes, (ii) metals like Mo and Mn have a larger contamination in the native oxide after the N_2 anneal, (iii) the remaining metal concentration in the oxide is temperature dependent and decreases with increasing oxidation temperature (for the same oxide thickness, not shown in the figure) [129]. It can be noticed that Pt has

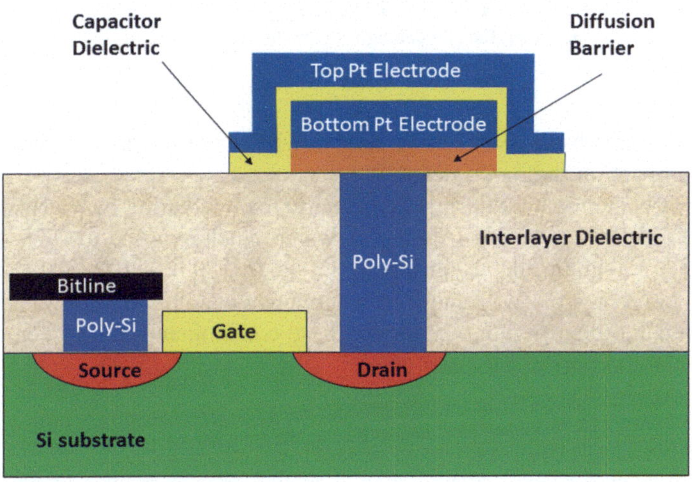

Fig. 3.22 Schematic illustration of a ferroelectric memory

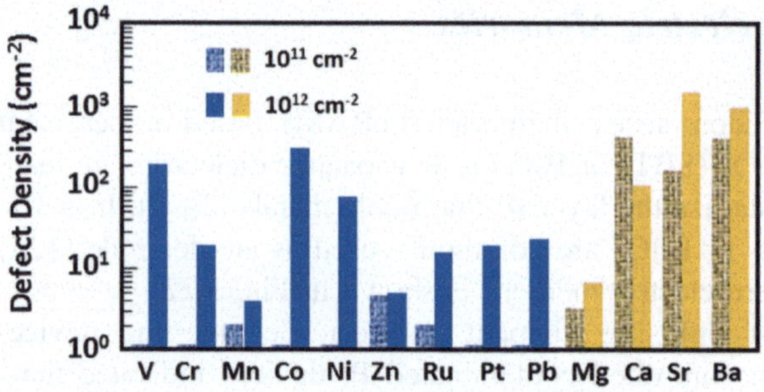

Fig. 3.23 Defect density in 4.5 nm gate oxides grown on intentionally contaminated samples. The data are obtained at 11 MV/cm from ramped voltage stress of capacitors with an area of 3.84 mm^2 [128]. In the case of Mg, Ca, Sr and Ba an area of 0.13 mm^2 was used

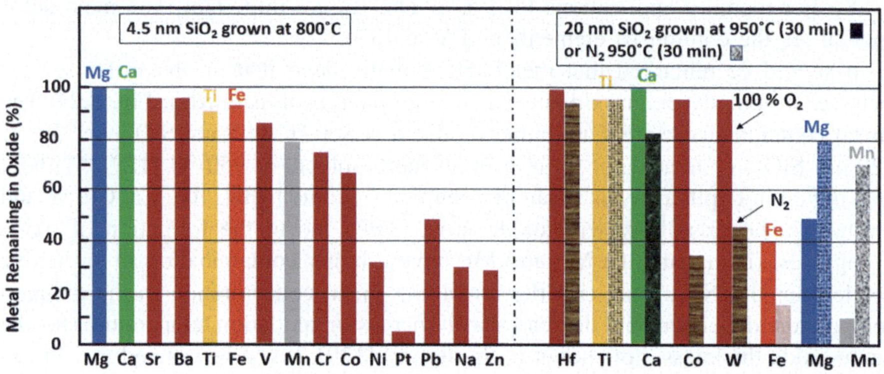

Fig. 3.24 Intentionally metal contaminated wafers to a level of about 10^{12} at/cm^2 are annealed in an oxidizing (4.5 nm SiO_2 at 800 °C or 20 nm SiO_2 at 950 °C) or inert ambient (950 °C for 30 min). The graph indicates the percentage of metal remaining in the oxide (data from Borde et al. [129])

no large affinity with the oxide, in agreement with the observation in Fig. 3.24 that Pt has no impact on the oxide defect density.

The cross-contamination during ferroelectric manufacturing has been studied for metals such as Pb, Zr, Ti and Ir [130]. A wafer chuck exposed to PZT or Ir films resulted in a maximum metal transfer of 1.1×10^{11}, 9.0×10^{10}, 5.1×10^{10} and 1.0×10^{9} cm^{-2} for Pb, Zr, Ti and Ir, respectively. Overall PZT films have a greater transfer potential than Ir films, however, these contaminants can be removed efficiently by an appropriate Si wafer clean.

References

1. K. Harada, H. Tanaka, J. Matsubara, Y. Shimanuki, H. Furuya, Origins of metal impurities in single crystal Czochralski silicon. J. Cryst. Growth **154**, 47–53 (1995). https://doi.org/10.1016/0022-0248(95)80028-X
2. A.A. Istratov, T. Buonassisi, R.J. McDonald, A.R. Smith, R. Schindler, J.A. Rand, J.P. Kalejs, E.R. Weber, Metal content of multicrystalline silicon for solar cells and its impact on minority carrier diffusion length. J. Appl. Phys. **94**, 6552–6559 (2003). https://doi.org/10.1063/1.1618912
3. P.F. Schmidt, C.S. Pearce, A neutron activation analysis study of the sources of transition group metal contamination in the silicon device manufacturing process. J. Electrochem. Soc. **128**, 630–637 (1981). https://doi.org/10.1149/1.2127472
4. A.A. Istratov, T. Buonassisi, M.D. Pickett, M. Heuer, E.R. Weber, Control of metal impurities in "dirty" multicrystalline silicon for solar cells. Mater. Sci. Eng. B **134**, 282–286 (2006). https://doi.org/10.1016/j.mseb.2006.06.023
5. E. Scheil, Bemerkungen zur Schichtkristallbildung. Z. Metallkd. **34**, 70–72 (1942)
6. D. Macdonald, A. Cuevas, A. Kinomura, Y. Nakano, L.J. Geerligs, Transition-metal profiles in a multicrystalline silicon ingot. J. Appl. Phys. **97**, 033523/1-7 (2005). https://doi.org/10.1063/1.1845584
7. R. Kvande, L.J. Geerligs, G. Coletti, L. Arnberg, M. Di Sabatino, E.J. Ovrelid, C.C. Swanson, Distribution of iron in multicrystalline silicon ingots. J. Appl. Phys. **104**, 064905/1-9 (2008). https://doi.org/10.1063/1.2956697
8. G. Allardyce, R. Barr, R. Chan, M. Moynihan, C. O'Connor, T. Ridler, Interaction between post wire saw cleaning and the subsequent cell fabrication saw damage etch and texturing process, in *35th IEEE Photovoltaic Specialists Conference* (2010), pp. 003494–003497. https://doi.org/10.1109/pvsc.2010.5614741
9. J. Rip, K. Wostyn, P. Mertens, S. De Gendt, M. Claes, Methodology for measuring trace metal surface contamination on PV silicon substrates. Energy Procedia **27**, 154–159 (2012). https://doi.org/10.1016/j.egypro.2012.07.044
10. W. Kern, D.A. Puotinen, Cleaning solution based on hydrogen peroxide for use in silicon semiconductor technology. RCA Rev. **31**, 187–205 (1970)
11. M. Itano, F.W. Kern, M. Miyashita, T. Ohmi, Particle removal from silicon wafer surface cleaning process. IEEE Trans. Semicond. Manufact. **6**, 258–267 (1993). https://doi.org/10.1109/66.238174
12. H. Morinaga, M. Suyama, T. Ohmi, Mechanism of metallic particle growth and metal-induced pitting on Si wafer surface in wet chemical processing. J. Electrochem. Soc. **141**, 2834–2841 (1994). https://doi.org/10.1149/1.2059240
13. G. Quimet, D.L. Rath, S.L. Cohen, E.E. Fisch, G.W. Gale, Defect reduction and cost savings through re-inventing RCA cleans, in *Proceedings of IEEE/SEMI Advanced Semiconductor Manufacturing Conference (ASMC)* (1996), pp. 308–313. https://doi.org/10.1109/asmc.1996.558026
14. H.F. Schmidt, M. Meuris, P.W. Mertens, A.L.P. Rotondaro, M.M. Heyns, T.Q. Hurd, Z. Hatcher, H_2O_2 decomposition and its impact on silicon surface roughening and gate oxide integrity. Jpn. J. Appl. Phys. **34**, 727–731 (1995). https://doi.org/10.1143/JJAP.34.727
15. K. Yamamoto, A. Nakamura, U. Hase, Control of cleaning performance of an ammonia and hydrogen peroxide mixture (APM) on the basis of a kinetic reaction model. IEEE Trans. Semicond. Manufact. **12**, 288–294 (1999). https://doi.org/10.1109/66.778192
16. C. Lee, N. Jo, C. Hwang, H.J. Kim, W. Lee, SC1 cleaning effect on electrical characteristics of 256 Mbit mobile DRAM with dual gate oxide. J. Electrochem. Soc. **151**, G683–G687 (2004). https://doi.org/10.1149/1.1786931
17. A. Siddiqui, M. Keswani, B. Broks, A. Fuerst, S. Raghavan, A study of hydrogen peroxide decomposition in ammonia-peroxide mixtures (APM). Microelectr. Eng. **102**, 68–71 (2013). https://doi.org/10.1016/j.mee.2012.04.003

18. O. Anttilla, M.V. Tilli, M. Schaekers, C. Claeys, Effect of chemicals on metal contamination on silicon wafers. J. Electrochem. Soc. **139**, 1180–1185 (1992). https://doi.org/10.1149/1.2069362
19. H.F. Schmidt, M. Meuris, P.W. Mertens, S. Verhaverbeke, M. Heyns, L. Hellemans, J. Snauwaert, K. Dillenbeck, Silicon surface roughening by the decomposition of hydrogen peroxide. Proc. Electrochem. Soc. **94**(7), 102–110 (1994)
20. M. Baeyens, W. Hub, B.O. Kolbesen, A.R. Martin, P.W. Mertens, Single step alkaline cleaning solution for advanced semiconductor cleaning. Solid-State Phemon. **56–66**, 23–26 (1999). https://doi.org/10.4028/www.scientific.net/SSP.65-66.23
21. H. Saloniemi, T. Visti, S. Eränen, A. Kiviranta, O. Anttilla, Improvement of SC-1 bath stability by complexing agents. Physica Scripta **101**, 226–229 (2002). https://doi.org/10.1238/Physica.Topical.101a00226
22. T.M. Pan, T.L. Lei, T.S. Chao, M.C. Liaw, F.S. Ko, C.P. Lu, One-step cleaning solution to replace the conventional RCA two-step recipe for pre-gate oxide cleaning. J. Electrochem. Soc. **148**, G315–G320 (2001). https://doi.org/10.1149/1.1369374
23. D. Sinha, Behavior of nickel deposition on silicon wafers from TMAH and ammonia based SC-1 cleaning process. ECS Trans. **69**(8), 77–83 (2015). https://doi.org/10.1149/06908.0077ecst
24. T.M. Pan, T.L. Lei, C.C. Chen, T.S. Chao, M.C. Liaw, W.L. Yang, M.S. Tsai, C.P. Lu, W.H. Chang, Novel cleaning solution for polysilicon film post chemical mechanical polishing. IEEE Electron Dev. Lett. **21**, 338–340 (2000). https://doi.org/10.1109/55.847373
25. M. Pourbaix, *Atlas of Ectrochemical Equilibria in Aqueous Solutions* (Pergamon Press, London, 1974)
26. T. Umeda, A. Tsuzuki, R. Kokuun, K. Yoneda, Further reduction of trace level ion from ultra-pure water and its effect on electrical property of device. Solid-State Phenom. **103–104**, 233–236 (2005). https://doi.org/10.4028/www.scientific.net/SSP.103-104.233
27. Y. Hashimoto, M. Amari, M. Komatsu, K. Fujiwara, Purification of trace amount of metal impurity from ultra pure wafer using membrane purifier/filter. Solid-State Phenom **103–104**, 265–268 (2005). https://doi.org/10.4028/www.scientific.net/SSP.103-104.265
28. F. Tardif, T. Lardin, B. Sandrier, P. Boelen, R. Matthews, L. Kaskoush, R. Novak, Performance of the 'DDC': diluted dynamic clean before 4.5 nm gate oxide. Proc. Electrochem. Soc. **97**(35), 15–22 (1997)
29. F. Pipia, E. Bellandi, B. Crivelli, M. Alessandri, Fe and Cu removal efficiency in HF-DIW/O_3 cleaning sequence. Solid-State Phenom. **65–66**, 109–112 (1999). https://doi.org/10.4028/www.scientific.net/SSP.65-66.109
30. P. Garnier, H. Fontaine, Metal removal efficiency in high aspect ratio structures. Solid-State Phenom. **255**, 313–318 (2016)
31. M. Meuris, P.W. Mertens, A. Opdebeeck, H.F. Schmidt, M. Depas, G. Vereecke, M. Heyns, A. Philipossian, IMEC clean: a new concept for particle and metal removal on Si surface. Solid State Techn. **38**, 109–114 (1995)
32. J.D. Joo, J.S. Kim, H. Morita, T. Ohmi, Comparison of cleaning efficiencies of noble metals on Si surface between O_3-UPW and SPM. Proc. Electrochem. Soc. **97**, 280–22 (1997)
33. T. Ohmi, Total room temperature wet cleaning for Si substrate surface. J. Electrochem. Soc. **143**, 2957–2964 (1996). https://doi.org/10.1149/1.1837133
34. J.R. Moody, S.E. Beary, D.S. Bushee, P.J. Paulsen, High purity fluoropolymer materials: trace element content and leaching. J. Cryst. Growth **89**, 43–48 (1988). https://doi.org/10.1016/0022-0248(88)90070-X
35. H. Ryssel, L. Frey, N. Streckfuss, R. Schork, F. Kroninger, T. Falter, Contamination control and ultrasensitive chemical analysis. Appl. Surf. Sci. **63**, 79–87 (1993). https://doi.org/10.1016/0169-4332(93)90068-M
36. D. Sinha, Nickel contamination from caustic etching of silicon wafers. ECS Trans. **80**(2), 223–232 (2017). https://doi.org/10.1149/08002.0223ecst
37. P.R. Camp, A study of etching rate of single-crystal germanium. J. Electrochem. Soc. **102**, 586–593 (1955). https://doi.org/10.1149/1.2429917

38. B.W. Batterman, Hillocks, pits and etch rate in germanium. J. Appl. Phys. **28**, 1236–1241 (1957). https://doi.org/10.1063/1.1722624
39. M.F. Ehman, J.W. Faust, W.B. White, The influence of the complexing agent concentration on the etch rate of germanium. J. Electrochem. Soc. **118**, 1143–1447 (1971). https://doi.org/10.1149/1.2408350
40. A. Kagawa, T. Mikawa, T. Kaneda, Chemical etching of germanium with H_2PO_4–H_2O_2–H_2O solution. Jpn. J. Appl. Phys. **21**, 1616–1618 (1982). https://doi.org/10.1143/JJAP.21.1616
41. D.P. Brunco, B. De Jaeger, G. Eneman, J. Mitard, G. Hellings, A. Satta, V. Terzieva, L. Souriau, F.E. Leys, G. Pourtois, M. Houssa, G. Winderickx, E. Vrancken, S. Sioncke, K. Opsomer, G. Nicholas, M. Caymax, A. Stesmans, J. Van Steenbergen, P.W. Mertens, M. Meuris, M.M. Heyns, Germanium MOSFET devices: advances in materials understanding, process development and electrical performance. J. Electrochem. Soc. **155**, H552–H561 (2008). https://doi.org/10.1149/1.2919115
42. B. Onsia, T. Conard, S. De Gendt, M. Heyns, I. Hoflijk, P. Mertens, M. Meuris, G. Raskin, S. Sioncke, I. Teerlinck, A. Theuwis, J. Van Steenbergen, C. Vinckier, A study of the influence of typical wet chemical treatments on the germanium wafer surface. Solid-State Penom. **103–104**, 27–30 (2005). http://doi.org/10.4028/www.scientific.net/SSP.103-104.27
43. S. Sioncke, B. Onsia, K. Struys, J. Rip, R. Vos, M. Meuris, P. Mertens, A. Theuwis, Metal deposition behavior on Ge surfaces. ECS Trans. **1**, 220–227 (2005)
44. D. Hellin, J. Rip, R. Bonzom, D. Nelis, S. Sioncke, G. Brammertz, M. Caymax, M. Meuris. S. De Gendt, C. Vinckier, How trace-analytical techniques contribute to the research and development of Ge and III/V semiconductor devices. ECS Trans. **3**(7), 173–181 (2006). https://doi.org/10.1149/1.2355806
45. Y. Limb, B.Y. Nguyen, P. Tobin, Removal of Fe and Al by pyrochemical cleaning. Proc. Electrochem. Soc. **94**(7), 409–415 (1997)
46. C. Elsmore, R. Gluck, P. Carr, M. Meuris, P.W. Mertens, M.M. Heyns, UV activated cleaning using NO, HCl and NO/HCl, in *Proceedings of the 2nd International Symposium on Ultra-Clean Processing of Silicon Surfaces* (Leuven, Belgium, 1994), pp. 125–129
47. J.A. Crighton, J. Carroll, B. Fairman, J. Haines, M. Hinds, Atomic spectrometry update—industrial analysis: metals, chemical and advanced materials. J. Anal. At. Spectr. **11**, 461–508 (1996). https://doi.org/10.1039/JA996110461R
48. M.Y. Wang, F.H. Ko, T.K. Wang, C.C. Yang, T.Y. Huang, Characterization and modeling of out-diffusion of manganese and zinc impurities from deep ultraviolet photoresist. J. Electrochem. Soc. **146**, 3448–3454 (1999). https://doi.org/10.1149/1.1392495
49. T.S. Yang, N.T. Hsu, K.S. Chen, F.H. Ko, Analysis of experimental data for metal impurity out-diffusion from deep-UV photoresist. J. Electrochem. Soc. **151**, G149–G154 (2004). https://doi.org/10.1149/1.1640631
50. S. Fujimura, H. Yano, Heavy metal contamination from resists during plasma stripping. J. Electrochem. Soc. **135**, 1995–2001 (1988). https://doi.org/10.1149/1.2095922
51. C. Curran, J.M. Lee, K.G. Watkins, Ultraviolet laser removal of small metallic particles from silicon wafers. Optics Laser Eng. **38**, 405–415 (2002). https://doi.org/10.1016/S0143-8166(02)00022-2
52. F. Kroninger, N. Streckfuss, L. Frey, T. Falter, C. Ryzlewicz, L. Plitzner, H. Ryssel, Application of advanced contamination analysis for qualification of wafer handling systems and chucks. Appl. Surf. Sci. **63**, 93–98 (1993). https://doi.org/10.1016/0169-4332(93)90070-R
53. A. Shimazaki, H. Sakurai, M. Iwase, R. Yoshimura, T. Tada, Metallic contamination control in leading-edge ULSI manufacturing. Solid-State Phenom. **145–146**, 115–116 (2009). https://doi.org/10.4028/www.scientific.net/SSP.145-146.115
54. K. Tomita, T. Nigita, S. Shimonishi, T. Shibata, T. Ohmi, T. Nitta, Eliminating metal sputtering contamination in ion implanter for low temperature annealed, low-reverse-bias-current junctions. J. Electrochem. Soc. **142**, 1692–1698 (1995). https://doi.org/10.1149/1.2048641

55. M.L. Polignano, C. Bresolin, F. Cazzaniga, A. Sabbadini, G. Queirolo, Investigation of metal contamination by photocurrent measurement: validation and application to ion implantation processes. Proc. SPIE **2638**, 14–26 (1995). https://doi.org/10.1117/12.221187
56. M.L. Polignano, D. Caputo, A. Giussani, V. Soncini, G. Di Toma, Metal contamination reduction in the evolution of ion implantation technology, in *Proceedings of the 10th International Conference on Ion Implantation Technology* (IEEE, 2000), pp. 686–689. https://doi.org/10.1109/.2000.924246
57. S.S. Todorov, A.F. Bertuch, M.L. Polignano, D. Caputo, Correlation of screen oxide thickness and iron levels introduced during high-current implants, in *Proceedings of the International Conference on Implantation Technology* (1999), pp. 646–649. https://doi.org/10.1109/iit.1999.812199
58. K. Saga, R. Ohno, D. Shibata, S. Kobayashi, K. Sueoka, Behavior of transition metals penetrating silicon substrate through SiO_2 and Si_3N_4 films by arsenic ion implantation and annealing. ECS J. Solid State Sci. Techn. **4**(5), P131–P136 (2015). https://doi.org/10.1149/2.0061505jss
59. M.L. Polignano, D. Codegoni, A. Galbiati, S. Grasso, I. Mica, G. Moccia, G. Nardone, F. Russo, Characterization of metal-contamination effects in silicon. ECS J. Solid State Sci. Techn. **5**, P3048–P3058 (2016). https://doi.org/10.1149/2.0081604jss
60. A. Cubina, M. Frost, Effect of molybdenum contamination resulting from BF_2 implantation. Nucl. Instr. Phys. Meth. Res. B **55**, 160–165 (1991). https://doi.org/10.1016/0168-583X(91)96154-D
61. E. Colelli, A. Galbiati, D. Caputo, M.L. Polignano, V. Soncini, G. Salva, Metal contamination monitoring in ion implantation technology, in *Proceedings of the International Symposium on Plasma and Process-Induced Damage* (2003), pp. 81–84. https://doi.org/10.1109/ppid.2003.1200923
62. J.P. Colinge, *Silicon-on-Insulator Technology: Materials to VLSI*. 2nd edn. (Kluwer Acad Press, 1997), chapter 2
63. A.C. Ipri, L. Jastrzebski, D. Peters, The effect of heavy metal contamination in SIMOX on radiation hardness of MOS transistors. IEEE Electron Dev. Lett. **10**, 571–573 (1989). https://doi.org/10.1109/55.43143
64. L. Frey, N. Kroninger, N. Streckfuss, H. Ryssel, Characterization of metal impurities in silicon-on-insulator materials. Mat. Sci. Eng. B **12**, 195–198 (1992). https://doi.org/10.1016/0921-5107(92)90285-H
65. M. Tachimori, S. Masui, T. Nakajima, K. Kawamura, I. Hamaguchi, T. Yano, Y. Nagatake, Quality improvement of SIMOX wafers by low-dose and high temperature oxidation techniques. Proc. Electrochem. Soc. **96**(3), 52–62 (1996)
66. P.F. Schmidt, Furnace contamination and its remedies. Solid-State Techn. **6**, 147–152 (1983)
67. P.H. Robinson, F.P. Heiman, Use of HCl gettering in silicon device processing. J. Electrochem. Soc. **118**, 141–143 (1971). https://doi.org/10.1149/1.2407929
68. M. Itsumi, Contamination determination for silicon carbide cantilever forks in diffusion furnaces. J. Electrochem. Soc. **141**, 1304–1308 (1994). https://doi.org/10.1149/1.2054914
69. T. Buonassisi, A.A. Istratov, S. Peters, C. Ballif, J. Isenberg, S. Riepe, W. Warta, R. Schindler, G. Willeke, Z. Cai, B. Lai, E.R. Weber, Impact of metal silicide precipitate dissolution during rapid thermal processing of multicrystalline silicon solar cells. Appl. Phys. Lett. **87**, 121918-1/3 (2005). https://doi.org/10.1063/1.2048819
70. T. Buonassisi, A. Istratov, M.A. Marcus, B. Lai, Z. Cia, S.M. Heald, E.R. Weber, Engineering metal-impurity nanodefects for low-cost solar cells. Nature Mat. **4**, 676–679 (2005). https://doi.org/10.1038/nmat1457
71. D.R. Sparks, N.S. Alvi, K. Sanders, Long-range diffusion of transition metals in silicon during rapid thermal annealing. MRS Proc. **146**, 385–390 (1989). https://doi.org/10.1557/PROC-146-385
72. D. Mathiot, D. Barbier, Solubility enhancement of metallic impurities in silicon by rapid thermal annealing. J. Appl. Phys. **69**, 3878–3881 (1991). https://doi.org/10.1063/1.348444

73. C.K. Tang, E. Lund, E.V. Monakhov, J. Mayandi, A. Holt, B.G. Svensson, Electrically active centers introduced in p-type Si by rapid thermal processing. Phys. Status Solidi C **3**, 725–725 (2012). https://doi.org/10.1002/pssc201000263
74. D. Barbier, M. Remram, J.F. Joly, A. Laugier, Defect-state generation in Czochralski-grown (100) silicon rapidly annealed with incoherent light. J. Appl. Phys. **61**, 156–160 (1987). https://doi.org/10.1063/1.338848
75. D. Mathiot, Quenched-in defect removal through silicide formation by rapid thermal processing. Appl. Phys. Lett. **58**, 131–133 (1991). https://doi.org/10.1063/1.104950
76. M. Ada-Hanifi, A. Chantre, D. Levy, J.P. Gonschond, Ph. Delpech, A. Nouailhat, Leakage mechanisms of titanium n^+ p junctions fabricated using thermal processing. Appl. Phys. Lett. **58**, 1280–1282 (1991). https://doi.org/10.1063/1.104336
77. N.E. Chabane-Sati, L. Thibaud, S. Kaddour, M. Berenguer, D. Barbier, Hole trap level generation in silicon during rapid thermal annealing: influence of substrate type and process conditions. J. Appl. Phys. **71**, 3320–3324 (1992). https://doi.org/10.1063/1.350952
78. L. Jastrzebski, Heavy metal contamination during integrated-circuit processing: measurement of contamination level and internal gettering efficiency by surface photovoltage. Mat. Sci. Eng. B **4**, 113–121 (1989). https://doi.org/10.1016/0921-5107(89)90226-2
79. R.B. Bergmann, L. Oberbeck, T.A. Wagner, High-quality and low-temperature epitaxial Si films deposited at very high deposition rate. J. Crys. Growth **225**, 335–339 (2001). https://doi.org/10.1016/S0022-0248(01)00906-X
80. N. Krause, H. Soltau, D. Hauff, J. Kemmer, D. Stötter, L. Strüder, J. Weber, Metal contamination analysis of the epitaxial starting material for scientific CCDs. Nucl. Instr. Meth. A **439**, 228–238 (2000). https://doi.org/10.1016/S0168-9002(99)009171
81. S.E. Holland, C.J. Bebek, P.J. Daniels, K.S. Dawson, J.H. Emes, D.E. Groom, S.R. Jelinski, A. Karcher, W.F. Kolbe, N.P. Palaio, C.H. Tran, N.A. Roe, G. Wang, Technology development for 4 k $\times$ 4 k, back-illuminated, fully depleted scientific CCD imagers, in *Proceedings of 2007 IEEE Nuclear Science Symposium Conference* (2007), pp. 2220–2225 https://ieeexplore.ieee.org/iel5/4436263/4436479/04436592
82. A.P.D. Nguyen, A. Stesmans, D. Hiller, M. Zacharias, Near-interface substrate 3*d* metal contamination during atomic layer deposition processing detected by electron spin resonance. J. Appl. Phys. **111**, 114308-1/6 (2012). https://doi.org/10.1063/1.4709445
83. S.P. Murarka, Transition metal silicides. Ann. Rev. Mater. Sci. **13**, 117–183 (1983). https://doi.org/10.1146/annurev.ms.13.080183.001001
84. K. Maex, Silicides for integrated circuits: $TiSi_2$, $CoSi_2$. Mater. Sci. Eng. Reports **11**, 53–153 (1993). https://doi.org/10.1016/0927-796X(93)90001-J
85. J.P. Gambino, E.G. Colgan, Silicides and Ohmic contacts. Mater. Chem. Phys. **52**, 99–146 (1998). https://doi.org/10.1016/S0254-0584(98)80014-X
86. N. Breil, C. Lavoi, A. Ozcan, F. Baumann, N. Klymko, K. Nummy, B. Sun, J. Jordan-Sweet, J. Yu, F. Zhu, S. Narasimha, M. Chudzik, Challenges of nickel silicidation in CMOS technology. Microelectron. Eng. **137**, 79–87 (2015). https://doi.org/10.1016/j.mee.2014.12.013
87. S.P. Murarka, Silicide thin films and their applications in microelectronics. Intermetallics **3**, 173–186 (1995). https://doi.org/10.1016/0966-9795(95)98929-3
88. P.R. Besser, C. Lavoi, C. Murray, C. D'Emic, K. Ohuchi, Silicide challenges for 22 nm technology and beyond. ECS Trans. **13**(1), 277–388 (2008). https://doi.org/10.1149/1.2911520
89. T.L. Lin, J. Park, S.D. Gunapala, E.W. Jones, H.M. Del Castillo, M.M. Weeks, P.W. Pellegrini, 7-μm-Cutoff PtSi infrared detector for high-sensitivity MWIR applications. IEEE Electron Dev. Lett. **16**, 94–96 (1995). https://doi.org/10.1109/55.363236
90. S. Gaudet, C. Detavernier, A.J. Kellock, P. Desjardins, C. Lavoi, Thin film reactions of transition metals with germanium. J. Vac. Sci. Techn. A **24**, 474–485 (2006). https://doi.org/10.1116/1.2191861

91. M. Shayesteh, C.L.M. Daunt, D. O'Connell, V. Djara, M. White, B. Long, R. Duffy, NiGe contacts and junction architectures for P and As doped germanium devices. IEEE Trans. Electron Dev. **38**, 3801–3807 (2011). https://doi.org/10.1109/ted.2011.2164801
92. C. Claeys, A. Firrincieli, K. Martens, J.A. Kitl, E. Simoen, Contact technology schemes for advanced Ge and III-V technology, in *Proceedings of 8th International Caribbean Conference Devices, Circuits and Systems* (2012), pp. 1–6. https://doi.org/10.1109/iccdcs.2012.6188889
93. N. Duan, J. Luo, G. Wang, J. Liu, E. Simoen, S. Mao, H. Radamson, X. Wang, J. Li, W. Wang, C. Zhao, T. Ye, Reduction of NiGe/n and p-Ge specific contact resistivity by enhanced dopant segregation in the presence of carbon during nickel germanidation. IEEE Trans. Electron Dev. **63**, 4546–4549 (2016). https://doi.org/10.1109/TED.2016.2610461
94. A. Sakata, M. Tomita, M. Koike, M. Koyama, I. Kunishima, Anomalous junction leakage behavior of Ti self-aligned silicide contacts on ultra-shallow junctions. Jpn. J. Appl. Phys. **36**, 1558–1562 (1997). https://doi.org/10.1143/JJAP.36.1558
95. H.D. Lee, J.M. Hwang, Accurate extraction of reverse leakage current components of shallow silicided p^+n junctions for quarter and sub-quarter-micron MOSFETs. IEEE Trans. Electron Dev. **45**, 1848–1850 (1998). https://doi.org/10.1109/16.704389
96. H.D. Lee, S.G. Lee, S.H. Lee, Y.J. Lee, J.M. Hwang, Characterization of corner-induced leakage current of a shallow silicided n^+p juntion for quarter-micron MOSFETs. Jpn. J. Appl. Phys. **37**, 1179–1183 (1998). https://doi.org/10.1143/JJAP.37.1179
97. H.D. Lee, M.S. Bae, H.H. Ji, K.M. Lee, S.H. Park, M.J. Jang, J.H. Lee, K.S. Yoon, J.H. Choi, G.S. Park, K.K. Kang, Y.J. Park, Characterization of the co-silicide penetration depth in the junction area for 0.15 and sub-0.15 micron CMOS technology. Jpn. J. Appl. Phys. **41**, 2445–2449 (2002). https://doi.org/10.1143/JJAP.41.2445
98. M. Tsuchiaki, C. Hongo, A. Takashima, K. Ohuchi, Intrinsic junction leakage generated by cobalt in-diffusion during $CoSi_2$ formation. Jpn. J. Appl. Phys. **41**, 2437–2444 (2002). https://doi.org/10.1143/JJAP.41.2437
99. D. Codegoni, G.P. Carnevale, C. De Marco, I. Mica, M.L. Polignano, Leakage current and deep levels in $CoSi_2$ silicided junctions. Mat. Sci. Eng. **124–125**, 349–353 (2005). https://doi.org/10.1016/j.mseb.2005.08.125
100. M. Tsuchiaki, A. Murakoshi, C. Hongo, Systematic investigation of leakage current suppression by pre-silicidation implantation for $CoSi_2$ formation on shallow n^+p Si diodes. Jpn. J. Appl. Phys. **42**, 1847–1854 (2003). https://doi.org/10.1143/JJAP.42.1847
101. C. D'Emic, K. Ohuchi, C. Murray, C. Lavoi, C. Scerbo, R. Carruthers, P.R. Besser, B. Yang, Material and integration issues for rare-earth silicides as gate and diffusion contacts in advanced CMOS. ECS Trans. **13**(1), 389–396 (2008). https://doi.org/10.1149/1.2911521
102. E. Simoen, K. Opsomer, C. Claeys, K. Maex, C. Detavernier, R.L. Van Meirhaeghe, S. Forment, P. Clauws, Point-defect generation in Ni-, Pd- and Pt-germanide Schottky barriers on n-type germanium. J. Electrochem. Soc. **154**, H857–H861 (2007). https://doi.org/10.1149/1.2759832
103. E. Simoen, K. Opsomer, C. Claeys, K. Maex, C. Detavernier, R.L. Van Meirhaeghe, S. Forment, P. Clauws, A deep level transient spectroscopy study of nickel germanide Schottky barriers on-type germanium. Appl. Phys. Lett. **88**, 183506/1-3 (2006). https://doi.org/10.1063/1.2199615
104. D. Edelstein, J. Heidenreich, R. Goldblatt, W. Cote, C. Uzoh, N. Lustig, P. Roper, T. McDevitt, W. Motsiff, A. Simon, J. Dukovic, R. Wachnik, H. Rathore, R. Schulz, L. Su, S. Luce, J. Slattery, Full copper wiring in a sub-0.25 μm CMOS ULSI technology. Proc. IEDM 773–776 (1997). https://doi.org/10.1109/iedm.1997.650496
105. S.M. Merchant, G. Gibson, M. Gross, D. Ibbotson, W. Josephson, S. Lytle, M. Oh, F. Stevie, S. Vatkavage, Recent developments in copper interconnects. Proc. Electrochem. Soc. **99** (31), 91–98 (1999)
106. H.A. Bakar, Z. Awang, W.A.A. Razali, Minimizing Cu contamination risk and establishment of control limit for floor surface when running both Al and Cu in a shared production

line, in *Proceedings of the IEEE International Symposium on Integrated Circuits* (2007), pp. 259–262. https://doi.org/10.1109/isicir.2007.4441847
107. M. Inohara, H. Sakurai, T. Yamaguchi, H. Tomita, T. Iijima, H. Oyamatsu, T. Nakayama, H. Yoshimura, Y. Toyoshima, Copper contamination induced degradation of MOSFET characteristics and reliability. Digest VLSI Techn. 26–27 (2000). https://doi.org/10.1109/vlsit.2000.852755
108. S. Prasad, How serious is the issue of backside Cu contamination in CMOS integrated circuits. Proc. Optoelectron. Microelectron. Mat. Dev. 25–28 (2000). https://doi.org/10.1109/commad.2000.1022884
109. G. Van den bosch, S. Demuynck, Z. Tokai, G. Beyer, M. Van Hove, G. Groeseneken, Impact of copper contacts on front-end yield and reliability. Digest IEDM 1–4 (2006). https://doi.org/10.1109/iedm.2006.346967
110. T. Kauerauf, G. Butera, K. Croes, S. Demuynck, C.J. Wilson, P. Roussel, C. Drijbooms, H. Bender, M. Lofrano, B. Vandevelde, Z. Tokei, G. Groeseneken, Degradation and failure analysis of copper and tungsten contacts under high fluence stress, in *Proceedings of the 2010 IEEE International Reliability Physics Symposium (IRPS)*—IRPS 712–716 (2002). https://doi.org/10.1109/irps.2010.5488744
111. G. Banerjee, R.L. Rhoades, Chemical mechanical planarization. ECS Trans. **13**(4), 1–19 (2008). https://doi.org/10.1149/1.2912973
112. M. Krishnan, J.W. Nalaskowski, L.M. Cook, Chemical mechanical planarization: slurry chemistry, materials and mechanisms. Chem. Rev. **110**, 178–204 (2009). https://doi.org/10.1021/cr900170z
113. D. Zhao, X. Lu, Chemical mechanical polishing: theory and experiment. Friction **1**, 306–326 (2013). https://doi.org/10.1007/s40544-013-0035-x
114. D. Ballutaud, P. de Mierry, A. Aucouturier, E. Darque-Ceretti, Influence of thermal treatments on the distribution of contaminating copper near the surface of silicon: a comparative SIMS and XPS study. Appl. Surf. Sci. **47**, 1–8 (1991). https://doi.org/10.1016/0169-4332(91)90096-3
115. Th. Prescha, T. Zundel, J. Weber, H. Prigge, P. Gerlach, Fast diffusing defects induced by copper in silicon. Mater. Sci. Eng. B **4**, 79–82 (1989). https://doi.org/10.1016/0921-5107(89)90220-1
116. M. Weling, V. Dunton, L. Zhang, R. Annapragada, CMP integration issues for sub 0.15 μm process technologies. Proc. Electrochem. Soc. **99–37**, 517–527 (1999)
117. N. Yarykin, J. Weber, Nickel in silicon: room temperature in-diffusion and interaction with radiation defects. Phys. Status Solidi C, 1700005/1–7 (2017). https://doi.org/10.1002/pssc.201700005
118. J. Lindroos, D.P. Fenning, D. Backlund, E. Verlage, A. Gorgulla, S.K. Estreicher, H. Savin, T. Buonassisi, Nickel: a very fast diffuser in silicon. J. Appl. Phys. **113**, 204906-1/7 (2013). https://doi.org/10.1063/1.4807799
119. J.P. Gambino, S.A. Adderly, J.U. Knickerbocker, An overview of through-silicon-via technology and manufacturing challenges. Microelectron. Eng. **135**, 73–106 (2015). https://doi.org/10.1016/j.mee.2014.10.019
120. N. Watanabe, M. Aoyago, D. Katagawa, T. Bandoh, T. Itsui, E. Yamamoto, Evaluation of a TSV reveal process using direct Si/Cu grinding and residual metal removal, in *Proceeding of IEEE 65th Electronic Components and Technology Conference* (2015), pp. 1452–1456. https://doi.org/10.1109/ectc.2015.7159788
121. N. Watanabe, M. Aoyago, D. Katagawa, T. Bandoh, T. Itsui, E. Yamamoto, Improvement of a TSV reveal process comprising direct Si/Cu grinding and residual metal removal, in *Proceeding of IEEE 66th Electronic Components and Technology Conference* (2016), pp. 1259–1264. https://doi.org/10.1109/ectc.2016.381
122. T. Kauerauf, A. Branka, K. Croes, A. Redolfi, Y. Civale, C. Torregiani, G. Groeseneken, E. Beyne, Effect of TSV presence on FEOL yield and reliability, in *Proceedings of Reliability Physics Symposium—IPRS* (2010), pp. 712–716. https://doi.org/10.1109/iedm.2006.346967

123. C. Croes, J. De Messemaeker, Y. Li, W. Guo, V. Pedreira, V. Cherman, M. Stucchi, I. De Wolf, E. Beyne, Reliability challenges related to TSV integration and 3-D stacking. IEEE Des. Test **33**, 37–45 (2016). https://doi.org/10.1109/MDAT.2015.2501302
124. M.A. Asiatici, A.C. Fischer, G. Stemme, F. Niklaus, Through silicon vias with invar metal conductor for high-temperature applications. J. Microelectron. Syst. **26**, 158–168 (2017). https://doi.org/10.1109/JMEMS.2016.2624423
125. H. Ishiwara, Ferroelectric random access memories. J. NanoSci. Nanotechn. **12**, 7619–7627 (2012). https://doi.org/10.1166/jnn.2012.6651
126. H. Boubekeur, T. Mikolajick, W. Pamler, J. Hopfner, L. Frey, H. Ryssel, Platinum contamination issues in ferroelectric memories. J. Appl. Phys. **92**, 3257–3265 (2002). https://doi.org/10.1063/1.1500414
127. H. Boubekeur, T. Mikolajick, N. Nael, C. Dehm, W. Pamler, A. Bauer, L. Frey, H. Ryssel, Impact of platinum contamination on ferroelectric memories. Int. Ferroelectr. **37**, 75–81 (2001). https://doi.org/10.1080/10584580108015667
128. T. Bearda, S. De Gendt, L. Loewenstein, M. Knotter, P. Mertens, M. Heyns, Behaviour of metallic contaminants during MOS processing. Solid-State Phenom. **65–66**, 11–14 (1999). https://doi.org/10.4028/www.scientific.net/SSP.65-66.1
129. Y. Borde, A. Maurel, A. Danel, A. Roche, M. Veillerot, Impact of metallic contamination on Si: short loop issues. ECS Trans. **11**(2), 151–158 (2007). https://doi.org/10.1149/1.2779374
130. S.R. Gilbert, D. Ritchey, M. Tavassoli, J. Amano, L. Colombo, S.R. Summerfelt, Cross-contamination during ferroelectric non-volatile memory fabrication. J. Electrochem. Soc. **184**, G195–G199 (2001). https://doi.org/10.1149/1.1353581

Chapter 4
Characterization and Detection of Metals in Silicon and Germanium

This chapter outlines a large variety of analytical techniques for the detection of metal contamination on and in silicon or germanium wafers. Trace analysis of metals in a semiconductor needs to answer three basic questions:

- What is the total amount of the species? This requires an integral sensitivity of the method, which should be smaller than the tolerable amount of metal contamination.
- Where are the metal atoms located? If in the active area, then a small amount can already kill the device or circuit yield; when in the gettering zone of the wafer, they are in principle harmless. This puts some demands on the lateral and depth resolution of the method.
- What is the chemical state of the metals? In other words, is the metal dissolved in interstitial or substitutional form, is it complexed or precipitated? This will determine the energy levels in the band gap and the electrical activity of the metal contamination. This addresses the chemical resolution of the analysis tool.

Generally, not all requirements can be fulfilled by the more common analysis methods, so that roughly a distinction can be made between techniques allowing for (1) chemical analysis, (2) the structural characterization of metal-related defects or (3) which are based on the impact on the electrical properties of the semiconductor material or simple device structures, i.e., Schottky barriers (SB) or metal-semiconductor junctions, p-n junctions and Metal-Oxide-Semiconductor (MOS) capacitors. Depending on the penetration depth of the used probe (wavelength of laser light, electron beam energy, X-ray energy,…) some methods will be more sensitive to surface contamination, including the native oxide, while others are more suitable for the bulk detection inside a wafer. In case of non-uniformities in the metal contamination, like for multicrystalline silicon (mc-Si) wafers for Photovoltaic (PV) applications, a high spatial resolution may be necessary in order to enable a mapping of the distribution of the metal atoms or defects across the wafer surface. Special attention will be given to tools, developed for the non-destructive and fast in-line metrology of contamination. Throughout this

C. Claeys and E. Simoen, *Metal Impurities in Silicon- and Germanium-Based Technologies*, Springer Series in Materials Science 270,
https://doi.org/10.1007/978-3-319-93925-4_4

chapter, examples will be given for the different methods, which are generally based on Fe-contaminated silicon—iron being the most prominent metal contaminant in silicon processing [1]. The chapter is concluded by a section, which is dealing with some examples of metal contamination studies in a process environment, in order to come up with a protocol for efficient detection of unknown metal contamination. As will be seen, depending on the properties of the respective metals (diffusivity, solubility, segregation, electrical activity,...) described in Chap. 2, one specific detection method will be more suitable than another. In general, a combination of different techniques is required to cover the full range of possible contaminants. A final remark is that in many cases, the finished devices may experience the detrimental effect of metals, when present even below the detection limit of most analytical tools. An example is a CMOS imager, where the dark current at room temperature can be sensitive to the presence of a single metal atom in a pixel. This implies that device-based metal detection can provide the ultimate detection limit – this will be the topic of Chap. 7.

4.1 Chemical Analysis of Metals

The advantage of chemical analysis is that it relies on a fingerprint of the specific element/atom, e.g., its mass, a characteristic X-ray or photon, etc. Most of the more standard techniques like Secondary Ion Mass Spectrometry (SIMS), Atomic Absorption Spectroscopy (AAS), Inductively-Coupled Plasma Atomic Emission Spectroscopy (ICP-AES) or Inductively-Coupled Plasma Mass Spectrometry (ICP-MS) are destructive in nature and usually detect the total amount of the specific metal atom, without making a distinction of the form it is present in or on the wafer. Although metal atoms can be dissolved in the lattice as individual interstitial (i) or substitutional (s) atoms, or can be present in the form of a defect complex, a small cluster or a precipitate, all atoms contribute equally to the detected signal in that case. As will be seen in Chaps. 5 and 6, the electrical impact of a metal in a semiconductor relies heavily on the way it is incorporated in the lattice, so that this is a potential drawback of the chemical analysis techniques. Another handicap is that for some more popular techniques like SIMS, the detection limit for metal atoms is rather poor —on the order of 10^{15} cm^{-3} [2], which is not sufficient to meet the requirements from the International Technology Roadmap for Semiconductors (ITRS). An example is ICP-MS, which is rather insensitive to Fe, due to the formation of ArO^+ in the plasma, which interferes with the mass of $^{56}Fe^+$ [1, 3].

In this section, the focus is on specific methods, which have proven suitable for the detection of metal contamination on silicon (or germanium) wafers. A first method which will be described is based on Total Reflection X-Ray Fluorescence (TXRF) analysis, which is nowadays the reference in-line monitoring technique for surface contamination [3]. In a second part, nuclear techniques for the detection of metals in the bulk of a semiconductor will be outlined. Next, Electron Paramagnetic Resonance (EPR) also sometimes called Electron Spin Resonance (ESR) will be briefly described. It has been the key technique for defect analysis since the early days of semiconductor technology, enabling the identification of a rich gamma of metal-related point defects in silicon [4], whereas its application to germanium has been limited to only a few specific metals. Finally, a brief description of Mössbauer spectroscopy is given, which provides information on the chemical state of some metals (Fe, Co, Sn,...) in semiconductors. For a description of other chemical analysis techniques like SIMS, ICP-MS, AAS, the Readers are referred to other text books [5, 6] or Review Articles [1].

4.1.1 Elemental Analysis of Surface Metal Contamination

X-ray Fluorescence is based on the ionization of an impurity atom in a semiconductor matrix by the absorption of an X-ray photon by a core electron. The ionized atom in a nonequilibrium state returns to its normal state within an extremely short time via the transition of electrons from the outer shells to the inner shells, giving rise to the emission of X-rays with wavelengths which are characteristic for the element. XRF is a spectroscopic method which can analyze simultaneously the concentration of a wide range of elements in a single measurement. Typical escape depths of the fluorescent X-rays in silicon are less than 100 μm so that the method is well suited for near surface element analysis [1]. Light elements (atomic number $Z_a < 11$) are usually hard to detect, since their principal XRF emission lines are in the range of long wavelengths (>1 nm), where most X-ray detectors lose sensitivity. This can be overcome by performing the analysis with a window-free detector in vacuum, so that lighter elements down to C ($Z_a = 6$) can be observed [1].

In the case of TXRF, the incident X-ray beam is at grazing incidence (Fig. 4.1), so that a condition of total reflection at the surface is reached. The critical angle for total reflection on a polished silicon surface is 0.21° at an energy of 8.4 keV and reduces with increasing beam energy [1]. In order to avoid too small critical angles, an X-ray primary energy in the range of 8–30 keV is employed. The information depth typically amounts from 2 to 10 nm, depending on the angle of incidence of the primary X-ray beam. In other words, TXRF is sensitive to metal atoms at the surface of a wafer. As shown schematically in Fig. 4.1, the fluorescence radiation is detected by a Si(Li) X-ray energy dispersive detector, perpendicular and close to the wafer surface. TXRF requires calibration by measuring a standard sample with a known contamination level. The detection limit of XRF is inversely proportional to the square root of the counting time, so that a compromise between sensitivity and throughput is made for in-line trace analysis [1]. Mapping of elemental contamination at the surface is possible, with a typical spot size of 0.5 cm^2 of the primary X-ray beam.

The reason for its popularity in the semiconductor industry is the fact that it can be applied for in-line monitoring without tedious sample preparation. It can detect metals down to the 10^{10} at/cm^2 level on the surface of polished silicon wafers—with the exception of light elements, like Al. It provides real-time multi-elemental

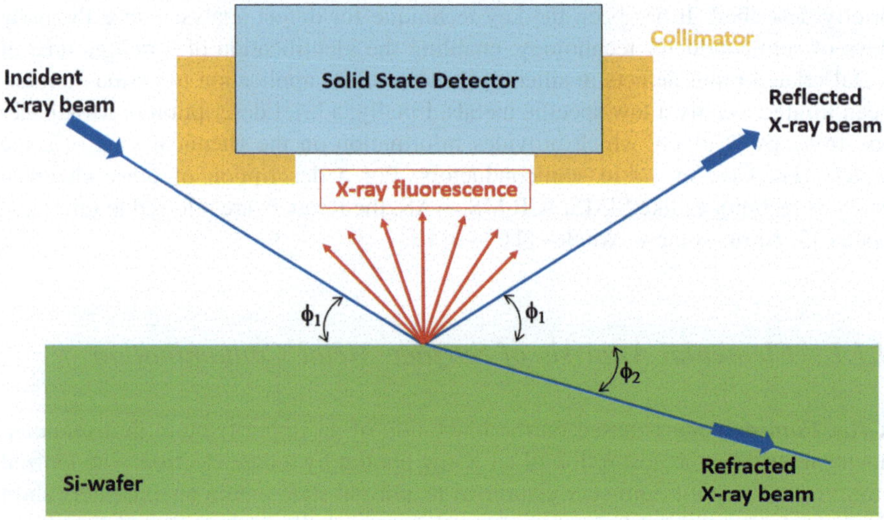

Fig. 4.1 The basic design of a TXRF experiment

data on wafers for elements with atomic number $20 < Z_a < 35$ and $56 < Z_a < 83$ with a high throughput [3]. In the case of Fe, the relevant fluorescence occurs at $K_{\alpha 1} = 6.403$ keV and $K_{\beta 1} = 7.057$ keV. If both peaks appear in the spectrum, it is considered an unambiguous proof of the presence of Fe [1]. A detection limit for Fe in the range of 10^9–10^{10} at/cm^2 for standard TXRF has been quoted [1]. This should be compared with the requirements for Fe contamination of the ITRS (<10^9 cm^{-2}).

The sensitivity of TXRF can be enhanced in combination with the Vapor Phase Decomposition (VPD) technique [3]. This consists of the vapor phase reaction of HF with native silicon oxide, forming water droplets that contain most of the soluble impurities. Next, an ultra-pure water droplet or an aqueous solution of H_2O_2 and/or HCl and/or HF and/or HNO_3 scans the surface to collect the soluble contaminants. A typical VPD procedure takes several hours [1]. The collected droplet can subsequently be analyzed by TXRF (or any other chemical technique like AAS, ICP-MS,…). In this way, information of the native oxide can be obtained. It has been quoted that VPD-TXRF improves the detection limit for most common metals by two orders of magnitude from typically ~10^{10} to ~10^8 at/cm^2 [3]. The enhancement factor is related to the ratio of the wafer area (VPD mode) to the typical spot size of 0.5 cm^2 [1]. A further lowering of the detection limit by a factor 20–100 can be achieved by using a high-intensity synchrotron radiation source (SR-TXRF) [3, 7, 8]. It also allows the detection of light elements like Al. Although developed for metal contamination in silicon, TXRF can be applied to the surface analysis of Ge wafers as well, provided the necessary calibration has been performed [9].

Time-Of-Flight ***(TOF)-SIMS*** has been recommended for the detection of metals segregated at the oxide/silicon interface [8, 10, 11]. In TOF-SIMS, the time of flight of an ion of known energy over a certain distance is measured in vacuum—which is proportional to the mass of the species. The sensitivity for surface Fe is quoted to be 6×10^8 at/cm^2 [11].

The chemical composition of structural defects, e.g., metal precipitates or decorated extended defects, can be inspected on the surface in a Scanning Electron Microscope (SEM) by ***Energy Dispersive X-ray (EDX)*** analysis. In that case, atomic core electrons are excited to a higher energy level by the primary electron beam, with typical energy in the range of a few keV to 20–30 keV. However, due to the large background of Bremstrahlung of the electron beam, the detection limit for most metal precipitates is rather low, on the order of 0.1–1 at% [12].

4.1.2 Elemental Analysis of Bulk Metal Contamination

Neutron Activation Analysis (NAA) measures the complete metal concentration in a certain volume of the material, without distinguishing the chemical state (isolated, complexed or precipitated) [1, 13]. The principle of NAA is that nominal thermal neutrons are absorbed by the semiconductor material during some time, which can take up several weeks. The activated nuclei undergo a radiative decay with the emission of specific γ-rays, which are then analyzed by a γ-ray Ge detector. The method is calibrated by standards with a known concentration of relevant isotopes. The counting efficiency is optimal if a radio-isotope with a half-life $t_{1/2}$ in the range of several days to weeks is produced [1]. In the case of Fe, ^{59}Fe is the only suitable isotope ($t_{1/2}$ = 45 days) and is formed by the activation of ^{58}Fe. The decay reaction is given by:

$$^{59}\mathrm{Fe} \rightarrow {}^{58}\mathrm{Co} + \gamma(1.2915\ \mathrm{MeV\ and}\ 1.0986\ \mathrm{MeV}) \tag{4.1}$$

The detection limit of NAA depends on several factors, including the weight of the sample, the natural abundance of the isotopes which can be detected after neutron exposure, the half-lifetimes of the isotopes, the neutron flux during the irradiation, the time between neutron irradiation and the beginning of the counting, the counting time,... [13]. Detection limits for Fe and Ni of 4.5×10^{14} and 2.3×10^{14} cm^{-3} have been mentioned in [13], respectively. The reason for the rather poor sensitivity of Fe detection by NAA is the small isotopic abundance of ^{58}Fe, which amounts to 0.31%, i.e., only 0.31% of the total number of Fe atoms can become activated with a reasonable $t_{1/2}$. The detection limit for other metals, like, e.g., Co and Au can be up to three decades smaller.

For semiconductor analysis, instrumental NAA is usually performed, which allows a simultaneous multi-element analysis after a single neutron exposure: up to 50 elements can be detected using multichannel γ-ray spectroscopy [14, 15]. Relative measurements are normally employed: the amount of impurity present in a sample is determined by comparing the γ-ray intensity at an element-specific energy with a co-irradiated standard.

While this technique has been utilized to evaluate the metal content of silicon and mc-silicon ingots, its application for process metal contamination monitoring is rather limited, given the lengthy procedure, the limited availability of thermal neutron sources and the high cost associated with it. NAA can, therefore, be considered as a tool for more fundamental studies, not for routine in-line process monitoring.

In more recent years, high intensity, focused X-ray beams have become available in several synchrotron facilities [7, 8, 10, 12, 16–18]. This has opened the door for different forms of X-ray analysis with high spatial resolution, enabling a mapping of the metal content over a wafer. Besides ***X-ray fluorescence microscopy (XRF)***, also structural information can be achieved based on X-ray Absorption Spectroscopy (XAS), while the local recombination activity can be investigated by the X-ray Beam-Induced Current (***XBIC***) mode of operation. The latter will be discussed in a following section. XRF employs the characteristic X-ray fluorescence radiation produced when core electrons of a certain atom are excited by the focused mono-energetic X-ray beam and electrons from higher orbitals fill the empty core state again.

XRF enables the detection of dissolved metal impurities and nm-scale precipitates with a 1–2 μm spatial resolution, which is orders of magnitude higher than for EDX, SIMS or Auger Electron Spectroscopy (AES) [12]. The sensitivity of the method, schematically shown in Fig. 4.2, depends on the primary beam energy and is for Fe ($K_{\alpha 1}$ and $K_{\beta 1}$ fluorescent signal) in the range of 5×10^{13} at/cm^2 at 8.5 keV beam energy and slightly higher ($\sim 10^{14}$ at/cm^2) at 12.4 keV. Energy-dispersive micro-analysis can thus be performed in the 10^{16} cm^{-3} range for 3*d* transition metals in Si and intense beams [17]. The minimum detectable size for a precipitate at the surface amounts to [18]:

$$diameter = 2 \times \left[\frac{3}{4\pi}\frac{N_{Prec}\,(atoms)}{atomic\ density}\right]^{1/3} \tag{4.2}$$

In (4.2), N_{Prec} (atoms) is the total number of atoms in one precipitate located near the sample surface. This can be determined according to [16]:

$$N_{Prec}\,(atoms) = \frac{C_{Prec}}{C_{Std}}\left\{[Fe]_{Std}\right\} \times [D_{Std}][A] \tag{4.3}$$

C_{Prec} and C_{Std} correspond to the measured XRF Fe $K_{\alpha 1}$ counts per second, normalized to the acquisition time, for the particle of interest and a standard sample, respectively. $[Fe]_{Std}$ is the areal Fe density in the standard material (in g/cm^2) and D_{Std} represents the number of Fe atoms per unit of mass of the standard material. Finally, A is the Full Width at Half Maximum (FWHM) of the X-ray spot size in cm^2. For typical experimental conditions with a 200 nm diameter beam size and an accumulation time of 1 s a cluster size of radius 23 ± 5 nm can be detected [16]. The theoretical limit is ~16 ± 3 nm assuming an atomic density of Fe in $FeSi_2$ of 4×10^{22} cm^{-3} [16].

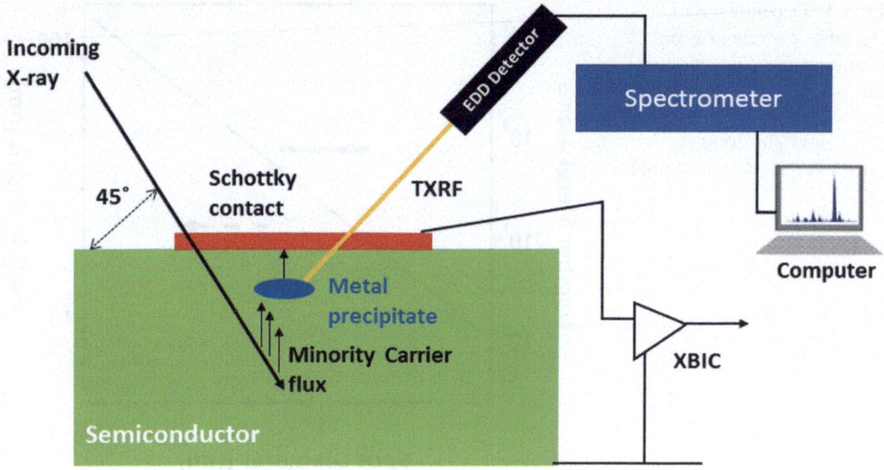

Fig. 4.2 A schematic diagram of the XBIC/XRF experiment. The focused X-ray beam from a synchrotron is incident at an angle of 45° to the sample surface. X-ray fluorescence arising when the beam hits the metal precipitate is measured by a Si:Li energy dispersive detector (EDD). Minority carriers induced due to the absorption of X-ray photons are collected with a Schottky diode. A reduction of minority carrier diode current due to enhanced recombination at a defect gives rise to the appearance of the XBIC contrast (after Vyvenko et al. [12])

The detectable size increases for precipitates deeper in the material, considering that the Fe $K_{\alpha1}$ fluorescence at 6.4 keV decays exponentially with depth in the silicon matrix with an attenuation length of 36 μm [16]. The depth of the precipitate can be derived by comparing the relative amplitude of the $K_{\alpha1}$ and $K_{\beta1}$ peaks for the standard measurement configuration and with the wafer rotated over 180° [18, 19]. The minimum detectable size is mainly governed by the X-ray beam diameter, as illustrated by Fig. 4.3 [17]. The high spatial resolution and fast detection has been exploited to perform mapping of metal precipitates (and their elemental analysis) in mc-Si wafers [12, 13, 16–19]. The acquisition time per spot depends on the intensity of the fluorescent signal and amounts typically to 1–2 s. The total time for a full XRF map can take up to a few hours. A primary beam energy well above 10 keV is necessary to detect all transition metals up to Cu [17].

Information on the chemical state of sub-micrometer sized metal precipitates in semiconductors can be obtained by ***XAS*** analysis [16, 17, 19]. It is based on the changes in the excitation of the core level electrons into empty valence band states by X-ray absorption. The valence band states are sensitive to the chemical state changes of the metal atoms. The μXAS beamline is similar to the μXRF beamline except a four-crystal channel-cut silicon monochromator to produce monochromatic radiation, with an energy resolution of 0.5 eV. An important aspect is to tune the energy of the impinging X-ray beam with the silicon monochromator while maintaining the beam position. The system is less sensitive than μXRF but has a higher energy resolution. The beam, positioned at the precipitate under analysis, is

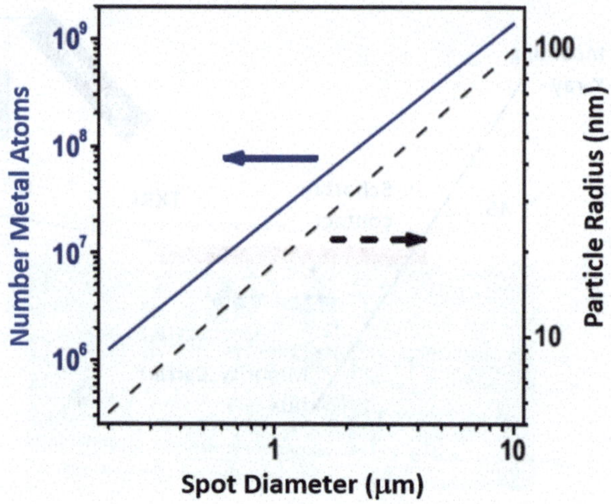

Fig. 4.3 Dependence of detectable particle size on beam diameter (detection limit of 10^{14} cm^{-2} assumed). Reproduced from Seifert et al. [17]. Copyright 2009 WILEY-VCH Verlag GmbH & Co

changed in small energy steps and the X-ray absorption is measured at each energy [17]. For small particles, the fluorescence is measured instead of the absorption. One can distinguish two energy regions [17]. The near-edge structure (***XANES***) depends on the oxidation state and local bonding and characterizes the chemical state of a metal atom in a particle. It is comparable to X-ray Photo-Electron Spectroscopy (XPS), where the photo-ionized electrons are detected. In the case of Extended X-ray Absorption Fine Structure (***EXAFS***) analysis a range of a few hundreds of eV is covered, starting from the absorption edge [17]. From the EXAFS spectra, information can be gathered of the nearest neighbor distances and atomic weights.

The detection of trace metal contamination in heavily p^+-doped silicon has shown to be challenging, particularly for Cu and Ni below 1×10^{13} cm^{-3} in volume production, because of the interaction of Cu with B and the formation of Ni silicides [20]. Dedicated methods have been developed which were described by Fabry et al. [20] and requiring a specific sample preparation in combination with VPD-TXRF and VPD-ICP-MS.

4.1.3 Electron Paramagnetic Resonance

EPR (or ESR) is based on the Zeeman splitting ΔE_Z in an external magnetic field with strength B of the energy level of an ***unpaired*** (or paramagnetic) electron, with spin $S = 1/2$ bound to an impurity atom or a defect in a material. The resonance condition in Fig. 4.4 is met at a field B_r when the energy of an electromagnetic wave of frequency f equals:

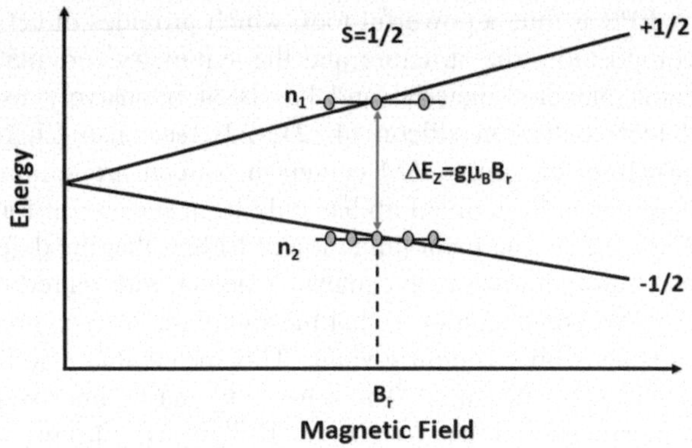

Fig. 4.4 Energy levels for a single electron ($S = 1/2$; $m_S = \pm 1/2$) in an applied magnetic field B, illustrating a magnetic resonance transition at the field B_r

$$hf = g_L \mu_B B \tag{4.4}$$

In (4.4), h is Planck's constant, g_L is the Landé factor or the effective g value of the detected electron and μ_B the Bohr magneton. Usually, microwaves in specific frequency bands are employed for ESR spectroscopy (X-band ~8.9 GHz, K-band ~20.5 GHz or Q-band ~34 GHz) at fixed magnetic field, to probe the resonance condition. The signal-to-noise ratio is largely improved by employing an ac magnetic field at ~100 kHz in combination with lock-in amplifier detection. The resulting spectrum consists of a series of resonances as a function of B, with each resonance corresponding with a specific g_L, i.e., with a specific defect center, which differs from the free electron value of $g_{free} = 2.002319$. Quantification of the concentration of spins happens by measuring at the same time a sample with a known concentration of a marker impurity, which, for silicon often is Si doped with phosphorus.

In the case of a paramagnetic defect in a crystalline semiconductor, the g value depends on the orientation of the magnetic field with respect to the major crystallographic axes. This means that g_L is in fact a matrix with different values according to the x, y and z axis. As a consequence, the symmetry of the defect and its orientation can be derived by studying the spin resonance at fixed magnitude B, while rotating the magnetic field in a certain crystal plane, i.e., studying the g map versus the angle ϕ_B the field makes with a reference crystallographic direction. Finally, information on the local atomic environment (nearest neighbor atoms) can be obtained from the so-called hyperfine (hf) interaction, resulting from the interaction of the studied unpaired electron with nuclei of spin angular momentum I different from 0. Under the influence of the hf interaction, the Zeeman level splits up into $(2I + 1)$ levels and as a result will give rise to $(2I + 1)$ hyperfine lines in the ESR spectrum. Based on the derived I value, one can identify the atoms involved in the defect center, especially if there are several isotopes with different I contributing. In that case, the ratio of the distinguished resonances will correspond with the abundance of the different isotopes.

EPR is thus a powerful tool, which provides direct information on the chemical composition, the structure and the symmetry (orientation) of a (point) defect in a semiconductor material and has been intensively used to identify metal-related defect centers in silicon [4, 21, 22] (see also Chaps. 5 and 6). The electronic structure for simple TM centers in silicon are summarized in Fig. 4.5 [21]. The application to germanium has only been successful for a few transition metals, like Ni [23, 24]. The main limitation of EPR is that the defect needs to have an unpaired electron (paramagnetic center). A defect with paired electrons cannot be detected. Another consideration is that the signal intensity is proportional to the total number of spins with a certain *g* value. This means that a sufficiently large sample volume needs to be measured. The sensitivity can be improved by measuring at cryogenic temperatures, down to 10^9 or 10^{10} spins. Moving the Fermi level beyond the ionization level by lowering the temperature can also transform a deep level defect in the appropriate unpaired electron state for EPR detection. It is clear that EPR is not directly suitable for in-line inspection, as it requires dedicated sample preparation, cooling and lengthy measurement time. Also the data analysis may not be straightforward in case of an unknown spectrum. On the other hand, once a defect has been identified by EPR, the spectrum can be considered as its unambiguous fingerprint. In spite of that, EPR has occasionally been used to identify metal contamination in silicon processing [25–27]. For further details on EPR, the Reader is referred to text books [28, 29].

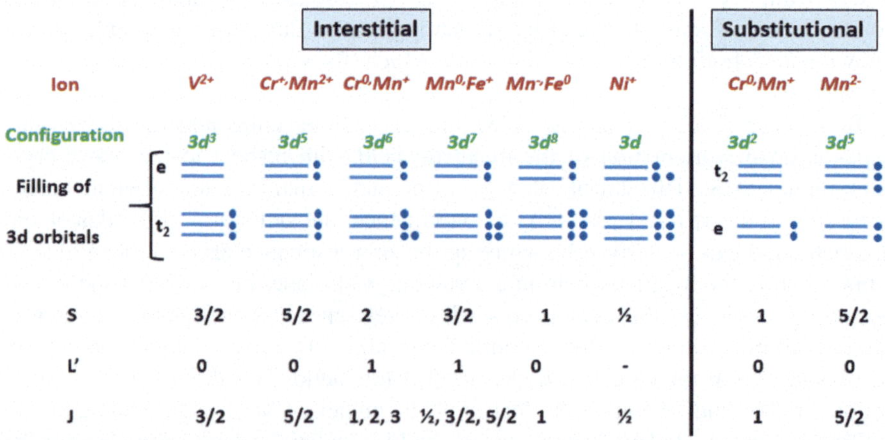

Fig. 4.5 Electronic structure of 3*d* metal ions in silicon, according to Ludwig and Woodbury [21]. The symbols *S*, *L'* and *J* represent the total spin, the effective orbital angular momentum and the effective total angular momentum, respectively. Reproduced from Ludwig and Woodbury [21] with permission from Elsevier

4.1.4 Mössbauer Spectroscopy

Mössbauer spectroscopy is a hyperfine interaction technique, using radio-active probe atoms [30]. It can be applied in a very broad temperature range, to study for example the diffusion of certain metals in semiconductors [30, 31]. In addition, the recoil energy of the decay can be sufficient to kick the daughter atom from its substitutional lattice site and so create interstitials or pairs between the metal atom and a vacancy. Decomposing the obtained γ-ray spectrum provides information on the chemical state and possible configuration of the impurity atoms in a semiconductor. Hyperfine interactions give rise to isomer shifts and quadrupole splitting of the emitted γ's. These parameters probe the shape of the electron density at nuclear sites, depending both on core and valence states and correspond with the electric monopolar and the quadrupolar interaction of a recoilless emission and absorption of γ-radiation from and by the nuclei of bound atoms, respectively. The isomer shift, for example, is proportional to the electron density at the nucleus and corresponds to the energy shift of the nuclear level for a Mössbauer component on a Doppler velocity scale of mm/s against the spectral center of α-Fe at room temperature [32].

The technique is, however, only possible for elements with suitable radio-isotopes. Fortunately, Fe is one among these elements and given its relevance for device yield in semiconductor processing, the method has been applied to iron-implanted silicon, germanium and silicon/germanium [30]. It relies on the decay of ^{57}Mn to the stable ^{57}Fe daughter by γ-emission with 14.4 keV energy and a half-life of about 100 ns, depicted in Fig. 4.6. Four lattice sites have been identified for iron in silicon: Fe in a damaged environment, in interstitial (Fe_i) and substitutional sites (Fe_s) and Fe_i paired with a vacancy [30]. In addition, the charge state of the Fe components in a silicon matrix has been investigated as well [32–34]. The diffusion behavior of Fe can be studied by the line broadening caused by diffusive jumps during the lifetime of the Mössbauer state. Mapping of the different Fe configurations in mc-Si has been developed as well [35]. Co is another 3*d* TM which lends itself to Mössbauer Spectroscopy; the results will be discussed in Chap. 6.

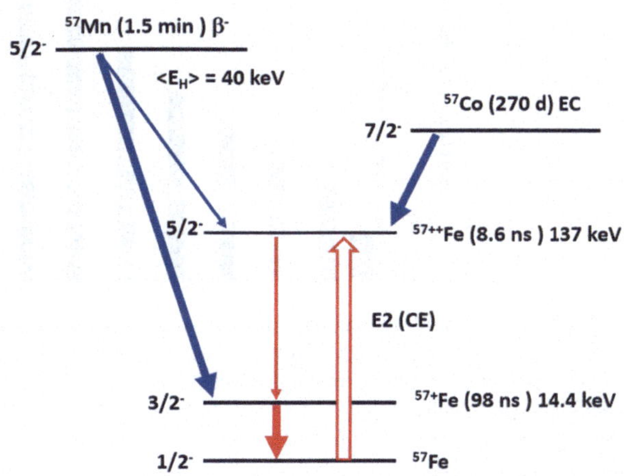

Fig. 4.6 Simplified decay scheme to the Mössbauer state of ^{57}Fe including Coulomb excitation of the 137 keV level. Reproduced from Weyer [30] by permission of Springer

4.2 Structural Analysis

Structural analysis of metal-related defects in semiconductors can be performed by different types of 'microscopies' whereby the lateral spatial resolution in the first instance depends on the wavelength of the used probe light or particle beam (electrons, ions,…). Figure 4.7 gives the lateral resolution for a range of techniques, comprising also chemical analysis tools. In principle, optical microscopy, including interference contrast Nomarski microscopy has a lateral resolution in the range of 0.25 μm [36] and is mainly suited for the detection of extended type of defects, like metal precipitates of sufficient size. At the other end of the spectrum, electron microscopies have an excellent lateral resolution approaching 0.05 nm, allowing imaging of individual atoms [36]. At the same time, the electron beam can be employed to ionize the atoms, thereby creating fluorescent X-rays, which can be analyzed by a detector to yield information on the composition of the defect or the material under investigation [36]. The back-scattered electrons can also be analyzed in so-called *Z*-contrast imaging, yielding chemical information of the sample.

While Transmission Electron Microscopy (TEM) has currently reached unprecedented resolution allowing both atomic level imaging and chemical analysis, it is usually not the most straightforward technique for the study of metal-related defects and contamination issues in semiconductor processing. TEM requires a dedicated and tedious sample preparation and is only successful if a high density of precipitates is present in the material, on the order of 10^8 cm^{-2} or higher. The focus here is on structural characterization techniques, which allow fast in-line monitoring of metal precipitates of low to medium densities (<10^8 cm^{-2}). They usually rely on a preferential wet etching of the extended defects in order to 'amplify' its size and increase the resolving power of optical or scanning electron

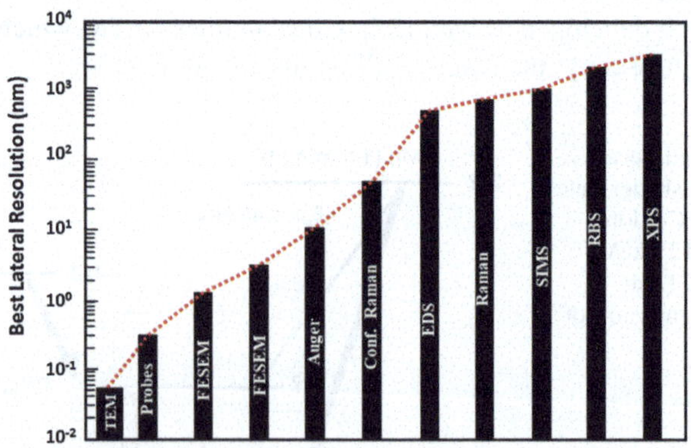

Fig. 4.7 Best lateral resolution for various characterization techniques. TEM: transmission electron microscope; FESEM: field emission secondary electron microscope; Conf: confocal; EDS: energy dispersive spectroscopy; SIMS: secondary ion mass spectrometry; RBS: Rutherford backscattering spectroscopy; XPS: X-ray photo-electron spectroscopy. Reproduced from Schroder [36] by permission of ECS-The Electrochemical Society

microscopy. This is also the basis for the well-known haze test to detect metal contamination in silicon processing. In a last paragraph, a brief introduction will be given to β-channeling, which provides detailed information on the lattice site of single metal atoms.

4.2.1 *Structural Analysis of Metal Precipitates*

Defect etching is a simple, fast and low-cost technique to assess the crystal quality of semiconductor materials. Wet etching of semiconductors is usually based on a three-component mixture:

- an oxidizing agent that oxidizes the surface of the wafer; examples are: HNO_3, H_2O_2, O_3, Br_2, CrO_3, $K_2Cr_2O_7$
- a complexing agent that dissolves the oxide that is formed on the surface (HF)
- a solvent for the dilution (H_2O, CH_3COOH)

In the case of defect etching, the overall reaction has to be sensitive to the difference in stress level and/or composition variation so that the etching proceeds faster or slower at the defect site than in the perfect matrix. As a result, etch pits (or hillocks) are formed where the extended defects cross the wafer surface, as sketched in the bottom Fig. 4.8, which can then be visualized under an optical microscope, an SEM or TEM or an Atomic Force Microscope (AFM). The extent of the etch features is usually much larger than the size of the underlying extended defect, rendering it more easy for observation and counting at a lower magnification. This is related to the fact that the stress field surrounding a defect covers a wider range than the defect dimensions.

Defect etching is often based on CrO_3/HF/H_2O mixtures [37], which achieve a satisfactory selectivity and etch rate to reveal dislocations and other types of extended defects. One major drawback is that it contains carcinogenic Cr^{VI}, which is not user friendly and also causes a waste handling issue. Another drawback is that many of the well-known defect etchants work well for one crystallographic direction, revealing clear etch pits, but are not at all effective for other planes. Therefore, other compositions have been developed in order to cope with these problems. Sopori has proposed an etchant based on 36 parts HF (49%), 20 parts of CH_3COOH (glacial) and x parts of HNO_3 (70%) where x may be varied between 1 and 2 [37]. It has been shown that this etch works very well for defect delineation in polycrystalline silicon samples. In other words, it can be applied to different crystallographic orientations of interest (100), (111) and (110).

In principle, the same etchants can be applied for defect inspection in Ge wafers, although no HF is needed to dissolve GeO_2, since this occurs easily in H_2O or H_2O_2 [38]. The situation is different when applying it to thin ($\sim$100 nm) hetero-epitaxial Ge layers on a silicon substrate. In that case, a solution is desired with a low etch rate, in order not to etch through the layer and into the underlying substrate, with

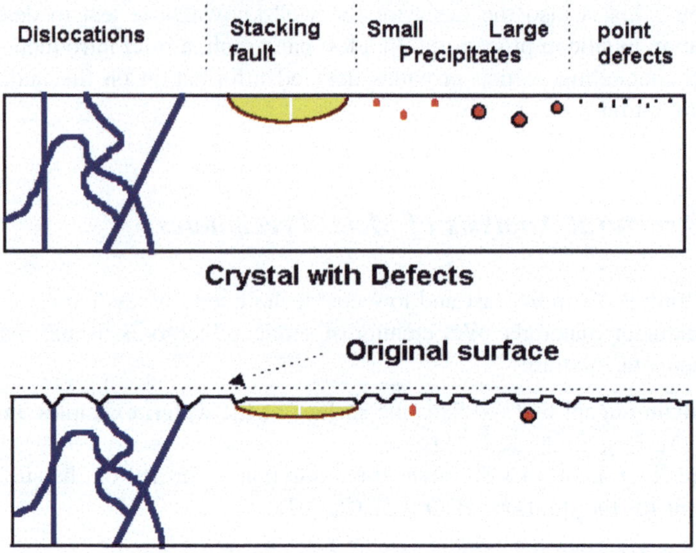

Fig. 4.8 Wafer containing extended and point defects before (upper) and after preferential wet etching (lower)

sufficient selectivity towards the defects. It has been shown that a Ce^{IV}-based solution ($(NH_4)_2Ce(NO_3)_6/H_2O$) and a permanganate based solution ($KMnO_4/HF/H_2O$) successfully reveal threading dislocations in thin Ge epitaxial layers [38], using AFM or SEM inspection.

An application of defect etching and optical inspection is the so-called haze test for metal contamination in combination with rapid thermal annealing (RTA) [39, 40]. It relies on the fact that many metals are fast diffusers in silicon and have the tendency to precipitate at the wafer surface, where they form so-called s(mooth) pits. These can be rendered visible by a defect etching (30 s of Sirtl etch [41]), where illumination with a bright lamp reveals a haze on the surface, due to the scattering of the light by the s-pits. TMs easily produce haze on the wafer frontside, when applied to the backside, followed by an RTA at 1000–1100 °C for 30 s [39]. The same method can be used to investigate the efficiency of certain (internal) gettering treatments (see Chap. 8), where a backside metal contamination is applied, followed by an RTA step and a haze test [42]. The ratio of the metal concentration at the front surface, detected by TXRF for example (in at/cm^2) to the applied contamination at the backside can be considered as a measure for the gettering efficiency.

4.2.2 Structural Analysis of Metal-Related Point Defects

Besides Mössbauer spectroscopy and EXAFS, there are few structural characterization techniques which provide direct evidence of the precise lattice site of individual metal atoms. One method which has gained some popularity in recent years due to the availability of suitable radio-isotopes in certain research facilities is the β-emission channeling [43–45]. It relies on the fact that the emission of β's from a nuclear decay experiences channeling or blocking effects along low-index crystal directions, similar as in Rutherford Back-Scattering (RBS). This leads to anisotropy in the emission yield from the crystal surface which depends in a characteristic way on the lattice site occupied by the emitting atom. In the case of Fe, ^{59}Mn ($t_{1/2}$ = 4.6 s) precursor atoms are implanted. They decay into ^{59}Fe, which is an unstable β emitter with a half-life of 44.6 days. The emission spectrum can be studied in function of the sample temperature.

For the case of Fe in high-resistivity Si, different sites have been identified [43]. From room temperature to 200 °C, as-implanted Fe was found in near-substitutional sites (similar as for Cu [44]), progressively disappearing above that temperature, with an activation energy of 2.0 eV. The displacement from the perfect-*S* site is 0.04–0.07 nm [44]. Vacuum annealing between 300 and 600 °C transfers Fe mainly into near-tetrahedral interstitial sites, with a displacement of 0.03–0.08 nm from the ideal near-*T* site. This could correspond with the Fe_i-V pair determined by Mössbauer and EPR [43]. Further annealing causes it increasingly incorporated on ideal substitutional sites, on which it is stable around 800 °C [44]. In heavily doped p^+-Si, the interstitial fraction dominates (probably due to the Fe_i-B pairs) [45], while it is reduced in n^+-Si.

4.3 Electrical Analysis

The main reason why metal contamination is highly undesired in semiconductor processing is because most metals give rise to energy levels in the band gap, irrespective of their chemical form in the lattice. As seen in Chap. 2, deep levels affect the recombination and generation of excess carriers, through the Shockley-Read-Hall (SRH) mechanism, thereby degrading the electrical performance of devices, particular the ones where the active region is contained in a significant part of the silicon or germanium wafer (solar cells; p-n junctions; power devices) or which require a low dark current level (Dynamic Random Access Memory (DRAM); Complementary Metal-Oxide-Semiconductor (CMOS) imagers; Charge-Coupled Devices (CCDs)). Metals precipitated at extended defects like dislocations, oxide precipitates or at the Si/SiO_2 interface will enhance the electrical activity and conductivity of these defects, causing excess leakage currents, shunts and early breakdown. It explains the huge efforts which have been undertaken since the early days of semiconductor electronics to develop techniques for the evaluation of the electrical activity of metals and whenever possible, the identification of the dominant lifetime killing centers. In the following section different techniques will be described which either evaluate the recombination (or generation) lifetime or the diffusion length L_{diff}, which is connected with τ_r through the relationship $L_{diff} = \sqrt{D\tau_r}$, with D the diffusivity for electrons or holes, respectively (D_n = 33.5 cm^2/s and D_p = 12.4 cm^2/s in Si at room temperature). In principle, lifetime measurements are far more sensitive than any other analytical technique and are only limited by the intrinsic band-to-band recombination lifetime, which amounts to 0.5 s in 10 Ω cm doped p-type Si [1]. This is equivalent to a detection limit for interstitial Fe in the range of 10^7 cm^{-3}.

Before detailing some of the more popular recombination lifetime techniques, like Surface PhotoVoltage (SPV), PhotoConductance Decay (PCD) or ELYMAT, we will take a closer look at the basic SRH theory underlying them. The assumptions where the model is valid will be highlighted and some pitfalls and limitations discussed. At the same time, the possibility for defect identification, based on so-called injection-level and temperature-dependent lifetime spectroscopy will be pointed out.

While there exist probably hundreds of different methods to measure the carrier lifetime, here, the focus is on techniques which enable a fast and in-line application, with an as high as possible spatial resolution. As will be seen, excess carriers are mostly created by a laser (light) pulse, while the material parameter measured to monitor the carrier recombination can vary widely. For example, in recent years, PL-based lifetime mapping has gained popularity in the field of mc-Si characterization. It is based on the complementarity between high non-radiative recombination and high PL intensity, i.e., high radiative recombination. Diffusion length mapping can also be achieved in an SEM, based on Electron-Beam-Induced Current (EBIC), Light- (or Laser-)Beam-Induced Current (LBIC) or by XBIC sketched in Fig. 4.2. This requires usually a depletion region to separate electron-hole pairs, which is for example present in a Schottky Barrier (SB) or p-n junction.

While in most cases the recombination lifetime is measured, extending over a diffusion length in the material, it is also of interest to address the generation lifetime in a narrow depletion region close to the surface of the wafer. As will be seen, this can be achieved by measurements on Metal-Oxide-Semiconductor (MOS) capacitors.

In a final section, Deep-Level Transient Spectroscopy (DLTS) will be briefly outlined. Since its introduction in 1974 by Lang [46], it is the standard technique for deep-level spectroscopy in semiconductor materials. Hence, it has also been intensively used for the characterization of metal-related deep states in both Si [47] and Ge [48]. For further details, the Readers are referred to reviews and text books on lifetime measurements [47, 49] and DLTS [47, 50].

4.3.1 Theoretical and Practical Considerations for Lifetime Measurements

Carrier lifetime measurements are based on the SRH theory, which, for a single level recombination center with concentration N_T is given by [51, 52]:

$$\tau_{SRH} = \tau_n = \tau_p = \frac{\tau_{p0}(n_0 + n_1 + \Delta n) + \tau_{n0}(p_0 + p_1 + \Delta n)}{n_0 + p_0 + \Delta n} \tag{4.5}$$

In (4.5), τ_{n0} and τ_{p0} are related to the capture cross section σ_n and σ_p for electrons and holes, respectively, through:

$$\tau_{n0} = \frac{1}{N_T \sigma_n v_{th}} \quad \text{and} \quad \tau_{p0} = \frac{1}{N_T \sigma_p v_{th}} \tag{4.6}$$

with v_{th} the thermal velocity of the carriers. The assumption of nondegeneracy is required to provide an unambiguous meaning to the capture cross section, since they represent "average" cross sections for carrier capture from all possible band states [53]. Further, n_0 and p_0 are the free electron and hole concentration in the semiconductor material, which depend on the doping density; for an n-type semiconductor $n_0 \gg p_0$ and vice versa. Δn, Δp are the excess electron and hole density created in the material for example by a light pulse of suitable wavelength (energy > band gap E_g). Finally, n_1 and p_1 are statistical factors and correspond with the equilibrium concentrations of electrons and holes when the Fermi level coincides with the energy level of the recombination center E_T:

$$n_1 = N_C \exp\left(\frac{E_T - E_C}{k_B T}\right) \quad \text{and} \quad p_1 = N_V \exp\left(\frac{E_C - E_g - E_T}{k_B T}\right) \tag{4.7}$$

with E_C the bottom of the conduction band (CB), E_V the top of the valence band (VB); N_C the density of states of the CB and N_V the density of states of the valence band. A crucial assumption in (4.6) is that $\Delta n = \Delta p$. In other words, no trapping by other defect centers, which creates an unbalance in the excess carrier densities, exists. Trapping effects will be important at low injection levels $\eta = \Delta n/n_0 \ll 1$ (n-type material) but can be suppressed by using a back-ground illumination to saturate the charge state of the traps.

Equation (4.5) is valid for arbitrary injection level, but small N_T and forms the basis for so-called Injection-Dependent Lifetime Spectroscopy (IDLS) [53–56], which is a very sensitive method to determine the parameters of the dominant recombination center in a semiconductor, i.e., the activation energy E_T and the capture cross section ratio $k = \sigma_n/\sigma_p$. It consists of determining the optimal fitting parameters in (4.5) to the experimental injection-level dependent recombination lifetime data at fixed T. For narrowing down the possible range of likely fitting parameters, usually one has to perform the IDLS measurements on wafers with different doping density. Alternatively, one can apply a related technique, namely, the so-called Temperature-Dependent Lifetime Spectroscopy (TDLS), where the low-level injection recombination lifetime is measured at constant Δn as a function of T. Assuming temperature independent capture cross sections enables accurate determination of the energy level and of $k = \tau_{n0}/\tau_{p0}$ of the dominant recombination center by fitting the complete T-dependence of the recombination lifetime. Combining with IDLS on the same material (same doping) allows for an unambiguous identification of the dominant recombination center [56]. The advantages with respect to DLTS is the fact that information not only about the majority capture cross section, but also on the minority capture cross section (vital for the minority lifetime) can be obtained for the dominant recombination center at room temperature.

An important boundary condition for the validity of (4.5) is that the density of the dominant recombination center is sufficiently small, as to not affect the condition $\Delta n = \Delta p$. In other words, at too high N_T, the deep level can cause preferential trapping of one of the carriers so that the equality condition is no longer fulfilled. A detailed study of the critical concentration N_{crit} for the case of Fe_i and FeB in silicon has been performed by MacDonald and Cuevas [53]. The result of this analysis is shown in Fig. 4.9. It is clear that trapping by the deep centers will be important at low injection levels and for $[Fe_i] \sim 1.4 \times 10^{12}$ cm^{-3} at a p-type doping of 1×10^{15} cm^{-3}, while it is 3.6×10^{15} cm^{-3} for FeB [53]. N_{crit} can be expressed in terms of the doping density of the substrate and the trapping lifetimes, according to [53]:

$$N_{crit} \approx N_a \frac{\tau_{n0}}{\tau_{p0}} = kN_a \tag{4.8}$$

at low injection. It turns out that energy levels close to mid gap have the lowest N_{crit}. As is seen in Fig. 4.9a, deep levels always correspond with a lifetime that increases with Δn (or remains approximately constant), while the shallow acceptor level of FeB (E_C k− 0.26 eV) can give rise to either an increasing or reducing τ_r with Δn, depending on the doping density N_A (p-type) and its activation energy [49]. Comparing Fig. 4.9a with 4.9b, it is evident that Fe_i is a much more efficient recombination center at low injection than FeB pairs.

Under certain conditions, information on the temperature dependence of the capture cross section can be gained from TDLS [57]. For p-type silicon, $n_0 \ll p_0$ and when Δn is sufficiently smaller than p_0 the τ_{SRH} can be rewritten as:

$$\tau_{SRH} = \frac{\tau_{n0}(p_1 + p_0) + \tau_{p0}(n_1 + n_0)}{p_0} + \frac{\tau_{p0}}{p_0}\Delta n \tag{4.9}$$

At sufficiently low T, where the emission from the defect level is negligible and $n_1(T) + n_0(T) \ll \Delta n$ and $p_1(T) \ll p_0(T)$, (4.9) can be further simplified to:

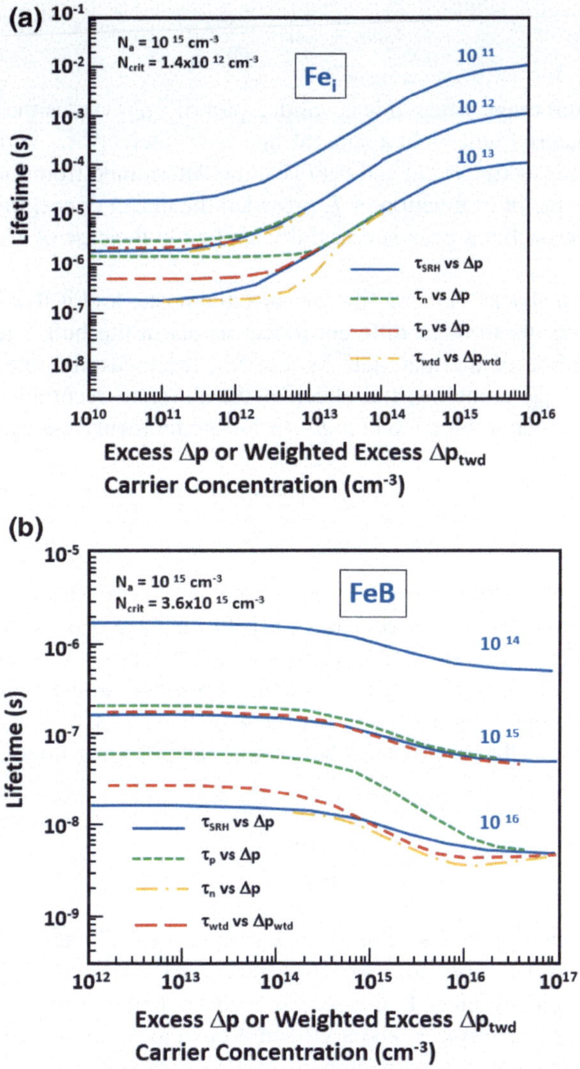

Fig. 4.9 **a** Comparison between the injection-level dependence of the simplified SRH lifetime measured by quasi-steady-state photoconductance (QSSPC) and the electron and hole lifetimes as calculated from the general steady-state solution, using the recombination parameters of interstitial iron. Curves are shown for three values of the iron concentration. The electron and hole lifetimes are plotted against the excess hole concentration Δp. Also shown for each iron concentration is a fourth curve, which shows the mobility-weighted combination of τ_n and τ_p plotted against a similarly weighted excess carrier concentration Δp_{wtd}. **b** Comparison between the injection-level dependence of the simplified SRH lifetime and the electron and hole lifetimes as calculated from the general steady-state solution, using the recombination parameters of the FeB pairs. Reprinted from MacDonald and Cuevas [53]. Copyright (2003) by The American Physical Society

$$\tau_{SRH} = \tau_{n0} + \frac{\tau_{p0}}{p_0} \Delta n \tag{4.10}$$

In the temperature range where this is valid, a plot of τ_{SRH} versus the injection level in the p-type material will yield a straight line with intercept τ_{n0} and a slope given by τ_{p0}/p_0. In other words, $\sigma_n(T)$ and $\sigma_p(T)$ can be determined from the intercept and the slope of the line fit in function of T, provided the defect concentration in (4.6) is known. This approach has been successfully applied to the case of Mo in crystalline Si [57].

One of the drawbacks of using lifetime as a diagnostic tool is that in many cases, recombination occurs through different defect levels in the bulk and from defects states at the surface of the material. As a result, the measured effective recombination lifetime (τ_{eff}) combines the effect of the different recombination channels (neglecting Auger recombination at high Δn for the moment) and can be written as:

$$\tau_{eff}^{-1} = \tau_s^{-1} + \sum_{i=1}^{n} \tau_i^{-1} \tag{4.11}$$

The surface recombination lifetime τ_s is determined by the surface recombination velocity s_r in cm/s by the surface states [47]. In the case of a good quality wafer with thickness d, the minority carrier diffusion length L_{diff} will be on the same order or even higher than d, so that surface recombination both at the front and the back surface will take place during a lifetime measurement. Assuming equal s_r at the front and the back, the relationship between τ_s and s_r is in good approximation given by [58]:

$$\tau_s \approx \frac{d^2}{\pi^2 D_{min}} + \frac{d}{2s_r} \tag{4.12}$$

with D_{min} the minority carrier diffusion coefficient. Typical values for s_r of silicon wafers are up to 10^7 cm/s for lapped surfaces, 10^3–10^5 cm/s for mechanically polished or chemically etched surfaces covered by native oxide, 1–100 cm/s for thermally oxidized Si surfaces, about 20 and 4 cm/s after immersion of the sample in bromine in methanol and iodine in methanol, respectively, and 1–0.25 cm/s after immersion in HF. Using a proper surface passivation method is thus essential to reduce the surface recombination lifetime contribution to the overall effective lifetime in (4.11). With respect to the application to Ge wafers it should be remarked that thermal GeO_2 usually is not a good passivation layer, due to the presence of a high density of surface states and the impact of humidity on the quality of the GeO_2. This implies that for Ge, a deposited SiO_2 or Si_3N_4 layer can be employed as surface passivation [49]. Alternatively, a surface passivation based on a solution of iodine in polyvinyl acetate enables accurate lifetime measurements on low resistivity p-type Ge wafers [59].

4.3.2 Surface Photo Voltage Lifetime Analysis

The Surface PhotoVoltage (SPV) technique is based on the spectral dependence of the surface photovoltage, which is created by the surface or insulator charge or work function difference [60] and is most commonly detected with a non-contacting probe, as illustrated in Fig. 4.10 [61]. Two types of probes are used, as shown in Fig. 4.11: a Kelvin probe, where the electrode vibrates vertically, changing the capacitance between the probe and the sample and the Monroe probe. In the latter case, the electrode is fixed and a grounded shutter, mounted in front of the electrode, is vibrating horizontally, thereby modulating the probe to wafer capacitance. Often, a corona charge is used to deposit a well-controlled charge on the semiconductor surface [61]. The contact potential difference V_{cp} is then given by the relationship: $I_{cp} = V_{cp}\ \mathrm{d}C/\mathrm{d}t$. When the current $I_{cp} = 0$ the contact potential difference equals the potential drop in the substrate V_B, which gives rise to a depletion of the near surface region, as can be seen in the band diagram of Fig. 4.10b.

In order to measure the diffusion length of minority carriers in a wafer, the surface is conditioned so that there exists a depletion. This surface is illuminated by monochromatic light of energy higher than the band gap, with the back surface kept in the dark. The light is chopped to enhance the signal/noise ratio using lock-in techniques. The wavelength is changed with a monochromator or a series of filters transmitting selected wavelengths [61]. Electron-hole pairs (ehp) are generated both in the space-charge region (scr) and in the p-type substrate by absorbed photons. The ehp generated in the scr are very quickly separated by the electric field with the minority carriers drifting toward the surface. Those minority carriers in the quasi-neutral substrate, within approximately a minority carrier diffusion length of the scr edge, diffuse and drift toward the surface, establishing a surface potential or

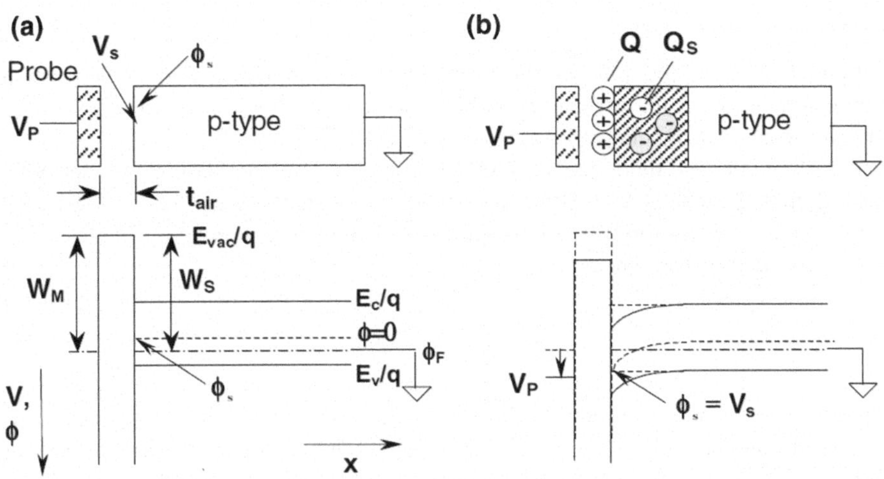

Fig. 4.10 Cross section and band diagram of a metal-air-semiconductor system with zero work function difference: **a** no surface charge, **b** with surface charge. Reproduced with permission from Schroder [61]. Copyright 2001 IOP Publishing

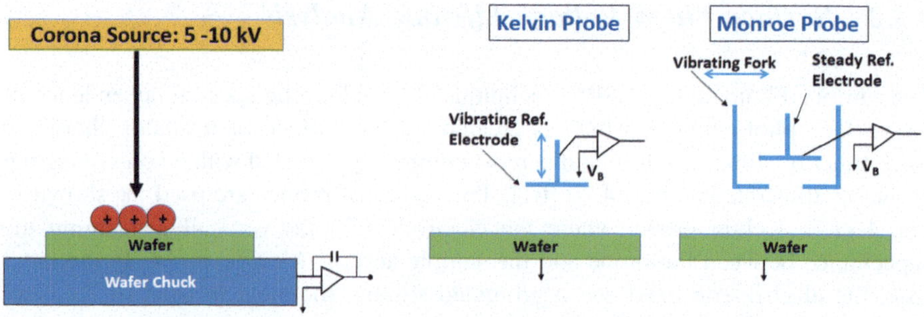

Fig. 4.11 Schematic illustration of a point electrode corona charging method, depositing positive charge on the surface for a negative Corona potential. The deposited charge is precisely measured with the operational amplifier charge meter. The contact potential difference is measured with either a Kelvin or a Monroe probe. Reproduced with permission from Schroder [61]. Copyright 2001 IOP Publishing

surface photovoltage relative to the grounded back surface. The SPV is proportional to the excess minority carrier density at the edge of the scr, $\Delta n(w)$, which is related in good approximation to the surface photovoltage V_s (for $qV_s \ll k_BT$) by [61]:

$$\Delta n(w) \approx \frac{n_0 q V_s}{k_B T} \tag{4.13}$$

with n_0 the equilibrium minority carrier density. The surface photovoltage is equal to the probe voltage V_P, given by [60]:

$$V_P = \frac{k_B T}{q} \frac{(1-R)\Phi}{n_0\left(s_{r1} + \frac{D_n}{L_n}\right)} \frac{L_n}{L_n + 1/\alpha_o} \tag{4.14}$$

In (4.14), Φ is the photon flux density, α_o is the absorption coefficient at wavelength λ_o; s_{r1} is the surface recombination velocity at $x = w$ (at the edge of the scr) and R is the frontside reflectivity of the wafer.

Based on (4.14), there are two straightforward implementations of the SPV method: first, a constant V_P method and second, a constant photon flux method [62]. The latter is now routinely used for diffusion length measurements. In that case, (4.14) can be transformed into [61]:

$$\frac{1}{V_P} = C_2(L_n + 1/\alpha_o) \tag{4.15}$$

where C_2 is a constant. A plot of V_P (usually in the mV range) versus $1/\alpha_o$ gives L_n, as illustrated schematically in Fig. 4.12 [61]. This is achieved by measuring the SPV at different λ_o. For (4.15) to hold, the minority carrier diffusion length L_n should be much larger than w, which is easily satisfied for a silicon wafer [61]. A so-called 'enhanced' SPV method has been developed to overcome this limitation [63].

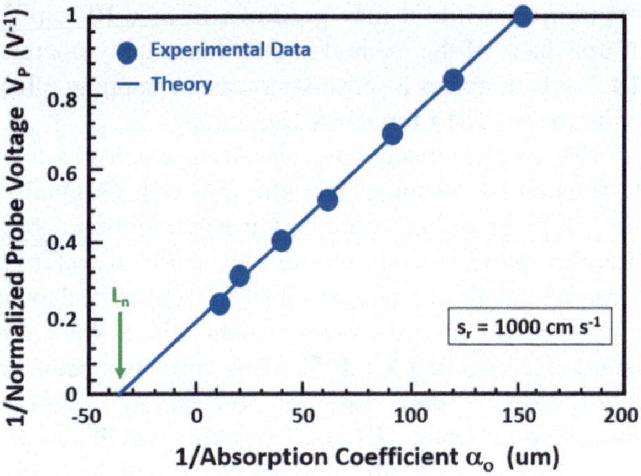

Fig. 4.12 Normalized SPV plot for constant photon flux density. Reproduced with permission from Schroder [61]. Copyright 2001 IOP Publishing

Another constraint is that for the undepleted wafer, its thickness should be much larger than the minority carrier diffusion length L_n; usually $L_n < d/3$ is considered [1]. The absorption coefficient should be sufficiently low that $\alpha_o w \ll 1$ but sufficiently high that $\alpha_o(d - w) \gg 1$, implying that light must be absorbed in the top part of the wafer. Equations (4.13) to (4.15) are based on a 1D solution of the transport equations. This means that the beam spot of the light should be larger than the wafer thickness. In addition, (4.13) requires a low injection condition to be valid, i.e., $\Delta n < p_0$, so that SPV is usually measured at low η [1].

SPV is an attractive diffusion length characterization technique, because it is nondestructive and contactless, the sample preparation is simple (no contacts, junctions or high temperature processing required), it is a steady-state method and the equipment is commercially available. It is thus suitable for fast in-line monitoring and wafer mapping of the diffusion length [61]. As an example, the effective minority carrier diffusion length detected for interstitial Fe_i and FeB at room temperature is given in Fig. 4.13, as a function of the bulk iron concentration [61]. The detection limit for $[Fe_i]$ has been quoted to be 2×10^9 cm^{-3} in routine IC

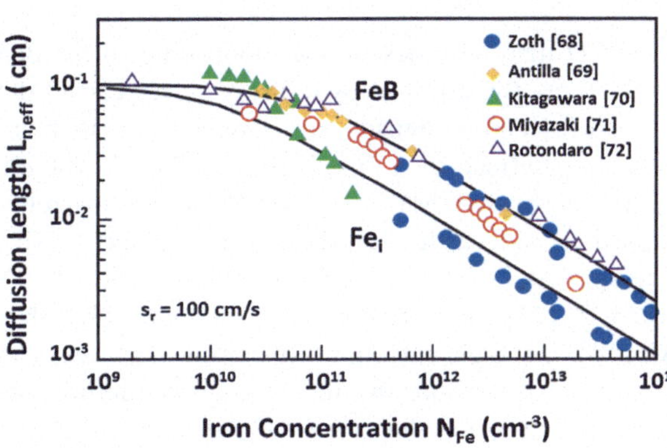

Fig. 4.13 Effective diffusion lengths versus iron concentration for FeB and Fe_i, collected from data in the literature [68–72]. s_r is the surface recombination velocity for the theoretical curves (lines). Reproduced with permission from Schroder [61]. Copyright 2001 IOP Publishing

monitoring, while it may go down to 8×10^7 cm^{-3} in laboratory conditions [1]. A drawback of the method is that it is usually operated at η in the range of 10^{-5} to 10^{-7}, which makes it sensitive to carrier trapping effects and it also does not allow a wide range of Δn for IDLS [1].

SPV measurements have also been employed for monitoring heavy metal contamination by chemical cleaning [64, 65]. Originally, surface contamination from wet cleaning by for example Cu was monitored by standard SPV, after an RTA drive-in of the metal in the bulk of a silicon wafer [64]. In situ monitoring during cleaning has been made possible by frequency dependent ac-SPV, which has been employed on n-type silicon wafers which were contaminated with an aqueous solution containing Cr [65]. This has also been proposed for the detection of low-level, on the order of 10^{10} at/cm^2 surface Cu contamination during Back-End-Of-Line (BEOL) copper metallization [66]. It is based on a short-wavelength (0.46 μm) surface excitation condition appropriate for measuring the surface lifetime τ_s rather than the bulk lifetime [65]. This eliminates the need to drive in the copper with a high-temperature heat treatment, when using standard SPV measurements. Modulated optical excitation is applied with a frequency f (angular frequency $\omega = 2\pi f$). At low frequencies, the measured ac-SPV is given by [66]:

$$\frac{1}{V^2} = \frac{1}{V_0^2} + \frac{\tau_s}{V_0^2}\omega^2 \tag{4.16}$$

with V_0 a constant. Alternatively, one can measure τ_s at a single frequency by measuring the phase shift between the illumination and the SPV signal [66], which can then be used for a fast wafer mapping.

A review of the use of Surface Voltage mapping in non-visual defect monitoring can be found in [67]. An isothermal DLTS type of analysis has been proposed for surface defects on a SiC wafer, whereby the voltage transient following a light pulse is monitored (non-steady-state method).

4.3.3 PhotoConductance Decay (PCD) and QSS-PC Method

The PCD method relies on the monitoring of the decay or transient in the sample conductivity (determined by the excess free carrier density) following a laser pulse. The excess carrier density $\Delta n(t)$ will decay in first approximation exponentially with time, defined by a time constant given by the effective recombination lifetime τ_{eff}. In microwave (MW) detected PCD, it is the reflected microwave power which is used as a measure for the excess carrier density decay with time, since the reflection coefficient varies proportionally with Δn. A schematic set-up is given in Fig. 4.14 [73]. In principle, one can also utilize the microwave absorption mode [48, 74]. In high injection, InfraRed (IR) absorption by free carriers can be employed to measure the recombination lifetime [49, 74]. The injection level is varied by changing the power of the laser pulse over a wide range, allowing for

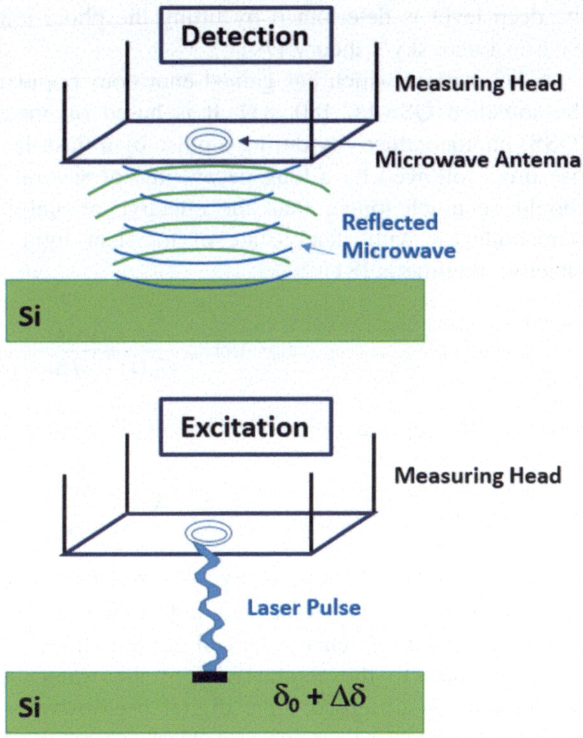

Fig. 4.14 **a** Laser pulse excitation of a silicon wafer with δ_0 conductivity. The excess conductivity generated by the illumination is indicated as $\Delta\delta$. **b** Detection of the conductive decay. Reproduced from Horányi et al. [73] by permission of ECS-The Electrochemical Society

IDLS. Using a heated wafer stage, one can perform the MW-PCD at elevated temperatures, as has been carried out for several melt-doped metals in Czochralski silicon crystals [75]. An extension of this method is to measure the PCD time constant of the carrier recombination decay as a function of T by MW reflection, applying the rate window concept like in DLTS (see Sect. 4.3.9) [76]. In this way, a spectrum can be generated, which allows for the determination of the deep level activation energy. This has been demonstrated for metal-doped silicon, showing reasonable agreement with the activation energy derived from standard DLTS [76].

Detection of the PCD by radio-frequency waves has also been proposed [77, 78], where the probing of the excess carriers relies on the reflected power because of the Eddy currents. It has been developed for in-situ monitoring of metallic contamination during aqueous HF cleaning of silicon wafers. The method was turned surface sensitive by using a wafer with sufficiently high bulk lifetime (no thermal in-diffusion of the metals performed), so that $\tau_{eff} \sim \tau_s$.

Another form of lifetime spectroscopy is based on the measurement of the MW-PC signal using photo-ionization of the deep levels present in the band gap by monochromatic light [49]. The photo-ionization spectrum consists of a number of steps, each step corresponding with the photo-ionization of a deep level, whenever the wavelength of the incident light equals E_C-E_T or E_T-E_V to excite a carrier to the CB or VB. The signal saturates for increasing energy and the activation energy of the deep level is determined by fitting the photo-ionization cross section, based, e.g., on Lucovsky's theory [79].

A PC method which has gained enormous popularity in the solar cell world is the so-called QSS-PC [80, 81]. It is based on measuring the quasi-steady-state (QSS) photoconductivity during a pulse by a flash-lamp light source, having a fast rise time, followed by a long decay time of several ms. The length of the decay should be much longer than the effective recombination lifetime. For a p-type semiconductor with steady-state of transient light incident on the sample, the effective lifetime is [81]:

$$\tau_{eff}(\Delta n) = \frac{\Delta n(t)}{G_o(t) - d\Delta n(t)/dt} \tag{4.17}$$

with $G_o(t)$ the ehp generation rate. Under QSS conditions, (4.17) transforms into [75]:

$$\tau_{eff}(\Delta n) = \frac{\Delta n}{G_o} \tag{4.18}$$

The generation rate G_o is given by $[(1 - R)\Phi]/d$, whereby the light flux at a specific power or lamp intensity and the conductance can be measured using a calibrated instrument and a reference solar cell or photodiode [80]. Usually, a coil in a bridge circuit couples to the conductivity of the wafer. A signal proportional to this conductivity is observed on a digital oscilloscope and sent to a computer for analysis. Compared to a transient decay approach, the QSS method allows measuring low τ_{eff} without the need for fast electronics or short light pulses. IDLS and TDLS can be easily performed by changing the light intensity and the temperature of the wafer, respectively. The illuminated area in QSS-PC is usually on the order of a few cm^2, so that the technique is less amenable to lifetime mapping.

4.3.4 ELYMAT

The Electrolytic Metal Tracer (ELYMAT) technique has been introduced by Lehmann and Föll [82, 83] and uses the wafer as the counter-electrode in an electrolytic cell. Diluted HF is used as the electrolyte, which, at the same time, assures a good passivation of the wafer surface on the wafer side where the current is not extracted. On the opposite side, a reverse bias is applied using the electrolyte as a full wafer area Schottky contact. The wafer is illuminated on one side, which creates a photo-induced diffusion current that is collected by the depletion region. The method can be operated in the usual backside photocurrent (BPC) mode (Fig. 4.15a) [84] or in the frontside photocurrent (FPC) mode. The most important constraint in the FPC mode is that the wafer thickness should greatly exceed the diffusion length. The main requirement in the BPC mode is that the penetration depth of the light is much smaller than the minority carrier diffusion length [84].

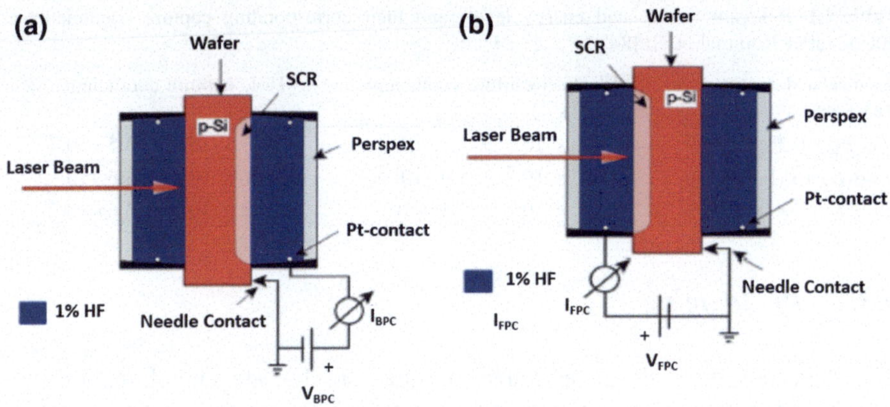

Fig. 4.15 **a** Elymat setup in BPC configuration and in **b** FPC configuration. Reproduced from Rommel et al. [84] by permission of ECS-The Electrochemical Society

These conditions are in many practical cases not fulfilled, so that one has to rely on an exact numerical extraction of L_{eff} (or τ_{eff}) [84].

As the current level depends directly on the number of recombination centers present in the material, it is a direct measure of τ_{eff}. The BPC mode is used routinely for material quality assessment (τ_{eff} mapping) or process monitoring [7, 8, 84–86], because in the FPC mode, mainly the region near the front surface affects the measured current. In BPC mode, the whole bulk as well as defects at the illuminated surface determine the collected photocurrent. It should be remarked that the extracted lifetime is to be considered an effective one, as the carrier density is not uniform across the wafer thickness, considering for example the depletion region at the collecting electrode [84]. By changing the power of the scanning laser, the injection level can be varied. Superimposed on the photocurrent is a dark current I_{dark}, generated in the scr without illumination. In order to determine the optimal operation point a dark and light current-voltage characteristic is measured for the actual lifetime mapping [84].

As the passivation by HF may not be perfect, one can use the ELYMOS variant of the method [84], whereby an insulator (commonly SiO_2) is present on the illuminated side. In that case, only the BPC mode is used, whereby the recombination centers at the SiO_2/Si interface also contribute to I_{BPC}. Using an accumulation bias at the electrolyte/oxide contact eliminates completely the contribution of the front interface states to τ_{eff}, so that only bulk defects are assessed [84]. As an illustration of the method, Table 4.1 summarizes the relevant Fe-related energy levels and capture coefficients at 300 K [84].

Table 4.1 Relevant iron-related energy levels and their corresponding capture coefficients at 300 K (after Rommel et al. [84])

Iron-related energy level (eV)	Electron capture coefficient ($cm^3\ s^{-1}$)	Hole capture coefficient ($cm^3\ s^{-1}$)
$Fe_i\ E_V + 0.38$	$5.45 \times 10^{-7} \pm 4.5{\times}10^{-8}$	$1.07 \times 10^{-9} \pm 3.5{\times}10^{-10}$
FeB $E_V + 0.10$	$1.15 \times 10^{-6} \pm 5.0{\times}10^{-7}$	$1.23 \times 10^{-6} \pm 3.6{\times}10^{-7}$
FeB $E_C - 0.29$	$5.31 \times 10^{-8} \pm 2.40{\times}10^{-8}$	$2.25 \times 10^{-7} \pm 3.6{\times}10^{-8}$

4.3.5 PL Imaging

The characterization of silicon wafers by quasi-steady-state PhotoLuminescence (PL) has been shown to be very sensitive at small excess carrier densities, as low as $10^9\ cm^{-3}$ [87]. It relies on the band-to-band optical recombination, whereby the PL intensity gives direct access to the excess carrier density Δn and, hence, to τ_r [87, 88]. It is unaffected by the so-called depletion region modulation (DRM) effect, where the presence of a depletion region distorts the excess carrier density and hence the recombination lifetime and is also not sensitive to minority carrier trapping [89]. PL intensity mapping is achieved in practice by illumination with a laser diode, while the luminescence light is detected by an one megapixel silicon CCD camera, cooled thermo-electrically to −20 °C in combination with a standard 28 mm lens [87]. In this case, the whole field of view is imaged at once by the camera instead of a site by site evaluation of the local lifetime as is performed in SPV, μW-PCD or ELYMAT. This reduces the evaluation time tremendously from the order of one hour to a matter of seconds.

Mapping the PL intensity or yield provides local information on the diffusion length or on τ_{eff}. Good quality regions normally correspond with a high intensity (high pixel count), while the opposite is true for poor quality defective regions. This is associated with the fact that optical recombination of carrier pairs is in competition with non-radiative SRH recombination. A spatial resolution on the order of 100 μm can be achieved for a total data acquisition time of 1.5 s. In addition, it is a contactless method, which can be used for wafer evaluation after different process steps in the manufacturing of solar cells. The injection level can be varied by a change in the laser power. Absolute light intensities are measured with a calibrated silicon sensor. Signal detection can proceed from the illuminated or from the other side of the wafer [87]. The relative pixel count is converted into a local effective minority carrier lifetime through (4.18). The conversion of counts per pixel into a local excess carrier density Δn is performed by comparing with the spatially averaged QSS-PL lifetime measurement taken at the same light intensity with a large-area p-i-n detector in a $2 \times 2\ cm^2$ area in the center of the wafer [87].

Performing the PL imaging in function of the injection level and the temperature enables a closer identification of the dominant recombination centers [90]. This is based on the fact that the so-called impurity parameter ξ, which is defined as the ratio of the derivative of the SRH lifetime with respect to the parameter X and the limit value of τ_{SRH} for $X \to 1$ only depends on the equilibrium hole concentration p_0, the SRH densities n_1 and p_1 and the symmetry factor k (= σ_n/σ_p), whereas it is independent on the defect concentration N_T [91]. For p-type material, X is defined as:

$$X = \frac{n}{p} \approx \frac{\Delta n}{p_0 + \Delta n} \tag{4.19}$$

Determining the temperature dependence of the impurity parameter can be exploited for the identification of the lifetime limiting defect, as is demonstrated by the calculated values in Fig. 4.16 for a number of metal-related recombination centers in silicon [90]. Moreover, it can also be applied to derive impurity maps from spatially resolved PL imaging [90]. The procedure for Fe imaging in mc-Si silicon has been worked out by Macdonald et al. [92], relying on the FeB pair breaking methodology (see Chap. 5). It also includes corrections for Auger recombination at high injection levels and for the fact that the carrier density is not uniform with depth [93].

A protocol has been proposed to identify several so-called metastable defects in silicon in PL imaging [94]. They comprise Fe_i/FeA (A = acceptor: B, Al, Ga, In,), Cr_i/CrB or the boron-oxygen related complex, which causes light-induced degradation (LID) of the bulk recombination lifetime in oxygen-containing Si [95]. A careful pre-treatment has been proposed in order to prepare well-defined states for either of these metastable defects, including the fact that usually incomplete separation of the FeA or CrA pairs is achieved after thermal and/or optical decomposition treatments [94]. At the same time, one should select carefully the illumination conditions in order to account for the non-uniform excess carrier density along the depth of the wafer [94].

Based on the same principles, the μPL imaging method using a confocal microscope has been developed to investigate the local lifetime around metal precipitates [95]. It has been shown that the band-to-band (BB) energy position of the defect PL peaks between 0.8 and 0.99 eV at room temperature or below, which

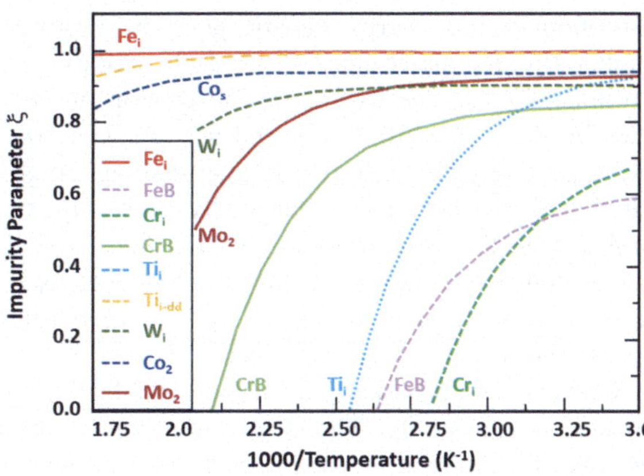

Fig. 4.16 Temperature dependence of the impurity parameter ξ for several defects and a p-type doping concentration of 1.5×10^{16} cm^{-3}. Reproduced with permission from Mundt et al. [90]. Copyright 2015 IEEE

may yield additional spectroscopic information on the nature of the recombination centers, with a high spatial resolution. The confocal microscope collects the PL light with a lateral resolution of 0.8 μm and an energy resolution of 0.5 nm. The depth resolution depends on the wavelength of the exciting laser, but is usually quite small, on the order of 1 μm for $\lambda_o = 532$ nm [95]. In the case of an Fe-contaminated sample, defect-specific PL at 1.3 μm is enhanced when normalized to the BB peak, while the opposite is true for Cu-contaminated Si [95].

A related technique is the so-called confocal micro-Raman spectroscopy (μRS), enabling the analysis of the carrier lifetime with submicron resolution [96]. The theoretical limit of a confocal microscope at 532 nm is 190 nm. Detailed analysis enables to measure both τ_r and the carrier density in the vicinity of an extended defect. Moreover, it allows to determine the local stress profile, demonstrating a complex stress field around metal precipitates, which affects the carrier lifetime [96, 97]. A decrease in the spatial extent of the region of maximum shear stress is correlated with an increase in the BB PL intensity. A direct correlation was established between a higher tensile residual stress and increased PL [98].

Application of PL analysis to the study of metal contamination and extended defects in mc-Si for solar cells can be found among others in [99, 100].

4.3.6 Carrier Lifetime by IR Imaging

Very fast lifetime mapping (a few minutes) with high spatial resolution can be achieved by monitoring the free carrier density after an excitation light pulse by its associated absorption (or emission) of long wavelength infrared (IR) light. This method has also been called Carrier Density Imaging (CDI) [56] and overcomes the necessity for a step by step illumination by a laser diode (at high injection density to limit the integration time), by using a CCD camera sensitive in the mid-IR. It measures absolute (and not differential) values of the lifetime under low-level injection conditions, similar as for QSS-PC, but with a higher spatial resolution. CDI is based on a measurement in the dark and under illumination of the free carrier absorption of low-energy mid-IR photons. It gives a values of Δn and of the absolute lifetime under quasi-steady-state conditions by taking the difference of the two images [56]. The free carrier absorption in silicon was found experimentally given by $\alpha_{on} \sim 1 \times 10^{-18}\ \lambda_o^2 n$ and $\alpha_{op} \sim 2.7 \times 10^{-18}\ \lambda_o^2 p$ in n- and p-type material, respectively [101]. In IR lifetime imaging, emission and absorption of thermal radiation always occurs simultaneously. One of the two processes will, however, be dominant. In the case of emission CDI, the wafer temperature is higher than the background, which offers the advantage of increased signal strength. This reduces the measurement time. It can be applied in production and allows for a quick assessment of technological problems [56].

A typical set-up is sketched in Fig. 4.17 [101]: a lock-in detection is applied to enhance the sensitivity. The IR transmittivity is measured in two different states: with illumination by a source $E_{ph} > E_g$, generating excess free carriers, and without

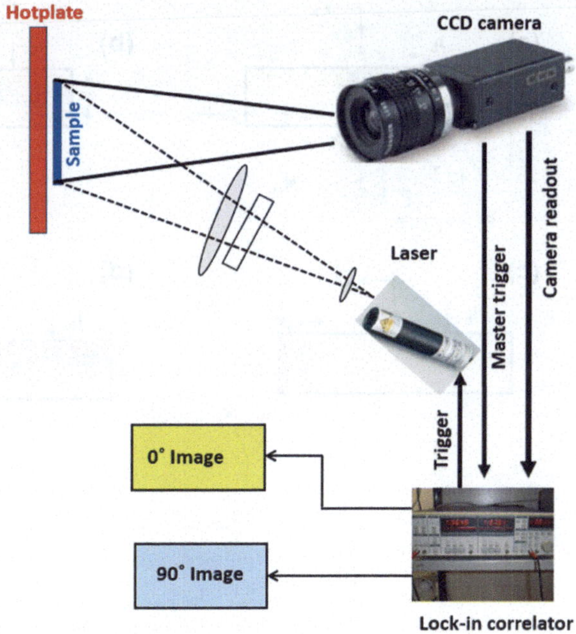

Fig. 4.17 Principle setup for infra-red carrier density measurements using a lock-in system for increased sensitivity (after Isenberg et al. [101])

illumination. If the two obtained images are subtracted from each other the difference is proportional to the IR absorption of the free carriers. A lateral resolution between 50 and 350 μm can be achieved using the set-up described in Fig. 4.17. Other modes of photocarrier radiometry have been developed over the years [102–104], enabling the detection of other relevant semiconductor material parameters besides the recombination lifetime, like the carrier mobility [105].

4.3.7 Lifetime Mapping of Extended Defects

As already indicated in Fig. 4.2, one can combine chemical or structural characterization of defects with electrical detection of the local diffusion length around a defect. The first technique which had such capabilities is the so-called Electron-Beam-Induced Current (***EBIC***), originally developed in the context of scanning electron microscopy [106]. In this case, the electron beam with energy in the range of a few keV up to several tens of keV is used to create ehps in the semiconductor. It has later also been extended to the case of TEM [107]. A schematic setup is depicted in Fig. 4.18. The principle is based on the minority carrier current which is collected by a space charge region in a semiconductor and created by the interaction of the scanning electron beam with the lattice. In general, a SB or p-n junction is required to provide the necessary electric field. The height of the collected current will depend on the local minority carrier diffusion length: for a

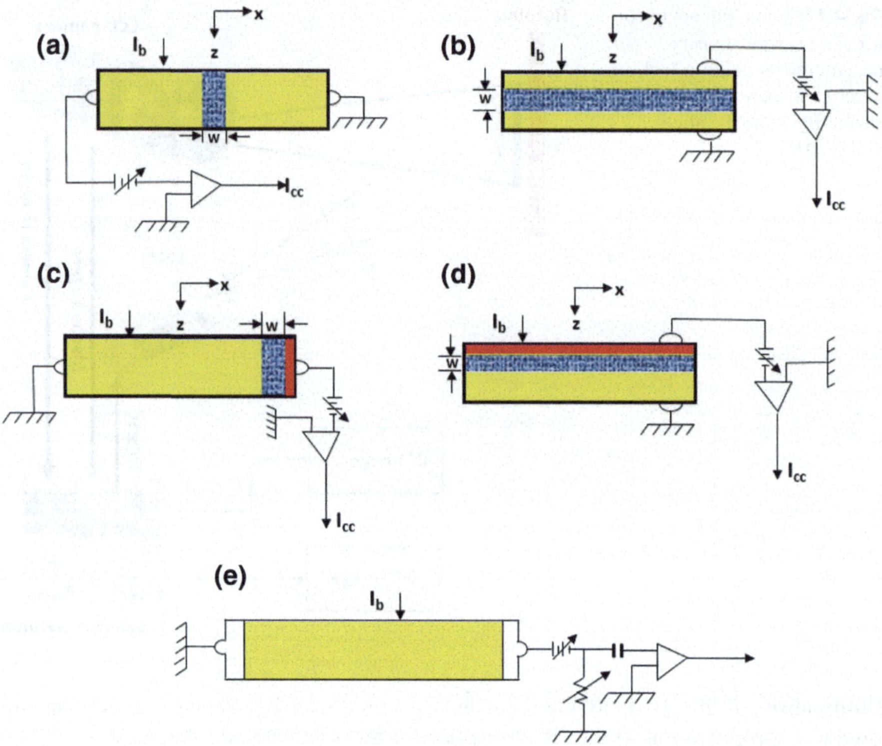

Fig. 4.18 Principle of EBIC. The charge collection area of the diode can be parallel or perpendicular to the electron beam as shown in (**a**, **b**) for a planar and in (**c**, **d**) for a Schottky barrier diode. The schematics for the beta conductivity technique is shown in (**e**)

high L_{diff}, a large current will be collected, while the opposite goes when a recombination active extended defect is present. This leads to a current intensity which varies over the wafer and is usually lower at the site of an extended defect. In this way, the recombination activity of defects can be visualized in an SEM, leading to an image with varying contrast, depending on the local L_{diff}. Usually, the relative change in contrast C_e (in %) with respect to a defect-free matrix is considered. In a first approximation, one has that $C_e \sim L_{diff}^{-2} \sim (D\tau_r)^{-1}$, under the assumption that SRH statistics are valid for the extended defect. The minority carrier diffusion length around an extended defect can be derived by measuring the collected current versus the electron beam energy [108–110].

Further information regarding the activation energy of an extended defect can be gathered by studying the contrast of a defect as a function of temperature [111]. Assuming that the SRH model is valid, the temperature dependence of the contrast follows from the one for the recombination lifetime and is in first instance given by [111]:

$$\tau_r = A\sigma_{min}^{-1}T\left[T^{-3/2} + BN_{dop}^{-1}\exp\left(-\frac{E_T}{k_B T}\right)\right] \quad (4.20)$$

The activation energy of the recombination level can thus be derived from the linear part of an $\ln(C_e T^{0.5})$ versus $1/k_B T$ plot [111]. Based on this method, it has been demonstrated that undecorated oxygen precipitates in Cz Si are little active at RT but exhibit clear contrast below 200 K, suggesting a rather shallow E_T ($\sim$0.1 eV in Si). On the other hand, when decorated with iron contamination, they become active at RT, with a contrast showing only a weak temperature dependence and a high activation energy ($\sim$mid gap). A systematic study of extended defects with different degrees of metal decoration has led to a model for the recombination activity of extended defects, which considers both the intrinsic shallow stress-induced states and the mid-gap impurity-related states [112].

A similar method consists of replacing the electron-beam injection of ehps by using optical excitation with an energy larger than the band gap of the semiconductor: Optical- or Light–Beam-Induced Current (***O-/LBIC***) [113, 114]. In this case, the image can be made by a CCD camera [114]. A procedure to identify recombination-active defects in silicon by means of focused ion-beam (FIB) sample preparation and TEM has been proposed by Stolze et al. [115].

As mentioned before, ***XBIC*** employs a mono-energetic synchrotron X-ray as excitation source (Fig. 4.2) [12, 16, 19, 116, 117]. The use of an X-ray beam which is inclined with respect to the wafer surface leads to a shift of the XBIC image with respect to μXRF in the direction of the beam projection on the sample surface [12]. This is because the maximum in XRF yield is achieved when the X-ray beam hits the precipitate, whereas the minimum in XBIC current is reached when the X-ray beam goes below the precipitate, so that the generated carriers can get trapped and recombine on their way to the p-n junction. The excitation regime in XBIC is equivalent to the low beam intensity regime in EBIC [12]. Its lateral resolution is on the order of 10 μm which is poorer than for EBIC, on the order of 1 μm [12]. This is related to the inclined incidence of the X-rays and the larger penetration depth. This implies that carrier pairs are created at a larger depth, on the order of 100 μm, which leads to their diffusive lateral scattering. The depth position of the defect can also change the contrast in XBIC [118].

While ***Cathodoluminescence*** (CL) has been proposed as a technique for local diffusion length determination in silicon [119] it is far more popular for direct gap III-V and III-N semiconductors, which have more efficient direct optical recombination. CL is an imaging mode in an SEM where the luminescent light created by the impinging scanning electron beam is measured by a photodetector.

4.3.8 MOS Generation Lifetime Techniques

All the foregoing techniques monitor the recombination lifetime, probing a depth in the semiconductor material on the order of the diffusion length or, in case $L_{diff} > d$, the wafer thickness. This means in practice that for the lifetime measurements described above, one usually obtains a spatially averaged value with respect to the sample depth. However, for specific cases, like, e.g., a thin epitaxial layer on a p^+-substrate or the denuded zone in an internally gettered (IG) silicon substrate or the Si film in a Silicon-On-Insulator (SOI) wafer, one is interested in the lifetime properties of a thin layer, on the order of a few μm. A suitable technique for that purpose is to measure the generation lifetime in a pulsed MOS capacitor [47, 120]. It is based on the so-called Zerbst technique [121], whereby the transient response of the capacitance is monitored after a bias pulse on the gate from accumulation into deep depletion. When biased in deep depletion, the capacitance will evolve to a steady-state inversion situation by the generation and/or in-diffusion of minority carriers, which will be collected in the inversion layer at the Si/SiO_2 interface. The capacitance is, thereby, measured by a small amplitude ($\sim$10 mV) ac modulation at 100 kHz on top of the dc bias.

The MOS capacitor can be processed by thermal oxidation and gate deposition, using lithography or it can be induced by a corona charge and a non-contact Kelvin probe. Also a contactless metal-air gap-oxide-semiconductor method has been developed [122]. The basic operation principle is illustrated by Fig. 4.19, representing the different minority carrier generation mechanisms, which help restoring the equilibrium condition after the pulse in deep depletion [120]. Five mechanisms have been distinguished in Fig. 4.19:

(1) bulk scr generation characterized by the generation lifetime τ_g;
(2) lateral surface scr generation characterized by the surface generation velocity s_g;
(3) surface scr generation under the gate characterized by the surface generation velocity s'_g;

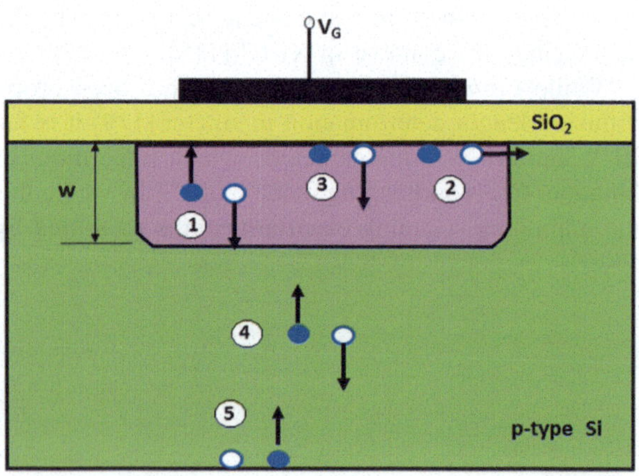

Fig. 4.19 Space-charge region and quasi-neutral bulk thermal generation mechanisms of a pulsed MOS capacitor (after Tan et al. [120])

(4) quasi-neutral bulk generation characterized by the minority carrier diffusion length L_n (for a p-type substrate);
(5) back surface generation characterized by the generation velocity s_c.

It is clear that the method is sensitive to defects at the Si/SiO$_2$ interface and to impurities in the scr. Usually, ehp generation in the scr is dominating when the device is pulsed into deep depletion at RT. Quasi-neutral bulk generation and diffusion of carriers to the scr dominates at high T [120]. The thermal generation rate of the various mechanisms is the negative of the change in the inversion charge density (dQ_n/dt) and given by [47]:

$$G_{th} = \frac{qn_i w}{\tau_g} + \frac{qn_i s_g A_S}{A_G} + qn_i s_g' + \frac{qn_i^2 D_n}{N_a L_{eff}} \tag{4.21}$$

with A_G the gate area, A_S the area of the lateral depletion region and L_{eff} the effective minority carrier diffusion length.

A so-called Zerbst plot represents $-d(C_{ox}/C(t)^2)/dt$ versus $(C_f/C - 1)$, with C_{ox} the oxide capacitance and C_f the final steady-state capacitance. This leads to the expression

$$-\frac{d}{dt}\left(\frac{C_{ox}}{C(t)}\right)^2 = \frac{2\varepsilon_{ox} n_i s_{g,eff}}{\varepsilon_s t_{ox} N_a} + \frac{2n_i}{\tau_{g,eff} N_a} \frac{C_{ox}}{C_f}\left(\frac{C_f}{C(t)} - 1\right) \tag{4.22}$$

The slope $(2n_i C_{ox})/(\tau_{g,eff} N_a C_f)$ enables to calculate the generation lifetime and the intercept is proportional with the surface generation velocity, as illustrated in Fig. 4.20. The effective generation lifetime is given by [47]:

$$\tau_{g,eff} = \frac{\tau_g}{1 + \tau_g/\tau_{sg}} \tag{4.23}$$

Fig. 4.20 Illustration of the Zerbst technique for the determination of the generation lifetime and the surface recombination velocity

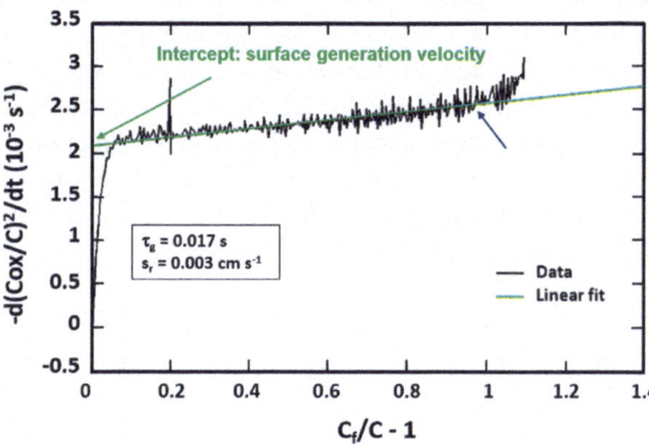

with $\tau_{sg} = r/2s_g$ the lateral surface generation lifetime and r the radius of a circular gate. In the limit of a well-passivated surface (thermal SiO_2, with an s_g in the range of 0.1–1 cm/s) the bulk scr generation lifetime will dominate τ_s, so that $\tau_{g,eff} \sim \tau_g$.

This method has been applied to study interstitial Fe_i and FeB pairs in silicon [120, 123] and to evaluate the quality of thin p/p^+ epitaxial layers [124]. In the latter case, τ_g in the scr is on the order of 10–50 ms, with a bulk trap density on the order of 10^{10} cm^{-3}, so that the surface generation can no longer be neglected. Combining measurements at different temperatures enables to separate the surface generation effect from the bulk one in thin, high-quality silicon epi-layers [124]. A method has also been proposed which is closely related to DLTS and measures an isothermal DLTS peak as a function of pulse frequency, whereby the capacitor starts from equilibrium in inversion and is repetitively pulsed into accumulation with a frequency f [125]. Using the resolving power of lock-in DLTS enables to measure a minority carrier generation peak, where the maximum corresponds with a predefined τ_g. Measurement of the DLTS peak at different T allows to construct an Arrhenius plot (see next section) wherefrom the activation energy and the capture cross section can be derived [125]. A higher lateral resolution for defect monitoring can be obtained in a scanning probe tool. For example, scanning capacitance microscopy has been applied to image surface Fe contamination in p-type silicon [126, 127]. It is based on the positive surface charge induced by the iron atoms and the associated negative shift in the flat-band voltage of the MOS capacitance [126].

Finally, it has been proposed to employ optical excitation through a transparent gate electrode (Kelvin probe) to measure the recombination lifetime in thin epitaxial layers [128, 129].

4.3.9 Deep-Level Transient Spectroscopy

Since its introduction in 1974 [46], Deep-Level Transient Spectroscopy (DLTS) is the reference technique for the fundamental study of deep-level defects in semiconductor materials. It replaces Hall effect measurements as a function of temperature, which was initially used in the fifties and sixties for the identification of metal levels in silicon and germanium [50]. The principle of DLTS has been described in several reviews and text books (see, e.g., [47, 50]) and will be briefly summarized here. Standard DLTS is a so-called junction space charge technique, relying on the capacitance transient of a p-n junction or SB, following a reverse-bias pulse, as sketched in Fig. 4.21. It can be shown that for a single deep level E_T and not too high trap concentrations N_T ($\ll N_a$ in p-type material) the transient is exponential, which can be represented by:

$$C(t) - C(\infty) = -\Delta C\left[1 - \exp\left(-\frac{t}{\tau_{ep}}\right)\right] \tag{4.24}$$

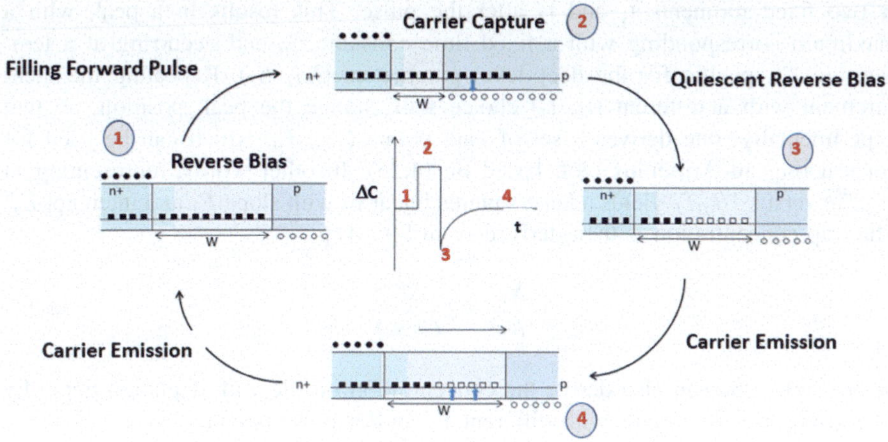

Fig. 4.21 Principle of DLTS: from 1 to 2, a bias pulse is applied to a reversely biased n^+p junction (SB or MOS capacitor), which reduces the space charge width w, so that majority carriers become available for capture by the empty deep levels in part of the depletion region. At 3, the original reverse bias V_R is restored, but the trapped holes do not respond immediately, so that the space charge density is reduced by the compensation of the ionized shallow acceptors (negative charge) and the trapped holes (positive charge). This results in a wider scr and a smaller capacitance immediately after the pulse. From 3 to 4, the capacitance relaxes by the thermal emission of the trapped holes, giving rise to a capacitance transient (middle figure). This is the basic signal for DLTS. In first approximation, this transient is exponential with t, with a relaxation time constant given by the inverse hole emission rate e_p. The amplitude ΔC of the transient is proportional with the deep-level concentration N_T

$C(\infty)$ is the steady-state reverse capacitance at $t \rightarrow \infty$ and corresponding with case 1 in Fig. 4.21. ΔC is the transient amplitude, which is proportional with the trap density N_T if a sufficiently long bias pulse is employed. The hole emission time constant τ_{ep} is given by:

$$\tau_{ep} = K_T^{-1} T^{-2} \exp\left(\frac{E_T}{k_B T}\right) \tag{4.25}$$

with (K_T, E_T) the trap signature. K_T is proportional with the majority carrier capture cross section. It is clear from (4.25) that the emission time constant is thermally activated, so that the capacitance transient exhibits a strong variation with T. This means that at low temperature, the trapped charge will remain on the deep levels, so that $C(t) \sim C(0)$. On the other hand, at high T, the transient will proceed very fast, so that $C(t) \sim C(\infty)$. For intermediate temperatures, an exponential transient will result.

The key feature of DLTS is the application of an analog (or nowadays digital) filtering of the repetitive transient signal (the pulse is periodically applied), so that a peak is obtained for each deep level, when varying the temperature. Originally, the double boxcar technique was employed [46], whereby the capacitance was sampled at two fixed moments t_1 and t_2 after the pulse. This results in a peak with a maximum corresponding with a fixed time constant τ_{ep} and occurring at a temperature T_p specific for the deep-level parameters (K_T, E_T). Repeating the measurement with a different (t_1, t_2) choice will change the peak position, so that experimentally, one derives a set of data points (τ_{ep}, T_p) which can be used for constructing an Arrhenius plot, based on (4.25). In other words, representing ln ($\tau_{ep}T_P^2$) versus $1/k_BT_P$ yields a least-squares linear fit with slope E_T and intercept K_T. The trap concentration is then derived from [46, 47]:

$$\frac{N_T}{N_a} = 2\frac{\Delta C}{C(\infty)} \tag{4.26}$$

In principle, one can also derive the concentration profile with depletion depth by performing measurements with different V_R and/or pulse height.

In practice, one has to vary the temperature of the diode over a certain range, in order to cover the accessible majority traps in the upper (n-type) or lower half (p-type) of the band gap. For silicon, this means a range between 30 and 300 K, while for Ge, this is shifted from 4 to 200 K, based on the lower ionization level of shallow dopants. From (4.26) one can intuitively derive that as long as there is a shallower doping level available to create the necessary back-ground space charge (qN_{dop}), it will be possible to get a DLTS peak, for sufficiently large N_T. The practical detection limit of DLTS is often quoted in the range 10^{-5}–10^{-4} of the back-ground doping density, which explains the large success of the technique. However, it is also clear that it is not possible to use it as a routine, in-line monitoring tool for defect-related problems, since one needs to prepare a diode and the recording of the spectrum itself takes a few hours. This is among others related to the fact that a sufficiently small scan rate $\mathrm{d}T/\mathrm{d}t$ is required, so that the temperature sensor tracks the real sample temperature. In addition, for deriving an Arrhenius plot, one has to record several spectra, i.e., perform several temperature scans. In more recent commercial systems, based on Fourier-Transform (FT) DLTS [130], spectra with different rate windows can be calculated based on one set of measured transients, so that in principle one T-scan suffices to generate an Arrhenius plot. Alternatively, one may change the frequency of the bias pulse at a fixed temperature (e.g., RT) in a so-called isothermal DLTS scan [50]. However, in order to construct an Arrhenius plot, measurements at different temperatures are still required. Consequently, DLTS is either used for more fundamental defect studies or in order to firmly identify an unknown contamination, once detected by faster in-line monitoring tools.

The use of a SB or MOS capacitor limits DLTS to scan the upper or the lower band half for majority carrier traps. This implies that in order to fully characterize all deep levels of a metal, one has to combine measurements on a p- and n-type substrate. This can be circumvented by measuring on a p-n junction (usually, asymmetric p^+n or n^+p junctions are preferred to facilitate the analysis), where minority carriers can be injected by a bias pulse in forward operation. As a result, minority carrier traps will give rise to a DLTS peak with opposite sign. This is based on the fact that the trapped charge has now the same sign as the fixed ionized charge in the scr, resulting in a positive capacitance transient. An alternative is the use of an optical pulse at fixed V_R in order to create ehps by absorption of photons with energy above the band gap. This can be applied in case a transparent metal contact is fabricated to obtain a SB or MOS capacitor. In any case, quantification of the minority trap concentration is more complicated than for majority traps. Of course, the advantage is that one can in principle detect all deep levels of a center with multiple charge states, as is typical for many metal-related defects, in one and the same sample.

It is important to remark that the value derived for the majority carrier capture cross section from the intercept value K_T provides at best an order of magnitude estimate at $T \rightarrow \infty$. For accurate determination of the majority capture cross section, one has to perform DLTS measurements with different bias pulse time t_p, to monitor the carrier capture. In principle, this yields an exponential increase of the DLTS amplitude with t_p, corresponding with a time constant given by $(nv_{th}\sigma_n)^{-1}$ for an n-type semiconductor. However, slow capture at long t_p can occur due to the existence of free carrier Debye tails at the edge of the depletion region [50]. Procedures for correcting this slow capture can be found in the literature [50], resulting in an accurate determination of the capture cross section. When performed at different T, one can derive the barrier for capture ΔE_B in case of a repulsive center for the majority carriers. Attractive centers will generally exhibit another type of temperature dependence. In this way, the nature (charge state) of the deep level in question can be derived, i.e., whether it is a deep donor or acceptor (or neutral when empty). Establishing the T-dependence of the capture cross section is also relevant from the viewpoint of practical device applications at room temperature, since the measured values in DLTS are usually obtained at T_P, which can be well below 300 K.

One serious limitation for DLTS is that it is not straightforward to measure the *minority carrier* capture cross section of a deep level. According to the SRH model, this is the cross section which will determine the low-level injection recombination lifetime and thus essential for estimating its impact on device performance. Double pulse biasing schemes for a p-n junction (one in forward injection and another one in reverse bias) have been proposed to overcome this limitation. However, some traps with a minority capture cross section which is much higher than the majority capture cross section may even escape DLTS observation, although they dominate the recombination lifetime. This is related to the fact that for an efficient recombination center, the trap occupation may not be changed during a bias pulse so that no measurable capacitance transient is obtained.

The energy resolution of DLTS is determined by the Full Width at Half Maximum of a peak (FWHM) and typically amounts to the order of k_BT [50]. This implies that the peak width increases at higher T_p, i.e., for deeper activation energy. A higher resolution of 1 or a few meV can be achieved by performing Laplace-Transform DLTS at constant temperature [131]. In that case, the DLTS signal is obtained versus the emission rate.

Another limitation of DLTS is that it requires the fabrication of contacts. Therefore, several proposals for non-contact DLTS have been reported, using either RF detection [76], Kelvin probe detection [67] or a metal-air gap-oxide-semiconductor scheme [122, 132]. The lateral resolution can be improved by using a scanning capacitance instrument for monitoring emission transients from deep levels [133].

Perhaps the biggest limitation of DLTS is that it is a quasi-spectroscopic technique, i.e., it provides and electrical signal but no direct chemical or structural data on the underlying defect. In order to obtain more information on the nature of the deep levels, one has to combine it with other analysis techniques. However, as the detection limit can be much lower than for most chemical and structural characterization tools, one usually has to rely on other strategies to identify the deep levels. One way is to study reference samples with well-known impurities, which has been the case for metal impurities in silicon, SiGe and Ge (see Chaps. 5 and 6). The metals were introduced either in the melt or by in-diffusion at high temperature, followed by quenching or by ion implantation and thermal anneal. For the case of native defects, like vacancy- and interstitial-related complexes, electron or gamma irradiation can be employed. Combined with the thermal stability of the defect levels upon isothermal or isochronal annealing helps to identify their nature. Finally, hydrogen can be introduced quite easily in semiconductors, either by a wet etch or a plasma treatment. In this way, the passivation of the deep levels by association with hydrogen can be studied. This means that the deep levels of a metal are removed from the band gap. In other cases, new deep levels will be found, which follow the in-diffusion profile of hydrogen and represent the binding with one of more hydrogen atoms. This will be detailed in Chaps. 5 and 6.

4.4 Strategy for Metal Contamination Monitoring

As shown in Chap. 2, the diffusion, segregation and solubility properties of metals in semiconductors vary over a wide range. This implies that the behavior of metals will be strongly different. As will be seen in Chaps. 5 and 6 in more detail, fast diffusing 3*d* TMs like Co, Ni and Cu are difficult to be kept in the bulk of the material and will tend to precipitate at the surface after a high-temperature treatment, irrespective of the cooling speed (quench). An intermediate diffusor like iron can be kept in the silicon bulk, when applying a fast quench after the heat treatment. However, interstitial Fe is not the lowest energy species in p-type silicon at room temperature. Fe_i will tend to pair with B, driven by Coulomb attraction, so that FeB pairs are the main Fe-related defects in steady state. Slow diffusors, like, Ti, Mo or W in silicon will not be able to diffuse very deep or precipitate at the surface even after slow cooling. This difference in behavior is also reflected to some extent in their impact on the electrical properties of materials and devices.

In the foregoing, a wide range of analysis methods has been described, with their strengths and weaknesses. Some tools will be more suitable for monitoring fast diffusing, surface-precipitating metals, while others will be more relevant for studying impurities in the bulk of a wafer. In case of process monitoring of usually unknown metal contamination, one needs to apply a combination of different techniques, in order to cover all the different types of metals and the associated defect states. Some guidelines for effective contamination analysis are in place and have been reported by Polignano and co-workers for a number of different cases [7, 8, 10]. In silicon device processing, it is recommended to investigate surface metal contamination by TXRF or TOF-SIMS [7, 8]. The electrical impact of for example copper can be monitored by a technique which is most sensitive to near surface effects, like MOS generation lifetime (see Sect. 4.3.8), which also enables the measurement of possible interface states with density D_{it} [8]. The electrical impact of slow diffusors like Ti, Mo or W is best investigated by DLTS, as it typically probes a depletion region on the order of 1 to a few μm beneath the surface, which contains the in-diffusion profile of such slow diffusors [7, 8, 10]. On the other hand, Fe is best detected by ELYMAT or SPV, which probes most part of the wafer. DLTS is less recommended in this case, as the concentration of Fe_i (or FeB) is lower in the near-surface region, by segregation at the Si/SO_2 interface [8].

When the properties of the substrate are non-uniform, like for an IG wafer with a defect-lean Denuded Zone (DZ) and a highly defective interior of the Cz Si wafer, again different techniques are more appropriate to investigate the gettering capability for metals, depending on the species [134]. Lifetime/diffusion length techniques like SPV or ELYMAT will mainly probe the poor lifetime in the defective bulk, while revealing no information on the DZ, where the active devices are being fabricated. In that case, MOS-based Zerbst lifetime measurements and DLTS can provide the desired information on the presence of metals in the DZ and, hence, on the gettering efficiency of the oxide precipitates in the bulk of the Cz Si wafer.

In the case of contamination monitoring during or after wet cleaning, most of the 'bulk' electrical techniques like ELYMAT, SPV or μPCD require in-diffusion of the surface-deposited metal by a heat treatment, for example, a Rapid Thermal Annealing (RTA) step [64, 66]. Non-contact techniques which can avoid this treatment have been developed [66] and are based on ac-SPV or Contact Potential Difference (CPD) measurements, which probe the surface charge induced by the deposited metal atoms.

Overall, the monitoring of in-line metal contamination is a challenging task, which generally requires a multi-method analysis approach. It is expected to become even more challenging for the upcoming technology nodes, where the active layers are confined in narrow regions between Shallow Trench Isolation (STI) with sub-10 nm width. This puts a tremendous demand on the lateral resolution of the analysis techniques. For metal detection in solar cell processing on mc-Si substrates, PL imaging (or CDI) appears to be the most performant analysis technique, providing a fast idea about the distribution of the dominant metal-related defect species in the material.

References

1. A.A. Istratov, H. Hieslmair, E.R. Weber, Iron contamination in silicon technology. Appl. Phys. A **70**, 489–534 (2000). https://doi.org/10.1007/s003390000458
2. D. Macdonald, A. Cuevas, A. Kinomura, Y. Nakano, L.J. Geerligs, Transition-metal profiles in a multicrystalline silicon ingot. J. Appl. Phys. **97**, 033523/1-7 (2005). https://doi.org/10.1063/1.1845584
3. L. Fabry, L. Köster, S. Pahlke, L. Kotz, J. Hage, Diagnostic and monitoring tools of large scale Si-manufacturing: trace-analytical tools and techniques in Si-wafer manufacturing. IEEE Trans. Semicond. Manufact. **9**, 428–436 (1996). https://doi.org/10.1109/66.536113
4. H.H. Woodbury, G.W. Ludwig, Spin resonance of transition metals in silicon. Phys. Rev. **117**, 102–108 (1960). https://doi.org/10.1103/PhysRev.117.102
5. K.C. Thompson, R.J. Reynolds, *Atomic Absorption Fluorescence and Flame Emission Spectroscopy: A Practical Approach* (Wiley, New York, 1978)
6. A. Benninghoven, F.G. Rudenauer, H.W. Werner, *Secondary Ion Mass Spectrometry: Basic Concepts, Instrumental Aspects, Applications and Trends* (Wiley-Interscience, New York, 1987)
7. M.L. Polignano, D. Codegoni, A. Galbiati, S. Grasso, I. Mica, G. Moccia, G. Nardone, F. Russo, Review - characterization of metal-contamination effects in silicon. ECS J. Solid State Sci. Technol. **5**, P3048–P3058 (2016). https://doi.org/10.11492/2.0081604jss
8. M.L. Polignano, D. Codegoni, S. Grasso, I. Mica, G. Borionetti, A. Nutsch, A comparative analysis of different measurement techniques to monitor metal and organic contamination in silicon device processing. Phys. Status Solidi A **212**, 495–505 (2015). https://doi.org/10.1002/pssa.201400082
9. D. Hellin, S. De Gendt, J. Rip, C. Vinckier, Total reflection X-ray fluorescence spectrometry for the introduction of novel materials in clean-room production environments. IEEE Trans. Device Mater. Reliab. **5**, 639–651 (2005). https://doi.org/10.1109/TDMR.2005.861351
10. M.L. Polignano, A. Galbiati, S. Grasso, D. Magni, I. Mica, Molybdenum and tungsten contamination in MOS capacitors. ECS J. Solid State Sci. Technol. **5**, P203–P210 (2016). https://doi.org/10.1149/2.0011605jss
11. P.K. Chu, B.W. Schueler, F. Reich, P.M. Lindley, Determination of trace metallic impurities on 200-mm silicon wafers by time-of-flight secondary-ion-mass-spectroscopy. J. Vac. Sci. Technol. B **15**, 1908–1912 (1997). https://doi.org/10.1116/1.589577
12. O.F. Vyvenko, T. Buonassisi, A.A. Istratov, E.R. Weber, X-ray beam induced current/microprobe X-ray fluorescence: synchrotron radiation based X-ray microprobe techniques for analysis of the recombination activity and chemical nature of metal impurities in silicon. J. Phys. Condens. Matter **16**, S141–S151 (2004). https://doi.org/10.1088/0953-8984/16/2/017
13. A.A. Istratov, T. Buonassisi, R.J. McDonald, A.R. Smith, R. Schindler, J.A. Rand, J. P. Kalejs, E.R. Weber, Metal content of multicrystalline silicon for solar cells and its impact on minority carrier diffusion length. J. Appl. Phys. **94**, 6552–6559 (2003). https://doi.org/10.1063/1.1618912
14. A.R. Smith, R.J. McDonald, H. Manini, D.L. Hurley, E.B. Norman, M.C. Vella, R.W. Odom, Low-background instrumental neutron activation analysis of silicon semiconductor materials. J. Electrochem. Soc. **143**, 339–346 (1996). https://doi.org/10.1149/1.1836433
15. A. Huber, G. Böhm, S. Pahlke, Industrial applications of instrumental neutron activation analysis. J. Radioanal. Nucl. Chem. **169**, 93–104 (1993). https://doi.org/10.1007/BF02046787
16. S.A. McHugo, A.C. Thompson, C. Flink, E.R. Weber, G. Lamble, B. Gunion, A. MacDowell, R. Celestre, H.A. Padmore, Z. Hussain, Synchrotron-based impurity mapping. J. Cryst. Growth **210**, 395–400 (2000). https://doi.org/10.1016/S0022-0248(99)00718-6
17. W. Seifert, O.F. Vyvenko, T. Arguirov, A. Erko, M. Kittler, C. Rudolf, M. Salome, M. Trushin, I. Zizak, Synchrotron microscopy and spectroscopy for analysis of crystal defects in silicon. Phys. Status Solidi C **6**, 765–771 (2009). https://doi.org/10.1002/pssc.200880717

18. T. Buonassisi, A.A. Istratov, M. Heuer, M.A. Marcus, R. Jonczyk, J. Isenberg, B. Lai, Z. Cai, S. Heald, W. Warta, R. Schindler, G. Willeke, E.R. Weber, Synchrotron-based investigations of the nature and impact of iron contamination in multicrystalline silicon solar cells. J. Appl. Phys. **97**, 074901/1-11 (2005). https://doi.org/10.1063/1.1866489
19. T. Buonassisi, A.A. Istratov, M.A. Marcus, M. Heuer, M.D. Pickett, B. Lai, Z. Cai, S.M. Heald, E.R. Weber, Local measurements of diffusion length and chemical character of metal clusters in multicrystalline silicon. Solid State Phenom. **108–109**, 577–584 (2005). https://doi.org/10.4028/www.scientific.net/SSP.108-109.577
20. L. Fabry, R. Hoelzl, A. Andrukhiv, K. Matsumoto, J. Qiu, S. Koveshnikov, M. Goldstein, A. Grabau, H. Horie, R. Takeda, Test methods for measuring bulk copper and nickel in heavily doped p-type silicon wafers. J. Electrochem. Soc. **153**, G566–G571 (2006). https://doi.org/10.1149/1.2186799
21. G.W. Ludwig, H.H. Woodbury, Electron spin resonance in semiconductors. Solid State Phys. **13**, 223–304 (1962). https://doi.org/10.1016/S0081-1947(08)60458-0
22. E.R. Weber, Transition metals in silicon. Appl. Phys. A **30**, 1–22 (1983). https://doi.org/10.1007/BF00617708
23. H.H. Woodbury, H.W. Ludwig, Spin of Ni^{61}. Phys. Rev. Lett. **1**, 16–17 (1958). https://doi.org/10.1103/PhysRevLett.1.16
24. G.W. Ludwig, H.H. Woodbury, Electron spin resonance in nickel-doped germanium. Phys. Rev. **113**, 1014–1018 (1959). https://doi.org/10.1103/PhysRev.113.1014
25. Y.H. Lee, R.L. Kleinhenz, J.W. Corbett, EPR of a thermally induced defect in silicon. Appl. Phys. Lett. **31**, 142–144 (1977). https://doi.org/10.1063/1.89630
26. F.C. Rong, G.J. Gerardi, W.R. Buchwald, E.H. Poindexter, M.T. Umlor, D.J. Keeble, W.L. Warren, Electrically detected magnetic resonance of a transition metal related recombination center in Si p-n diodes. Appl. Phys. Lett. **60**, 610–612 (1992). https://doi.org/10.1063/1.106569
27. A.P.D. Nguyen, A. Stesmans, D. Hiller, M. Zacharias, Near-interface Si substrate 3*d* metal contamination during atomic layer deposition processing detected by electron spin resonance. J. Appl. Phys. **111**, 114308/1-6 (2012). https://doi.org/10.1063/1.4709445
28. C. Poole, *Electron Spin Resonance: A Comprehensive Treatise on Experimental Techniques* (Dover Publications, New York, 1996)
29. J.A. Weil, J.R. Bolton, *Electron Paramagnetic Resonance* (Wiley, Hoboken, 2006)
30. G. Weyer, Defects in semiconductors—results from Mössbauer spectroscopy. Hyperfine Interact. **177**, 1–13 (2007). https://doi.org/10.1007/s10751-008-9607-y
31. P. Schwalbach, S. Laubach, M. Hartick, E. Kankeleit, B. Keck, M. Menningen, R. Sielemann, Diffusion and isomer shift of interstitial iron in silicon observed via in-beam Mössbauer spectroscopy. Phys. Rev. Lett. **64**, 1274–1277 (1990). https://doi.org/10.1103/PhysRevLett.64.1274
32. Y. Yoshida, Y. Ino, K. Tanaka, Mössbauer spectroscopy on Fe impurities in Si materials. Solid State Phenom. **242**, 211–217 (2016). https://doi.org/10.4028/www.scientific.net/SSP.242.211
33. Y. Yoshida, Y. Kobayashi, K. Hayakawa, K. Yukihira, A. Yoshida, H. Ueno, F. Shimura, F. Ambe, In situ observation of substitutional and interstitial Fe atoms in Si after GeV-implantation: an in-beam Mössbauer study. Physica B **376**, 69–72 (2006). https://doi.org/10.1016/j.physb.2005.12.019
34. Y. Yoshida, Y. Suzuki, A. Matsushita, K. Suzuki, K. Sakata, Fermi level dependence of Mössbauer spectroscopic components corresponding to iron interstitials and their clusters in silicon. Physica B **401–402**, 167–170 (2007). https://doi.org/10.1016/j.physb.2007.08.138
35. Y. Yoshida, Y. Tsukamoto, M. Ichino, K. Tanaka, Direct observation of carrier trapping processes on Fe impurities in mc-Si solar cells. Solid State Phenom. **205–206**, 40–46 (2014). https://doi.org/10.4028/www.scientific.net/SSP.205-206.40
36. D.K. Schroder, Nano characterization of materials. ECS Trans. **33**(11), 3–22 (2010). https://doi.org/10.1149/1.3485679

37. B.L. Sopori, A new defect etch for polycrystalline silicon. J. Electrochem. Soc. **131**, 667–672 (1984). https://doi.org/10.1149/1.2115670
38. L. Souriau, T. Atanasova, V. Terzieva, A. Moussa, M. Caymax, R. Loo, M. Meuris, W. Vandervorst, Characterization of threading dislocations in thin germanium layers by defect etching: toward chromium and HF-free solution. J. Electrochem. Soc. **155**, H677–H681 (2008). https://doi.org/10.1149/1.2953495
39. D.R. Sparks, R.G. Chapman, The use of rapid thermal annealing for studying transition metals in silicon. J. Electrochem. Soc. **133**, 1201–1205 (1986). https://doi.org/10.1149/1.2108819
40. P. Alpern, W. Bergholz, R. Kakoschke, Detection of fast diffusing metal impurities in silicon by haze test and by modulated optical reflectance: a comparison. J. Electrochem. Soc. **136**, 3841–3848 (1989). https://doi.org/10.1149/1.2096559
41. E. Sirtl, A. Adler, Chromsäure-Flußsäure als spezifisches System zur Ätzgrubenentwicklung auf Silizium. Z. Metallkde **52**, 231–529 (1961)
42. K. Graff, Transition metals in silicon and their gettering behaviour. Mater. Sci. Eng. B **4**, 63–69 (1989). https://doi.org/10.1016/0921-5107(89)90217-1
43. U. Wahl, J.G. Correia, E. Rita, J.P. Araújo, J.C. Soares, The ISOLDE Collaboration, Lattice sites of implanted Fe in Si. Phys. Rev. B **72**, 014115/1-8 (2005). https://doi.org/10.1103/physrevb.72.014115
44. U. Wahl, J.G. Correia, E. Rita, J.P. Araújo, J.C. Soares, The ISOLDE Collaboration, Fe and Cu in Si: lattice sites and trapping at implantation-related defects. Nucl. Instrum. Meth. Phys. Res. B **253**, 167–171 (2006). https://doi.org/10.1016/j.nimb.2006.10.053
45. D.J. Silva, U. Wahl, J.G. Correia, J.P. Araújo, Influence of n^+ and p^+ doping on the lattice sites of implanted Fe in Si. J. Appl. Phys. **114**, 103503/1-9 (2013). https://doi.org/10.1063/1.4819210
46. D.V. Lang, Deep-level transient spectroscopy: a new method to characterize traps in semiconductors. J. Appl. Phys. **45**, 3023–3032 (1974). https://doi.org/10.1063/1.1663719
47. D. Schroder, *Semiconductor Material and Device Characterization* (Wiley, New York, 1998)
48. C. Claeys, E. Simoen (eds.), *Germanium-Based Technologies—From Materials to Devices* (Elsevier, 2007)
49. E. Gaubas, E. Simoen, J. Vanhellemont, Review—carrier lifetime spectroscopy for defect characterization in semiconductor materials and devices. ECS J. Solid State Sci. Technol. **5**, P3108–P3137 (2016). https://doi.org/10.1149/2.0201604jss
50. E. Simoen, J. Lauwaert, H. Vrielinck, Analytical techniques for electrically active defect detection, in *Semiconductors and Semimetals*, vol. 91, ed. by L. Romano, V. Privitera, C. Jagadish (Elsevier, 2015), pp. 205–250. https://doi.org/10.1016/bs.semsem.2014.12.001
51. W. Shockley, W.T. Read Jr., Statistics of the recombination of holes and electrons. Phys. Rev. **87**, 835–842 (1952). https://doi.org/10.1103/PhysRev.87.835
52. R.N. Hall, Electron-hole recombination in germanium. Phys. Rev. **87**, 387 (1952). https://doi.org/10.1103/PhysRev.87.38
53. D. Macdonald, A. Cuevas, Validity of simplified Shockley-Read-Hall statistics for modeling carrier lifetimes in crystalline silicon. Phys. Rev. B **67**, 075203/1-7 (2003). https://doi.org/10.1103/physrevb.67.075203
54. S. Rein, T. Rehrl, W. Warta, S.W. Glunz, Lifetime spectroscopy for defect characterization: systematic analysis of the possibilities and restrictions. J. Appl. Phys. **91**, 2059–2070 (2002). https://doi.org/10.1063/1.1428095
55. J. Schmidt, Temperature- and injection-dependent lifetime spectroscopy for the characterization of defect centers in semiconductors. Appl. Phys. Lett. **82**, 2178–2180 (2003). https://doi.org/10.1063/1.1563830
56. W. Warta, Advanced defect and impurity diagnostics in silicon based on carrier lifetime measurements. Phys. Status Solidi A **203**, 723–746 (2006). https://doi.org/10.1002/pssa.200564510

57. B.B. Paudyal, K.R. McIntosh, D.H. Macdonald, G. Coletti, Temperature dependent carrier lifetime studies of Mo in crystalline silicon. J. Appl. Phys. **107**, 054511/1-5 (2010). https://doi.org/10.1063/1.3309833
58. T.S. Horányi, T. Pavelka, P. Tüttõ, In situ bulk lifetime measurement on silicon with a chemically passivated surface. Appl. Surf. Sci. **63**, 306–311 (1993). https://doi.org/10.1016/0169-4332(93)90112-O
59. D. Poelman, P. Clauws, B. Depuydt, Chemical surface passivation of low resistivity p-type Ge wafers for solar cell applications. Solar Energy Mater. Solar Cells **76**, 167–173 (2003). https://doi.org/10.1016/S0927-0248(02)00216-7
60. A.M. Goodman, A method for the measurement of short minority carrier diffusion lengths in semiconductors. J. Appl. Phys. **32**, 2550–2552 (1961). https://doi.org/10.1063/1.1728351
61. D.K. Schroder, Surface voltage and surface photovoltage: history, theory and applications. Meas. Sci. Technol. **12**, R16–R31 (2001). https://doi.org/10.1088/0957-0233/12/3/202
62. J. Lagowski, P. Edelman, M. Dexter, W. Henley, Non-contact mapping of heavy metal contamination for silicon IC fabrication. Semicond. Sci. Technol. **7**, A185–A192 (1992). https://doi.org/10.1088/0268-1242/7/1A/036
63. J. Lagowski, A.M. Kontkiewicz, E.P. Jastrzebski, Method for the measurement of long minority carrier diffusion lengths exceeding wafer thickness. Appl. Phys. Lett. **63**, 2902–2904 (1993). https://doi.org/10.1063/1.110292
64. L. Jastrzebski, O. Milic, M. Dexter, J. Lagowski, D. DeBusk, K. Nauka, R. Witowski, M. Gordon, E. Persson, Monitoring of heavy metal contamination during chemical cleaning with surface photovoltage. J. Electrochem. Soc. **140**, 1152–1159 (1993). https://doi.org/10.1149/1.2056215
65. H. Shimizu, E. Omori, M. Ikeda, Anomalous behavior of Schottky barrier-type surface photovoltages in chromium-contaminated n-type silicon wafers exposed to air. Jpn. J. Appl. Phys. **45**, 4982–4984 (2006). https://doi.org/10.1143/JJAP.45.4982
66. M. Wilson, A. Savtchouk, J. D'Amico, I. Tarasov, L. Jastrzebski, J. Lagowski, Manifestation of Cu impurities on silicon surfaces, implications for monitoring Cu contamination. ECS Trans. **11**(3), 347–361 (2007). https://doi.org/10.1149/1.2778676
67. A. Findlay, D. Marinskiy, P. Edelman, M. Wilson, A. Savtchouk, J. Lagowski, Non-visual defect monitoring with surface voltage mapping. ECS J. Solid State Sci. Technol. **5**, P3087–P3095 (2016). https://doi.org/10.1149/2.0161604jss
68. G. Zoth, W. Bergholz, A fast, preparation-free method to detect iron in silicon. J. Appl. Phys. **67**, 6764–6771 (1990). https://doi.org/10.1063/1.345063
69. O.J. Anttila, M.V. Tilli, Metal contamination removal on silicon wafers using dilute acidic solutions. J. Electrochem. Soc. **139**, 1751–1756 (1992). https://doi.org/10.1149/1.2069488
70. Y. Kitagawara, T. Yoshida, T. Hanaguchi, T. Takenaka, Evaluation of oxygen-related Czochralski-grown Si crystals by the bulk lifetime measurements. J. Electrochem. Soc. **142**, 3505–3509 (1995). https://doi.org/10.1149/1.2050012
71. M. Miyazaki, S. Miyazaki, T. Kitamura, T. Aoki, Y. Nakashima, M. Hourai, T. Shigematsu, Influence of Fe contamination in Czochralski-grown silicon single crystals on LSI-yield related crystal quality characteristics. Jpn. J. Appl. Phys. **34**, 409–413 (1995). https://doi.org/10.1143/JJAP.34.409
72. A.L.P. Rotondaro, T.Q. Hurd, A. Kaniava, J. Vanhellemont, E. Simoen, M.M. Heyns, C. Claeys, G. Brown, Impact of Cu and Fe contamination on the minority carrier lifetime of silicon substrates. J. Electrochem. Soc. **143**, 3014–3019 (1996). https://doi.org/10.1149/1.1837141
73. T.S. Horányi, P. Tüttõ, C. Kovacsics, Identification possibility of metallic impurities in p-type silicon by lifetime measurement. J. Electrochem. Soc. **143**, 216–220 (1996). https://doi.org/10.1149/1.1836411
74. E. Gaubas, J. Vanhellemont, Comparative study of carrier lifetime dependence on dopant concentration in silicon and germanium. J. Electrochem. Soc. **154**, H231–H238 (2007). https://doi.org/10.1149/1.2429031

75. F. Shimura, T. Okui, T. Kusama, Noncontact minority-carrier lifetime assessment at elevated temperatures for metal-doped Czochralski silicon crystals. J. Appl. Phys. **67**, 7168–7171 (1990). https://doi.org/10.1063/1.344548
76. Y. Kirino, A. Buczkowski, Z.J. Radzimski, G.A. Rozgonyi, F. Shimura, Noncontact energy level analysis of metallic impurities in silicon crystals. Appl. Phys. Lett. **57**, 2832–2834 (1990). https://doi.org/10.1063/1.103756
77. J. Michel, A.J. Reddy, G.J. Norga, M. Platero, L.C. Kimerling, In-situ determination of Si wafer contamination using photoconductance decay measurements. Proc. Electrochem. Soc. **97–12**, 212–221 (1997)
78. G.J. Norga, M. Platero, K.A. Black, A.J. Reddy, J. Michel, L.C. Kimerling, Detection of metallic contaminants on silicon by surface sensitive minority carrier lifetime measurements. J. Electrochem. Soc. **145**, 2602–2607 (1998). https://doi.org/10.1149/1.1838688
79. G. Lucovsky, On the photoionization of deep impurity centers in semiconductors. Solid State Commun. **3**, 299–302 (1965). https://doi.org/10.1016/0038-1098(65)90039-6
80. R.A. Sinton, A. Cuevas, Contactless determination of current-voltage characteristics and minority-carrier lifetimes in semiconductors from quasi-steady-state photoconductance data. Appl. Phys. Lett. **69**, 2510–2512 (1996). https://doi.org/10.1063/1.117723
81. D.K. Schroder, Some recent advances in contactless silicon characterization. ECS Trans. **3** (4), 321–337 (2006). https://doi.org/10.1149/1.2355767
82. V. Lehmann, H. Föll, Minority carrier diffusion length mapping in silicon wafers using a Si-electrolyte-contact. J. Electrochem. Soc. **135**, 2831–2835 (1988). https://doi.org/10.1149/1.2095442
83. H. Föll, Properties of silicon-electrolyte junctions and their application to silicon characterization. Appl. Phys. A **53**, 8–19 (1991). https://doi.org/10.1007/BF00323428
84. M. Rommel, A.J. Bauer, H. Ryssel, Detailed carrier lifetime analysis of iron-contaminated boron-doped silicon by comparison of simulation and measurement. J. Electrochem. Soc. **155**, H117–H123 (2008). https://doi.org/10.1149/1.2819628
85. J. Wittmann, W. Bergholz, H. Hoffmann, Elymat measurement of intentionally contaminated and dry etched wafers. J. Electrochem. Soc. **146**, 313–320 (1999). https://doi.org/10.1149/1.1391606
86. D. Caputo, P. Bacciaglia, C. Carpanese, M.L. Polignano, P. Lazzeri, M. Bersani, L. Vanzetti, P. Pianetta, L. Moro, Quantitative evaluation of iron at the silicon surface after wet cleaning treatments. J. Electrochem. Soc. **151**, G289–G296 (2004). https://doi.org/10.1149/1.1668993
87. T. Trupke, M. Bardos, M.C. Schubert, W. Warta, Photoluminescence imaging of silicon wafers. Appl. Phys. Lett. **89**, 044107/1-3 (2006). https://doi.org/10.1063/1.2234747
88. T. Roth, M. Rüdiger, W. Warta, S.W. Glunz, Electronic properties of titanium in boron-doped silicon analyzed by temperature-dependent photoluminescence and injection-dependent photoconductance lifetime spectroscopy. J. Appl. Phys. **104**, 074510/1-6 (2008). https://doi.org/10.1063/1.2996252
89. R.A. Bardos, T. Trupke, M.C. Schubert, T. Roth, Trapping artifacts in quasi-steady-state photoluminescence and photoconductance lifetime measurements on silicon wafers. Appl. Phys. Lett. **88**, 053504/1-3 (2006). https://doi.org/10.1063/1.2165274
90. L.E. Mundt, M.C. Schubert, J. Schön, B. Michl, T. Niewelt, F. Schindler, W. Warta, Spatially resolved impurity identification via temperature- and injection-dependent photoluminescence imaging. IEEE J. Photovolt. **5**, 1503–1509 (2015). https://doi.org/10.1109/JPHOTOV.2015.2447837
91. J.D. Murphy, K. Bothe, R. Krain, V.V. Voronkov, R.J. Falster, Parameterisation of injection-dependent lifetime measurements in semiconductors in terms of Shockley-Read-Hall statistics: an application to oxide precipitates in silicon. J. Appl. Phys. **111**, 113709/1-10 (2012). https://doi.org/10.1063/1.4725475
92. D. Macdonald, J. Tan, T. Trupke, Imaging interstitial iron concentrations in boron-doped crystalline silicon using photoluminescence. J. Appl. Phys. **103**, 073710/1-7 (2008). https://doi.org/10.1063/1.2903895

93. M.C. Schubert, M.J. Kerler, W. Warta, Influence of heterogeneous profiles in carrier density measurements with respect to iron concentration measurements in silicon. J. Appl. Phys. **105**, 114903/1-6 (2009). https://doi.org/10.1063/1.3138805
94. M.C. Schubert, H. Habenicht, W. Warta, Imaging metastable defects in silicon. IEEE J. Photovolt. **1**, 168–173 (2011). https://doi.org/10.1109/JPHOTOV.2011.2169942
95. P. Gundel, M.C. Schubert, W. Kwapil, J. Schön, M. Reiche, H. Savin, M. Yli-Koski, J.A. Sans, G. Martinez-Criado, W. Seifert, W. Warta, E.R. Weber, Micro-photoluminescence spectroscopy on metal precipitates in silicon. Phys. Status Solidi RRL **3**, 230–232 (2009). https://doi.org/10.1002/pssr.200903221
96. P. Gundel, M.C. Schubert, F.D. Heinz, J. Benick, I. Zizak, W. Warta, Submicron resolution carrier lifetime analysis in silicon with Fano resonances. Phys. Status Solidi RRL **4**, 160–162 (2010). https://doi.org/10.1002/pssr.201004170
97. P. Gundel, M.C. Schubert, W. Warta, Simultaneous stress and defect luminescence study on silicon. Phys. Status Solidi A **207**, 436–441 (2010). https://doi.org/10.1002/pssa.200925368
98. V. Pogue, S.N. Melkote, B. Rounsaville, S. Danyluk, The effect of residual stress on photoluminescence in multi-crystalline silicon wafers. J. Appl. Phys. **121**, 085701/1-9 (2017). https://doi.org/10.1063/1.4976328
99. H. Sugimoto, K. Araki, M. Tajima, T. Eguchi, I. Yamaga, M. Dhamrin, K. Kamisako, T. Saitoh, Photoluminescence analysis of intragrain defects in multicrystalline silicon wafers for solar cells. J. Appl. Phys. **102**, 054506/1-5 (2007). https://doi.org/10.1063/1.2776003
100. M. Tajima, M. Ikebe, Y. Ohshita, A. Ogura, Photoluminescence analysis of iron contamination effect in multicrystalline silicon wafers for solar cells. J. Electron. Mater. **39**, 747–750 (2010). https://doi.org/10.1007/s11664-010-1131-6
101. J. Isenberg, S. Riepe, S.W. Glunz, W. Warta, Imaging method for laterally resolved measurement of minority carrier densities and lifetimes: measurement principle and first applications. J. Appl. Phys. **93**, 4268–4275 (2003). https://doi.org/10.1063/1.1555688
102. M.E. Rodríguez, A. Mandelis, G. Pan, J.A. García, V. Gorodokin, Y. Raskin, Minority carrier lifetime and iron concentration measurements on p-Si wafers by infrared photothermal radiometry and microwave photoconductance decay. J. Appl. Phys. **87**, 8113–8121 (2000). https://doi.org/10.1063/1.373506
103. A. Mandelis, M. Pawlak, C. Wang, I. Delgadillo-Holtfort, J. Pelzl, Time-domain and lock-in rate-window photocarrier radiometric measurements of recombination processes in silicon. J. Appl. Phys. **98**, 123518/1-13 (2005). https://doi.org/10.1063/1.2148631
104. D. Shaughnessy, A. Mandelis, Electronic defect and contamination monitoring in Si wafers using spectrally integrated photocarrier radiometry. J. Electrochem. Soc. **153**, G283–G290 (2006). https://doi.org/10.1149/1.2168052
105. J. Batista, A. Mandelis, D. Shaughnessy, Temperature dependence of carrier mobility in Si wafers measured by infrared photocarrier radiometry. Appl. Phys. Lett. **82**, 4077–4079 (2003). https://doi.org/10.1063/1.1582376
106. H.J. Leamy, Charge collection scanning electron microscopy. J. Appl. Phys. **53**, R51–R80 (1982). https://doi.org/10.1063/1.331667
107. C.J. Wu, D.B. Wittry, Investigation of minority-carrier diffusion lengths by electron bombardment of Schottky barriers. J. Appl. Phys. **49**, 2827–2836 (1978). https://doi.org/10.1063/1.325163
108. C. Donolato, Theory of beam induced current characterization of grain boundaries in polycrystalline solar cells. J. Appl. Phys. **54**, 1314–1322 (1983). https://doi.org/10.1063/1.332205
109. M. Kittler, W. Seifert, G. Morgenstern, Annealing-induced changes of recombination in cast polycrystalline silicon. J. Electrochem. Soc. **140**, 556–560 (1993). https://doi.org/10.1149/1.2221087
110. C. Cabanel, J.Y. Laval, Localization of the electrical activity of structural defects in polycrystalline silicon. J. Appl. Phys. **67**, 1425–1432 (1990). https://doi.org/10.1063/1.345673

111. W. Seifert, M. Kittler, J. Vanhellemont, EBIC study of recombination activity of oxygen precipitation related defects in Si. Mater. Sci. Eng. B **42**, 260–264 (1996). https://doi.org/10.1016/S0921-5107(96)01718-7
112. V. Kveder, M. Kittler, W. Schröter, Recombination activity of contaminated dislocations in silicon: a model describing electron-beam-induced current contrast behavior. Phys. Rev. B **63**, 115208 (2001). https://doi.org/10.1103/PhysRevB.63.115208
113. B. Sopori, C. Wei, Z. Yi, J. Madjdpour, High-speed mapping of grown-in defects and their influence in large-area silicon photovoltaic devices. J. Cryst. Growth **210**, 346–350 (2000). https://doi.org/10.1016/S0022-0248(99)00708-3
114. B. Moralejo, M.A. González, J. Jiménez, V. Parra, O. Martínez, J. Gutiérrez, O. Charro, LBIC and reflectance mapping of multicrystalline Si solar cells. J. Electron. Mater. **39**, 663–670 (2010). https://doi.org/10.1007/s11664-010-1174-8
115. L. Stolze, P. Saring, C. Rudolf, M. Seibt, Localisation and identification of recombination-active extended defects in crystalline silicon by means of focused ion-beam preparation and transmission electron microscopy. Phys. Status Solidi C **6**, 1862–1867 (2009). https://doi.org/10.1002/pssc.200881471
116. O.F. Vyvenko, T. Buonassisi, A.A. Istratov, H. Hieslmair, A.C. Thompson, R. Schindler, E. R. Weber, X-ray beam induced current—a synchrotron radiation based technique for the in situ analysis of recombination properties and chemical nature of metal clusters in silicon. J. Appl. Phys. **91**, 3614–3617 (2002). https://doi.org/10.1063/1.1450026
117. M.C. Schubert, J. Schön, P. Gundel, H. Habenicht, W. Kwapil, W. Warta, Imaging of metal impurities in silicon by luminescence spectroscopy and synchrotron techniques. J. Electron. Mater. **39**, 787–793 (2010). https://doi.org/10.1007/s11664-010-1114-7
118. M. Trushin, O. Vyvenko, W. Seifert, M. Kittler, I. Zizak, A. Erko, M. Seibt, C. Rudolf, Combined XBIC/μ-XRF/μ-XAS/DLTS investigation of chemical character and electrical properties of Cu and Ni precipitates in silicon. Phys. Status Solidi **6**, 1868–1873 (2009). https://doi.org/10.1002/pssc.200881430
119. K.J. Fraser, R.J. Falster, P.R. Wilshaw, Cathodoluminescence assessment of annealed silicon and a novel technique for estimating minority carrier lifetime in silicon. Mater. Sci. Eng. B **159–160**, 194–197 (2009). https://doi.org/10.1016/j.mseb.2008.05.006
120. S.-E. Tan, D.K. Schroder, M. Kohno, M. Miyazaki, Iron contamination in silicon wafers measured with the pulsed MOS capacitor generation lifetime technique. IEEE Trans. Electron. Devices **47**, 2392–2398 (2000). https://doi.org/10.1109/16.887027
121. M. Zerbst, Relaxation effects at semiconductor-insulator interfaces (in German). Z. Angew. Phys. **22**, 30–33 (1966)
122. H. Yoshida, S. Kuge, Characterization of generation lifetime and surface generation velocity of semiconductor wafers by a contactless Zerbst method. J. Electron. Mater. **39**, 773–776 (2010). https://doi.org/10.1007/s11664-010-1126-3
123. G. Obermeier, D. Huber, Iron detection in polished and epitaxial silicon wafers using generation lifetime measurements. J. Appl. Phys. **81**, 7345–7349 (1997). https://doi.org/10.1063/1.365440
124. A.E. Khorasani, T.L. Alford, D.K. Schroder, Modified pulsed MOS capacitor for characterization of ultraclean thin p/p^+ silicon epitaxial layers. IEEE Trans. Electron. Devices **60**, 2592–2597 (2013). https://doi.org/10.1109/TED.2013.2267209
125. A.E. Khorasani, D.K. Schroder, T.L. Alford, A fast technique to screen carrier generation lifetime using DLTS on MOS capacitors. IEEE Trans. Electron. Devices **61**, 3282–3288 (2014). https://doi.org/10.1109/TED.2014.2337898
126. M.N. Chang, T.Y. Chang, F.M. Pan, B.W. Wu, T.F. Lei, An investigation of scanning capacitance microscopy on iron-contaminated p-type silicon. Electrochem. Solid-St Lett. **4**, G69–G71 (2001). https://doi.org/10.1149/1.1389877
127. M.N. Chang, C.Y. Chen, F.M. Pan, T.Y. Chang, T.F. Lei, Observation of differential capacitance images on slightly iron-contaminated p-type silicon. Electrochem. Solid-State Lett. **5**, G69–G71 (2002). https://doi.org/10.1149/1.1486820

128. A.E. Khorasani, D.K. Schroder, T.L. Alford, Optically excited MOS-capacitor for recombination lifetime measurement. IEEE Electron Device Lett. **35**, 986–988 (2014). https://doi.org/10.1109/LED.2014.2345058
129. A.E. Khorasani, D.K. Schroder, T.L. Alford, Carrier recombination lifetime measurement in silicon epitaxial layers using optically excited MOS capacitor technique. IEEE Trans. Electron Devices **62**, 1553–1560 (2015). https://doi.org/10.1109/TED.2015.2409291
130. S. Weiss, R. Kassing, Deep level transient Fourier spectroscopy (DLTFS)—a technique for the analysis of deep level properties. Solid-State Electron. **31**, 1733–1742 (1988). https://doi.org/10.1016/0038-1101(88)90071-8
131. L. Dobaczewski, A.R. Peaker, K. Bonde Nielsen, Laplace-transform deep-level spectroscopy: the technique and its application to the study of point defects in semiconductors. J. Appl. Phys. **96**, 4689–4728 (2004). https://doi.org/10.1063/1.1794897
132. H. Yoshida, R. Nakanishi, S. Kishino, Sensitivity of contactless transient spectroscopy and actual measurement of localized states in oxidized Si wafer. J. Cryst. Growth **210**, 379–383 (2000). https://doi.org/10.1016/S0022-0248(00)00715-0
133. N. Chinone, R. Kosugi, Y. Tanaka, S. Harada, H. Okumura, Y. Cho, Local deep level transient spectroscopy using super-higher-order scanning nonlinear dielectric microscopy. Microelectron. Reliab. **64**, 566–569 (2016). https://doi.org/10.1016/j.microrel.2016.07.088
134. M.L. Polignano, P. Bacciaglia, D. Caputo, C. Clementi, B. Padovani, F. Priolo, T. Simpson, Detection of metal contamination in internally gettered wafers. Proc. Electrochem. Soc. **2002–20**, 206–217 (2002)

Chapter 5
Electrical Activity of Iron and Copper in Si, SiGe and Ge

The main reason for the interest in metal impurities in semiconductors is their electrical activity, resulting in device performance and yield degradation (to be discussed in Chap. 7). In this chapter, the electrical properties of the 3*d* transition metals iron and copper will be discussed in detail. Iron is perhaps the best studied metal impurity in Group IV semiconductors because it is the most prominent contaminant in silicon processing, relying mainly on stainless steel equipment (Chap. 3). It is characterized by an intermediate diffusivity and solubility, so that upon quenching, Fe can be kept in the silicon lattice as an interstitial atom. As this is a non-equilibrium state, iron tends to evolve into a lower energy state, which, given its mobility at room temperature, happens within a few hours in p-type silicon by pairing with ionized shallow acceptors. The fact that iron occurs in different bonding states also affects its electrical (recombination) properties. Therefore, the different possible configurations of iron in silicon (and Ge) will be discussed, based on the observations from chemical analysis techniques, like EPR and Mössbauer. Besides different types of Fe-related point defects and complexes with other impurities, there will be a tendency to form pairs, small clusters and at higher temperatures precipitates. In a second section, the electrical properties of the different types of Fe-related defects are discussed, as they have been revealed by for example DLTS. In a third section, the detection and identification of Fe and Fe-related species in silicon by lifetime techniques will be outlined.

A similar scheme will be followed for the description of the properties of copper in Si and Ge. The interest in copper is mainly triggered by the fact that Back-End-Of-Line (BEOL) interconnects are based on Cu metallization. Also in solar cells, expensive Ag metallization is being replaced by copper. In order to prevent the in-diffusion of copper in the silicon substrate, one has to employ diffusion barriers. This is even more the case for the three-dimensional (3D) integration of ICs, using copper-filled plugs. Copper is the prototype of an extremely fast diffuser in silicon (like Ni and Co) which is difficult to keep in solution even after a fast quench from high temperature. It has a tendency to precipitate at the bare

C. Claeys and E. Simoen, *Metal Impurities in Silicon- and Germanium-Based Technologies*, Springer Series in Materials Science 270,
https://doi.org/10.1007/978-3-319-93925-4_5

silicon surface, so that precipitated copper will mainly affect the surface recombination properties. However, it can be kept in solution at room temperature as interstitial species in p-type silicon, covered by SiO_2, where it interacts with the dopant atoms to form CuB pairs. Eventually, copper will form larger precipitates, which gives rise to the so-called Light-Induced-Degradation (LID) of silicon solar cells. The high mobility of copper has also led to an unique identification method based on copper drift in the electric field present in a depletion region. Copper is also one of the best-studied 3*d* transition metal impurities in germanium. In contrast to the case of silicon, TMs in germanium have a tendency to occupy predominantly substitutional (lattice) sites. This also explains the triple acceptor character of Cu_s in Ge, which is based on the hypothesis that the s^1 metal impurities try to form tetrahedral bonds [1, 2]. In this and the following chapter, recent DLTS results obtained on metal-doped germanium will be reviewed; a more complete review of the pre-existing data can be found in Chap. 5 of [3].

The electrical properties of other metal impurities will be the subject of Chap. 6. The results from ab initio quantum mechanical calculations of the properties of metals in Si and Ge will be covered by Chap. 9.

5.1 Iron

Iron is probably the most studied metal impurity in silicon and its properties have been the subject of Review papers, covering more than 1000 papers until the year 2000 [4, 5]. In silicon, at least 30 different Fe-related complexes have been identified by EPR and at least 20 different deep levels associated with Fe [4]. Details can be found in Appendix A and B of [4], respectively. In the following, an overview of the different types of Fe-related point defects, complexes and clusters/precipitates will be discussed, both from a structural and from an electrical viewpoint. In a third section, some lifetime-based methods for the identification of the presence of Fe in silicon will be described.

5.1.1 Configurations of Fe

5.1.1.1 Interstitial and Substitutional Fe

The electronic configuration of ***interstitial iron*** in silicon has been identified by the pioneering EPR work of Woodbury and Ludwig [6]. The electronic structure derived for the 3*d* TM from their model (Fig. 4.5) indicates that the 4*s*-shell electrons are transferred to the 3*d* shell. Then, all the N_e valence electrons ($N_e = m + n$ in $3d^m4s^n$) fill the levels in such a way that a maximum electron spin S is achieved. The tetrahedral symmetry of the crystal field splits the tenfold degenerate atomic 3*d* orbital into a sixfold degenerate t_2 orbital and a fourfold degenerate e orbital, separated by the crystal-field energy. EPR has identified Fe_i as the dominant thermal defect introduced in silicon after a high-temperature heat treatment, followed by a fast quench [7, 8]. The ionization energy of the Fe_i^0 donor level has been established at E_V + 0.375 eV by comparing the intensity of the EPR signal as a function of temperature with Hall effect measurements [9]. Detailed electron nuclear double resonance of Fe_i has revealed that hyperfine interactions could be resolved with six shells of neighbor nuclei comprising 42 silicon atoms [10]. The localization of the two paramagnetic 3*d* electrons at Fe_i^0 was estimated between 80 and 95%. Interstitial Fe has also been identified by Mössbauer spectroscopy [11], while a predominant near-tetrahedral interstitial Fe has been identified in ion-implanted Si after vacuum annealing in the 300–600 °C range by β-emission channeling, with a displacement of 0.03–0.08 nm from the ideal near-*T* site [12–14].

Evidence for the occurrence of ***substitutional Fe*** is more controversial. The only EPR spectrum that can be interpreted as a substitutional Fe in a distorted configuration is the Si-NL19 center [15] in irradiated iron-doped silicon. Since the second component of the complex in irradiated silicon is most probably a vacancy, it was suggested that this is a singly positively charged interstitial iron-vacancy complex, $(Fe_iV)^+$. Because the center has ⟨111⟩ axial symmetry, the iron atom cannot be exactly at the lattice position but has to be shifted along a trigonal axis toward the tetrahedral interstitial position [15]. Photo-EPR places the energy level at 0.51 eV above the valence band [16]. The formation of Fe_iV pairs, with an activation energy of 0.6–0.9 eV has been confirmed by Mössbauer spectroscopy [17], which is consistent with the diffusion activation energy of Fe_i. In Ge, a similar Fe_iV pair is formed at 350 K [17]. Iron in near substitutional sites has been confirmed by β-emission channeling in ion-implanted silicon, with a displacement from the perfect *S*-site of 0.04–0.07 nm [12–14]. Vacuum annealing between 300 and 600 ° C transforms this Fe-species into a near interstitial *T*-site. This is in agreement with the pair stability derived from EPR [16].

Iron on an ideal *S*-site has been found by Mössbauer (see e.g. [18]) and by β-emission channeling [12–14], where it is dominant after high-temperature annealing around 800 °C. It has been argued that substitutional iron does not introduce a deep level in the band gap of silicon [4, 18], which may explain why its observation in DLTS remains elusive. On the other hand, reports in the literature of Fe-donor pairing (see also the next paragraph) would suggest an acceptor character of Fe_s. This was derived from a significant enhancement of iron in highly n^+-doped silicon [19]. However, as will be seen, this could also be explained by the formation of Fe_i-V pairs, so that the matter is far from resolved.

Mössbauer studies in implanted Ge indicate that after annealing at ~350 K, Fe is mainly incorporated on substitutional lattice sites [20]. Also interstitial Fe and Fe_iV pairs have been detected, which are stable up to ~700 K. Similar studies by β-emission channeling confirm the substitutional preference. Surprisingly, a fraction of these TMs have also been found in sixfold coordinated bond-centered sites [21]. According to density functional theory (DFT) calculations, this results from the trapping of a vacancy, spontaneously forming a metal-vacancy complex in the split-vacancy configuration. Similar observations hold for Cu in Ge [21]. Data on the configuration of individual Fe atoms in $Si_{1-x}Ge_x$ are scarce. One EPR report describes the spectrum of neutral Fe_i in dislocation-free single crystals with x up to 2.4% [22]. At higher Ge contents, line broadening prohibits the detection of Fe_i. Consequently, it is for example not at all clear at which composition the transition from the predominantly interstitial site in Si changes into predominantly substitutional for pure Ge.

5.1.1.2 Dopant-Iron Pairs

When Fe-doped silicon is quenched from high temperature, interstitial iron will be kept in solution, as the diffusivity is not high enough for full precipitation at the surface (or in the interior of the wafer at extended defects). Fe_i is, however, not the lowest energy state at room temperature in p-type material. For a sufficiently high acceptor (A) concentration, Fe_i will be ionized (positively charged) and long-range Coulomb attraction by negatively charged dopant atoms will cause ion pairing. This can be described by the Reiss' theory [23], as already mentioned in Sect. 2.2.1. The formation and dissociation reactions of FeB pairs are determined by three energies schematically represented in Fig. 5.1 [4]: the equilibrium binding energy E_b, the diffusion barrier of interstitial iron E_m and the dissociation energy of the pairs E_{diss}. E_b can be determined by the temperature dependence of the equilibrium fraction of Fe_i paired with an acceptor, according to the law of mass action:

$$\frac{N(FeB)}{N(Fe_i^+) \times N(B_s^-)} = \frac{Z}{D_i}\exp\left(\frac{E_b}{k_B T}\right) \tag{5.1}$$

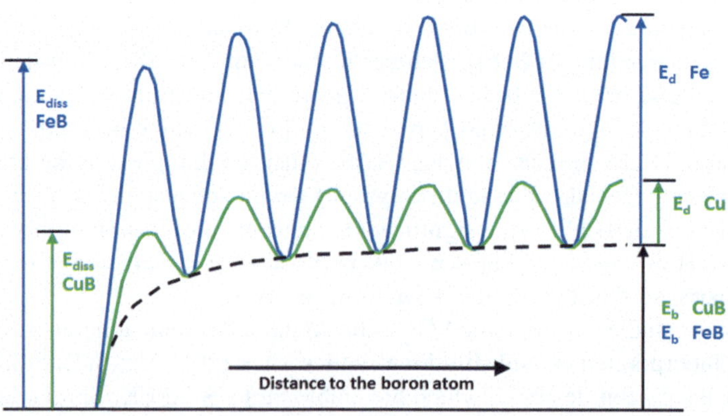

Fig. 5.1 Energy diagram of the FeB interaction in silicon, E_{diss} is the dissociation energy of the pair, E_b the equilibrium binding energy (a difference in equilibrium energy in the paired and unpaired states) and E_m is the diffusion barrier for migration of iron. Note that $E_{diss} < E_b + E_m$ and that there is no straightforward manner to calculate one of the three parameters from the other two. Reproduced from Istratov et al. [4] by permission of Springer

D_i is the density of interstitial sites in the lattice (5 × 10^{22} cm^{-3}), Z the number of the possible orientations of the pairs with the same symmetry around one acceptor atom (4 for the tetrahedral symmetry). The concentration of ionized Fe_i is usually smaller than the total Fe_i concentration, which depends on the doping density and the temperature, i.e., on the Fermi level position with respect to the ionization level of Fe_i. Experimental binding energies for FeB over a range from 0.53 to 0.65 eV have been reported [4], close to the simple theoretical estimate for a close ionic pair of 0.52 eV given by (2.6). This has been a strong argument in favor of the ion pair formation model for the FeA pairs [4]. However, the deviations from the simple estimate indicate that besides an ionic character, there is also a partial covalent character of the bonding. The different E_b for the different acceptors from Al to In furthermore suggests a strain component in E_b, which is supported by theoretical calculations [4]. For example, values for E_b of 0.60 eV (FeB); 0.70 eV (FeAl) and 0.47 eV (FeGa) have been reported by Lemke [24], showing the chemical effect of the acceptor on the pair binding energy.

The association kinetics can be described by the theory of diffusion-limited precipitation or trapping of mobile donors by immobile acceptors (Ham's theory) [25] according to [4]:

$$\tau_{ass} = \frac{1}{4\pi D(Fe_i)N_a R_C} = \frac{566.7 \times T}{D(Fe_i)N_a} \quad (5.2)$$

with $D(Fe_i)$ the diffusivity of interstitial iron and R_C the capture radius, which can be approximated by [4]

$$R_C \approx \frac{q^2}{4\pi\varepsilon_0\varepsilon_{Si}k_B T} \quad (5.3)$$

which is valid for $N_a \leq 1 \times 10^{17}$ cm^{-3}. This yields a capture radius on the order of 4.5 nm, which is close to the experimentally observed values at room temperature [4]. The association rate has recently been re-assessed [26], showing that on the basis of (5.2) the rate is 14% faster than in reality. The improved association rate equals then:

$$\tau_{ass} = \frac{5.7 \times 10^5}{N_a} Texp\left(\frac{0.66}{k_B T}\right) \quad (5.4)$$

Pair dissociation occurs when an Fe_i ion jumps away from the closest nearest neighbor interstitial position to B_s across a potential barrier E_{diss}. The kinetics of this process can be described by a time constant:

$$\tau_{diss}^{-1} = vexp\left(\frac{E_{diss}}{k_B T}\right) \quad (5.5)$$

with ν an attempt frequency. Experimental dissociation barriers have been found in the range of 1.2 eV, with a $\nu = 5 \times 10^{10}\ s^{-1}$ for FeB [4]. It has been observed that the pair dissociation can be accelerated by injection of minority carriers using an $n^{+}p$ junction or by light with sufficient energy [4]. This minority-carrier-injection stimulated enhancement of the pair dissociation can be explained as a recombination-enhanced defect reaction. In that case, the electron capture energy on Fe_i^+ can be converted into defect vibrational modes, which lowers the E_{diss} by about 0.77 eV [4] and thus facilitating pair dissociation. The attempt frequency in that case would correspond with the electron-hole recombination rate.

More detailed studies on the dynamics of light-induced FeB dissociation has indicated that the dissociation rate evolves with $G_{min}^2/[FeB]^2$, with G_{min} the minority carrier generation rate and [FeB] the concentration of FeB pairs. This suggests that 2 recombination or electron capture events are required to break the pair [27]. A possible scenario consists of the neutralization of Fe_i^+ by a first electron capture. However, this does not suffice to separate the pair. This requires a second electron to be captured. The negatively charged Fe_i will experience repulsion from the immobile B_s and moves away. Alternatively, a second electron-hole recombination can supply the necessary vibrational energy to break the pair [27]. Enhanced pair dissociation has also been reported for wet chemical etched p-type silicon, containing Fe [28, 29]. Iron becomes reactivated by a hydrogen stimulated FeB pair dissociation. It has been shown that only one hydrogen atom participates in the dissociation reaction, whereby hydrogen replaces Fe_i and passivates the B atom [28].

The pairing kinetics has been further studied in co-doped n-type silicon [30, 31]. In that case, it becomes important to consider the electron capture and emission rates of Fe_i as a function of the charge carrier densities. This permits to determine the charge carrier density dependent dissociation rate [31] and provides an experimental way to derive the compensating acceptor concentration in n-type silicon [31]. While the formation of FeB pairs in n-type silicon is at first sight not expected, there is a weakened Coulomb attraction caused by the permanent switch of Fe_i^0 and Fe_i^+ by electron emission, about 1 time every µs at room temperature. Considering the emission and capture balance, a time ratio between Fe_i^+/Fe_i^0 of 0.06 exists, yielding a finite probability for pairing with B_s^- in compensated n-type Si [31]. Based on the FeB association kinetics in compensated p-type silicon, a method has been proposed to derive the ionized donor density [32].

An important remark in the context of pair association and dissociation is that this happens under dynamic equilibrium. In other words, during dissociation, there is always a finite chance of association and vice versa. This implies in practice that the time constant measured under most experimental conditions is given by $\tau_{eff}^{-1} = r_{ass} + r_{diss}$ with r_{ass} and r_{diss} the association and dissociation rates [4].

The existence of $\langle 111 \rangle$ oriented FeB pairs was originally reported by Ludwig and Woodbury [6]. The paramagnetism arises from three unpaired electrons of the $3d^7$ iron impurity in the crystal field of the negative acceptor. According to their model, the spin density of the ionized acceptor is small and does not markedly contribute to the EPR signal of the pair [6]. There have been EPR reports on Fe-donor pairs [33, 34], although in one case, distant pairs in a diluted random distribution were found [33]. In the case of quenched Fe-doped n-type FZ material, irradiated by 3 MeV electrons and annealed in the range 100–550 °C for 30 min, an EPR signal below 12 K was detected, showing monoclinic-*I* symmetry. It has been suggested that this could correspond with an FeVP pair [33], with the Fe atom displaced from the exact substitutional position, remaining in the same (100)-type plane with P_s. These centers anneal out above 400 °C.

5.1.1.3 Small Fe-Related Clusters and Fe-Related Complexes

While FeA pairs are the energetic stable form of Fe_i in p-type silicon at room temperature, they easily dissociate by minority carrier injection or by heating at 200 °C. Iron clustering will occur from a supersaturated solution, based on the chemical driving force given by (2.12). By means of EPR it has been shown that clustering of Fe starts with the formation of small entities $(Fe)_n$, with $n = 2$ and $n = 4$ [4]. $(Fe)_2$ pairs have been found directly after quenching from 1200 °C but disappear slightly above 300 K [35]. Based on photo-EPR measurements the energy levels of the transitions from Fe_2^+ to Fe_2^0 and from Fe_2^0 to Fe_2^+ were determined as $E_V + 0.61$ eV and $E_C - 0.67$ eV, suggesting a pronounced lattice relaxation on electron capture [36].

The formation of $(Fe)_4$ clusters starts at 120 °C and goes hand in hand with a drop in the Fe_i concentration [15]. The $(Fe)_4$ concentration achieves a maximum at 140 °C, reaches a plateau when $[Fe_i]$ has dropped below 1% of its initial value, at about 160 °C. The maximum of $[(Fe)_4]$ indicates a competition between $(Fe)_4$ formation and the growth to larger clusters of silicides by Fe_i incorporation [4]. The $(Fe)_4$ clusters survive up to 250 °C and then dissociate to form larger clusters or precipitates. No higher order clusters have been detected by EPR, but DLTS studies have been applied to monitor the initial further clustering of Fe_i in silicon [37]. This will be discussed in the next section on the electrical activity.

Centers with two Fe atoms and an acceptor have been reported in a number of studies [1, 36]. They exhibit monoclinic or orthorhombic symmetry. A new modification of an Fe_2B center was found by Strube et al. [36]. EPR has thus identified at least 9 species with two to four Fe atoms. The donor level of the "old" Fe_2B center was found at $E_V + 0.57$ eV [36]. In addition, EPR centra with two equivalent Fe atoms and one or two vacancies have been reported by Muller et al., following quenching and electron irradiation [15]. Photo-EPR studies have shown that the Fe_2V complex is less stable than the FeV pair and anneals out at about 300–350 °C [16].

Besides pairing with acceptors, Fe-Metal pairs have also been identified in EPR; details can be found in Appendix A of [4].

5.1.1.4 Fe Precipitation

Below the eutectic point of the Fe–Si phase diagram at 1210 °C, the equilibrium solid solubility of Fe reduces exponentially with reducing temperature (Fig. 2.1). This means that upon cooling, there is a chemical driving force for precipitation, which increases with increasing supersaturation (2.12). However, since the solubility of iron is much lower than for Cu or Ni, it will be more difficult to find Fe-silicide precipitates in bulk silicon after heat treatment and slow cool down [5]. This is because the solubility is much lower and, so the density of precipitates will also be smaller. Besides Fe-silicide, other types of precipitates (silicates,...) can be formed in silicon; Table 5.1 shows the formation energies and Gibbs free energies at 1053 and 1273 K [38], indicating that Fe_2SiO_4 has the most negative values and is, therefore, the most stable precipitate form.

Rod-like $FeSi_2$ precipitates, with a length up to 1.5 μm have been found in bulk silicon [5]. Their thermal stability has been systematically investigated by Ramappa and Henley [39], showing that at low temperatures, orthorhombic β-$FeSi_2$ is the stable phase ($T < 800$ °C), while tetragonal α-$FeSi_2$ is stable at higher temperatures (Fig. 5.2). Below 915 °C, α-$FeSi_2$ films decompose eutectoidally to form β-$FeSi_2$. Between 760 and 900 °C no Fe-silicide precipitates are formed. This establishes a temperature window for the dissolution of Fe-silicide precipitates as part of a gettering treatment. Precipitate dissolution occurs quite fast at 900 °C [39], which can be repeatedly performed, i.e. Fe can be cycled between Fe_i and $FeSi_2$.

Given the relatively low stability of the $FeSi_2$ precipitates, iron will bind much more easily to pre-existing extended defects, like dislocations or oxide precipitates [5]. For example, an electrically detected magnetic resonance signal has been reported for Fe atoms decorating an oxide precipitate [40]. Iron also decorates oxidation-induced stacking faults (OISF) [5] and even assists in their growth. Combining data on Fe-diffused crystalline p-type Si (Cz and FZ substrates) and edge-defined, film-fed ribbon growth (EFG) solar cell material, a correlation has been revealed between the diffusion length of the starting material and the iron precipitation rate, measured after heating at 235 °C by the DLTS concentration of

Table 5.1 Formation enthalpies ΔH_T and Gibbs free energies ΔG_T at 1053 and 1273 K (after Hackl et al. [38])

Compound	ΔH_{1053} [kJ/mol]	ΔG_{1053} [kJ/mol]	ΔH_{1273} [kJ/mol]	ΔG_{1273} [kJ/mol]
FeO	−230	−260	−218	−264
Fe_2O_3	−745	−737	−724	−733
Fe_3O_4	−1009	−1026	−980	−1026
SiO_2	−879	−863	−867	−863
FeSi	−38	−80	−29	−80
$FeSi_2$	−33	−75	−17	−75
Fe_3Si	−21	−105	+2	−109
Fe_2SiO_4	−1382	−1386	−1352	−1398

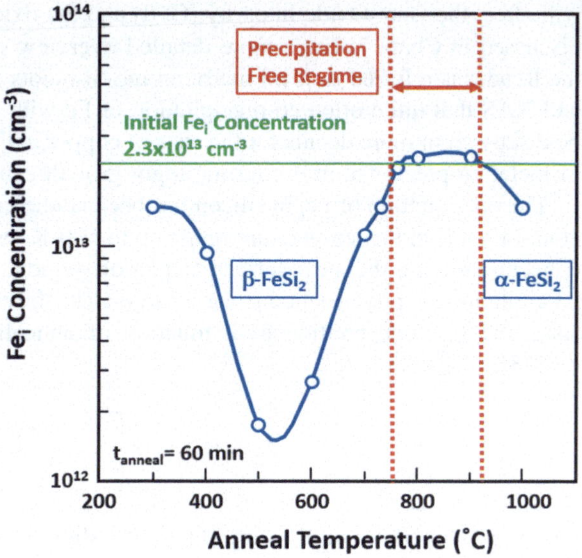

Fig. 5.2 Precipitation behavior of iron in silicon. Iron does not precipitate in the temperature interval 760–920 °C. Reproduced from Ramappa and Henley [39] by permission of ECS-The Electrochemical Society

the FeB donor level [41, 42]. This trend is represented in Fig. 5.3 [42], showing a stronger than linear tendency. For the case of EFG silicon, it was concluded that the same intragrain defects (probably SiC precipitates) were responsible for the carrier diffusion length and the iron precipitation in the material [41]. In addition, it was derived that these defects exhibit a higher level of impurity decoration (probably already in the as-grown state) than in high-lifetime silicon. In the latter case, oxygen precipitates (and associated secondary extended defects) were thought responsible for the Fe precipitation.

The surface of a wafer also forms an efficient sink for Fe. Particularly when Fe is present at the surface during silicon oxidation, very strong chemical bonds may be formed between oxygen, iron and silicon, yielding iron oxides or iron silicides. This

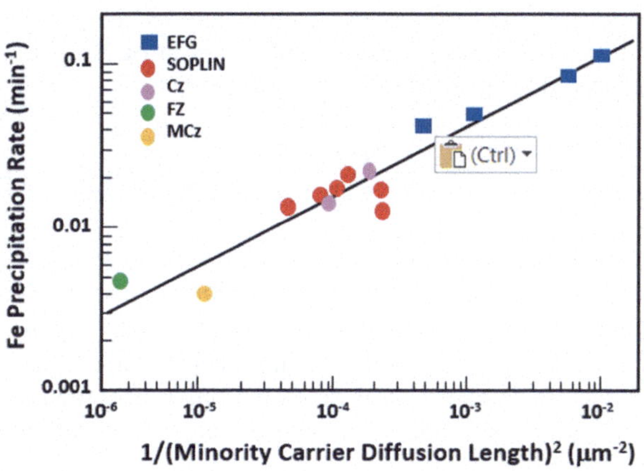

Fig. 5.3 Iron precipitation rate (inverse of precipitation time constant) versus as-grown minority-carrier diffusion recombination rate for different types of p-type silicon substrates. Reproduced from Bailey et al. [42] by permission of Springer

will affect the Gate-Oxide Integrity (GOI) and the oxide breakdown field as will be discussed in Chap. 7. For a more detailed overview on Fe precipitation in silicon, the Reader is referred to [5]. Finally, in mc-Si ingots it has been observed by XRF and XAS that quite often co-precipitation of Fe with other metals occurs; this will be discussed in more details for the case of copper in the next part. A brief summary of metal co-precipitation during mc ingot growth can be found in [37].

The precipitation of Fe_i in silicon has been studied more fundamentally by Mesli et al. [43, 44] in the temperature range up to 500 K, using DLTS to monitor the Fe_i concentration in the near-surface depletion region. It was concluded that iron precipitation in p-type silicon was charge-state dependent. In the ***neutral*** charge state, the Fe concentration as a function of annealing time could be described by [43]:

$$C = C(t_i)\exp\left[-\frac{(t - t_i)^{\frac{3}{2}}}{\tau_{prec}^{3/2}}\right] \tag{5.6}$$

with t_i an incubation time where $[Fe_i]$ even slightly increases. It could correspond with the time delay necessary to redistribute the nucleation centers before the onset of precipitation [43]. Equation (5.6) is in agreement with Ham's theory [25], where in general, the time dependence is defined by an exponent n. The exponent $n = 3/2$ is valid for a spherical precipitate, which keeps its shape during growth.

In the case of Fe_i^+ precipitation, exponential first order kinetics was found [43]. This means $n = 1$, which applies to rodlike defects (or dislocations). The corresponding precipitation rate is given by [25]:

$$\frac{1}{\tau_{prec}} = 2D\left[\left(\frac{4\pi}{3}\right)N_{prec}\right]^{2/3}(V_{eff}C_0)^{1/3} \tag{5.7}$$

with N_{prec} the precipitate density, V_{eff} the effective volume of iron in a precipitate and C_0 the initial concentration of electrically active iron.

Interest in the precipitation modeling of Fe in silicon has in recent years mainly been triggered by PV applications and by the presence of Fe precipitates in mc-Si [45–48]. The heterogenous precipitation of Fe at for example oxygen precipitates can be described by the equation [45]:

$$\frac{\partial[Fe]}{\partial t} = -4\pi r_{ox}N_{ox}D\left([Fe] - [Fe]_{eq}\right) \tag{5.8}$$

with $[Fe]_{eq}$ the equilibrium solid solubility of Fe at the precipitation temperature and [Fe] the actual iron concentration. It is assumed that heterogeneous precipitation occurs by spherical oxide precipitates with radius r_{ox} and density N_{ox}. However, such a traditional approach cannot describe that only a fraction of the oxide precipitates causes precipitation, as observed experimentally [45]. In addition, there appears to be a critical supersaturation level above which precipitation of iron can occur. These issues have been tackled in a more refined approach based on a set of Fokker-Planck equations, establishing the Chemical Rate Equation (CRE) for each precipitate size with n iron atoms [45, 46]. Key to the model is to start from $n = 0$, which stands in contrast to the homogeneous nucleation case [45] and, secondly, a fraction f_0 of oxide precipitates is assumed which do not contain iron. The corresponding growth and dissolution rates for a precipitate with n iron atoms are then given by Haarahiltunen et al. [45]:

$$g_n = 4\pi r_{ox} D[Fe] \quad \text{and} \quad d_n = 4\pi r_{ox} D[Fe]_{eq} \tag{5.9}$$

with $[\mathrm{Fe}]_{\mathrm{eqn}}$ a size-dependent equilibrium concentration. This takes into account the fact that a large oxide precipitate more easily becomes an effective iron precipitation site than a small one [45]. In other words, the chemical potential of iron will be higher in a small than in a large cluster. Both experimentally and from simulations based on the described model, it has been concluded that Fe precipitation starts at about 650 °C, corresponding with a critical supersaturation level $k_B T\ln([\mathrm{Fe}]/[\mathrm{Fe}]_{\mathrm{eq}}) \sim 0.34$ eV [45].

For the case of homogeneous precipitation of Fe in mc-silicon, where β-$\mathrm{FeSi_2}$ is already present, one can rephrase (5.8) in the form [47]:

$$\frac{\partial [Fe]}{\partial t} = -4\pi r_{FeSi_2} D \frac{[Fe]_P V_{Fe/FeSi_2}}{4/3\pi r_0^3}\left([Fe] - [Fe]_{eq}\right) \tag{5.10}$$

with $[\mathrm{Fe}]_{\mathrm{P}}$ the as-grown precipitated iron concentration, $V_{Fe/FeSi_2}$ the volume of one iron atom in β-$\mathrm{FeSi_2}$ and r_0 the average radius of the as-grown β-$\mathrm{FeSi_2}$ precipitates.

Finally, the trapping of iron by extended defects (dislocations) in mc-Si has also been modeled, considering the impact of the local strain field on the solubility of iron [48]. It was proposed to represent this by adding an additional term in the iron solubility:

$$[Fe]_{eqSD}(SD) = [Fe]_{eq}(1 + C_1 \times SD) \tag{5.11}$$

with:

$$C_1 = 0.5\pi r_{SD}^2 \exp\left(\frac{E_{i0}}{k_B T}\right) \tag{5.12}$$

In (5.11) SD is the density of extended defects, r_{SD} is the radius where the lattice distortion acts and E_{i0} is an interaction energy between dislocations and iron atoms, which is inhomogeneous, reaching more than 1 eV in a small core region and less than 0.8 eV in the vicinity of the dislocation. In fact, where the strain field is unfavorable for Fe (below or above the extra half plane of a dislocation), a negative E_{i0} is present [47].

It is clear from the above that the equilibrium solid solubility of Fe in silicon can vary locally, depending whether extended defects, surrounded by a strain field are present. In fact, the equilibrium solid solubility of a metal is defined by the equilibrium of silicon and a specific silicide phase (see also Chap. 2). This has been successfully demonstrated by the experiments of Murphy and Falster [49], showing that the relaxation behavior of iron in silicon between 500 and 750 °C depends strongly on the fact whether a β-$FeSi_2$ phase is present or not during the heat treatment. This has important consequences for the gettering performance in Si (See Chap. 8).

5.1.2 Electrical Properties of Fe

As mentioned before, at least 20 deep levels have been identified in Fe-doped silicon. An overview of the parameters can be found in Appendix B of [4]. In this chapter, we will mainly focus on the most relevant Fe-related species, important for recombination lifetime, i.e., Fe_i and the FeA pairs and on data which have been published since 2000.

5.1.2.1 Fe_i and FeA Pairs in Silicon

According to Istratov et al. [4], the energy position of the **interstitial iron donor level** is at $\boldsymbol{E_V + 0.38}$ **eV**, derived from Hall effect and EPR. The average hole capture cross section obtained by DLTS can be represented by [4]:

$$\sigma_p(Fe_i) = (3.9 \pm 0.5) \times 10^{-16} \times \exp\left(-\frac{0.045 \pm 0.005\mathrm{eV}}{k_B T}\right)(\mathrm{cm}^2) \quad (5.13)$$

This implies that the **emission enthalpy** derived from an Arrhenius plot in DLTS should correspond with an average value of 0.38 eV + 0.045 eV = **0.425 eV**. The average hole capture cross section at 300 K (important for minority carrier lifetime modeling) is assumed 7.0×10^{-17} cm^2. A much less accurate electron capture cross section of $\sigma_n = 4.0 \times 10^{-14}$ cm^2 has been derived from lifetime experiments at room temperature [50]. This corresponds approximately with a lifetime asymmetry factor $k = 540$ of the Fe_i donor. This asymmetry indicates that at low level injection, Fe_i will be a much more efficient lifetime killer in p- than in n-type silicon [51].

As outlined in Chap. 4, recent Injection- and Temperature-Dependent Lifetime Spectroscopy allows for a more precise determination of the recombination lifetime parameters of a specific metal impurity [50]. This has been applied to the case of Fe_i, yielding a donor level at $E_T = E_V + (0.394 \pm 0.005)$ eV and a k-factor equal to 51 ± 5, lower than for previous data. The poorly defined electron capture cross section has been re-determined and a value for $\sigma_n = (3.6 \pm 0.4) \times 10^{-15}$ cm^2 at 300 K was obtained. It has furthermore been derived that σ_n increases exponentially with temperature, with an activation energy $E_{\infty} = 0.024$ eV, which identifies the multiphonon emission as the dominant electron capture mechanism [50].

The ionization energy of interstitial silicon is a key parameter for the understanding of the behavior of mobile Fe_i in silicon, as it defines the dominant charge state in p-type material with respect to the Fermi level, which is depending on the doping density N_a and T. As seen before, the charge state will determine the diffusivity or mobility of Fe_i at a given temperature, which also drives its pairing and precipitation behavior. For the modeling of iron gettering, it would be of high value to know the position of the Fe_i donor level in the band gap as a function of T. The early results have been summarized by Istratov et al. [4], showing that up to 800 °C, the level stays constant, while for $T > 900$ °C, it dives towards the valence band, where it merges at $T > 1100$ °C. This is shown in Figs. 5.4 [52], where also more recent data (from Gilles et al. [19] and McHugo et al. [53]) derived in the T-range of 685–853 K have been indicated.

Interstitial iron has also been studied in ***$Si_{1-x}Ge_x$ epitaxial layers*** up to $x = 0.28$ [54]. As can be seen in Fig. 5.5, the activation energy derived from an Arrhenius plot reduces with increasing x, both for strained and relaxed layers. A slightly different behavior is found for the Cr_i donor level, showing a reduction for relaxed and a slight increase for strained epi layers. The fact that the deep donor levels do not follow either of the valence or conduction band is in agreement with the fact that the impurity-related deep states are formed by wave functions originating from the entire Brillouin zone [54]. The strain dependence can be interpreted in terms of a uniaxial deformation potential of the deep level [54]. This parameter only weakly depends on the Ge content in the range up to 28% and amounts to ~5 eV for Fe_i. Such a low value is consistent with the fact that interstitial iron is a simple point defect, which is less sensitive to a stress change.

It is now well-established that the ***FeB pair*** has both a donor level, close to the VB and an acceptor level near the CB [4]. A simplified explanation for this is given in Fig. 5.6 [4] showing the upward shift of the $Fe_i^{0/+}$ donor level by about 0.52 eV (the Coulomb binding energy of the closest pair) to the $(FeB)^{0/-}$ acceptor level. Likewise,

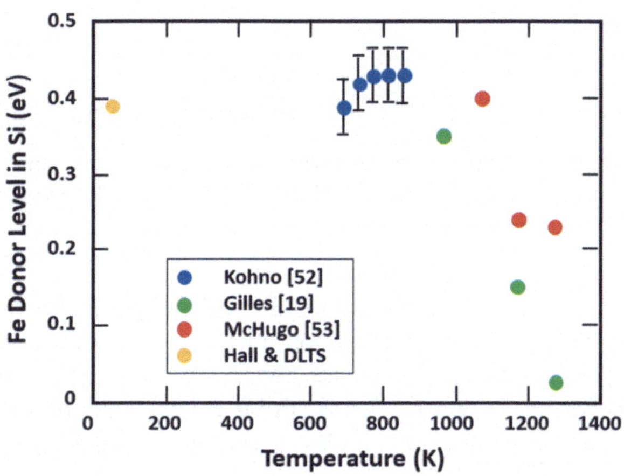

Fig. 5.4 Temperature dependence of the iron donor level in silicon. Reproduced from Kohno et al. [52]. Copyright 2000 AIP Publishing

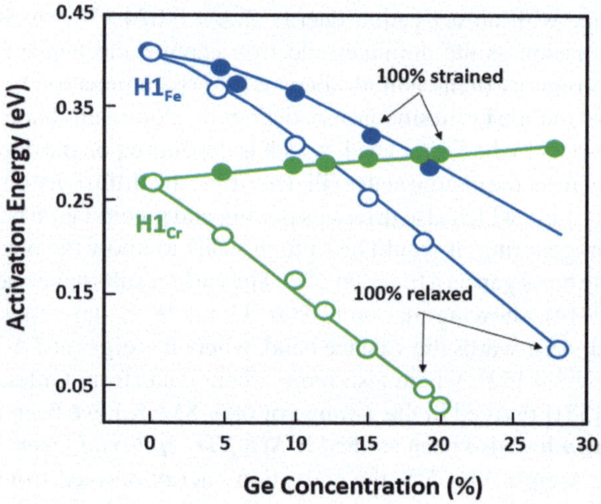

Fig. 5.5 Evolution of the Fe- and Cr-related deep states as a function of Ge concentration. Two extreme cases are shown: 100% relaxed (open symbols) and 100% strained (closed symbols). Reproduced from Nauka and Kamins [54] with permission of Elsevier

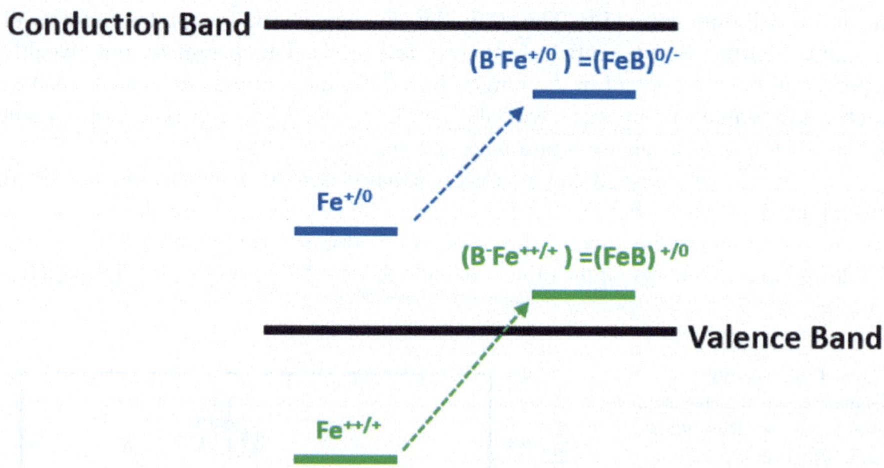

Fig. 5.6 A diagram of the energy levels of interstitial iron in the electrical field of a shallow acceptor. This diagram is based on the ionic model of iron-boron pairs. Reproduced from Istratov et al. [4] by permission of Springer

the $Fe^{+/++}$ level is shifted inside the band gap to form the $(FeB)^{+/0}$ donor level. The generally accepted activation energy for the donor level is $E_V + 0.100$ eV, corresponding with a σ_p in the range of 1.5×10^{-13} cm^2 (55 K) to 2×10^{-14} cm^2 between 50 and 65 K [4]. From a viewpoint of recombination lifetime at room temperature, the acceptor level is more relevant and its recombination properties have been studied, therefore, to greater detail [50, 55–57]. The results of four lifetime-based assessments for the stable FeB acceptor level are summarized in Table 5.2 [50, 56], converging at an average activation energy of 0.26 eV, a σ_p at 300 K in the range of 2×10^{-15} cm^2, while the values for the electron capture cross section are more scattered.

While the FeB pair is formed by an Fe_i in the nearest tetrahedral interstitial site to B_s, along the ⟨111⟩ direction, the energy difference for a pair along ⟨100⟩, where Fe_i occupies a second nearest neighbor site is not that large. A quick estimate based on the Coulomb pairing model indicates that the additional energy is only 0.071 eV [58]. This implies that below room temperature, the two configurations can coexist, with a fractional occupation defined by a Boltzmann factor given by $\exp[-0.071/k_BT]$. This idea has been confirmed by DLTS experiments [4]. In fact, depending on the bias on the diode during cooling, a metastable occupation of the second nearest neighbor ⟨100⟩ configuration can be frozen in [55], which depends on the charge state of the pair during cooling. This process is completely reversible and is driven by a single-jump motion of Fe_i in the vicinity of the ionized acceptor. The configuration-coordinate (CC) diagram of Fig. 5.7 sketches the general energy scheme for the bistable iron-acceptor pairs in silicon [4]. Based on the scheme of Fig. 5.6, one can state that each configuration of the FeB pair has a donor and acceptor level, but in the case of the ⟨100⟩ 2nd nearest neighbor pair, they are shifted 70 meV closer to the valence band edge. In fact, in n-type silicon, five energy levels have been detected in DLTS, corresponding with second, third and fourth nearest neighbor sites for the FeB pair [59]. These more distant pairs have been created by low-level minority carrier injection at low temperature, to break the close pairs by one or more Fe_i jumps, but having insufficient thermal energy to diffuse away.

High-resolution Laplace DLTS confirms an activation energy of 0.074 eV for the metastable 2nd nearest neighbor configuration of the FeB pair [60]. The impact of a magnetic field on the hole emission from the FeB level has also been investigated in order to make a direct link with EPR observations. It was found that at increasing magnetic field, the hole emission rate substantially slows down, which has been mainly attributed to the Zeeman splitting of the FeB level [61]. In addition, the optical activity of the FeB center in p-type silicon has been measured by PL, revealing a no-phonon line at 1.067 eV [61].

In ***relaxed*** $\boldsymbol{Si_{1-x}Ge_x}$ ***epi layers***, the FeB donor level has been derived from DLTS up to $x = 0.07$ [62]. A similar down-shift as for the Fe_i donor level is found in

Table 5.2 Electronic properties of the FeB acceptor level derived from IDLS and TDLS

Energy level (eV)	σ_n (cm^2)	σ_p (cm^2)	k	References
0.26 ± 0.03	2.5×10^{-15}	5.5×10^{-15}	0.45	[50]
0.23	3×10^{-14}	2×10^{-15}	–	[55]
0.26 ± 0.02	$(1.4 \pm 0.2) \times 10^{-14}$	$(0.5–2.5) \times 10^{-15}$	–	[56]
0.26	5×10^{-15}	3×10^{-15}	1.7	[57]

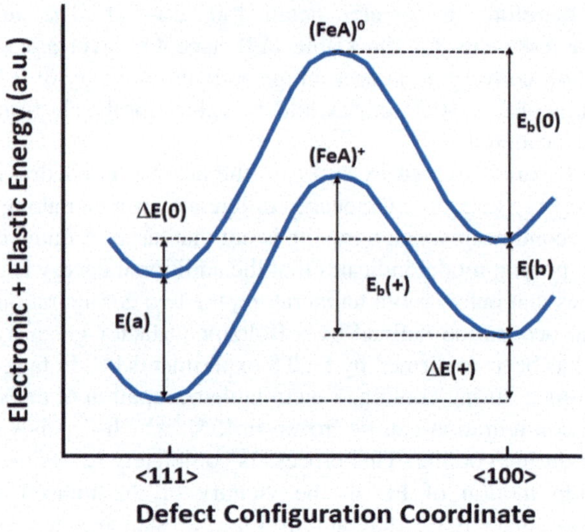

Fig. 5.7 A configuration-coordinate diagram of the bistable iron-acceptor pairs in silicon. The FeB, FeAl and FeGa pairs have a lower sum of the electronic and elastic energies in the ⟨111⟩ configuration than in the ⟨100⟩ configuration, whereas FeIn pairs have a lower equilibrium energy in the ⟨100⟩ configuration than in the ⟨111⟩ configuration. E_b is the binding energy of the pairs. *E(a)* and *E(b)* are the hole emission activation energy in the two states of the defect. $\Delta E(0)$ and $\Delta E(+)$ are the differences in equilibrium energies of the ⟨100⟩ and ⟨111⟩ configuration of the iron-acceptor pair in neutral $(FeA)^0$ and positive $(FeA)^+$ charge states; $E_b(0)$ and $E_b(+)$ are the potential barriers for the jump of a positively charged Fe_i^+ and the doubly positively charged Fe_i^{2+} ions from the metastable to the stable position. Values for the different pairs can be found in Table I from [4]

Fig. 5.8, which can be explained by considering that the trapped hole mainly resides in the delocalized *d* orbital of the Fe_i atom. The B atom will not markedly perturb the electronic structure. The shift with increasing *x* can be explained by acknowledging that the *d* orbital is sensitive to a disordered environment by adding Ge atoms, due to its delocalized nature [62]. The shift is faster than the shift of the valence band edge (or the band gap). The fact that the Fe_i donor level moves closer to the valence band indicates that more atoms will be in a neutral charge state and this should be accounted for when modeling the FeB pairing kinetics in $Si_{1-x}Ge_x$ [62].

A more systematic study of the alloy effect on the levels of Fe_i and FeB has been undertaken on relaxed bulk single crystals of $Si_{1-x}Ge_x$ with *x* up to 7.1% [63–65]. High-resolution Laplace DLTS was employed in order to investigate the energy level shift with Ge content. Figure 5.9 illustrates this for Fe_i at 240 K and for FeB at 48 K [63]. Obviously, increasing the Ge content gives rise to a peak broadening and at the same time satellite his explains the growing asymmetry of a standard DLTS peak for higher *x* [62, 63]. In addition, as in Fig. 5.8, the emission energies of the main and side peaks reduce with increasing *x* [62]. These alloying effects are related to the increasing disorder in the local atomic environment and in the local variation

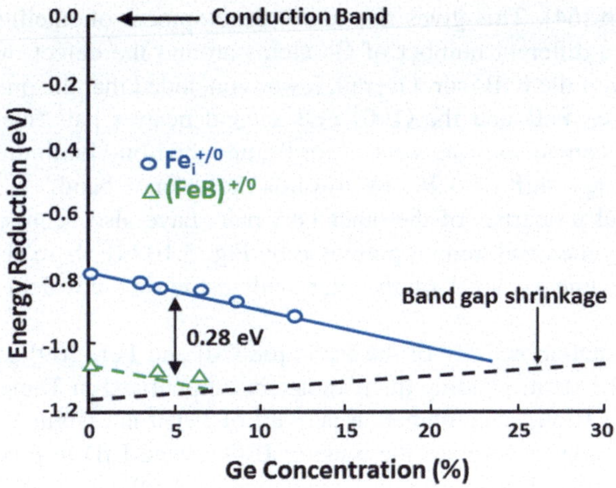

Fig. 5.8 Decrease of the hole ionization enthalpy at 0 K of both $Fe_i^{+/0}$ and $FeB^{+/0}$ donor levels, as a function of the Ge concentration. Reproduced from Mesli et al. [62]. Copyright (2002) by The American Physical Society

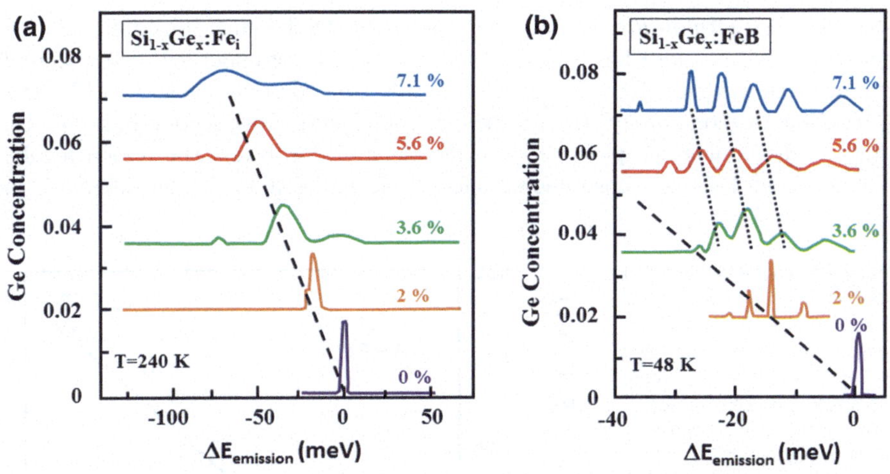

Fig. 5.9 **a** Laplace DLTS spectra observed for interstitial isolated iron in $Si_{1-x}Ge_x$ ($0 < x < 0.071$). The baseline of the spectra is displaced vertically according to the germanium content of the crystal. The peak magnitudes are plotted to an unmarked arbitrary scale. The dashed line connects the centres of gravity of the main peak in each of the spectra, demonstrating the linear alloy shift of the peak. **b** Laplace DLTS spectra observed for the iron-boron pair in $Si_{1-x}Ge_x$ ($0 < x < 0.071$). The dashed lines represent the same shift as in (**a**). The dotted lines connect the centers of gravity of the main peaks observed for the 3.6, 5.6 and 7.1% Ge samples (after Kolkovsky et al. [63])

of the band gap [64]. This gives rise to the development of satellite peaks corresponding with a different number of Ge atoms around the defect, which modifies the bond length of the FeB pair. Overall, it was concluded that the pure alloy shift is the same for Fe_i, FeB and the ⟨100⟩ FeB second nearest pair (metastable form) [66]. The same conclusion was reached for Fe_i and the iron-aluminum pair [66, 67], showing an energy shift of $0.98x$ eV towards the valence band.

The electrical properties of the other FeA pairs have also been studied [4, 23], showing a clear chemical trend represented by Fig. 5.10 [4]: there is a proportional increase of the energy level of the pair with increasing covalent radius of the acceptor.

The recombination activity of the FeGa [68–70] and FeIn [69] pairs have been studied to some extent, yielding the parameters summarized in Table 5.3. It can be derived that the strong recombination activity of FeGa is assigned to its large σ_n [69]. While Ga-doping resolves the issue of B-O related LID in p-type solar cells, the higher recombination activity of the pairs will make the material more sensitive to Fe-contamination effects.

5.1.2.2 Fe-Related Point Defects in Si and Ge

Starting with ***germanium***, it has been shown by DLTS that substitutional iron introduces two acceptor levels in the band gap, one in the upper half and a second one in the lower half [71, 72]. Detailed results are shown in Table 5.4 [72]. The non-zero activation energy ΔE_B for the electron capture cross section in n-type Ge supports the double acceptor nature of the electron trap level. The reason for the preference of TMs on substitutional sites in Ge is the fundamental difference in

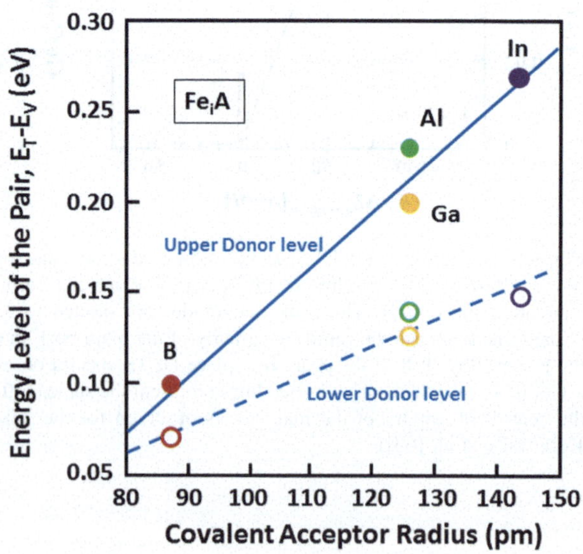

Fig. 5.10 The dependence of the hole ionization energy on the covalent radii of shallow acceptor impurities. The filled and open circles correspond to the upper and lower donor levels of the iron-acceptor pairs, respectively. Reproduced from Istratov et al. [4] by permission of Springer

Table 5.3 Energy levels and capture cross sections for iron-related recombination centers in crystalline silicon [69]

Recombination center	Energy level E_T (eV)	σ_n (cm^2)	σ_p (cm^2)	k
Fe_i	$E_V + 0.38$	4×10^{-14}	7×10^{-17}	571
FeB acceptor	$E_C - 0.26$	1.4×10^{-14}	1.1×10^{-15}	13
FeGa donor	$E_V + 0.20$	4×10^{-14}	2×10^{-14}	2
FeIn donor	$E_V + 0.15$	3.5×10^{-13}	1.5×10^{-14}	23

The parameters for Fe_i and FeB are taken from the literature [4, 56]

Table 5.4 Deep level parameters and assignment of the traps corresponding with Fe_s impurities in Ge

Level	E_T(eV)	K_T (s^{-1}K^{-2})	ΔE_B (eV)	σ_∞ (cm^2)	ΔH (eV)	ΔS (k_B)
$Fe^{-/2-}$	$E_C - 0.342$	7.5×10^7	0.050	3.7×10^{-15}	0.292	2.0
$Fe^{-/0}$	$E_V + 0.345$	2.0×10^9	0	2.3×10^{-14}	0.345	4.3

ΔE_B is the barrier for capture, σ_∞ is the temperature-independent capture cross section, ΔH is the free enthalpy of the deep level (=$E_T - \Delta E_B$) and ΔS is the entropy change upon emission of the carrier from the deep level to the corresponding band (after Lauwaert et al. [72])

diffusion mechanism, where vacancy-assisted diffusion (Chap. 2) dominates in Ge, due to the lower formation energy of vacancies compared with Si. This leads to a stabilization of TMs on substitutional lattice sites. As a consequence, the valence electrons of the TM impurity in Ge are involved in local bonding, allowing acceptor states to prevail, whereas in silicon, the TM atoms mainly stabilize in interstitial sites, impeding them from engaging their valence electrons in bonding with their neighbors, leading to donor states [71]. Application of Laplace DLTS has revealed that the shoulder to the left of the main Fe_s peak corresponds with another electron trap with a rather large electron capture cross section and activation energy of 0.287 eV (below E_C) [71]. It has been assigned to an Fe_sH level, with the hydrogen introduced during the wet etching of the sample prior to Schottky barrier fabrication.

In contrast, the possible energy levels of Fe_s acceptors ***in silicon*** are far from well-established even at this time. There have been indications that Fe_s may not have a level in the band gap [4]. One report places it at $E_C - (0.380 \pm 0.02)$ eV [73] (see Table 5.5). There is more evidence for the existence of iron in a non-ideal substitutional site, also considered as a nearest neighbor pair between a single vacancy and Fe_i, placed at $E_V + (0.500 \pm 0.020)$ eV [73]. A metastable iron-vacancy pair in aluminum-gettered p-type Si has been reported at $E_V + 0.33$ eV [74], based on the fact that the latter treatment injects vacancies in the material, which can become trapped by Fe_i, especially under white light illumination. This defect is stable up to 175 °C [74]. In irradiated and annealed silicon, trapping of Fe_i by a V_2 can occur, giving rise to a hole trap at $E_V + 0.29$ eV [75, 76]. Upon annealing at 225 °C a more stable VFeV configuration is reached, corresponding with a level at $E_V + 0.34$ eV [76]. Originally, this was assigned to a

Table 5.5 Deep level parameters and assignment of the traps corresponding with Fe-related point defects and complexes in silicon

E_T (eV)	K_T (s^{-1} K^{-2})	σ_n/σ_p (cm^2)	Identity	References
0.085 ± 0.03	9.0×10^7	–	$(Fe_i)_2V_2$, h	[73]
0.152	3.9×10^6	–	Fe-P-H, e	[79]
0.184		3.0×10^{-15}	$(Fe_i)V_2$, h	[77]
0.199	3.0×10^6	–	Fe-V-P, e	[79]
0.220 ± 0.01	6.0×10^8	–	$(Fe_i)_2V_2$, e	[73]
0.290		9×10^{-15}	FeV_2, h	[76]
0.305 ± 0.01	5.0×10^7	–	Fe_iO_i	[74]
0.31	–	–	Fe_iH, h	[80]
0.31		5.2×10^{-17}	Fe_iH, h	[81]
0.33	–	3×10^{-14}	$(Fe_i)V$, h	[74]
0.340		2×10^{-14}	VFeV, h	[75]
0.365		6.8×10^{-15}	FeVO, e	[77]
0.380 ± 0.02	3.0×10^6	–	Fe_s, e	[73]
0.410	2.6×10^7	–	$(Fe_i)_2$	[78]
0.500 ± 0.02	9.0×10^7	–	$(Fe_i)V$, h	[73]
0.530 ± 0.02	1.5×10^8	–	$(Fe_i)_2V$, h	[73]

The label e indicates an electron trap in the upper half of the band gap, h a hole trap in the lower half of the band gap

complex between VO and Fe [60], but later re-interpreted [76]. Another report places a V_2 + Fe hole trap at $E_V + 0.184$ eV [77] or at $E_V + 0.53$ [73]. The latter defect is stable up to 400 °C. The association of VO and Fe_i gives rise to a level at $E_C - 0.36$ eV [75, 77].

In n-type silicon, both FZ and Cz material doped with iron, two deep acceptors at $E_C - 0.12$ eV, $E_C - 0.21$ eV and a deep donor at $E_C - 0.41$ eV have been observed by DLTS [78]. The latter defect level has been assigned to the $(Fe_i)_2$ pair. P-related complexes with Fe have been identified in [79] (Table 5.5) in the context of Phosphorous-Diffusion-Gettering (PDG). It was argued that these complexes contribute to the trapping of Fe_i in heavily n^+ doped layers. One of the electron traps has been assigned to a Fe-P-H complex [79]. An FeH related hole trap at $E_V + 0.31$ eV has been observed both in n-type [80] and in p-type [81] silicon. The spectra for p-type silicon indicate that under reverse bias annealing the pair is stable up to 110 °C but begins dissociating at 120 °C. In n-type material, dissociation was found at 175 °C [80]. The corresponding hole capture cross section was not temperature-dependent and equal to 5.2×10^{-17} cm^2 [81]. No evidence for a complete passivation of Fe_i was found; the total concentration of electrically active species, i.e., Fe_i, FeB and FeH was found constant. Table 5.5 summarizes the DLTS results of the last two decades on Fe-doped silicon, with a possible identification, which in some cases is more firmly established than in other.

5.1.2.3 Fe-Related Clusters and Precipitates

There are some hints in the literature of Fe clustering [78] and precipitation-related bands [79], A more systematic DLTS study of the early phase of Fe clustering has been undertaken by the Göttingen group [37, 82, 83]. Annealing of silicon with $[Fe_i]$ of 1.4×10^{14} cm^{-3} at 250 °C for only 15 min, yielded Fe_i clusters, which have not been observed by EPR [37]. The corresponding spectrum in Fig. 5.11 shows a peak at the same position of the Fe_i donor level, i.e., corresponding with similar hole emission characteristics, while the capture kinetics follows a logarithmic law [82]. It implies that hole capture is strongly impeded by an occupation-dependent repulsive barrier. Furthermore, as can be seen in Fig. 5.11b, the lines coincide on the high-temperature side after normalization, a fingerprint of deep localized states at extended defects. The unusual feature for the DLTS-line assigned to the Fe-cluster is that it is almost not broadened compared with the Fe_i peak [82]. The fact that similar hole emission characteristics are obtained indicates that the Fe_i atoms in the cluster occupy the same T_d-sites as individual Fe_i. This leads to a picture where the iron atoms form a phase with T_d sites which are occupied (next nearest neighbors) and empty (nearest neighbors) in equal parts [82].

These clusters appear to be metastable and under conditions allowing further relaxation should be replaced by $FeSi_2$ precipitates. This happens upon annealing p-type silicon with an $[Fe_i] = 1.6 \times 10^{16}$ cm^{-3}, i.e., a higher supersaturation, at 200 or 250 °C for 15 min [83]. A DLT-spectrum like the one in Fig. 5.12 is then

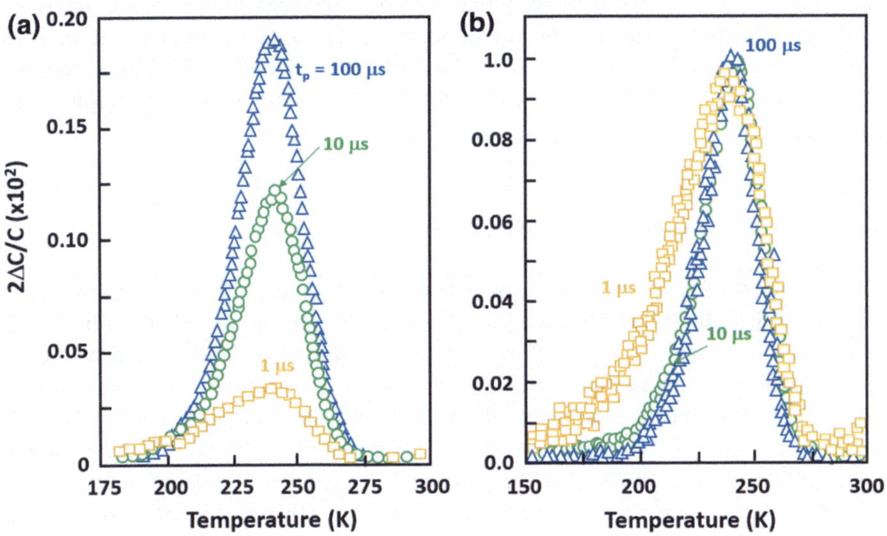

Fig. 5.11 DLT-spectra after iron was in-diffused at 950 °C for 1 h and annealed at 250 °C for 15 min. **a** Logarithmic capture kinetics, and **b** normalized spectra coinciding on the high temperature side. Such characteristics are the fingerprint of extended localized states. Reproduced from Khalil et al. [82]. Copyright 2005 WILEY-VCH Verlag GmbH & Co

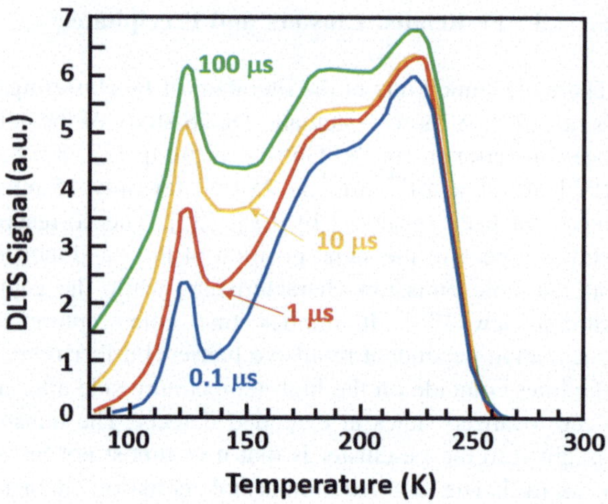

Fig. 5.12 Band-like states observed after iron in-diffusion at 1200 °C, quenching, and subsequent annealing at 250 °C, which has been associated with iron silicide precipitates. Reproduced from Seibt et al. [37] by permission of Springer

obtained, which indicates a band-like density of states spread over some 0.1 eV and located in the bottom half of the gap [37]. Possible candidates for this phase are evidently β-$FeSi_2$, the thermodynamically stable phase below 915 °C and γ-$FeSi_2$, which is a metastable disilicide isomorphous to $CoSi_2$ and $NiSi_2$ [37]. If the latter would be the case, a simple replacement reaction can transform the interstitial Fe clusters into γ-$FeSi_2$, whereby a Fe_i-atom pushes a Si atom onto a T_d-site.

When the size of the precipitate becomes sufficiently large that a kind of metallic $FeSi_2$ bulk forms, the electrical properties will be rather determined by the interface states between the silicide and the silicon matrix [84]. The recombination properties in that case can be described in terms of the metal-semiconductor Schottky barrier model, discussed in Chap. 2, which is generally valid for large metal precipitates in silicon.

5.1.2.4 Fe Activation of Extended Defects

The activation of extended defects in silicon by iron decoration has mainly been studied using EBIC-related techniques (Sect. 4.3.8). In one of the first reports by Miremadi and Morrison, it was shown that the conduction along a grain boundary was dramatically increased after Fe decoration [85]. This can be a possible cause of shunts in poly- or mc-Si solar cells. The EBIC contrast of a grain boundary (GB) is clearly enhanced when Fe is present [86, 87]. The carrier collection efficiency at the GB exhibits a minimum, indicating enhanced carrier recombination activity. The contrast reduces when reducing the temperature from 300 to 80 K [86]. Admittance measurements under e-beam irradiation in an SEM of an Fe-decorated grain boundary have been used to derive the conductance G_{GB} across the GB versus pulsed electron-beam frequency [87]. The steep increase of G_{GB} could be modeled by assuming a single energy level at $E_V + 0.504$ eV ($\sigma_p = 1.5 \times 10^{-13}$ cm^2) with a state density of 5.1×10^{11} cm^{-2}. Fermi level pinning at this energy level also determined the GB barrier height, which was lowered compared with the undecorated GB [87]. Likewise, the recombination velocity derived from EBIC versus the effective carrier generation rate was explained by an Fe-decoration-related level at $E_V + 0.504$ eV, a $\sigma_p = 8\times10^{-13}$ cm^2 and a $\sigma_n = 5\times10^{-17}$ cm^2, for an Fe-related state density of 6.8×10^{11} cm^{-2} [87].

The interaction between Fe and oxygen precipitates, important for internal gettering has been studied by EBIC [88, 89] and DLTS [89, 90]. It has been shown that the EBIC contrast exhibits a positive temperature coefficient when decorated with Fe [89], whereby the contrast at room temperature is high. This means that the oxide precipitates are strongly recombination active, with a corresponding level around mid-gap [89]. In a detailed study of the IG of Fe by oxygen precipitates in p-type Si, single-step thermal treatments at 850 °C for 16 h or 950 °C for 2, 8 or 16 h have been performed, followed by a slow cooling [89]. Such high-T annealing resulted in the disappearance of the FeB and Fe_i DLTS peaks and the appearance of other peaks at $E_V + 0.29$ eV ($\sigma_p = 1.0 \times 10^{-15}$ cm^2) and $E_V + 0.52$ eV. The first peak was interpreted in terms of an Fe-O complex, while the H(0.52) level has been associated with B [89]. After longer anneals at 950 °C, the peaks disappear, indicating that all iron is collected by the oxide precipitates. XRF maps indicate a mixture of $FeSi_2$ precipitates and Fe-oxide particles. This leads to the following picture for Fe-oxygen co-precipitation [89]: firstly, occurs the nucleation of iron-oxide or silicide cluster-embryos, which serve as nucleation sites for oxygen. At the second stage, the subsequent growth of oxygen precipitates takes place. These particles in turn serve as sites for further iron gettering giving rise to the production of the second generation of iron-silicide embryos and so on.

5.1.3 Detection and Identification of Fe in Silicon

Over the years, recombination lifetime measurements have proven to be an excellent tool for the identification and mapping of iron in silicon wafers. The method is based on the reversible cycling between Fe_i and FeA, whereby the pairs can be decomposed by application of intense white light or a heat treatment at 200 °C for example. Repairing occurs spontaneously at room temperature within a few hours or a day, determined by the association rate of (5.2). Considering the different recombination parameters (Sect. 5.1.2.1), the minority carrier lifetime (at low injection level) will be markedly different if iron is interstitial or in a pair with the acceptors.

The basic principle of the iron-detection method was first proposed by Zoth and Bergholz [91], using the SPV method and pair separation by a hot plate anneal at 210 °C for 3 min and cooled quickly to room temperature to suppress repairing as much as possible. Since SPV is operated at low-level injection (low light intensity), the recombination lifetime τ_r is smaller when iron is in the interstitial state, compared with when it is paired (Fig. 5.13). The iron concentration is then derived from the difference of the two recombination lifetime measurements [92]:

$$\frac{1}{\tau_{r0}} = \frac{1}{\tau_{FeB}} + \frac{1}{\tau_{Auger0}} + \frac{1}{\tau_{other0}} \tag{5.14a}$$

$$\frac{1}{\tau_{r1}} = \frac{1}{\tau_{Fe_i}} + \frac{1}{\tau_{Auger1}} + \frac{1}{\tau_{other1}} \tag{5.14b}$$

resulting in:

$$[Fe_i] = C'\left(\frac{1}{\tau_{r1}} - \frac{1}{\tau_{r0}}\right) \tag{5.15}$$

with C' a calibration factor which in general depends on the injection level η [92]. Equation (5.14a) includes a term for Auger recombination, which is necessary when measuring at high η or Δn, but not in the original method [91], since for SPV, $\eta \ll 1$. The contribution τ_{other} stands for the recombination lifetime determined by the other recombination centers. In case these are stable, not light-sensitive centers, one usually assumes that $\tau_{other1} = \tau_{other0}$ and thus cancels in (5.15). On the other hand, a dedicated procedure can be applied in order to separate the contribution of different metastable recombination centers like CrB pairs, Cu and BO light-induced centers [93].

If both measurements are applied at the same injection level (Δn), the Auger lifetimes also cancel—at low-level injection like for SPV or the PL imaging, these terms can obviously be neglected. A constant Δn for both measurements is in general not always easy to achieve, as the injection level Δn depends on the local recombination lifetime through G_o/τ_r with G_o the generation rate of excess carriers by illumination. Most of the popular lifetime techniques operate under a constant generation rate, as indicated in Fig. 5.13 [92]. Moreover, the injection level is not uniform across the thickness of the wafer, because the generation volume is defined by the penetration depth (absorption length) of the light and by the local recombination lifetime. In the ultimate case of infinitely fast recombination, the excess carrier density profile replicates the light absorption profile [94]. Correction procedures can be applied in combination with numerical solution of the 1D transport equations [94, 95].

The calibration factor C' depends on the injection level and the doping density [92, 96] and should also consider the fact that not all the iron is in the paired (before dissociation) or the separated state (after pair dissociation). Another consideration is that the recombination by FeB occurs in parallel through its donor and acceptor state. The SRH statistics for this set of coupled recombination channels has been worked out by Walz et al. [97]. However, it turns out that the recombination by FeB is mainly governed by the $E_C - 0.26$ eV acceptor level, so that the contribution of the $E_V + 0.1$ eV donor can be neglected [97]. The relevant recombination parameters for the other FeA pairs have been summarized in the foregoing (Sect. 5.1.2.1).

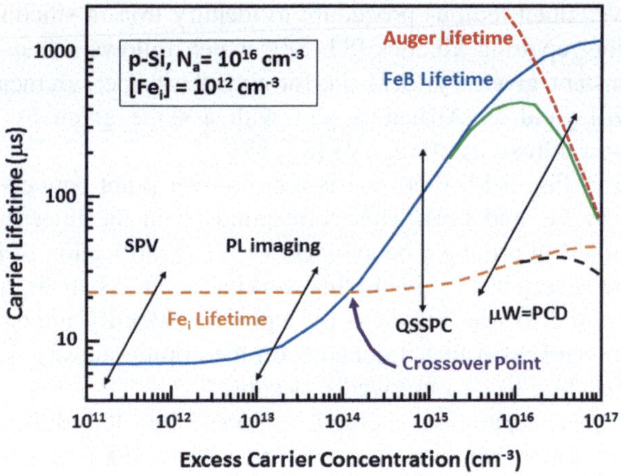

Fig. 5.13 The expected carrier lifetimes due to Fe_i and FeB pairs in boron-doped p-type silicon with $N_a = 10^{16}$ cm^{-3} and an iron concentration of 10^{12} cm^{-3} as a function of the excess carrier concentration Δn. Auger recombination is also shown at high injection and its impact on the lifetimes due to Fe_i and FeB. The injection regimes in which SPV, PL imaging, QSSS-PC and µW-PCD techniques typically operate are also indicated. Those shown with an angled arrow are constant generation rate (G_o) techniques. Reproduced from Macdonald et al. [92]. Copyright 2008 AIP Publishing

The paired fraction depends on the doping density and the temperature (Fermi level position). An important side remark is that during the lifetime measurement, pair dissociation can occur by the excitation light pulse or by the probing light (SPV and QSS-PC measurements). Keeping the exposure to the light as short as possible and the intensity as low as possible is recommended and should minimize the pair dissociation. In the case of the PL imaging technique, which allows wafer mapping in a few seconds, the following calibration factor has been proposed [92]:

$$C' = \frac{1}{\delta^{Fe_i} - \delta^{FeB}} \tag{5.16a}$$

with:

$$\delta^{Fe_i} = \frac{v_{th}(N_a + \Delta n_1)}{\frac{1}{\sigma_n^{Fe_i}}\left(N_a + p_1^{Fe_i} + \Delta n_1\right) + \frac{\Delta n_1}{\sigma_p^{Fe_i}}} \tag{5.16b}$$

$$\delta^{FeB} = \frac{v_{th}(N_a + \Delta n_0)}{\frac{1}{\sigma_n^{FeB}}(N_a + \Delta n_0) + \frac{1}{\sigma_p^{FeB}}\left(n_1^{FeB} + \Delta n_0\right)} \tag{5.16c}$$

with v_{th} the thermal velocity, σ is the capture cross section (for electrons: n, for holes: p) and p_1 and n_1 are the statistical factors in the SRH model (see (4.7)).

An alternative, albeit tedious procedure to identify iron in silicon is by lifetime monitoring of the repairing kinetics [83, 88], which follows a first-order kinetics, with a time constant given by (5.2). Performing these measurements at different temperatures will yield an Arrhenius plot with a slope given by the activation energy of the iron diffusivity (0.68 eV) [83, 88].

As indicated in Fig. 5.13, there exists a cross-over point between the recombination lifetime by Fe_i and FeB. This corresponds with the injection level where $\tau_{r0} = \tau_{r1}$ and should in principle be avoided for the identification of iron in silicon, since the method is very insensitive in this working point. As an alternative method, it has been proposed to use this cross-over point to identify interstitial iron [96]. Later studies have revealed that it depends on the doping density N_a, as shown in Fig. 5.14 [97], leading to an improved procedure.

In fact the pair separation method to determine the diffusion length in Fe-contaminated silicon has already been proposed in 1986 by Kitler et al., performing EBIC experiments [98]. A two-side surface photovoltage method has been developed as an approach for iron contamination monitoring—it provides information on the iron depth distribution and on the possible origin of the contamination [99].

Although DLTS is less suitable for routine inspection, it has been applied to the deeper analysis on the role of Fe and FeB pairs in mc-silicon for PV applications [100, 101]. Scanning DLTS in an SEM has also been performed, whereby the temperature of the sample is fixed and the electron-beam pulse is scanned across the sample [102]. This allows to map the Fe_i concentration in p-type mc-Si. A standard temperature-scan DLTS spectrum at a fixed position of the beam has revealed the presence of two minority carrier electron traps at $E_C - 0.29$ eV (FeB acceptor level) and at $E_C - 0.45$ eV [102]. In order to detect FeB pairs in heavily p^+-doped silicon, a hydrogen passivation procedure has been worked out, to reduce the active B doping concentration and to increase the sensitivity of the method [103].

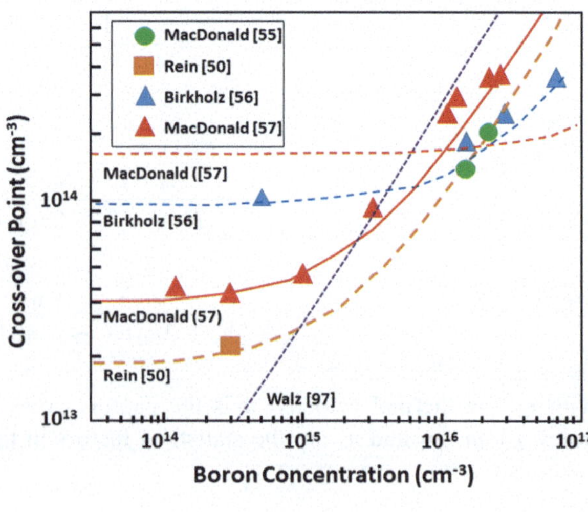

Fig. 5.14 Dependence of the cross-over point on the capture cross sections and boron dopant concentration. The fits are based on the energy levels summarized in Table 5.3. Reproduced from Macdonald et al. [57]. Copyright 2006 AIP Publishing

Lifetime analysis has also been applied to the case of iron precipitates [104, 105]. Combining μPL and μRaman measurements has established a correlation between the strain around a precipitate and its recombination activity [105]. A linear relationship was derived between the area of an iron precipitate and its recombination active area (which is found about a factor of 3 bigger). This trend can be explained if the precipitate has an infinite surface recombination velocity (or conversely a close to zero lifetime), leading to a capture cross section which is proportional to its surface area, independent of the metal element [105]. For the case of $FeSi_2$ no measurable stress could be detected (mismatch $\sim 1\%$), in contrast to the case of Cu_3Si, which corresponds with a lattice mismatch of 5% [105]. In this context, it is important to remark that the recombination activity of a certain concentration of iron will be much higher when dissolved interstitially, compared with the same amount of iron in precipitates. Moreover, a small density of large precipitates will be more favorable to obtain a high minority carrier lifetime than a high density small precipitates.

5.2 Copper

Copper in silicon is the fastest diffusing 3*d* TM, with a fairly high interstitial solubility at high temperatures [106]. The combination of a high diffusivity with a steeply decaying solubility creates a high driving force for precipitation upon cooling. One of the unique features of copper is that below the eutectic temperature of 820 °C, the phase in equilibrium with silicon is a copper-rich silicide, namely Cu_3Si [107], which has a large lattice mismatch. This means that compressive stress is built up during precipitation, which can be relaxed by the emission of one interstitial for every two precipitated Cu atoms. It implies that in spite of the high driving force upon cooling, given by [107]:

$$\Delta f_{chem} = k_B T ln\left(\frac{C}{C_0(T)}\right) = E_a \frac{\Delta T}{T + \Delta T} \tag{5.17}$$

with ΔT is the undercooling and E_a = 1.75 eV, there exists a high barrier for the ***homogeneous*** nucleation of Cu precipitates. This barrier can be lowered by the presence of pre-existing extended defects (dislocations, stacking faults, oxygen precipitates,…) favoring the ***heterogeneous*** nucleation of large precipitate colonies. However, in defect-free silicon substrates and for a fast cooling rate from high temperature, large amounts of copper can be frozen well above the solubility limit at 300 K. This means that at room temperature and considering the high mobility of interstitial copper ions (Cu_i^+) other reactions may take place, as indicated in Fig. 5.15 [106], whereby copper either out-diffuses to the surface or becomes trapped by other impurities (dopants, Cu_s, other metals,…).

In this part, the different configurations of copper in Si and (Si)Ge will be discussed. Particular emphasis is on the precipitation behavior of copper in silicon,

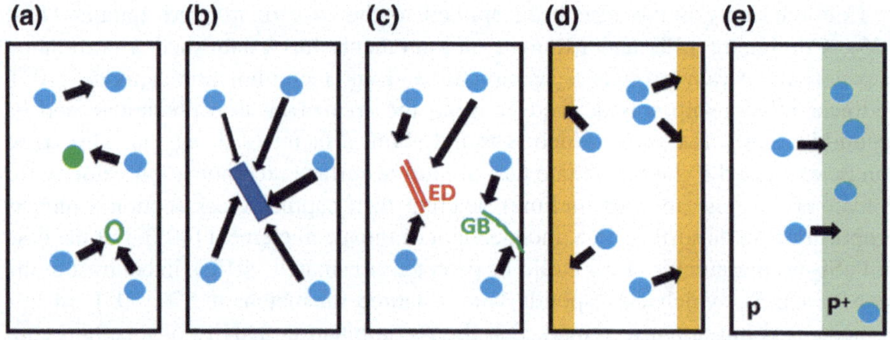

Fig. 5.15 Five main types of defect reactions of copper in silicon: **a** formation of point defects and their complexes; **b** formation of copper-silicide precipitates in the bulk; **c** decoration of existing defects, such as dislocations or grain boundaries; **d** out-diffusion to the surface; **e** segregation in the p^+ areas. Blue circles are interstitial copper ions Reproduced from Istratov and Weber [106] by permission of ECS-The Electrochemical Society

based on heterogeneous nucleation, assisted by other defects or homogeneous nucleation. A second section discusses the electrical properties of copper and copper-related complexes and precipitates. In the final section, the electrical detection and identification of copper in Si by lifetime and other dedicated techniques will be described.

5.2.1 Configurations of Copper

5.2.1.1 Cu-Related Point Defects

The minimum energy site for Cu in silicon is the interstitial tetrahedral (*T*-) site, where it will behave as a donor. As a substitutional impurity, Cu_s is expected to behave as a triple acceptor in Si and Ge [1, 2]. The lattice location of implanted Cu ***in silicon*** has been investigated by β-emission channeling, where up to 600 °C annealing the majority of Cu atoms was found close to substitutional (near-*S*) sites [108]. The displacement, however, was significant and amounted to 0.05 nm along the ⟨111⟩ direction towards the bond center position. It is believed that in this case, near-*S* Cu is related to copper in a multivacancy complex, stemming from the ion implantation. In intrinsic ***germanium***, a significant fraction already occupies ideal substitutional sites in the as-implanted state [109]. This fraction increases upon annealing up to 300 °C; after 400 °C the channeling effects disappear. A dissociation energy for Cu_s in Ge of 1.6–1.9 eV was derived from the thermal stability [109]. In the case of the displaced substitutional sites in silicon, the activation energy was found 1.8–2.2 eV [108].

In highly n^+ doped silicon, Cu was observed in ideal substitutional sites upon annealing between 200 and 600 °C [110]; a corresponding dissociation energy of 2.9 eV has been derived. The location in ideal *S*-sites could point to the pairing of copper with donor atoms by Coulomb attraction (Cu_s–D_s^+), in line with the pairing process described in Chap. 2. Evidence for pair formation between Cu_i and A_s^- has been found in perturbed gg angular correlation measurements of Cu-doped p-type silicon [111]. A dissociation energy for the pairs of 0.70 eV was established. Evidence for the formation of other Cu-related pairs and complexes was obtained from EPR measurements [6, 112, 113]. On the other hand, no EPR fingerprints of individual copper have been reported so far [113].

In contrast to other TMs in silicon, only a fraction of about 0.1% copper forms electrically active point defects and complexes, giving rise to deep level(s) in the band gap [106]. It is the high diffusivity of Cu that enhances the dissociation of its point-defect complexes, thus greatly reducing their stability and favoring the diffusion of copper to thermodynamically more stable sinks, such as the surface or extended defects, where it precipitates. The kinetics of defect complexes and ultimately their equilibrium concentration is determined by the balance between two reactions: association and dissociation [106]. This is schematically represented in Fig. 5.1, following the general idea of Metal-Acceptor pair formation by long-range Coulomb attraction (Chap. 2). The kinetic barrier that an impurity atom has to overcome to dissociate from a defect is determined by two components: (i) the depth E_b of the attractive potential between the impurity and the defect and (ii) the height of the diffusion barrier, E_d, for the impurity. Roughly speaking, the dissociation energy, $E_{diss,}$ becomes: $E_{diss} \approx E_d + E_b$. Assuming simple Coulomb pairing between nearest neighbors, E_b can be approximated by 0.52 eV in silicon [106]. CuB pairs are unstable, even at room temperature, and dissociate within ms. As the majority of complexes of Cu with other impurities and defects very easily dissociate, they cannot permanently trap Cu, but rather slow down its diffusion towards more stable sinks. These sinks can be extended defects in the bulk of the wafer, the wafer surface or a p^+ layer.

5.2.1.2 Heterogeneous Precipitation of Copper

Since the first observations of copper precipitation in silicon [114–116], many studies have been devoted to investigate the structure and composition of the extended colonies (see, e.g., [117–126] and Fig. 5.16 [119]), which generally nucleate on some extended defect. In fact, the tendency of copper to decorate extended defects has frequently been used to investigate grown-in defects in Czochralski [127–129] and SOI [130] substrates by defect etching and optical or SEM inspection.

Based on extensive TEM observations, a model for the formation of copper colonies has been developed [131, 132]. This mechanism is schematically represented in Fig. 5.17 [132]. Briefly summarized, the colonies consist of planar arrangements of copper silicide particles bounded by extrinsic dislocation loops.

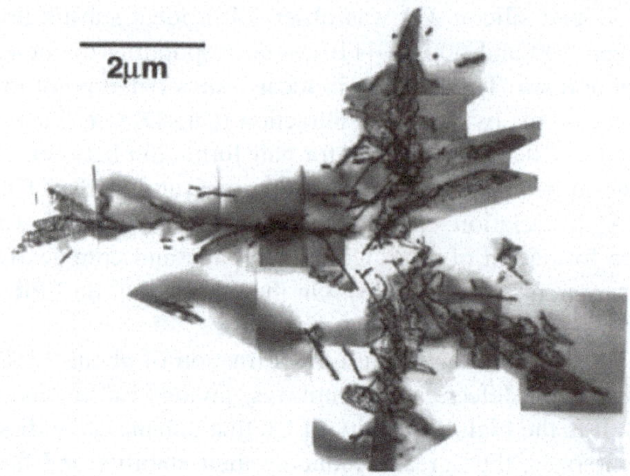

Fig. 5.16 Low-magnification TEM micrograph of a copper silicide particle colony after moderately fast cooling from 900 °C. Reproduced from Seibt et al. [107]. Copyright 1998 WILEY-VCH Verlag GmbH & Co

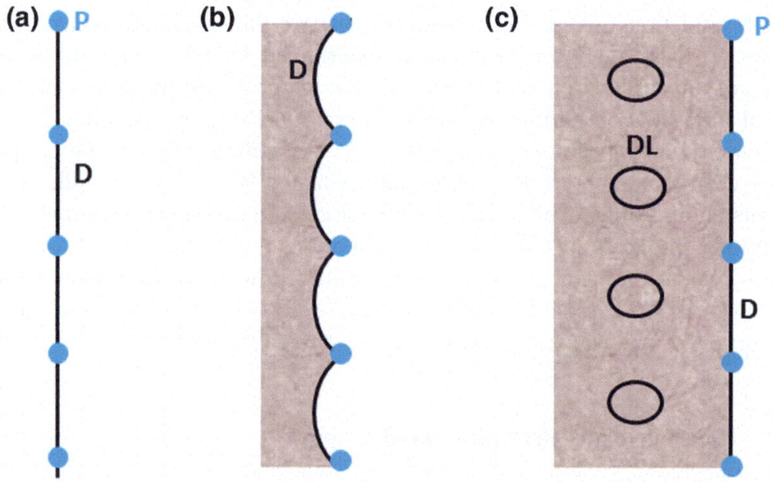

Fig. 5.17 Schematic representation of precipitate colony growth after Nes [131]: **a** heterogeneous nucleation of copper silicide precipitates at a pre-existing dislocation, **b** growth of the precipitates and climb of the dislocation due to incorporation of self-interstitials emitted by the growing particles, and **c** dislocation unpinning restoring of the situation in (**a**)

The phase of the particles has been established as η'-Cu_3Si (non-cubic, hexagonal lattice). Colony growth is usually observed in the presence of extended defects like grain boundaries [133–135], dislocations [136–139], stacking faults [138, 139] or oxygen precipitates in Cz silicon [133–135]. In defect-free silicon, such colonies form either at the surface (haze formation [143, 144]) at moderate cooling rates

($\sim$4 K/s) or as a consequence of Ostwald ripening after homogeneous nucleation (see next section).

When comparing the efficiency of the different types of extended defects as nucleation site for copper precipitation, it appears that the colonies develop at some special sites on dislocations (i.e., non-dissociated edge type dislocation segments) and do not decorate other parts of the dislocation if the wafers are cooled slowly [137]. This is in contrast with Fe in which case a homogeneous decoration of the dislocation occurs. Oxidation-Induced Stacking Faults (OISFs) are more efficient gettering sites for copper than oxygen precipitates [139]. Colonies of Cu precipitates develop irrespective of the cooling rate at punched-out dislocation loops around oxygen precipitates [139, 140], which is relevant for internal gettering of copper (see Chap. 8). In heavily contaminated samples cooled fast from the contamination temperature, Cu also precipitates on Frank partial dislocations bounding bulk stacking faults [139]. Given the strain associated with copper precipitation and the emission of self-interstitials, there can be an impact on the precipitation of oxygen in silicon, which is also accompanied by an internal volume expansion of SiO_2 versus silicon. Conflicting reports can be found in the literature: in a first study, no impact was observed on the oxygen precipitation in copper contaminated (111) Cz silicon, whereby the pre-existing large colonies where not affected by the formed oxide precipitates or other secondary defects (bulk SFs and punched-out dislocation loops) [145]. In other words, the copper precipitate colonies exhibit a high thermal stability against a two-step heat treatment in the range 600–1100 °C. More recent studies indicate just the opposite: the presence of Cu enhances significantly the precipitation of oxygen [142]. This is explained by considering that due to the Cu precipitation the high density of dislocation loops, formed by the interstitial emission, can serve as nucleation sites for oxygen precipitates. Furthermore, the loops can absorb the interstitials which are emitted during the precipitation of oxygen.

A detailed microscopic and theoretical study of copper precipitation at small-angle tilt boundaries in p-type Si revealed that in the initial stages, Cu agglomerates along the boundaries, forming coherent layers of Cu_3Si (less than $\sim$2 nm thick), with a bcc structure in a metastable state [146]. As the layer thickens, it becomes semi-coherent with misfit dislocations on the (220) interphase boundaries, reducing the strain. Subsequently, the metastable layers convert into polyhedrons or orthorhombic η”-Cu_3Si in equilibrium state [146]. The tendency of Cu to precipitate on dislocations also explains the difficulty in doping relaxed SiGe hetero-epitaxial layers on silicon with copper [147].

Under a moderately fast cooling from the contamination temperature, such as typically applied during RTP, copper will have sufficient time to diffuse to the surface and precipitate there. This is clearly a favorable site for precipitation when there is sufficient time during cooling, since the surface is highly reactive due to the presence of dangling bonds, there is less steric hindrance to form another phase and no stress is built up at the surface upon precipitate formation. In that case, large star-shaped precipitate colonies can form, consisting of a planar arrangement of small particles parallel to Si{110} planes and surrounded by extrinsic dislocation loops [143]. Upon preferential etching, large etch pits develop, as illustrated by the

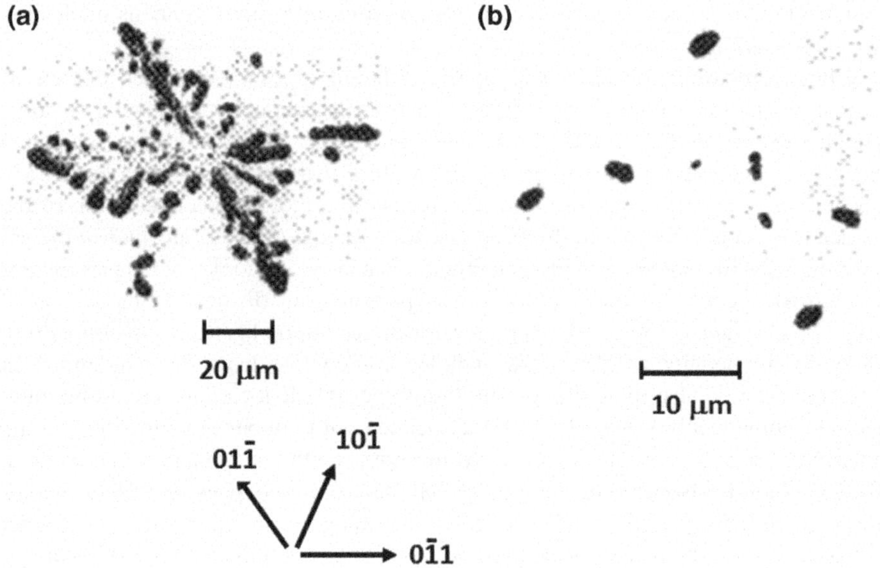

Fig. 5.18 Optical micrographs of etch pits obtained by preferential etching of Cu-diffused Si wafers: **a** starlike etch pits in regions of high Cu supersaturation; **b** small rodlike etch pits in regions of smaller Cu concentration. Reproduced from Seibt and Graff [143]. Copyright 1988 AIP Publishing

optical micrograph of Fig. 5.18 [143], while the surface of the wafer appears hazy under illumination. This is the basis of the so-called *haze test* for evaluating the efficiency of gettering treatments, whereby one usually employs Pd back-side contamination as the test vehicle, inspecting the front-side after RTP and defect etching (see Sect. 4.2.1).

It is clear that intrinsic point defects, present in as-grown wafers and controlled by the crystal pulling conditions, can have an impact on the precipitation behavior of copper: in the presence of excess self-interstitials (I-rich region), precipitation will be hindered and conversely, in vacancy-rich parts of a Cz silicon crystal, copper precipitate nucleation will be promoted, as the vacancies may serve as a sink for the emitted interstitials and in the case large agglomerates of vacancies (voids) are present, they can act as nucleation site. This principle is exploited in the ion-implantation-damage and void-gettering of Cu. Fundamental studies on the impact of intrinsic point defects on copper precipitation have been carried out in the past, either on material which was exposed before to radiation damage [148] and with grown-in [149, 150] or processing-induced point defects present [151]. It has been shown that in regions with A-swirl defects (interstitial clusters) a smaller density of large colonies is present, while in D-defect (void) regions a larger density of smaller individual copper precipitates with no colonies occurs after contamination at 1000 °C [149]. The suppression of the colony formation in vacancy-rich crystals is related to the fact that the emitted self-interstitials can recombine with O_2V

complexes, so that no extrinsic dislocation loops are formed for renewed Cu precipitation. In addition, the O_2V defects can compensate the compressive stress built up during copper precipitation [149]. In the interstitial-rich region, on the other hand, the repeated nucleation process on the dislocation loops is not hindered, explaining the growth of large colonies. At the same time, it has been observed that the minority carrier lifetime is more degraded in the D-zone because of the higher precipitate density [150]. A similar tendency was found in n-type Cz wafers after introduction of interstitials by a thermal oxidation step at 900 °C [148]: with increasing oxidation time, the Cu precipitate behavior changed from a high density of spherical particles to a smaller density of rod-like colonies, nucleated at the oxidation-induced dislocation loops.

An important question to answer regards the thermal stability of Cu precipitates. Before, it was already mentioned that pre-existing copper precipitates were not affected by subsequent oxygen precipitation, as it happens during an IG step [142]. Further studies focused on the dissolution of precipitated copper at 560 °C for 45 min followed by a rapid quench (1000 °C/s). A partial dissolution of copper precipitated on the Frank partials of a SF was observed, whereby the limiting factor for the dissolution is the diffusion of I's at low temperature [152]. The most critical factor determining the stability of a precipitate is its chemical state. According to the diagram of Fig. 5.19, Cu_2O is more stable, due to the higher binding energy of the metal atom, than a Cu film or Cu_3Si [152].

The role of the cooling rate on the precipitation behavior has been thoroughly investigated as well [154]. If a wafer is under thermodynamic equilibrium conditions (furnace cool) little precipitation takes place and the copper will have diffused to the surface, where so-called self-gettering by a Cu layer or a silicide will take place. Under an air cool (rate 25 °C/s), no interstitial copper above the detection

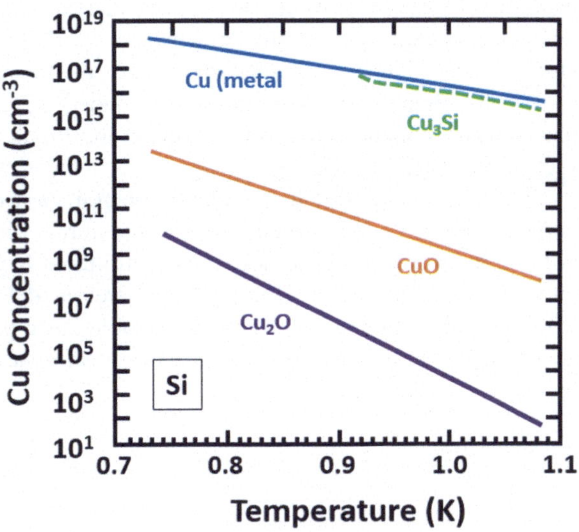

Fig. 5.19 Solid solubility of Cu in silicon with respect to a Cu, Cu_3Si, Cu_2O, and CuO phase. There is a great disparity in solubility between metal/silicide and oxides. Reproduced from McHugo et al. [153]. Copyright 2002 AIP Publishing

limit of the Transient Ion Drift (TID) technique (10^{11} cm^{-3}) is maintained in solution. Most copper is precipitated at the surface, while only a minor fraction is gettered by the internal heterogeneous sites (the Frank partials of the bulk SFs rather than the oxide precipitates). Following a rapid quench in ethylene glycol (1000°C/s), a Cu_i concentration equal to the B concentration is obtained, with no copper at the internal gettering sites or at the surface. In this case, copper is in a high degree of supersaturation. In addition, small platelet precipitates can be homogeneously nucleated as discussed next.

Finally, the co-precipitation of several TMs (Ni, Fe, Cu) has been studied in slowly cooled $Si_{1-x}Ge_x$/Si heterostructures, revealing that the chemical reactions between metals and silicon during precipitation may reduce the lattice mismatch compared to single-metal precipitates [154]. This renders mixed-metal-silicide precipitates more stable and energetically favorable. Such precipitates are frequently found in mc-Si wafers, which are subjected to contamination by different TMs during growth.

5.2.1.3 Homogeneous Precipitation of Copper

Homogeneous nucleation, without the assistance of an extended defect proceeds via the formation of plate-shaped precipitates surrounded by an extrinsic SF, as shown in Fig. 5.20 [107]. Such early-stage precipitates can be found after a ***rapid quench from high temperature*** (2000 K/s) and their habit plane is (111) or occasionally (113) [107, 155, 156]. The shape of the initial precipitates is determined by minimizing the elastic and surface energy, which is smallest for platelets or spheres, respectively. Plates are thus formed to reduce the elastic strain energy, resulting from the pseudomorphic growth of copper silicide [107]. The interfaces of the platelets are not atomically flat – they are wavy and occasionally change orientation from $\langle 111 \rangle$ to $\langle 113 \rangle$. At its border, the platelet is connected to an extrinsic SF, which has been formed by the condensation of I's injected by the growing platelet [150]. These platelets are metastable and are transformed into compact almost spherically shaped Cu_3Si particles surrounded by a SF [107, 156]. These may develop into colonies by precipitation on the Frank partials bordering the SF, as sketched in Fig. 5.21 [107].

These early stages of metal silicide precipitation in dislocation-free silicon are known to be related to non-equilibrium structures, which are kinetically selected in order to facilitate fast relaxation of extraordinarily large driving forces occurring in undercooled solutions of TMs [37]. The created defects are frequently associated with deep band-like states. These metastable early precipitates relax upon further annealing ('internal ripening') into the silicon-richest phase, with respect to the binary phase diagram [37].

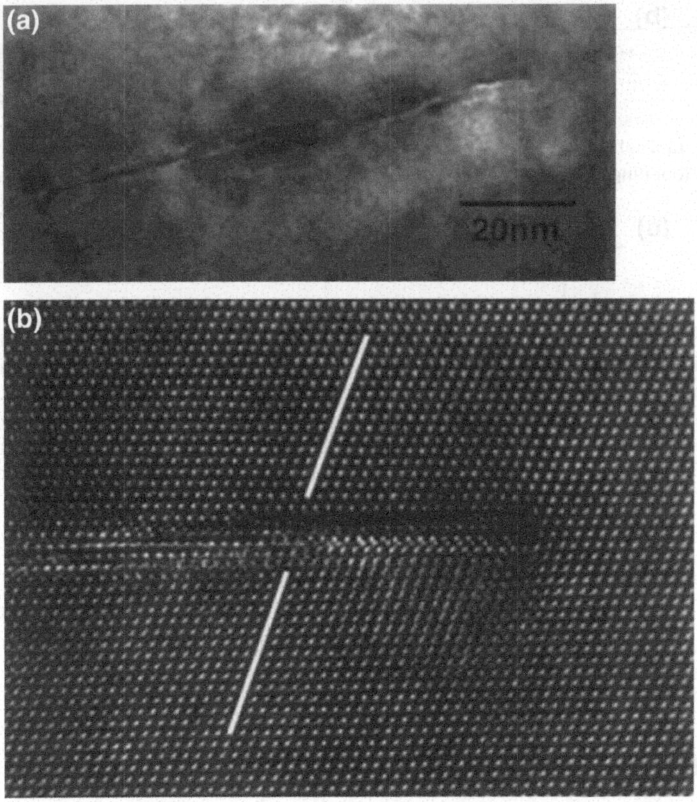

Fig. 5.20 Plate-shaped copper silicide precipitates after rapid quenching (cooling approximately 2000 K/s) from high temperatures. **a** Bright-field micrograph, **b** HRTEM image showing an extrinsic stacking fault surrounding the platelet. Reproduced from Seibt et al. [107]. Copyright 1998 WILEY-VCH Verlag GmbH & Co

For a somewhat lower quenching rate of 1000 K/s two additional effects may occur:

(i) Strain relaxation due to misfit dislocations formed in the interface between the platelet and silicon.
(ii) Heterogeneously nucleated spherical precipitates at the partial dislocation bounding the SF may be formed, which is a precipitate colony in its embryonic stage.

Finally, for a cooling rate of 250 K/s small precipitate colonies are observed, which contain larger platelets in their central parts [107].

The 'ripening' of these platelet precipitates, produced by a rapid quench has been studied in detail by Seibt et al. [107], whereby in principle two mechanisms can be at work: a so-called 'internal ripening' whereby the total defect density is preserved but the platelet is transformed into a spherical particle without long range

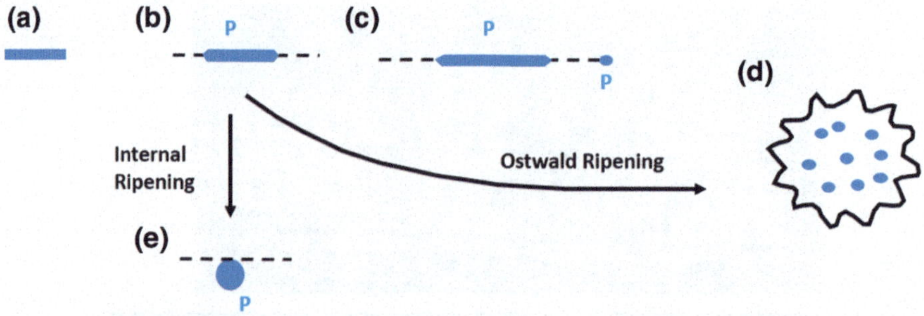

Fig. 5.21 Summary of precursor stages to colony growth explained in terms of misfit strain accommodation: **a** formation of plate-shaped precipitates in order to reduce elastic strain energies, **b** formation of an extrinsic stacking fault surrounding the platelet allowing strain relaxation by self-interstitial emission, **c** heterogeneous nucleation of precipitates at the partial dislocation bounding the SF, **d** colony growth. In **e** the formation of spherical precipitates as a result of internal ripening is shown (after Seibt et al. [107]. Copyright 1998 WILEY-VCH Verlag GmbH & Co)

diffusion after an anneal at 450 °C for 15 min (vertical arrow in Fig. 5.21). On the other hand, at higher annealing temperature (700 °C 15 min) the particle density is reduced by one decade, giving rise to small precipitate colonies (lateral process in Fig. 5.21), by Ostwald ripening.

5.2.1.4 Precipitation Versus Out-Diffusion

Perhaps surprisingly, it is concluded that after rapid cooling (>500 K/s), copper can be kept interstitially at room temperature for several hours, without immediately precipitating. During this time, Cu_i^+ diffuses large distances eventually reaching the surface, where it accumulates. The copper out-diffused at room temperature can easily be removed with an HF:H_2O_2 mixture, etching the native oxide. This suggest that no silicidation takes place, i.e., no bonds are formed, but that copper gets incorporated in the growing native oxide. In fact, the behavior of copper at the silicon surface will depend on whether or not there is already an oxide layer present, on its thickness, etc. This is related to the fact that an oxidized silicon surface forms a potential barrier for positive copper ions, due to the presence of fixed positive charge in SiO_2. Interaction of copper with native oxide is known, e.g., the catalyzation of oxidation in the presence of Cu [106]. Copper contamination during Chemical Mechanical Polishing (CMP) is a reversed out-diffusion. However, in this case, the 0.2–0.3 eV potential barrier at the surface now assists in the in-diffusion, which, therefore could happen much faster [106].

Alternatively, Cu may also precipitate or become trapped by a dopant or other metal impurities. However, the shallow acceptors do not form stable traps due to the low binding energy. Ionized acceptors can only slow down the copper diffusion to stable sinks, which Cu eventually finds within several hours after the quench.

Shabani et al. were the first to observe the out-diffusion of Cu at room temperature after several days in p-type Cz wafers, *after removing the native oxide* [157]. This implies that the positive charge in the native oxide forms a Coulomb barrier, preventing Cu from reaching the surface. In *n-type Si*, **no** out-diffusion was observed [157], indicating the impact of the Fermi level on the Cu_i out-diffusion. In order to explain these observations, an electrostatic model has been proposed by Flink et al. [158]. According to the data of Fig. 5.22, obtained on p-type FZ and Cz Si, B-doped between 10^{14} and 10^{16} cm^{-3}, there appears to exist a critical Cu concentration where above precipitation starts. As a rule of thumb, this concentration is approximately given by [158]:

$$[Cu]_{crit} = [B] + 10^{16} cm^{-3} \tag{5.18}$$

and thus increases linearly with the boron concentration. For copper concentrations above the critical value, the precipitated Cu is close to the solubility (Fig. 5.23). In other words, above $[Cu]_{crit}$, the main reaction path is precipitation. In addition, the precipitation happens during or immediately after the quench and is, therefore, much faster than the out-diffusion.

On the other hand, below $[Cu]_{crit}$, out-diffusion to the surface is the dominant reaction path. In further agreement with this picture is the observation that the decay rate of Cu_i at 300 K evolves non-exponentially and depends on the thickness of the wafer: it happens slower for thicker samples. This is typical for an out-diffusion process. The Cu_i decay curves versus time can be modeled assuming field-enhanced diffusion, including a repulsive surface potential of 50–100 meV as a boundary condition [158]. The data of Fig. 5.22 can be further transformed by deriving a critical Fermi level distinguishing between out-diffusion and copper precipitation (Fig. 5.24) [158]. The position of the Fermi level can be obtained by solving the charge balance equation, considering both fully ionized acceptors and Cu_i donors. It turns out that the critical Fermi level is at $E_C - 0.2$ eV, which coincides with the neutrality level of Cu_3Si (see next paragraph). In other words, besides the elastic strain barrier for precipitation, there exists a Coulomb barrier, where the

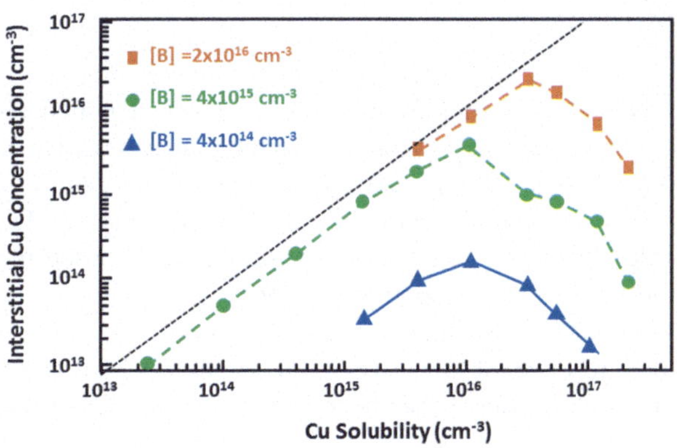

Fig. 5.22 Interstitial copper concentration as measured with TID 30 min after quench at room temperature versus the Cu solubility concentration at in-diffusion temperature. Reproduced from Flink et al. [158]. Copyright (2000) by The American Physical Society

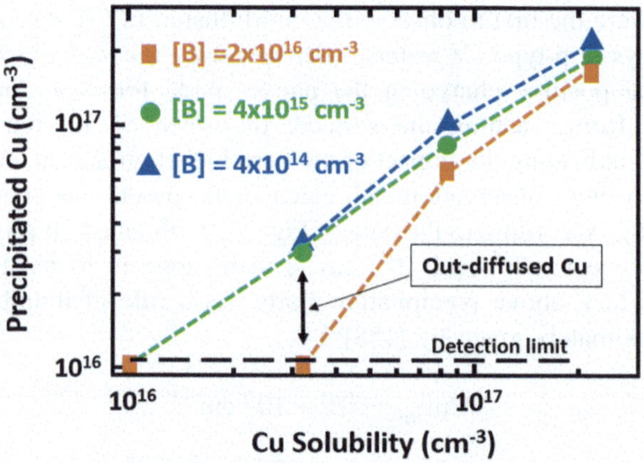

Fig. 5.23 Copper concentration measured with X-ray Fluorescence (XRF) versus the solubility concentration of copper at the in-diffusion temperature. Reproduced from Flink et al. [158]. Copyright (2000) by The American Physical Society

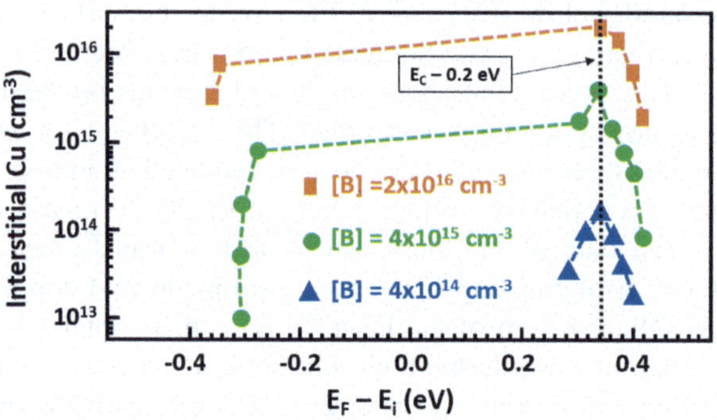

Fig. 5.24 Interstitial copper concentration as measured with TID 30 min after quench at room temperature versus the Fermi level position immediately after quench. The Fermi level was calculated from the acceptor concentration and the initial copper donor concentration at the in-diffusion temperature. Reproduced from Flink et al. [158]. Copyright (2000) by The American Physical Society

precipitation can start for a Fermi level at which the acceptor levels of the precipitates become occupied, resulting in a negative charge, attractive for positive copper ions. This model self-consistently explains the absence of out-diffusion in n-type Si, where precipitation is favored, once the Fermi level is above the critical value [158]. More detailed thermodynamics-based modeling has largely confirmed the Fermi level dependence [159].

It is also clear that the out-diffusion of copper depends on the surface condition: the presence of a native oxide or not, which modifies the surface charge and potential barrier. It has been shown that the Cu surface precipitation is accelerated by employing a solution of low HF concentration or by an H_3BO_3 wet treatment [160]. Also adsorbed organic molecules can lower the surface potential, so that long-term storage of wafers in a shipping box increases the Cu precipitation in p-type Cz wafers [160]. Alternatively, the application of a deposited positive corona charge keeps interstitial copper in solution as long as it is present, while a negative charge accelerates the out-diffusion.

5.2.2 Electrical Activity of Cu

From the foregoing it is clear that copper in silicon can exist in different forms and, correspondingly, it is expected that this will impact on the electrical activity. Roughly speaking, only a minor fraction of copper is normally incorporated as simple point defects (PDs) or point defect complexes, giving rise to deep levels in the band gap of silicon. These can be studied by sensitive techniques like DLTS or, in case of optical recombination activity, by PL spectroscopy. Considering that Cu_s is a triple acceptor in silicon, it may act as an efficient recombination center, determining the lifetime when present in a sufficiently high concentration. It should be reminded that interstitial copper is always positively charged, implying a shallow donor level in the upper half of the band gap (or coinciding with the CB). This means that there is no impact on the minority carrier lifetime of Cu_i^+, while, of course, it contributes to the ionized charge (and electrical field) in the material. Copper can also form precipitates in the bulk, on extended defects or at the surface. In the former two cases, a strong effect on the minority carrier lifetime has been demonstrated [107, 156, 161], which can be explained by considering the distribution of states induced in the silicon band gap. In addition, when precipitated on extended defects, copper will modify their electrical activity. Finally, copper at the surface of silicon (or at the Si/SiO_2 interface) will react with the dangling bonds present there, modifying the surface recombination and generation velocity. In the following, the electrical activity of copper-related point defect complexes, of Cu_3Si precipitates and of copper decoration on extended defects will be discussed.

5.2.2.1 Copper-Related Point Defects

Early-day spectroscopic studies of copper-related PDs have been based on photoconductivity [162, 163] infrared absorption [164], Hall effect versus temperature [162, 165, 166] or space-charge limited current analysis [167]. However, most accurate and reliable data have been obtained by capacitance-based measurements [168–175]. Reviews on the DLTS results of copper-related deep levels can be found in [176] and [177].

Figure 5.25 shows the spectrum of the double acceptor level (−/2− state) of Cu_s in n-type silicon. Two more acceptor levels can be found in p-type Si (see also Table 5.6). Another well-known copper level is the $E_V + 0.1$ eV donor, represented

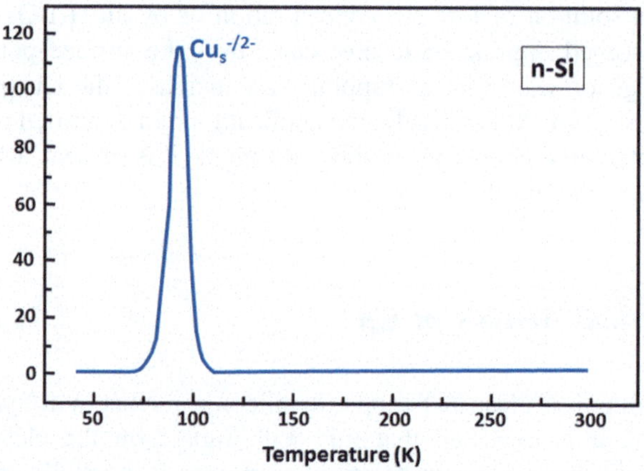

Fig. 5.25 Double-acceptor level of substitutional copper in n-type silicon measured by DLTS. Two more levels are found in p-type silicon. Reproduced from Knack [177] with permission of Elsevier

Table 5.6 Summary defect properties of Cu-related defects in silicon: trap level, activation energy, electron capture cross section, barrier for thermally activated electron capture, hole capture cross section, level assignment and radius of hydrogen capture (after Knack [177])

Level	E_a (eV)	c_n (cm^3 s^{-1})	ΔE_B (eV)	c_p (cm^3 s^{-1})	Assignment	r_H (nm)
E(100)	0.167(3)	3.2×10^{-10} (105 K)	0.025(3)	6.5×10^{-6} (105 K)	$Cu_s^{-/2-}$	0.7
E(140)	0.254(4)	4.6×10^{-11} (143 K)	0.050(4)	7.0×10^{-6} (144 K)	$CuH_2^{-/2-}$	0.9
E(180)	0.360(3)	7.5×10^{-11} (183 K)	0.050(4)	1.3×10^{-5} (185 K)	$CuH_1^{-/2-}$	1.8
H(100)	0.207(4)			$2.4 \times 10^{-5}/8.9 \times 10^{-7}$ (97 K)	$Cu_s^{+/0}$	–
H(125)	0.27(1)				$CuH_2^{0/-}$	1.0
H(200)	0.478(5)			2.9×10^{-7} (202 K)	$Cu_s^{0/-}$	0.3
H(260)	0.54(1)			1.4×10^{-7} (248 K)	$CuH_1^{0/-}$	0.3

in Fig. 5.26 and responsible for the 1.014 eV luminescence line [61, 178–185] represented in Fig. 5.27. Originally, this defect was associated with a Cu_s–Cu_i pair (see e.g. [176, 185]) but more recent experimental [186–193] and theoretical [194, 195] investigations have definitively shown that the center contains four copper atoms: a central Cu_s surrounded by three copper interstitials.

Another Cu-related deep level found after Cu diffusion and quenching is the so-called M-defect, represented by Fig. 5.28 [172, 177]. It can be passivated by pairing with Cu_i but the structure of the defect is so far not clear. In Table 5.6 the main deep level parameters of some Cu-related point defects are summarized [177].

The interaction of Cu with radiation-induced defects in p-type silicon has been studied by Hall effect [196] and DLTS [169, 197, 198]. Several Cu-related levels have been observed either as a result of low-temperature copper diffusion into

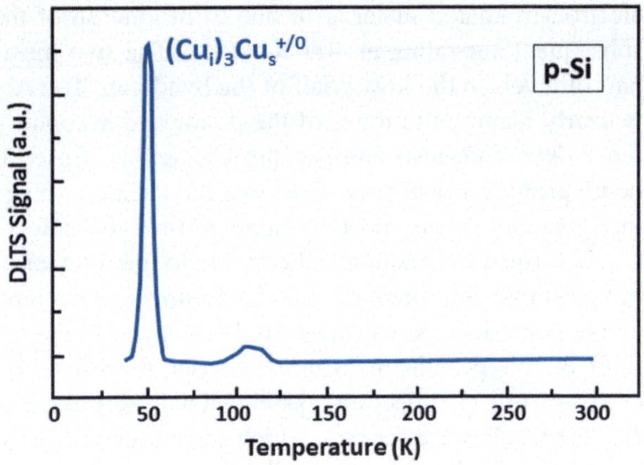

Fig. 5.26 The donor level in p-type silicon at $E_V + 0.1$ eV, which has been correlated with the 1.014 eV luminescence. Reproduced from Knack [177] with permission of Elsevier

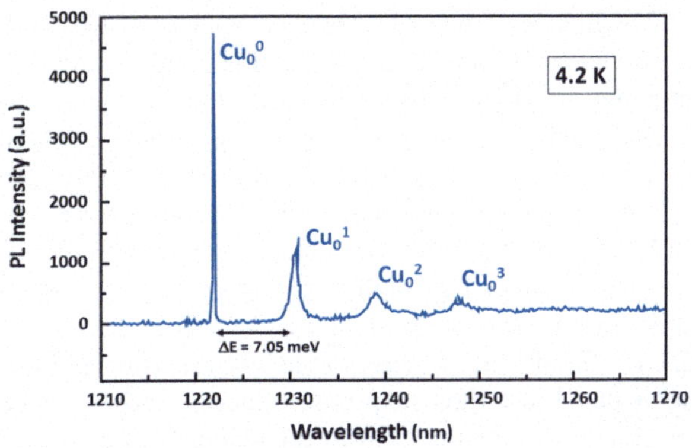

Fig. 5.27 Luminescence feature of the Cu-related complex measured at 4.2 K. The zero-phonon line and anti-Stokes phonon-replicas with a phonon energy of 7.05 meV can be seen. Reproduced from Knack [177] with permission of Elsevier

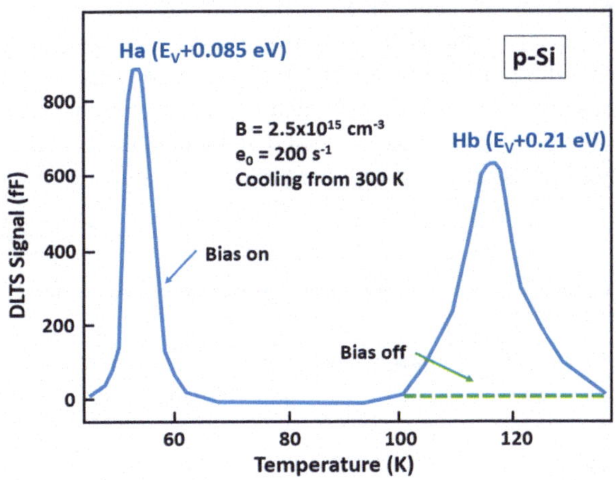

Fig. 5.28 DLT-spectra showing the Cu-related defects observed after Cu diffusion and fast quenching. Peak Hb corresponds with the M-defect. The emission rate window e_0 is 200 s^{-1}. Reproduced from Mesli and Heiser [172]. Copyright (2000) by The American Physical Society

electron-irradiated material or due to irradiation of the Cu-containing samples and subsequent annealing at 400 °C [198]. The two most prominent defects possess a pair of levels in the lower half of the band gap. The Arrhenius signatures of one pair is nearly identical to those of the donor and acceptor pair of Cu_s. The levels of the other defect are also close to the Cu_s levels. Based on the introduction rates and depth profiles, it was concluded that the vacancy-oxygen (VO or A) center serves as the precursor of the most abundant Cu-related defect [198]. It was concluded that Cu_s is formed in irradiated silicon due to the interaction with the VO centers via the rather stable intermediate CuVO complex. The interaction between copper and native point defects (vacancies and self-interstitials) was investigated by performing RTP on p-type silicon prior to copper in-diffusion, either under Ar atmosphere (Vs) or O_2 (I injection) [199]. Two new hole traps at $E_V + 0.35$ eV and $E_V + 0.32$ eV were formed, which were ascribed to the interaction between Cu and V and Cu and I, respectively [199].

As can be seen from Table 5.6, copper also interacts with hydrogen [177, 200–202] which modifies the electrical activity of Cu_s and even complete passivation is achieved when three hydrogen atoms bond to copper. As seen before, copper is also attracted by shallow acceptors and other metals in silicon. Taking into account the high diffusivity of copper at room temperature explains the limited stability of some of the associated deep levels even at room temperature storage [203].

The deep levels associated with Cu_s ***in germanium*** have been discussed in detail in Chap. 5 of [3]. What Fe is for Si, copper is for Ge: it is the pre-dominant thermal defect, introduced by high temperature heat treatments followed by a quench. It is, therefore, by far the most studied TM in germanium. The parameters derived from DLTS for the three acceptor levels are summarized in Table 5.7, corresponding with two hole traps in p-type material and one electron trap in n-type Ge. The presence of interstitial Cu in quenched Ge has been investigated by Kamiura and Hashimoto [207], showing the presence of two new Cu-related hole traps at $E_V + 0.09$ eV and at $E_V + 0.23$ eV. Based on the room temperature annealing behavior of the H(0.23) trap the level has been assigned to Cu_i, while the corresponding growth of the H(0.09) hole trap points to the existence of a Cu_s-Cu_i pair. Finally, it has been shown that Cu_s can be completely passivated by the attachment of three hydrogen atoms, with CuH and CuH_2 complexes giving rise to intermediate acceptor states in p-type material. The same can be achieved by pairing with Li.

Table 5.7 Deep-level parameters of the three acceptor levels of copper as derived from DLTS [204, 205]

Charge state	Energy level (eV)	Capture cross section (cm^2)		Temperature (K)
		Holes	Electrons	
$Cu_s^{0/-}$	$E_V + 0.037$	2.4×10^{-12}	–	22
$Cus^{-/2-}$	$E_V + 0.322$	2.0×10^{-13}	–	145
$Cu^{2-/3-}$	$E_C - 0.259$	4.0×10^{-16}	5.3×10^{-19}*	183

*Determined by Paramonova and Rzhanov [206], extrapolating from 300 K assuming a T^{-3} law

5.2.2.2 Electrical Activity of Precipitated Copper

The electrical activity of copper precipitates has been studied by different techniques, like EBIC [156, 208, 209], lifetime measurements [210–217] and DLTS [37, 107, 126, 218–220]. It has been reported that on large Cu precipitates, exceptionally strong contrast can be found in EBIC [156]. This has been assigned among others to the internal Schottky barrier present at the silicide/silicon interface [221]. This repels majority carriers and attracts minority carriers to the precipitate, strongly impacting on the recombination lifetime. The advantage of EBIC is that a direct correlation between an extended defect and its local impact on the lifetime can be established. However, the technique is only applicable to large precipitates and quantification is not always that straightforward. In p-Cz Si, diffusion lengths on copper colonies in the range of 0.3–0.4 μm have been derived from EBIC compared with 100 μm in uncontaminated substrates [209]. At the same time, large surface recombination velocities in the range of a few 10^5 cm/s were obtained, to be compared with $<10^4$ cm/s in uncontaminated regions [209].

There exists a strong barrier for homogeneous precipitation of copper. Therefore, copper prefers decorating existing extended defects in the interior of a wafer or, in absence of such sites, will tend to agglomerate at the surface. This means that in general the surface recombination lifetime will be more sensitive to copper contamination than the bulk minority carrier lifetime. This has for example been shown for p-type Cz Si, subjected to around 1×10^{12} at/cm^2 of copper and 950 °C, 35 min thermal oxidation [117]. The bulk lifetime is less affected, while $\tau_{eff} = 1/(1/\tau_s + 1/\tau_r)$ is dominated by the surface recombination, due to the segregation of the copper impurities at the Si/SiO_2 interface during the slow cooling from the oxidation temperature [210].

When copper is driven in at low annealing temperatures, typical for back-end metallization, little effect is found on the bulk diffusion length L_{diff} (Fig. 5.29a), while the surface recombination velocity s_r in Fig. 5.29b is degraded already after a 250 °C drive-in [216], as measured by ELYMAT. The degradation of the s_r at low thermal budget indicates the segregation of copper into the native oxide, while after a high-temperature anneal the surface recombination is further enhanced by the presence of copper at the Si/SiO_2 interface [216]. The degradation of the bulk lifetime after a 900 °C anneal indicates the effect of precipitation in the bulk [216].

Assuming that the minority carrier diffusion length L_{diff} is dominated by a single recombination center, one can derive an effective trap density N_{Teff} from it, through:

$$L_{diff}^2 = D_{min}\tau_r = \frac{D_{min}}{N_{Teff}\sigma_{min}v_{th}} \tag{5.19}$$

There is a strongly different behavior in the diffusion length of copper contaminated Si [212, 215]: in n-type Si, the density of recombination centers increases proportionally with the surface Cu concentration, whereas in p-type material, the density remains very low at low $[Cu]_{surf}$ and increases sharply at a Cu concentration of $\sim 10^{16}$ at/cm^2. This corresponds with the onset of the formation of Cu

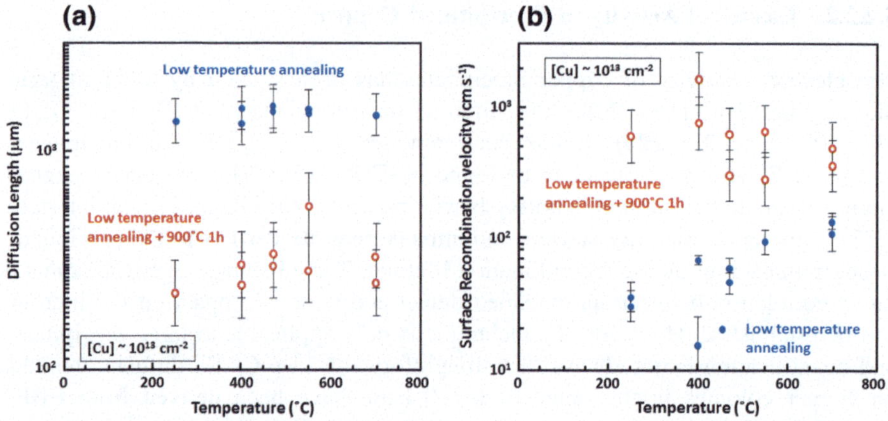

Fig. 5.29 Carrier bulk diffusion length (**a**) and surface recombination velocity (**b**) as a function of the first annealing temperature. Reproduced from Polignano et al. [216]. Copyright 2004 EDP Sciences

precipitates and depends on the p-type doping density. In other words, the recombination activity of copper in n-Si is much higher than in p-type silicon and this is related to the presence of copper precipitates, which start to form for a Fermi level at the charge neutrality level of the precipitates. This E_F is determined both by the acceptor concentration and by [Cu_i] (compensating donors).

In internally gettered (IG) p-type Cz Si, little impact is observed by the oxygen precipitates at copper concentrations above the critical level [215]. In other words, oxygen precipitates do not affect too much the precipitation of copper and, hence, are no favorable sites for heterogeneous nucleation. In n-type silicon copper recombination happens through multiple copper atoms as the trap concentration follows a power law on $[Cu]_{surf}$ [215]. It has, finally, been shown that the lifetime of copper-diffused Cz silicon also depends on the atmosphere during a high-temperature annealing, whereby an interstitial injecting anneal under O_2 creates less Cu-related recombination centers than a vacancy-injecting atmosphere like Ar or N_2 [216]. In addition, the presence of nitrogen in Cz:N material promotes the formation of a smaller density of large Cu precipitates, due to the existence of more grown-in oxygen precipitates in Cz:N material [217].

It has also been found that the lifetime of copper-doped p-type silicon is degraded after exposure to intense visible light [222]. This so-called Light-Induced Degradation is very important for photovoltaic applications and is usually ascribed to a defect complex related to oxygen and B. However, besides in B-doped material, it was also found in Ga-doped p-type Si [223] and in B- and Ga-doped mc-Si [224], which indicates that the origin is not related to the B-O complex. Instead, it is believed that in the case of copper-doped silicon, defect reactions can take place, transforming copper into a more recombination-active state under illumination (see also Sect. 5.2.3). The role of the light in this process is to shift the

Fermi level above the neutral level for precipitation or pairing (complexing) to overcome Coulomb repulsive forces.

As shown in 5.2.1.1, ***homogeneous precipitation*** of copper only takes place after ultra-fast quenching, yielding thin platelet precipitates. These correspond with a non-equilibrium structure, which are kinetically selected [37, 107, 156, 218–220]. The electrical properties of these early precipitates have been investigated in detail by DLTS, leading to a specific behavior. According to the spectra, represented in Fig. 5.30, they are broadened, suggesting the presence of a band of states in the band gap. The rather unusual trapezoidal shape of the peak has been explained by considering a two-dimensional density of states at the precipitate/matrix interface [37]. This can be represented by:

$$N_p(E) = N_{ip}/\Delta E,\ E_p - \Delta E_p \le E \le E_p \,\text{and}\, N_p(E) = 0, \text{else} \tag{5.20}$$

with N_{ip} being the total density of electrically active interface states, E_p the upper energy of the band and ΔE_p its width [37]. The dependence of the capture barrier is assumed as:

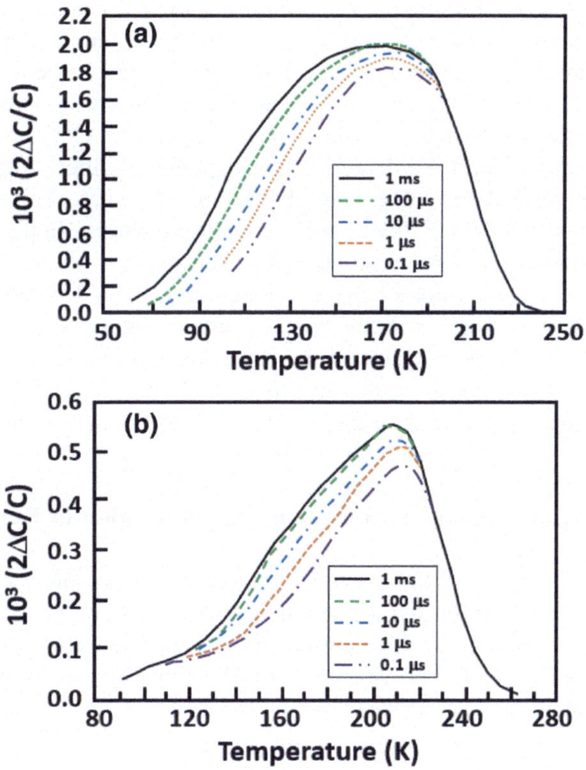

Fig. 5.30 Capacitance transient spectra plotted as the relative capacitance change $2\Delta C/C$ versus temperature. The curves show the filling pulse kinetics (pulse length t_p indicated) of deep levels associated with copper silicide precipitates **a** after quenching and **b** after additional ripening at 700 °C. The coincidence of the high temperature sides shows that band-like states have formed in both cases. Reproduced from Seibt et al. [107]. Copyright 1998 WILEY-VCH Verlag GmbH & Co

$$q\Phi = \alpha_C\left(F_p - F_p^{(N)}\right) \tag{5.21}$$

with $F_p^{(N)}$ the fraction of defect states occupied by electrons when the defect is electrically neutral, F_p the total occupation and α_C to be determined from a fit to the DLTS spectrum [37].

The ***band-like*** character of the states is derived from the evolution of the peak with trap filling time t_p in Fig. 5.30a. For a band of states, the high-temperature flank of the spectrum overlaps for all t_p, while the low-temperature part shifts towards lower T for increasing pulse duration [37], causing a corresponding shift of the peak maximum. Such behavior can be explained by considering the internal equilibration rate Γ_i. If Γ_i is much shorter than the inverse of the emission and recombination rates R_e^{-1}, R_c^{-1}, a band of states is present, like at the silicide/silicon interface. For very thin Cu_3Si platelets ($\sim$5 nm for a diameter of 240 nm) the DLTS line is only slightly broader than for a point defect [37]. At the same time, the DLTS-line rounds off and the line width decreases for increasing platelet diameter. This also occurs after a high-temperature ripening of the platelets; compare, e.g., Figure 5.30a (as-quenched) with Fig. 5.30b after ripening [107].

According to Fig. 5.30, both small, planar Cu precipitates (a), as well as copper colonies, formed after Ostwald ripening (b) exhibit a similar band-like behavior. It is the reason why the DLTS line has been assigned to the precipitate/silicon interface [37]. From DLTS, the band of states extends between $E_C - 0.15$ eV and $E_C - 0.45$ eV, with a neutral occupation fraction of 0.72 and a capture barrier $\delta E_C = 0.58(F\text{-}F^{(N)})$ which results from the Coulomb interaction of free charge carriers with the defect if its occupation F differs from the occupation $F^{(N)}$ of the neutral defect. The neutral level of $E_C - 0.2$ eV thus corresponds with a large occupation and, hence, a large repulsive barrier. In the band, the states change from donor-like (low F) to acceptor-like (above $F^{(N)}$, i.e., $E_C - 0.2$ eV) and, therefore, Cu_3Si precipitates behave as amphoteric extended defects. The large recombination activity stems from the band of states, making efficient recombination possible and from the barrier around the precipitates, forming a potential well for the minority carriers. As a result, the diffusion length in copper-precipitated silicon is only slightly higher than the average distance between the precipitates.

5.2.2.3 Copper Activation and Passivation of Extended Defects

As shown in the previous paragraph, copper precipitation prefers heterogeneous nucleation, so that it easily decorates pre-existing extended defects like dislocations and stacking faults. Consequently, the presence of copper precipitates on extended defects will modify their electrical activity by changing the states in the band gap. In most cases, enhanced recombination activity is found, following copper decoration.

This is shown by the different types of EBIC contrast behavior in function of the sample temperature, reported in Fig. 5.31 for dislocations with different degrees of copper contamination [156, 225]. In the case of heavily decorated (and activated) dislocations, a high contrast value is obtained from low to room temperature (type H1 and H2). For 'clean' dislocations, significant contrast is only found at low temperatures (type L1 and L2), while mixed behavior is observed for intermediate contamination levels. This also corresponds with a shift of the recombination levels from shallow, near band edge, to close to mid gap [156, 226]. This can also be observed in the temperature dependence of the EBIC contrast of Frank partials bounding a SF [138, 139]. For both uncontaminated samples and Cu-contaminated at 1000 °C samples slowly cooled down, the contrast decreases for increasing temperature. The opposite behavior is observed for contaminated samples after fast cooling. Similar activation has been revealed on grain boundaries [227–232], which is of concern in mc-Si solar cells. The impact of copper contamination on 'clean' GBs has been studied for bonded (110)/(100) interfaces [233]. It was derived that the presence of copper at the interface transforms the localized states at clean GBs into band-like states, with much larger hole capture cross sections.

Copper decoration also activates misfit dislocations at the relaxed SiGe/Si interface, resulting in a drastic increase in the EBIC contrast from low to high temperatures [234], depending on the amount of contamination.

Besides activation of extended defects, copper may also passivate both extended [235] and point defects [236–238]. In this sense, it behaves like other fast diffusing interstitial ions like hydrogen and lithium [236].

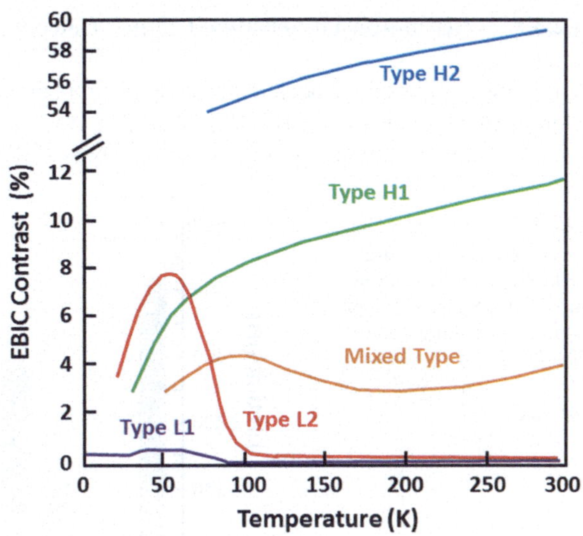

Fig. 5.31 Temperature dependence of the EBIC contrast of dislocations in Si. Different types of behavior reflect the effect of different concentrations of deep traps in the dislocation core (or its vicinity). The high concentration of deep traps in the dislocation is usually a result of dislocation contamination. Reproduced from Schröter et al. [156] with permission of Elsevier

5.2.3 Detection of Copper

Given the strong impact of copper on the materials' and device electrical characteristics, several chemical techniques have been developed for the sensitive detection of copper on/in silicon wafers [38, 239–243]. However, most of these techniques are destructive and can, therefore, not be implemented for in-line contamination monitoring. In many cases, the analysis is also lengthy, with a limited sensitivity and spatial resolution. Detection methods based on an electrical parameter are generally fast and can be non-destructive. In the case of lifetime measurements, there is the mapping feature, enabling local detection of copper with μm resolution on a wafer. However, lifetime measurements are integral and, therefore, sensitive to all types of recombination centers. Another issue is that in many cases, depending on the contamination level and the wafer history, only a small fraction (typically 10^{-4}) of the copper in the wafer is in an electrically active form. This lowers the detection limit and furthermore, makes a quantitative analysis rather cumbersome, unless dedicated techniques are developed. In addition, standard methods like DLTS reveal only a fraction of the copper present in the contaminated wafer, so that it is not at all very useful for copper detection. In the next, it will be shown that these problems have been tackled: the analysis of the injection level dependence of the minority carrier recombination lifetime and the impact of light enables identification of copper-related recombination centers. In addition, the TID method, in combination with an appropriate pre heat treatment allows a sensitive detection of copper contamination down to the 10^{11} cm^{-3} level.

5.2.3.1 Lifetime-Based Sensitive Copper Detection

Quite some studies have been reported on the impact of copper on the recombination lifetime of silicon wafers [210–217, 244–251]. As can be seen from a comparison of the high-injection lifetime in Fig. 5.32 for p-type Cz Si, doped in the melt with TM

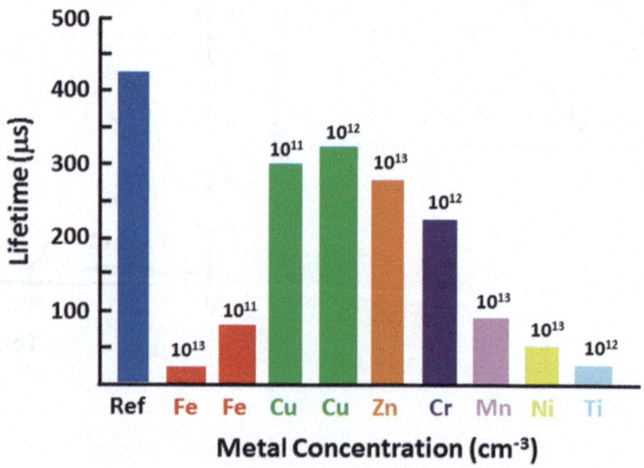

Fig. 5.32 High excitation level dependence of the recombination lifetime in metal-doped p-type silicon wafers (after Kurita and Shingyouji [246]. Copyright 1998 The Japan Society of Applied Physics)

atoms, copper is one of the least efficient recombination centers [246]. On the positive side, this means that silicon processing can tolerate relatively high levels of copper contamination ($\sim 10^{12}$–10^{13} cm^{-3}). At the same time, its detection by electrical techniques is less sensitive than for other metals (Fe, Ni, Ti,…), given its lesser impact.

A detailed SPV analysis by Ramappa and Henley [247] for p-type copper-contaminated Cz Si as a function of annealing temperature revealed a complex behavior of the diffusion length in function of T_a (Fig. 5.33a) and the sample history (Fig. 5.33b). While the interpretation of the responsible centers should be carefully reconsidered in view of the later insights, it is evident from Fig. 5.33 that the diffusion length first increases with annealing temperature up to about 150 °C, followed by a steep decrease. This increase could indicate either a passivation of other recombination centers or a removal of already-present Cu-related recombination centers. This can equally be by a further agglomeration with Cu_i or by a dissociation of copper-related PDs and PD complexes. The reduction of the lifetime indicates the transformation of Cu_i and other Cu-related PDs into a more recombination active form by further complexing or by the dissociation of the so-called (Cu_i–Cu_s) complex into Cu_s by removal of the three surrounding Cu_i ions.

As mentioned before, the injection level dependence of the recombination lifetime can be used to identify a specific metal impurity. As an example, Fig. 5.34 shows the correlation between the carrier lifetime in p-type and n-type Si for Cu, Fe and stainless-steel (SS) elements [245, 252]. In the case of Cu-contaminated Si, a ratio of

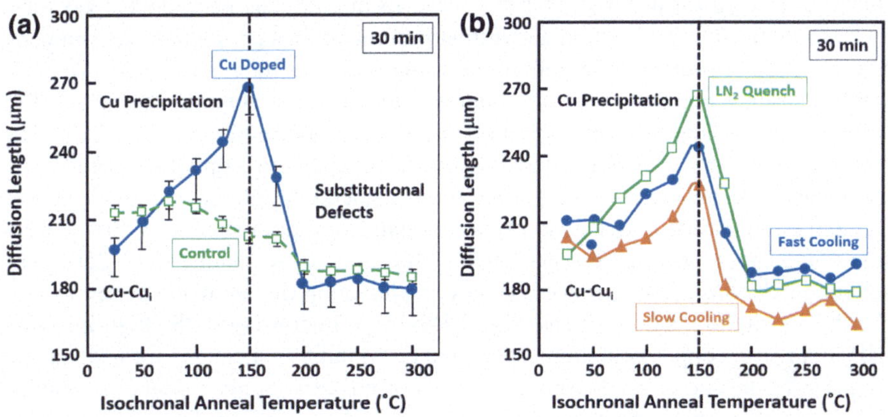

Fig. 5.33 Isochronal annealing results of copper contaminated Cz samples. **a** Minority carrier diffusion lengths measured after successive annealing for 30 min at each successive temperature. Increase in the diffusion length indicates copper passivation. Copper in-diffusion was performed at 600 °C. **b** Same samples cooled at different rates following in-diffusion at 600 °C. Liquid nitrogen quenching shows the greatest change in diffusion length following an isochronal anneal because of the presence of a higher concentration of Cu complexes. Reproduced from Ramappa and Henley [247]. Copyright 1998 AIP Publishing

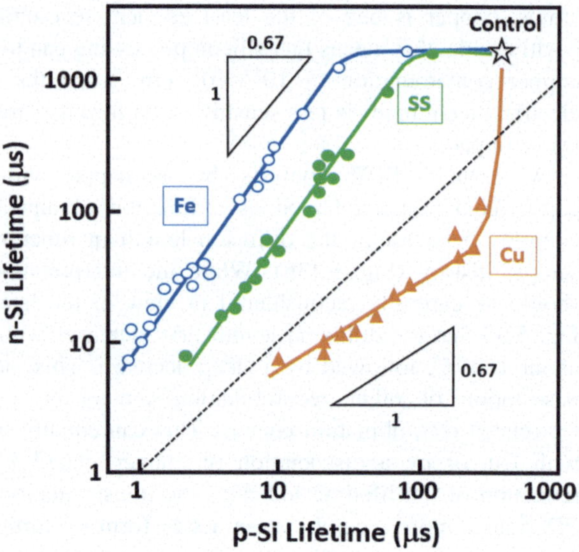

Fig. 5.34 N-type Si recombination lifetime versus p-type Si recombination lifetime. Reproduced from Itsumi et al. [245]. Copyright 1997 AIP Publishing

1:0.67 exists for the lifetime in p- and n-type Si. The opposite occurs for Fe: 0.67 (p) versus 1 (n-Si). This could help to identify the presence of copper contamination, with as a drawback the need for two types of wafers [245, 252].

This has spurred other groups to look for a sensitive and clear identification method for the detection of trace amounts of copper introduced in p-type Si during high temperature processing, based on carrier lifetime [248–250]. One such approach relying on the so-called *Q-V* SPV technique, has been developed for a routine monitoring of oxidation and diffusion furnaces in a production environment. The presence of copper in the bulk of the wafer and at its interface or in the oxide is detected at the same time. One of the aims is to use the total oxide charge as a fast and sensitive indicator for the presence of copper in oxidized p-Si [248–250]. The bulk lifetime of the oxidized wafers is assessed by open circuit photovoltage decay, while the oxide and interface properties are evaluated through the total oxide charge Q_{tot} and the interface trap density D_{it} at mid-gap from quasi-static *Q-V* SPV curves employing the noncontact corona-oxide semiconductor technique [248–250]. The resulting behavior versus surface copper contamination (in at/cm^2) is represented in Fig. 5.35a (high-injection τ_r) and Fig. 5.35b (D_{it}). Interestingly, the high-injection (HI) lifetime in copper-contaminated p-type Cz Si initially increases for low surface concentrations, due to the passivation effects of pre-existing recombination centers. This is a feature, typical for Cu. Only above 3×10^{13} cm^{-3}, τ_r becomes degraded by copper-related deep level centers [250].

It is also evident from Fig. 5.35b that n-type Si is more sensitive to surface diffusion of Cu than p-Si. In p-type material, the out-diffusion to the Si-SiO_2 interface is hindered by the near-surface repulsive electrical potential, induced by the positive oxide charges. The increase of the density of interface states occurs above a few times 10^{10} at/cm^2 in n-type Si, while in p-type, D_{it} at mid-gap only

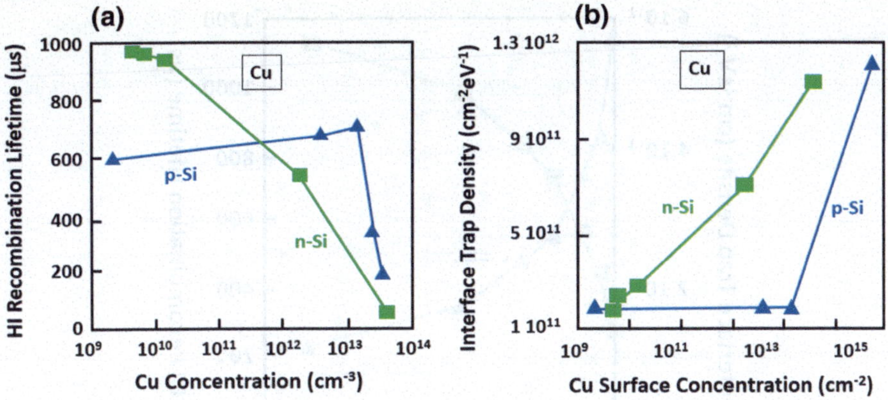

Fig. 5.35 **a** High-injection recombination lifetime HI-τ_r versus copper surface contamination prior to in-diffusion for p-type and n-type Si. **b** Interface trap density D_{it} at mid-gap versus copper concentration prior to in-diffusion for p- and n-type Si. Reproduced from Boehringer et al. [250] by permission of ECS-The Electrochemical Society

slightly increases up to a surface copper concentration of about 1×10^{13} at/cm^2 [250]. Note that τ_r is much more sensitive to the copper contamination level in n- than in p-type Si (Fig. 5.35a).

For the non-contact *Q-V* SPV measurements, one can also derive Q_{tot}, which captures charged defects in the oxide bulk. In this case, the response is one decade more sensitive in p-Si and can be used as a fast indicator of copper contamination [250]. The total oxide charge represents a sensitive, however, unspecific indicator for the presence of trace amounts of copper in oxidized p-Si. The more detailed analysis of silicon and interface properties reveals a characteristic copper fingerprint. Illumination with high-intensity light induces the transformation of copper into a defect type with considerably higher recombination activity. This effect was first described by Ramapa et al. [253, 254]. The transformation process has an activation energy of 0.419 eV in B-doped Si [254] and corresponds with the change of phase in the dominant copper-related recombination center. This light-induced change is permanent and irreversible in the case of copper. It also reduces the barrier at the surface, so that the properties of the Si/SiO_2 interface and Q_{tot} are affected too [250]. The evolution with illumination time is represented in Fig. 5.36 [249]. According to Fig. 5.37, the increase of both the total oxide charge and D_{it} at mid-gap increases much faster under strong illumination. In other words, the detection sensitivity of copper can be significantly enhanced by applying so-called optical activation of copper [250]. Furthermore, it is recommended to monitor simultaneously Q_{tot}, D_{it} and HI-τ_R to distinguish the different copper species present in the oxidized wafer. Moreover, discrimination between Fe and Cu can only be performed under high-injection lifetime conditions [250]. Indeed, FeB pairs are very recombination active under low-to-medium injection, but less under high-injection. Copper remains highly recombination active under HI conditions [250].

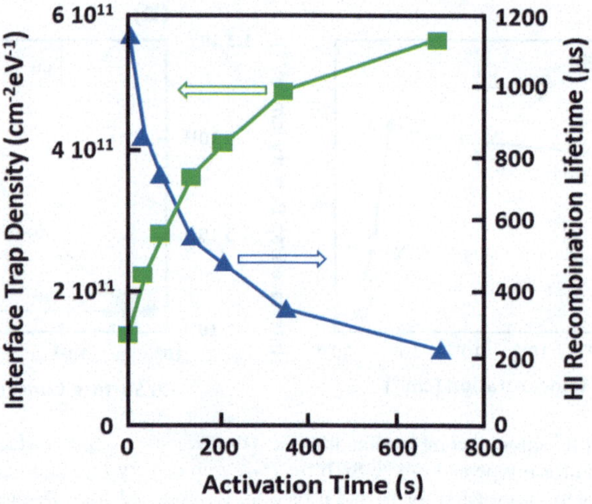

Fig. 5.36 Interface trap density D_{it} at mid-gap (left scale) and recombination lifetime HI-τ_r (right scale) versus activation time for oxidized p-type silicon with copper. Reproduced from Boehringer et al. [250] by permission of ECS-The Electrochemical Society

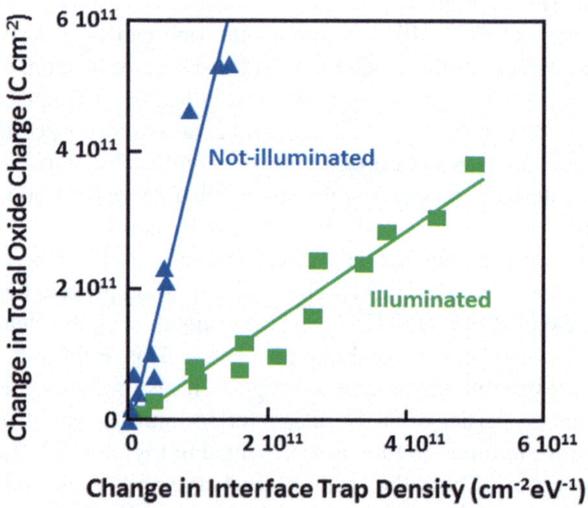

Fig. 5.37 Change of the total oxide charge ΔQ_{tot} versus the change of the interface trap density ΔD_{it} induced by illumination (36–600 s) of oxidized p-type Si contaminated with copper concentrations between 1×10^{10} at/cm^2 and 4×10^{12} at/cm^2. For comparison, ΔQ_{tot} versus ΔD_{it} for copper at distinct concentrations without illumination prior to the measurement is reproduced in the same graph (filled triangle). Reproduced from Boehringer et al. [250] by permission of ECS-The Electrochemical Society

The data represented in Figs. 5.35, 5.36 and 5.37 are consistent with the following scenario for the behavior of trace amounts of copper in oxidized p-Si. In the medium contamination range, after cooling to ambient temperature, copper resides at electrically weakly active interstitial sites in the substrate and as positive ions in the bulk of the oxide. Neither precipitation nor out-diffusion occurs to a significant extent. Therefore, the bulk recombination lifetime and the interface trap density are only weakly affected. However, the copper impurity can be recognized from the presence of bulk oxide charge. At ambient temperature, this distribution is stable over a long period of time (hours to days). Above a critical concentration, however, copper starts to out-diffuse to the Si-SiO_2 interface, which results in a simultaneous steep increase in D_{it} and Q_{tot} at surface concentrations exceeding 1×10^{13} at/cm^2 [250]. Simultaneous copper precipitation in the bulk causes the formation of defect complexes with high recombination activity and a drastic decrease in τ_r is observed. Most important for a sensitive copper detection in routine contamination monitoring, similar reaction paths can already be activated at considerably lower contamination levels by illumination with high intensity light. Whether out-diffusion or precipitation prevails upon optical activation depends on the detailed experimental conditions, for example, on the density and distribution of oxygen precipitates [251, 255–259]. On the other hand, it has been observed that optical activation of copper-related centers has little effect on the lifetime in n-type silicon. This can be explained by the fact that copper is already in a stable, precipitated phase, which operates at the same time as a very efficient recombination center [254].

A further optimization of the lifetime assessment of copper contamination in p-type Cz silicon wafers was worked out [255–259]. Oxygen precipitates are usually not very efficient gettering centers for copper due to the compressive strain of both SiO_2 and Cu_3Si precipitates. In the presence of a high density of small SiO_x precipitates, low levels of Cu_i do not precipitate at room temperature [255], but can be gettered at high T [141]. However, this process is stimulated at RT by exposure to intense illumination for some time, whereby it is believed that the optical activation may decrease the barrier for precipitation by rendering Cu_i neutral or by screening the positive charge of the oxide precipitates by the injected high free carrier density. It is shown that in order to achieve a strong precipitation effect under high-intensity illumination, the Fermi level has to be above $E_C - 0.23$ eV, close to the neutrality level for copper precipitation [257]. Copper precipitation at a high density of small oxygen precipitates may be exploited by recombination lifetime measurements, where an acceleration of the recombination rate is detected after optical activation even at low copper concentrations [255]. The small oxygen precipitates can be formed by a two-step annealing: a nucleation step at 625 °C for 4 h, followed by a ripening step at 800 °C for 4 h [255]. Based on Ham's theory for precipitation (5.5), (5.6), a value for t_{prec} of 1650 ± 50 s was derived [255]. The impact of both illumination time and the presence of a large density of oxide precipitates on the recombination lifetime for low levels of copper-contaminated p-type Cz Si is illustrated in Fig. 5.38 [259]. The correlation between the recombination lifetime measured by microwave absorption (μWA) and the copper concentration is represented in Fig. 5.39 [259]. As can be seen, the detection limit for

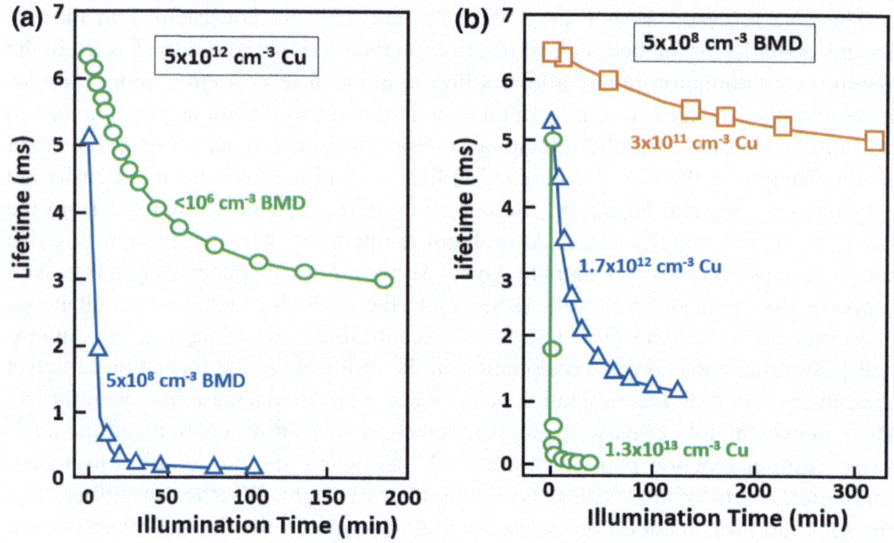

Fig. 5.38 Measured lifetime as a function of light illumination time. **a** The wafers have the same amount of copper (5×10^{12} cm^{-3}) but different density of oxide precipitates, called Bulk Micro Defect (BMD) density. **b** The wafers have the same density of oxide precipitates (5×10^{8} cm^{-3})

Fig. 5.39 Correlation between copper concentration and light activated lifetime in three different groups of wafers, with varying density of oxide precipitates: homogenized (circles), low BMD (triangles) and high BMD (squares) (after Väinölä et al. [259]. Copyright 2005 AIP Publishing)

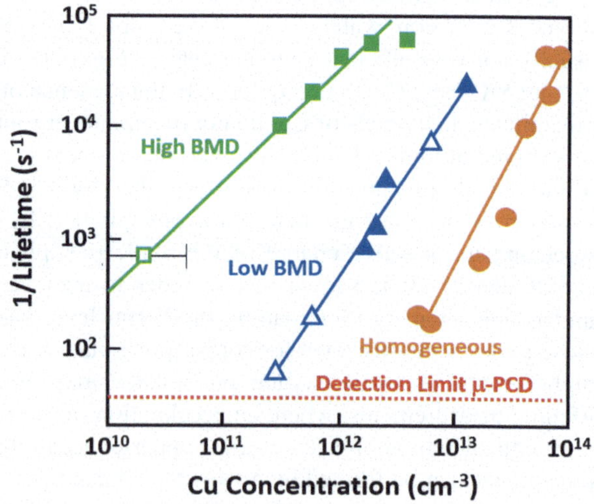

copper can be improved by increasing the bulk micro-defect density (BMD) to the level of about 1×10^{9} cm^{-3} in the case of Fig. 5.39.

Alternatively, one can use the out-diffusion of copper combined with, e.g., μWA recombination lifetime measurements to detect small amounts of copper in ***p-type*** silicon [251, 255]. On the one hand, deposition of a positive corona charge on the surface will create a barrier for out-diffusion of Cu_i, so that the lifetime remains unchanged for days, i.e., copper is kept in solution without out-diffusion. It has been estimated that a surface band bending of 0.2 eV gives a copper out-diffusion time constant of 1 day [251], with about +1 μC/cm^2 corona charge deposited on oxide (>5 nm). Changing the polarity of the corona charge allows copper out-diffusion to the surface, which can be monitored by μWA after optical activation for 1 h or several h [255]. Due to out-diffusion, a reduction of the recombination rate can be observed. Conversely, changing the sign of the corona charge reverses the situation: copper diffuses in again, increasing the recombination rate [255]. It also implies that copper at the surface does not form stable bonds (silicides) as the out-diffusion can be reversed by changing the corona surface charge. For low-level copper detection, a high intensity spot-like illumination is preferred [251]. The illumination time needed to complete the reactions depends on the copper contamination level, the oxide precipitate density and the light intensity; it can vary between 10 min and several h. The copper concentration can then be derived from [251]:

$$[Cu] = 5 \times 10^{11} \left(\frac{1}{\tau_{final}} - \frac{1}{\tau_{init}} \right)^{1/2} (\mathrm{cm}^{-3}) \quad (5.22)$$

In order to distinguish Cu from Fe contamination, a high-injection lifetime measurement like μWA is required [251]. The unique fingerprint of copper is that after light exposure, the HI lifetime only drops for Cu, while it increases for Fe. Moreover, the light-induced change is irreversible in the case of copper [251].

Based on an accelerated LID effect, a method has been proposed for the detection of copper in p-type silicon [260]. It consists of a combined heating at 120 °C and a 0.5 Sun Xenon lamp exposure. The responsible trap level for Cu LID has been investigated by I- and TDLS [261], placing the dominant recombination level at $E_C - E_T = 0.48 - 0.62$ eV with a moderate donor-like capture asymmetry ($k = 1.2 - 2.6$) [261]; an additional shallow energy level located at $E_C - E_T = 0.1 - 0.2$ eV was required for an optimal fit to the experimental data. From a detailed study of the formation kinetics of the Cu-LID centers in oxidized p-Si wafers, it has been concluded that the defect formation is limited by copper diffusion [262], further strengthening the idea that copper precipitation is the root cause of the phenomenon. Further confirmation of the relationship with copper precipitation in the bulk silicon has been derived from the fact that Cu-LID can be suppressed by applying a negative corona charge to the surface of oxidized p-type crystalline [263] and solar-grade silicon [264, 265]. Alternatively, the application of an Al_2O_3 passivation layer, containing negative fixed charge achieves the same goal [265]. The suppression can be explained by considering the fact that positive copper ions will be attracted by the negative surface charge, and, provided sufficient time for diffusion to the surface is allowed, no copper precipitation in the bulk will occur upon illumination. A detailed model for Cu LID in p-type Cz silicon has been established, considering Cu precipitation as the main mechanism [266] and yielding a satisfactory comparison with a wide range of experimental data [267].

5.2.3.2 Transient Ion Drift Analysis of Copper in Silicon

The high mobility of copper even at room temperature can be exploited to develop a dedicated analysis method, based on the ion drift in the electric field of a Schottky barrier on p-type silicon [268–273]. The ion transport is monitored through the capacitance transient as a function of time, obeying [269]:

$$\Delta C(t) = [C(0) - C(\infty)]\left(1 - \exp\left(-\frac{t}{\tau_d}\right)\right) \tag{5.23}$$

with the time constant τ_d given by:

$$\tau_d = \frac{k_B T \varepsilon_{Si} \varepsilon_0}{q^2 D N_a} \tag{5.24}$$

Equation (5.23) is valid under the following conditions:

(i) linearity of the electric field in the depletion region, valid only for low impurity (copper) concentrations ([Cu] $\ll N_a$)
(ii) negligible transfer of ions out of the depletion region
(iii) no precipitation in the space charge region
(iv) the ion and the majority carrier charge state are of the same sign (p-type Si)

In order to make the technique non-destructive one can use a mercury probe (after native oxide removal) instead of depositing a SB [272]. This results in a spatial resolution in the range of 1 mm. The detection limit is in the low 10^{11} cm^{-3} (depends on the substrate doping density).

For enhancing the sensitivity of the TID technique, one can optimize the pre-heat treatment in order to transform most of the copper in the highly mobile interstitial form, which can be detected by drift at room temperature [274]. From Fig. 5.40, it is observed that 100% activation (up to the B doping level of the p-type material) of interstitial copper can be achieved after 1 min RTP at 600 °C [274]. The sample cooling rate is between 50 and 100 K/s to avoid excessive bulk precipitation. Below 600 °C, the solid solubility limit of copper is followed. The activation is much less effective for furnace annealing In addition, exposure to UV light during RTP may accelerate the copper activation [274]. Figure 5.40 represents the in-diffusion kinetics from the surface [274]. In the same way, precipitate dissolution studies have been undertaken on samples which were contaminated at 900 °C for 1 min and subsequently annealed at 200 °C for 30 min. At this low temperature, the high copper supersaturation triggers precipitation or complex formation and reduces the TID signal by two orders of magnitude. The dissolution kinetics is then investigated by submitting the wafers to a final 1 min high temperature treatment and quench, with T varying from 450 to 900 °C. According to the results of Fig. 5.41, precipitate dissolution starts at 400 °C, which is consistent with what is expected from a diffusion limited model [274]. This also implies that copper which was

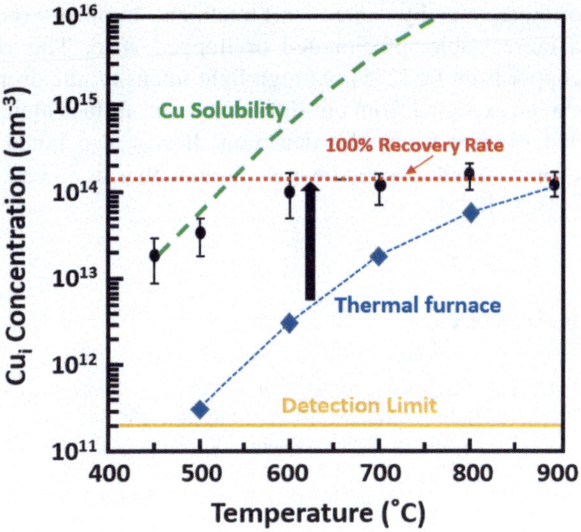

Fig. 5.40 Copper concentration after 1 min isochronal annealing as a function of annealing temperature (circles: RTP; diamonds: furnace anneal FA—10 min). Reproduced from Belayachi et al. [274] with permission of Elsevier

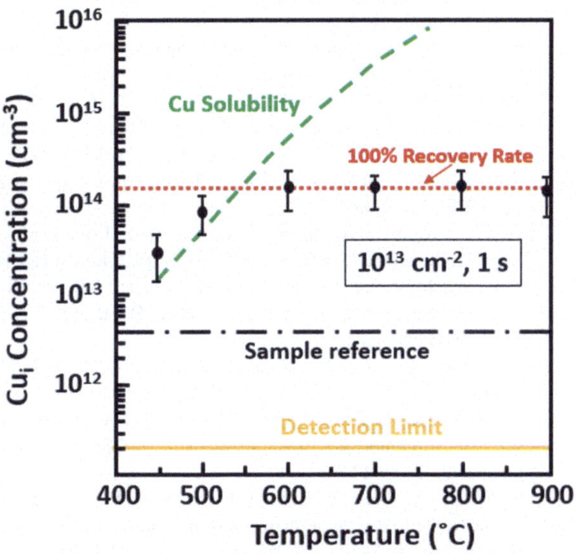

Fig. 5.41 Copper concentration as measured by TID 5 min after the quench as a function of the dissolution temperature. Reproduced from Belayachi et al. [274] with permission of Elsevier

precipitated during processing can be redissolved and measured by TID after an RTP step at 600 °C for 1 min.

Besides "thermal activation" of copper, which can be defined as a thermally assisted transfer of trapped or precipitated copper into its interstitial form, high-intensity light exposure can achieve the same goal [275]. It should be mentioned also that the effects induced by light exposure in low-level Cu contaminated wafers are irreversible, i.e., the recombination lifetime cannot be restored to its original value by some (heat) treatment. In other words, copper is transformed into a more stable, precipitated or trapped state. This differentiates the behavior of copper from Fe [275]. At high light intensity, the drop in mobile Cu_i is higher than can be expected from out-diffusion to the surface alone. A large fraction precipitates and escapes from TID detection. It was also noted that the thermal stability of copper complexes can be quite high, with a significant fraction surviving RTA up to 900 °C [275].

References

1. H.H. Woodbury, W.W. Tyler, Triple acceptors in germanium. Phys. Rev. **105**, 84–92 (1957). https://doi.org/10.1103/PhysRev.105.84
2. G.D. Watkins, P.M. Williams, Vacancy model for substitutional Ni^-, Pd^-, Pt^-, Au^0 in silicon. Phys. Rev. B **52**, 16575–16580 (1995). https://doi.org/10.1103/PhysRevB.52.16575
3. C. Claeys, E. Simoen (eds.), *Germanium-Based Technologies—From Materials to Devices* (Elsevier, 2007)
4. A.A. Istratov, H. Hieslmair, E.R. Weber, Iron and its complexes in silicon. Appl. Phys. A **69**, 13–44 (1999). https://doi.org/10.1007/s003399900059
5. A.A. Istratov, H. Hieslmair, E.R. Weber, Iron contamination in silicon technology. Appl. Phys. A **70**, 489–534 (2000). https://doi.org/10.1007/s003390000458
6. G.W. Ludwig, H.H. Woodbury, Electron spin resonance in semiconductors. Solid State Phys. **13**, 223–304 (1962). https://doi.org/10.1016/S0081-1947(08)60458-0
7. Y.H. Lee, R.L. Kleinhenz, J.W. Corbett, EPR of a thermally induced defect in silicon. Appl. Phys. Lett. **31**, 142–144 (1977). https://doi.org/10.1063/1.89630
8. E.R. Weber, Transition metals in silicon. Appl. Phys. A **30**, 1–22 (1983). https://doi.org/10.1007/BF00617708
9. H. Feichtinger, J. Waltl, A. Gschwandtner, Localization of the Fe^0-level in silicon. Solid State Commun. **27**, 867–871 (1978). https://doi.org/10.1016/0038-1098(78)90194-1
10. S. Greulich-Weber, J.R. Niklas, E.R. Weber, J.M. Spaeth, Electron nuclear double resonance of interstitial iron in silicon. Phys. Rev. B **30**, 6292–6299 (1984). https://doi.org/10.1103/PhysRevB.30.6292
11. G. Weyer, Defects in semiconductors—results from Mössbauer spectroscopy. Hyperfine Interact. **177**, 1–13 (2007). https://doi.org/10.1007/s10751-008-9607-y
12. U. Wahl, J.G. Correia, E. Rita, J.P. Araújo, J.C. Soares, The ISOLDE Collaboration, Lattice sites of implanted Fe in Si. Phys. Rev. B **72**, 014115/1-8 (2005). https://doi.org/10.1103/physrevb.72.014115
13. U. Wahl, J.G. Correia, E. Rita, J.P. Araújo, J.C. Soares, The ISOLDE Collaboration, Fe and Cu in Si: lattice sites and trapping at implantation-related defects. Nucl. Instr. Meth. Phys. Res. B **253**, 167–171 (2006). https://doi.org/10.1016/j.nimb.2006.10.0
14. D.J. Silva, U. Wahl, J.G. Correia, J.P. Araújo, Influence of n^+ and p^+ doping on the lattice sites of implanted Fe in Si. J. Appl. Phys. **114**, 103503/1-9 (2013) https://doi.org/10.1063/1.4819210
15. S.H. Muller, G.M. Tuynman, E.G. Sieverts, C.A.J. Ammerlaan, Electron paramagnetic resonance on iron-related centers in silicon. Phys. Rev. B **25**, 25–40 (1982). https://doi.org/10.1103/PhysRevB.25.25
16. T. Mchedlidze, M. Suezawa, Properties of an iron-vacancy pair in silicon. Jpn. J. Appl. Phys. **41**, 7288–7292 (2002). https://doi.org/10.1143/JJAP.41.7288

17. H.P. Gunnlaugsson, G. Weyer, N.E. Christensen, M. Dietrich, M. Fanciulli, K. Bharuth-Ram, R. Sielemann, A. Svane, The ISOLDE Collaboration, On the kinetics of the formation of interstitial Fe-vacancy pairs in silicon at high temperatures. Physica B **340-342**, 532–536 (2003). https://doi.org/10.1016/j.physb.2003.09.141
18. G. Weyer, S. Degroote, M. Fanciuli, V.N. Fedoseyev, G. Langouche, V.I. Mishin, M. van Bavel, A. Vantomme, Mössbauer spectroscopy of Fe in silicon with the novel laser-ionized $^{57}Mn^{+}$ ion beam at ISOLDE. Mater. Sci. Forum **258–263**, 437–442 (1997). https://doi.org/10.4028/www.scientific.net/MSF
19. D. Gilles, W. Schroter, W. Bergholz, Impact of the electronic structure on the solubility and diffusion of 3*d* transition metal elements in silicon. Phys. Rev. B **41**, 5770–5782 (1990). https://doi.org/10.1103/PhysRevB.41.5770
20. H.P. Gunnlaugsson, G. Weyer, N.E. Christensen, M. Dietrich, M. Fanciulli, K. Bharuth-Ram, R. Sielemann, A. Svane, The ISOLDE Collaboration, Mössbauer spectroscopy on Fe impurities in germanium. Physica B **340-342**, 537–540 (2003). https://doi.org/10.1016/j.physb.2003.09.140
21. S. Decoster, S. Cottenier, B. De Vries, H. Emmerich, U. Wahl, J.G. Correia, A. Vantomme, Transition metal impurities on the bond-centered site in germanium. Phys. Rev. Lett. **102**, 065502/1-4 (2009). https://doi.org/10.1103/physrevlett.102.065502
22. M. Höhne, U. Juda, J. Wollweber, D. Schulz, J. Donecker, A. Gerhardt, Electron paramagnetic resonance of phosphorus, platinum, and iron in Float Zone $Si_{1-x}Ge_x$ crystals. Mater. Sci. Forum **196–201**, 359–364 (1995). https://doi.org/10.4028/www.scientific.net/MSF.196-201.359
23. H. Reiss, C.S. Fuller, F.J. Morin, Chemical interaction among defects in germanium and silicon. Bell Syst. Techn. J. **3**, 535–636 (1956). https://doi.org/10.1002/j.1538-7305.1956.tb02393
24. H. Lemke, Energieniveaus und Bindungsenergien von Ionenpaaren in Silizium. Phys. Status Solidi A **76**, 223–234 (1983). https://doi.org/10.1002/pssa.2210760126
25. F.S. Ham, Theory of diffusion-limited precipitation. J. Phys. Chem. Solids **6**, 335–351 (1958). https://doi.org/10.1016/0022-3697(58)90053-2
26. J. Tan, D. Macdonald, F. Rougieux, A. Cuevas. Accurate measurement of the formation rate of iron-boron pairs in silicon. Semicond. Sci. Technol. **26**, 055019/1-5 (2011). https://doi.org/10.1088/0268-1242/26/5/055019
27. L.J. Geerligs, D. Macdonald, Dynamics of light-induced FeB pair dissociation in crystalline silicon. Appl. Phys. Lett. **85**, 5227–5229 (2004). https://doi.org/10.1063/1.1823587
28. O.V. Feklisova, A.L. Parakhonsky, E.B. Yakimov, J. Weber, Dissociation of iron-related centers in Si stimulated by hydrogen. Mater. Sci. Eng. B **71**, 268–271 (2000). https://doi.org/10.1016/S0921-5107(99)00388-8
29. C.K. Tang, L. Vines, B.G. Svensson, E.V. Monakhov, Interaction between hydrogen and the Fe-B pair in boron-doped p-type silicon. Appl. Phys. Lett. **99**, 052106/1-3 (2011). https://doi.org/10.1063/1.3619848
30. T. Bartel, F. Gibaja, O. Graf, D. Gross, M. Kaes, M. Heuer, F. Kirscht, C. Möller, K. Lauer, Dynamics of iron-acceptor-pair formation in co-doped silicon. Appl. Phys. Lett. **103**, 202109/1-4 (2013). https://doi.org/10.1063/1.4830227
31. C. Möller, T. Bartel, F. Gibaja, K. Lauer, Iron-boron pairing kinetics in illuminated p-type and in boron/phosphorus co-doped n-type silicon. J. Appl. Phys. **116**, 024503/1-8 (2014). https://doi.org/10.1063/1.4889817
32. D. Macdonald, A. Cuevas, L.J. Geerligs, Measuring dopant concentrations in compensated p-type crystalline silicon via iron-acceptor pairing. Appl. Phys. Lett. **92**, 202119/1-3 (2008). https://doi.org/10.1063/1.2936840
33. M. Höhne, U. Juda, H. Riemann, J.-M. Spaeth, S. Greulich-Weber, Distant iron-shallow-donor pairs in silicon detected by electron paramagnetic resonance. Phys. Rev. B **49**, 16999–17006 (1994). https://doi.org/10.1103/PhysRevB.49.16999
34. T. Mchedlidze, M. Suezawa, An iron-phosphorus pair in silicon. J. Phys. Condens. Matter **16**, L79–L84 (2004). https://doi.org/10.1088/0953-8984/16/8/L02

35. J.J. van Kooten, E.G. Sieverts, C.A.J. Ammerlaan. Electron paramagnetic resonance of an Fe-Fe pair in silicon. Solid State Commun. **64**, 1489–1494 (1987). https://doi.org/10.1016/0038-1098(87)90364-4
36. S. Strube, H. Vollmer, R. Labusch, Electron paramagnetic resonance (EPR) and photo-EPR studies of aggregate centers with two iron atoms in silicon. Appl. Phys. A **81**, 87–91 (2005). https://doi.org/10.1007/s00339-004-3031-1
37. M. Seibt, R. Khalil, V. Kveder, W. Schröter, Electronic states at dislocations and metal silicide precipitates in crystalline silicon and their role in solar cell materials. Appl. Phys. A **96**, 235–253 (2009). https://doi.org/10.1007/s00339-008-5027-8
38. B. Hackl, K.-J. Range, P. Stallhofer, L. Fabry, Correlation between DLTS and TRXFA measurements of copper and iron contaminations in FZ and CZ silicon wafers; application to gettering efficiencies. J. Electrochem. Soc. **139**, 1495–1498 (1992). https://doi.org/10.1149/1.2069438
39. D.A. Ramappa, W.B. Henley, Stability of iron-silicide precipitates in silicon. J. Electrochem. Soc. **144**, 4353–4356 (1997). https://doi.org/10.1149/1.1838192
40. T. Mchedlidze, K. Matsumoto, Electrically detected magnetic resonance signal from iron contaminated Czochralski silicon crystal. J. Appl. Phys. **83**, 4042–4048 (1998). https://doi.org/10.1063/1.367160
41. J. Bailey, E.R. Weber, Precipitation of iron in polycrystalline silicon. Phys. Status Solidi A **137**, 515–523 (1993). https://doi.org/10.1002/pssa.2211370221
42. J. Bailey, S.A. McHugo, H. Hieslmair, E.R. Weber, Efficiency-limiting defects in silicon solar cell material. J. Electron. Mater. **25**, 1417–1421 (1996). https://doi.org/10.1007/BF02655376
43. A. Mesli, T. Heiser, N. Amroun, P. Siffert, Charge-state-dependent iron precipitation in silicon. Appl. Phys. Lett. **57**, 1898–1900 (1990). https://doi.org/10.1063/1.104005
44. T. Heiser, A. Mesli, How far does the charge state affect the iron behavior in silicon? Appl. Phys. Lett. **58**, 2240–2242 (1991). https://doi.org/10.1063/1.104938
45. A. Haarahiltunen, H. Väinölä, O. Anttila, E. Saarnilehto, M. Yli-Koski, J. Storgårds, J. Sinkkonen, Modeling of heterogeneous precipitation of iron in silicon. Appl. Phys. Lett. **87**, 151908/1-1 (2005). https://doi.org/10.1063/1.2099531
46. A. Haarahiltunen, H. Väinölä, O. Anttila, M. Yli-Koski, J. Sinkkonen, Experimental and theoretical study of heterogeneous iron precipitation in silicon. J. Appl. Phys. **101**, 043507/1-6 (2007). https://doi.org/10.1063/1.2472271
47. D.P. Fenning, J. Hofstetter, M.I. Bertoni, G. Coletti, B. Lai, C. del Cañizo, T. Buonassisi. Precipitated iron: a limit on gettering efficacy in multicrystalline silicon. J. Appl. Phys. **113**, 044521/1-12 (2013). https://doi.org/10.1063/1.4788800
48. J. Schön, H. Habenicht, M.C. Schubert, W. Warta, Understanding the distribution of iron in multicrystalline silicon after emitter formation: theoretical model and experiments. J. Appl. Phys. **109**, 063717/1-8 (2011). https://doi.org/10.1063/1.3553858
49. J.D. Murphy, R.J. Falster, The relaxation behaviour of supersaturated iron in single-crystal silicon at 500 to 750 °C. J. Appl. Phys. **112**, 113506/1-7 (2012). https://doi.org/10.1063/1.4767378
50. W. Rein, S.W. Glunz, Electronic properties of interstitial iron and iron-boron pairs determined by means of advanced lifetime spectroscopy. J. Appl. Phys. **98**, 113711/1-12 (2005). https://doi.org/10.1063/1.2106017
51. D. Macdonald, L.J. Geerligs, Recombination activity of interstitial iron and other transition metal point defects in p- and n-type crystalline silicon. Appl. Phys. Lett. **85**, 4061–4063 (2004). https://doi.org/10.1063/1.1812833
52. H. Kohno, H. Hieslmair, A.A. Istratov, E. Weber, Temperature dependence of the iron donor level in silicon at device processing temperatures. Appl. Phys. Lett. **76**, 2734–2736 (2000). https://doi.org/10.1063/1.126459
53. A.S. McHugo, R.J. McDonald, A.R. Smith, D.L. Hurley, E.R. Weber, Iron solubility in highly boron-doped silicon. Appl. Phys. Lett. **73**, 1424–1426 (1998). https://doi.org/10.1063/1.121964

54. K. Nauka, T.I. Kamins, Deep state defects in strained and relaxed $Si_{1-x}Ge_x$ on Si introduced by 3*d* transition metal and 5*d* noble metal impurities. Physica B **273–274**, 603–607 (1999). https://doi.org/10.1016/S0921-4526(99)00584-0
55. D. Macdonald, A. Cuevas, J. Wong-Leung, Capture cross sections of the acceptor level of iron-boron pairs in p-type silicon by injection-level dependent lifetime measurements. J. Appl. Phys. **89**, 7932–7939 (2001). https://doi.org/10.1063/1.1372156
56. J.E. Birkholz, K. Bothe, D. Macdonald, J. Schmidt, Electronic properties of iron-boron pairs in crystalline silicon by temperature- and injection-level-dependent lifetime measurements. J. Appl. Phys. **97**, 103708/1-6 (2005). https://doi.org/10.1063/1.1897489
57. D. Macdonald, T. Roth, P.N.K. Deenapanray, T. Trupke, R.A. Bardos, Doping dependence of the carrier lifetime crossover point upon dissociation of iron-boron pairs in crystalline silicon. Appl. Phys. Lett. **89**, 142107/1-3 (2006). https://doi.org/10.1063/1.2358126
58. A. Chantre, D. Bois, Metastable-defect behavior in silicon: charge-state-controlled reorientation of iron-aluminum pairs. Phys. Rev. B **31**, 7979–7988 (1985). https://doi.org/10.1103/PhysRevB.31.7979
59. H. Nakashima, T. Sadoh, T. Tsurushima, Electrical and thermal properties of structurally metastable iron-boron pairs in silicon. Phys. Rev. B **49**, 16983–16993 (1994). https://doi.org/10.1103/PhysRevB.49.16983
60. L. Dobaczewski, P. Kamiński, R. Kozłowski, M. Surma, High resolution DLTS studies of transition-metal-related defects in silicon. Mater. Sci. Forum **196–201**, 669–676 (1995). https://doi.org/10.4028/www.scientific.net/MSF.196-201.669
61. H. Conzelmann, Photoluminescence of transition metal complexes in silicon. Appl. Phys. A **42**, 1–18 (1987). https://doi.org/10.1007/BF00618154
62. A. Mesli, B. Vileno, C. Eckert, A. Slaoui, C. Pedersen, A. Nylandsted Larsen, N.V. Abrosimov. Iron in relaxed $Si_{1-x}Ge_x$ alloy: band gap related levels, diffusion, and alloying effects. Phys. Rev. B **66**, 045206/1-12 (2002). https://doi.org/10.1103/physrevb.66.045206
63. VI. Kolkovsky, A. Mesli, L. Dobaczewski, N.V. Abrosimov, Z.R. Zytkiewicz, A.R. Peaker, Stable and metastable configurations of iron atoms in SiGe alloys. J. Phys. Cond. Matter **17**, S2267–S2272 (2005). https://doi.org/10.1088/0953-8984/17/22/014
64. Vl. Kolkovsky, A. Mesli, L. Dobaczewski, N.V. Abrosimov, Z.R. Żietkiwicz, A.R. Peaker, Interaction of iron with the local environment in SiGe alloys investigated with Laplace transform deep level spectroscopy. Phys. Rev. B **74**, 195204/1-8 (2006). https://doi.org/10.1103/physrevb.74.195204
65. A. Mesli, VI. Kolkovsky, L. Dobaczewski, A. Nylandsted Larsen, N.V. Abrosimov, Defects and impurities in SiGe: the effect of alloying. Nucl. Instrum. Meth. Phys. Res. B **253**, 154–161 (2006). https://doi.org/10.1016/j.nimb.2006.10.025
66. P. Kruszweski, VI. Kolkovsky, A. Mesli, L. Dobaczewski, N.V. Abrosimov, V. P. Markevich, A.R. Peaker, Alloy shift of “no-germanium” iron-related electronic levels in unstrained silicon-germanium alloys. Phys. Rev. B **76**, 233203/1-4 (2007). https://doi.org/10.1103/physrevb.76.233203
67. P. Kruszewski, A. Mesli, L. Dobaczewski, N.V. Abrosimov, V.P. Markevich, A.R. Peaker, Iron-aluminium pair reconfiguration processes in SiGe alloys. J. Mater. Sci. Mater. Electron. **18**, 759–762 (2007). https://doi.org/10.1007/10854-006-9104-5
68. T.F. Ciszek, T.H. Wang, Silicon defect and impurity studies using float-zone crystal growth as a tool. J. Cryst. Growth **237–239**, 1685–1691 (2002). https://doi.org/10.1016/S0022-0248(01)02325-9
69. J. Schmidt, D. Macdonald, Recombination activity of iron-gallium and iron-indium pairs in silicon. J. Appl. Phys. **97**, 113712/1-9 (2005). https://doi.org/10.1063/1.1929096
70. Y. Yoon, Y. Yan, N.P. Ostrom, J. Kim, G. Rozgonyi, Deep level transient spectroscopy and minority carrier lifetime study on Ga-doped continuous Czochralski silicon. Appl. Phys. Lett. **101**, 222107/1-4 (2012). https://doi.org/10.1063/1.4766337
71. Y. Gurimskaya, D. Mathiot, A. Sellai, P. Kruszewski, L. Dobaczewski, A. Nylandsted Larsen, A. Mesli, Spectroscopic studies of iron and chromium in germanium. J. Appl. Phys. **110**, 113707/1-7 (2011). https://doi.org/10.1063/1.3664761

72. J. Lauwaert, J. Vanhellemont, E. Simoen, H. Vrielinck, P. Clauws, Electronic properties of iron and cobalt impurity centres in germanium. J. Appl. Phys. **111**, 113713/1-7 (2012). https://doi.org/10.1063/1.4729037
73. P. Kaminski, R. Kozlowski, A. Jelenski, T. Mchedlidze, M. Suezawa, High-resolution photoinduced transient spectroscopy of electrically active iron-related defects in electron irradiated high-resistivity silicon. Jpn. J. Appl. Phys. **42**, 5415–5419 (2003). https://doi.org/10.1143/JJAP.42.5415
74. D. Abdelbarey, V. Kveder, W. Schröter, M. Seibt, Light-induced point defect reactions of residual iron in crystalline silicon after aluminum gettering. J. Appl. Phys. **108**, 043519/1-6 (2010). https://doi.org/10.1063/1.3474658
75. C.K. Tang, L. Vines, B.G. Svensson, E.V. Monakhov, Deep level transient spectroscopy on proton-irradiated Fe-contaminated p-type silicon. Phys. Status Solidi C **9**, 1992–1995 (2012). https://doi.org/10.1002/pssc.201200163
76. C.K. Tang, L. Vines, V.P. Markevich, B.G. Svensson, E.V. Monakhov, Divacancy-iron complexes in silicon. J. Appl. Phys. **113**, 044503/1-4 (2013). https://doi.org/10.1063/1.4788695
77. B.A. Komarov, Special features of radiation-defect annealing in silicon p-n structures: the role of Fe impurity atoms. Semiconductors **38**, 1041–1046 (2004). https://doi.org/10.1134/1.1797482
78. S. Tanaka, H. Kitagawa, Diffusion and electrical properties of iron-related defects in n-type silicon grown by Czochralski- and floating zone method. Jpn. J. Appl. Phys. **37**, 4656–4662 (1998). https://doi.org/10.1143/JJAP.37.4656
79. T. Mchedlidze, M. Kittler, Involvement of iron-phosphorus complexes in iron gettering for n-type silicon. Phys. Status Solidi A **203**, 786–791 (2006). https://doi.org/10.1002/pssa.200564512
80. T. Sadoh, K. Tsukamoto, A. Baba, D. Bai, A. Kenjo, T. Tsurushima, H. Mori, H. Nakashima, Deep level of iron-hydrogen complex in silicon. J. Appl. Phys. **82**, 3828–3831 (1997). https://doi.org/10.1063/1.365746
81. S. Leonard, V.P. Markevich, A.R. Peaker, B. Hamilton, J.D. Murphy, Evidence for an iron-hydrogen complex in p-type silicon. Appl. Phys. Lett. **107**, 032103/1-5 (2015). https://doi.org/10.1063/1.4927323
82. R. Khalil, V. Kveder, W. Schröter, M. Seibt, Early stages of iron precipitation in silicon. Phys. Status Solidi C **2**, 1802–1806 (2005). https://doi.org/10.1002/pssc.200460507
83. R. Khalil, V. Kveder, W. Schröter, M. Seibt, Electrical properties of clustered and precipitated iron in silicon. Solid State Phenom. **108–109**, 109–114 (2005). https://doi.org/10.4028/www.scientific.net/SSP.108-109.10
84. E.K. Evangelou, G.E. Giakoumakis, C.A. Dimitriadis, Deep levels in β-$FeSi_2$/n-Si heterojunctions. Solid State Commun. **86**, 309–312 (1993). https://doi.org/10.1016/0038-1098(93)90379-2
85. B.K. Miremadi, S. Roy Morrison, Conductance along iron-doped silicon grain boundaries. J. Appl. Phys. **55**, 3658–3663 (1984). https://doi.org/10.1063/1.332915
86. M. Kittler, W. Seifert, M. Stemmer, J. Palm, Interaction of iron with a grain boundary in boron-doped multicrystalline silicon. J. Appl. Phys. **77**, 3725–3728 (1995). https://doi.org/10.1063/1.358611
87. C. Hässler, G. Pensl, M. Schulz, A. Voigt, H.P. Strunk, Grain boundaries in multicrystalline silicon. Phys. Status Solidi A **137**, 463–484 (1993). https://doi.org/10.1002/pssa.2211370218
88. W. Seifert, M. Kittler, J. Vanhellemont, EBIC study of recombination activity of oxygen precipitation related defects in Si. Mater. Sci. Eng. B **42**, 260–264 (1996). https://doi.org/10.1016/S0921-5107(96)01718-7
89. M. Trushin, O. Vyvenko, W. Seifert, G. Jia, M. Kittler, Iron-oxygen interaction in silicon: a combined XBIC/XRF-EBIC-DLTS study of precipitation and complex building. Physica B **404**, 4645–4648 (2009). https://doi.org/10.1016/j.physb.2009.08.132

90. A. Castaldini, D. Cavalcoli, A. Cavallini, S. Binetti, S. Pizzini, Electronic transitions at defect states in Cz p-type silicon. Appl. Phys. Lett. **86**, 162109/1-3 (2005). https://doi.org/10.1063/1.1881788
91. G. Zoth, W. Bergholz, A fast, preparation-free method to detect iron in silicon. J. Appl. Phys. **67**, 6764–6771 (1990). https://doi.org/10.1063/1.345063
92. D. Macdonald, J. Tan, T. Trupke, Imaging interstitial iron concentrations in boron-doped crystalline silicon using photoluminescence. J. Appl. Phys. **103**, 073710/1-7 (2008). https://doi.org/10.1063/1.2903895
93. M.C. Schubert, H. Habenicht, W. Warta, Imaging of metastable defects in silicon. IEEE J. Photovolt. **1**, 168–173 (2011). https://doi.org/10.1109/JPHOTOV.2011.2169942
94. M.C. Schubert, M.J. Kerler, W. Warta, Influence of heterogeneous profiles in carrier density measurements with respect to iron concentration measurements in silicon. J. Appl. Phys. **105**, 114903/1-6 (2009). https://doi.org/10.1063/1.3138805
95. L.E. Mundt, M.C. Schubert, J. Schön, B. Michl, T. Niewelt, F. Schindler, W. Warta, Spatially resolved impurity identification via temperature- and injection-dependent photoluminescence imaging. IEEE J. Photovolt. **5**, 1503–1509 (2015). https://doi.org/10.1109/JPHOTOV.2015.2447837
96. D.H. Macdonald, L.J. Geerligs, A. Azzizi, Iron detection in crystalline silicon by carrier lifetime measurements for arbitrary injection and doping. J. Appl. Phys. **95**, 1021–1028 (2004). https://doi.org/10.1063/1.1637136
97. D. Walz, J.-P. Joly, G. Kamarinos, On the recombination behaviour of iron in moderately boron-doped p-type silicon. Appl. Phys. A **62**, 345–353 (1996). https://doi.org/10.1007/BF01594232
98. M. Kittler, W. Seifert, K. Schmalz, K. Tittelbach-Helmrich, Comparison of EBIC and DLTS measurements on boron-doped CZ silicon contaminated with iron. Phys. Status Solidi A **96**, K133–K137 (1986). https://doi.org/10.1002/pssa.2910960247
99. I. Rapoport, P. Taylor, J. Kearns, D.K. Schroder, Two-side surface photovoltage studies for implanted iron diffusion in silicon during rapid thermal anneal. J. Appl. Phys. **107**, 013518/1-6 (2010). https://doi.org/10.1063/1.3275045
100. E.H. Dahl, V. Osinniy, K. Friestad, A.-K. Søiland, Y. Safir, W. Skorupa, R. Tronstad, A. Nylandsted Larsen, Assessing the role of iron-acceptor pairs in solar grade multicrystalline silicon wafers from the metallurgical route. Phys. Status Solidi C **9**, 2017–2022 (2012). https://doi.org/10.1002/pssc.201200258
101. T. Mchedlidze, J. Weber, Radial distribution of iron in silicon crystals grown by Czochralski method from contaminated feedstock. Phys. Status Solidi RRL **8**, 228–230 (2014). https://doi.org/10.1002/pssr.201308327
102. K. Knobloch, W. Seifert, M. Kittler, Application of scanning deep level transient spectroscopy for characterisation of multicrystalline silicon. Mater. Sci. Eng. B **42**, 254–259 (1996). https://doi.org/10.1016/S0921-5107(96)01684-4
103. Y. Tokuda, T. Namizaki, T. Murase, T. Hasegawa, H. Shiraki, Deep-level transient spectroscopy detection of iron in hydrogenated p^+ silicon. Jpn. J. Appl. Phys. **40**, L533–L535 (2001). https://doi.org/10.1143/JJAP.40.L533
104. W. Kwapil, J. Schön, F. Schindler, W. Warta, M.C. Schubert, Impact of iron precipitates on carrier lifetime in as-grown and phosporus-gettered multicrystalline silicon wafers in model and experiment. IEEE J. Photovolt. **4**, 791–798 (2014). https://doi.org/10.1109/JPHOTOV.2014.2304355
105. P. Gundel, M.C. Schubert, F.D. Heinz, W. Kwapil, W. Warta, G. Martinez-Criado, M. Reiche, E.R. Weber, Impact of stress on the recombination at metal precipitates in silicon. J. Appl. Phys. **108**, 103707/1-5 (2010). https://doi.org/10.1063/1.3511749
106. A.A. Istratov, E.R. Weber, Physics of copper in silicon. J. Electrochem. Soc. **149**, G21–G30 (2002). https://doi.org/10.1149/1.1421348
107. M. Seibt, M. Griess, A.A. Istratov, H. Hedemann, A. Sattler, W. Schröter, Formation and precipitation of copper silicide precipitates in silicon. Phys. Status Solidi A **166**, 171–182 (1998). https://doi.org/10.1002/(sici)1521-396x(199803)166:1<171::aid-pssa171>3.0.c);2-2

108. U. Wahl, A. Vantomme, G. Langouche, J.P. Correia, ISOLDE Collaboration, Lattice location and stability of ion implanted Cu in Si. Phys. Rev. Lett. **84**, 1495–1498 (2000). https://doi.org/10.1103/PhysRevLett.84.1495
109. U. Wahl, J.G. Correia, J.C. Soares, the ISOLDE Collaboration, Lattice location and stability of implanted Cu in Ge. Physica B **340-342**, 799–802 (2003). https://doi.org/10.1016/j.physb.2003.09.216
110. U. Wahl, A. Vantomme, G. Langouche, J.P. Araújo, L. Peralta, J.G. Correia, Lattice location of implanted Cu in highly doped Si. Appl. Phys. Lett. **77**, 2142–2144 (2000). https://doi.org/10.1063/1.1314876
111. R. Keller, M. Deicher, W. Pfeiffer, H. Skudlik, D. Steiner, Th. Wichert, Copper in silicon. Phys. Rev. Lett. **65**, 2023–2026 (1990). https://doi.org/10.1103/PhysRevLett.65.2023
112. A.M. Frens, M.T. Bennebroek, J. Schmidt, W.M. Chen, B. Monemar, Zero-field optical detection of magnetic resonance on a metastable sulfur-pair-related defect in silicon: evidence for a Cu constituent. Phys. Rev. B **46**, 12316–12322 (1992). https://doi.org/10.1103/PhysRevB.46.12316
113. P.N. Hai, T. Gregorkiewicz, C.A.J. Ammerlaan, D.T. Don, Copper-related defects in silicon: electron-paramagnetic-resonance identification. Phys. Rev. B **56**, 4620–4625 (1997). https://doi.org/10.1103/PhysRevB.56.4620
114. W.C. Dash, Copper precipitation on dislocations in silicon. J. Appl. Phys. **27**, 1193–1195 (1956). https://doi.org/10.1063/1.1722229
115. G.H. Schwuttke, Study of copper precipitation behavior in silicon single crystals. J. Electrochem. Soc. **108**, 163–167 (1961). https://doi.org/10.1149/1.2428034
116. L. Fiermans, J. Vennik, The precipitation behaviour of copper in silicon single crystals. Phys. Status Solidi **12**, 277–289 (1965). https://doi.org/10.1002/pssb.19650120125
117. E. Nes, G. Lunde, Copper precipitate colonies in silicon. J. Appl. Phys. **43**, 1835–1837 (1972). https://doi.org/10.1063/1.1661405
118. S.M. Hu, M.R. Poponiak, Habit and morphology of copper precipitates in silicon. J. Appl. Phys. **43**, 2067–2074 (1972). https://doi.org/10.1063/1.1661455
119. E. Nes, J. Solberg, *In situ* transmission electron microscope investigation of the annealing of copper colonies in silicon. J. Appl. Phys. **44**, 486–487 (1973). https://doi.org/10.1063/1.1661910
120. E. Nes, J.K. Solberg, The dragging of precipitate particles by climbing dislocations in silicon. J. Appl. Phys. **44**, 488–489 (1973). https://doi.org/10.1063/1.1661911
121. G. Das, Precipitation of copper in silicon. J. Appl. Phys. **44**, 4459–4467 (1973). https://doi.org/10.1063/1.1661982
122. E. Nes, J. Washburn, Transmission electron microscope investigation of the growth of copper precipitate colonies in silicon. J. Appl. Phys. **44**, 3682–3688 (1973). https://doi.org/10.1063/1.1662820
123. J.K. Solberg, The crystal structure of η-Cu_3Si precipitates in silicon. Acta Cryst. **A34**, 684–698 (1978). https://doi.org/10.1107/S0567739478001448
124. Y. Chikaura, K. Kishimoto, Extinction X-ray topography images of copper precipitates in silicon single crystals. Jpn. J. Appl. Phys. **19**, L5–L8 (1980). https://doi.org/10.1143/JJAP.19.L5
125. W. Wijaranakula, S.S. Kim, Precipitation of 3*d* transition metal silicides in Czochralski silicon crystals. J. Appl. Phys. **76**, 6017–6019 (1994). https://doi.org/10.1063/1.358355
126. M. Seibt, V. Kveder, W. Schröter, O. Voß, Structural and electrical properties of metal impurities at dislocations in silicon. Phys. Status Solidi A **202**, 911–920 (2005). https://doi.org/10.1002/pssa.20046051
127. W.C. Dash, Generation of prismatic dislocation loops in silicon crystals. Phys. Rev. Lett. **1**, 400–402 (1958). https://doi.org/10.1103/PhysRevLett.1.400
128. T. Yamauchi, Y. Tsumori, T. Nakashizu, H. Esaka, S. Takao, S. Shinoyama, Application of copper-decoration method to characterize as-grown Czochralski-silicon. Jpn. J. Appl. Phys. **31**, L439–L442 (1992). https://doi.org/10.1143/JJAP.31.L439

129. J.-S. Shin, I.W. Lyo, Influence of Cu-decoration to individual crystal originated pits on Si wafer. Jpn. J. Appl. Phys. **42**, 4187–4192 (2003). https://doi.org/10.1143/JJAP.42.4187
130. M. Itsumi, Y. Omura, K. Imai, T. Ueki, H. Akiya, M. Tomita, M. Yamawaki, Copper decoration followed by TEM observation identifying defects in the buried oxides of SOI substrates. J. Electrochem. Soc. **143**, 2357–2361 (1996). https://doi.org/10.1149/1.1837007
131. E. Nes, The mechanism of repeated precipitation on dislocations. Acta Metall. **22**, 81–87 (1974). https://doi.org/10.1016/0001-6160(74)90127-8
132. J.K. Solberg, E. Nes, On the micromechanisms of repeated precipitation on edge dislocations. J. Mater. Sci. **13**, 2233–2340 (1978). https://doi.org/10.1007/BF00541679
133. M. Elkajbaji, J. Dessus, J. Thibault, Structure of copper precipitates in a symmetrical silicon tilt bicrystal: high-resolution electron microscopy and energy-dispersive X-ray analysis. Philos. Mag. A **66**, 873–888 (1992). https://doi.org/10.1080/01418619208247996
134. S.A. McHugo, Release of metal impurities from structural defects in polycrystalline silicon. Appl. Phys. Lett. **71**, 1984–1986 (1997). https://doi.org/10.1063/1.119762
135. T. Buonassisi, M.A. Marcus, A.A. Istratov, M. Heuer, T.F. Ciszek, B. Lai, Z. Cai, E.R. Weber, Analysis of copper-rich precipitates in silicon: Chemical state, gettering, and impact on multicrystalline silicon solar cell material. J. Appl. Phys. **97**, 063503/1-9 (2005). https://doi.org/10.1063/1.1827913
136. H. Gottschalk, Precipitation of copper silicide on glide dislocations in silicon at low temperature. Phys. Status Solidi A **137**, 447–461 (1993). https://doi.org/10.1002/pssa.2211370217
137. B. Shen, T. Sekiguchi, R. Zhang, Y. Shi, H. Shi, K. Yang, Y. Zheng, K. Sumino, Precipitation of Cu and Fe in dislocated Floating-Zone-grown silicon. Jpn. J. Appl. Phys. **35**, 3301–3305 (1996). https://doi.org/10.1143/JJAP.35.3301
138. B. Shen, R. Zhang, Y. Shi, Y.D. Zheng, T. Sekiguchi, K. Sumino, Precipitation of Cu and Fe on Frank-type partial dislocations in Czochralski-grown silicon. Appl. Phys. Lett. **68**, 214–216 (1996). https://doi.org/10.1063/1.116464
139. B. Shen, T. Sekiguchi, J. Jablonski, K. Sumino, Gettering of copper by bulk stacking faults and punched-out dislocations in Czochralski-grown silicon. J. Appl. Phys. **76**, 4540–4546 (1994). https://doi.org/10.1063/1.357285
140. W.K. Tice, T.Y. Tan, Nucleation of CuSi precipitate colonies in oxygen-rich silicon. Appl. Phys. Lett. **28**, 564–565 (1976). https://doi.org/10.1063/1.88825
141. A. Bazzali, G. Borionetti, R. Orizio, D. Gambaro, R. Falster, Oxygen precipitate precursors and size threshold for the preferential nucleation for copper and nickel precipitation in silicon: the detection of copper and nickel contamination by minority carrier lifetime methods. Mater. Sci. Eng. B **36**, 85–90 (1996). https://doi.org/10.1016/0921-5107(95)01298-2
142. J. Xu, N. Wang, D. Yang, Influence of oxygen precipitation on copper precipitation in Czochralski silicon. J. Appl. Phys. **111**, 094907/1-4 (2012). https://doi.org/10.1063/1.4705421
143. M. Seibt, K. Graff, Characterization of haze-forming precipitates in silicon. J. Appl. Phys. **63**, 4444–4449 (1988). https://doi.org/10.1063/1.340164
144. S. Sadamitsu, S. Sumita, N. Fujino, T. Shiraiwa, TEM observation of defects induced by Cu contamination on Si(100) surface. Jpn. J. Appl. Phys. **27**, L1819–L1821 (1988). https://doi.org/10.1143/JJAP.27.L1819
145. B. Shen, J. Jablonski, T. Sekiguchi, K. Sumino, Influences of Cu and Fe impurities on oxygen precipitation in Czochralski-grown silicon. Jpn. J. Appl. Phys. **35**, 4187–4194 (1996). https://doi.org/10.1143/JJAP.35.4187
146. Y. Ohno, K. Inoue, K. Kutsukake, M. Deura, T. Ohsawa, I. Yonenaga, H. Yoshida, S. Takeda, R. Taniguchi, H. Otubo, S.R. Nishitani, N. Ebisawa, Y. Shimizu, H. Takamizawa, K. Inoue, Y. Nagai, Nanoscopic mechanism of Cu precipitation at small-angle tilt boundaries in Si. Phys. Rev. B **91**, 235315/1-5 (2015). https://doi.org/10.1103/physrevb.91.235315

147. G. Kissinger, H. Grimmeiss, Difficulties in doping SiGe alloys with transition metal point defects. Phys. Status Solidi A **145**, K5–K9 (1994). https://doi.org/10.1002/pssa.2211540249
148. R. Gleichmann, Precipitation of copper in transmutation doped silicon. Mat. Res. Soc. Symp. Proc. **25**, 287–292 (1984). https://doi.org/10.1002/crat.2170180303
149. Z. Xi, D. Yang, J. Xu, Y. Ji, D. Que, H.J. Moeller, Effect of intrinsic point defects on copper precipitation in large-diameter Czochralski silicon. Appl. Phys. Lett. **83**, 3048–3050 (2003). https://doi.org/10.1063/1.1617377
150. Z. Xi, J. Chen, D. Yang, A. Lawerenz, H.J. Moeller, Copper precipitation in large-diameter Czochralski silicon. J. Appl. Phys. **97**, 094909/1-4 (2005). https://doi.org/10.1063/1.1875740
151. W. Wang, D. Yang, X. Ma, D. Que, Effect of silicon interstitials on Cu precipitation in n-type Czochralski silicon. J. Appl. Phys. **103**, 093534/1-4 (2008). https://doi.org/10.1063/1.1.2908215
152. S.A. McHugo, C. Flink, Thermal stability of copper precipitates in silicon. Appl. Phys. Lett. **77**, 3598–3600 (2000). https://doi.org/10.1063/1.1328769
153. S.A. McHugo, A. Mohammed, A.C. Thompson, B. Lai, Z. Cai, Copper precipitates in silicon: Precipitation, dissolution, and chemical state. J. Appl. Phys. **91**, 6396–6405 (2002). https://doi.org/10.1063/1.1471944
154. M. Heuer, T. Buonassisi, A.A. Istratov, M.D. Pickett, M.A. Marcus, A.M. Minor, E.R. Weber, Transition metal interaction and Ni-Fe-Cu-Si phases in silicon. J. Appl. Phys. **101**, 123510/1-6. (2007). https://doi.org/10.1063/1.2748346
155. M. El Kajbaji, J. Thibault, Platelet copper precipitates in silicon: a high-resolution electron microscopy study. Philos. Mag. Lett. **71**, 335–339 (1995). https://doi.org/10.1080/09500839508241016
156. W. Schröter, V. Kveder, M. Seibt, H. Ewe, H. Hedemann, F. Riedel, A. Sattler, Atomic structure and electronic states of nickel and copper silicides in silicon. Mater. Sci. Eng. B **72**, 80–86 (2000). https://doi.org/10.1016/S0921-5107(99)00499-7
157. M.B. Shabani, T. Yoshimi, H. Abe, Low-temperature out-diffusion of Cu from silicon wafers. J. Electrochem. Soc. **143**, 2025–2029 (1996). https://doi.org/10.1149/1.1836943
158. C. Flink, H. Feick, S.A. McHugo, W. Seifert, H. Hieslmair, T. Heiser, A.A. Istratov, E.R. Weber, Out-diffusion and precipitation of copper in silicon: an electrostatic model. Phys. Rev. Lett. **23**, 4900–4903 (2000). https://doi.org/10.1103/PhysRevLett.23.4900
159. H.-W. Guo, S.T. Dunham, Accurate modeling of copper precipitation kinetics including Fermi level dependence. Appl. Phys. Lett. **89**, 182106/1-3 (2006). https://doi.org/10.1063/1.2374689
160. Y. Ohkubo, K. Matsumoto, K. Nagai, Acceleration of Cu surface precipitation from bulk by adsorbed organic molecules. Jpn. J. Appl. Phys. **44**, 3793–3797 (2005). https://doi.org/10.1143/JJAP.44.3793
161. M. Seibt, H. Hedemann, A.A. Istratov, F. Riedel, A. Sattler, W. Schröter, Structural and electrical properties of metal silicide precipitates in silicon. Phys. Status Solidi A **171**, 301–310 (1999). https://doi.org/10.1002/(sici)1521-396x(199901)171:1<301::aid-pssa301>3.0.co;2-p
162. C.B. Collins, R.O. Carlson, Properties of silicon doped with iron and copper. Phys. Rev. **108**, 1409–1414 (1957). https://doi.org/10.1103/PhysRev.108.1409
163. A.A. Lebedev, M.M. Akhmedova, Photoconductivity of copper-doped silicon. Sov. Phys. Semicond. **10**, 1130–1132 (1976)
164. C.S. Chen, J.C. Corelli, Spectroscopy study of copper impurities in silicon. J. Appl. Phys. **44**, 5622–5623 (1973). https://doi.org/10.1063/1.1662209
165. N. Tōyama, Copper impurity levels in silicon. Solid-State Electron **26**, 37–46 (1983). https://doi.org/10.1016/0038-1101(83)90159-4
166. N. Tōyama, T. Otsuji, Strain-induced level in copper-doped silicon. J. Appl. Phys. **57**, 4623–4625 (1985). https://doi.org/10.1063/1.335371
167. H. Lemke, Properties of copper donor levels in silicon. Phys. Status Solidi A **1**, 283–286 (1970). https://doi.org/10.1002/pssa.19700010211

168. M.M. Akhmedova, L.S. Berman, L.S. Kostina, A.A. Lebedev, Investigation of the parameters of copper levels in silicon by capacitance methods. Sov. Phys. Semicond. **10**, 1400–1401 (1976)
169. S.J. Pearton, A.J. Tavendale, The electrical properties of deep copper- and nickel-related centers in silicon. J. Appl. Phys. **54**, 1375–1379 (1983). https://doi.org/10.1063/1.332160
170. H. Lemke, Störstellenreaktionen bei Cu-dotierten Siliziumkristallen. Phys. Status Solidi A **95**, 665–677 (1986). https://doi.org/10.1002/pssa.2210950237
171. S.D. Brotherton, J.R. Ayres, A. Gill, H.W. van Kesteren, F.J.A.M. Greidanus, Deep levels of copper in silicon. J. Appl. Phys. **62**, 1826–1832 (1987). https://doi.org/10.1063/1.339564
172. A. Mesli, T. Heiser, Defect reactions in copper-diffused and quenched p-type silicon. Phys. Rev. B **45**, 11632–11641 (1992). https://doi.org/10.1103/PhysRevB.45.11632
173. S. Koveshnikov, Y. Pan, H. Mollenkopf, Investigation of electronic states in copper doped p-type silicon. Electrochem. Soc. Proc. **96**(13), 473–479 (1996)
174. A.A. Istratov, H. Hieslmair, C. Flink, T. Heiser, E.R. Weber, Interstitial copper-related center in n-type silicon. Appl. Phys. Lett. **71**, 2349–2351 (1997). https://doi.org/10.1063/1.120026
175. S.D. Ganichev, E. Ziemann, W. Prettl, I.N. Yassievich, A.A. Istratov, E.R. Weber, Distinction between the Poole-Frenkel and tunneling models of electric-field-stimulated carrier emission from deep levels in semiconductors. Phys. Rev. B **61**, 10361–10365 (2000). https://doi.org/10.1103/PhysRevB61.10361
176. E.R. Weber, A.A. Istratov, Electrical properties and recombination activity of copper, nickel and cobalt in silicon. Appl. Phys. A **66**, 123–136 (1998). https://doi.org/10.1007/s003390050649
177. S. Knack, Copper-related defects in silicon. Mater. Sci. Semicond. Process. **7**, 125–141 (2004). https://doi.org/10.1016/j.mssp.2004.06.002
178. N.S. Minaev, A.V. Mudryĭ, V.D. Tkachev, Radiative recombination at thermal defects in silicon. Sov. Phys. Semicond. **13**, 233–234 (1979)
179. J. Weber, H. Bauch, R. Sauer, Optical properties of copper in silicon: excitons bound to isoelectronic copper pairs. Phys. Rev. B **25**, 7688–7699 (1982). https://doi.org/10.1103/PhysRevB.25.7688
180. H.B. Erzgräber, K. Schmalz, Correlation between the Cu-related luminescent center and a deep level in silicon. J. Appl. Phys. **78**, 4066–4068 (1995). https://doi.org/10.1063/1.359863
181. M. Nakamura, Dissociation of the 1.014 eV photoluminescence copper center in silicon crystal. Appl. Phys. Lett. **73**, 3896–3898 (1998). https://doi.org/10.1063/1.122928
182. M. Nakamura, Thermal equilibrium and stability of copper complexes in silicon crystal. Appl. Phys. Lett. **76**, 2089–2091 (2000). https://doi.org/10.1063/1.126264
183. M. Nakamura, Formation of the photoluminescence Cu center on in-diffusion and out-diffusion of Cu in dilute Cu-contaminated silicon crystals. J. Appl. Phys. **92**, 6625–6629 (2002). https://doi.org/10.1063/1.1521515
184. M. Nakamura, H. Ohno, S. Murakami, Formation of the 1.014 eV photoluminescence Cu center in Cu-implanted silicon crystals and the center's model. Jpn. J. Appl. Phys. **43**, L1466–L1468 (2004). https://doi.org/10.1143/JJAP.43.L1466
185. A.A. Istratov, H. Hieslmair, T. Heiser, C. Flink, E.R. Weber, The dissociation energy and the charge state of a copper-pair center in silicon. Appl. Phys. Lett. **72**, 474–476 (1998). https://doi.org/10.1063/1.120790
186. M. Nakamura, S. Murakami, N.J. Kawai, S. Saito, H. Arie, Diffusion-temperature-dependent formation of Cu centers in Cu-saturated silicon crystals studied by photoluminescence and deep-level transient spectroscopy. Jpn. J. Appl. Phys. **47**, 4398–4402 (2008). https://doi.org/10.1143/JJAP.47.4398
187. M.L.W. Thewalt, M. Steger, A. Yang, N. Stavrias, M. Cardona, H. Riemann, N.V. Abrosimov, M.F. Churbanov, A.V. Gusev, A.D. Bulanov, I.D. Kovalev, A.K. Kaliteevskii, O.N. Godisov, P. Becker, H.-J. Pohl, J.W. Ager III, E.E. Haller, Can highly enriched ^{28}Si reveal new things about old defects? Physica B **401–402**, 587–592 (2007). https://doi.org/10.1016/j.physb.2007.09.028

188. M. Steger, A. Yang, N. Stavrias, M.L.W. Thewalt, H. Riemann, N.V. Abrosimov, M.F. Churbanov, A.V. Gusev, A.D. Bulanov, I.D. Kovalev, A.K. Kaliteevskii, O.N. Godisov, P. Becker, H.-J. Pohl, Reduction of the linewidths of deep luminescence centers in ^{28}Si reveals fingerprints of the isotope constituents. Phys. Rev. Lett. **100**, 177402/1-4 (2008). https://doi.org/10.1103/physrevlett.100.177402
189. M. Nakamura, S. Murakami, N.J. Kawai, S. Saito, K. Matsukawa, H. Arie, Compositional transformation between Cu centers by annealing in Cu-diffused silicon crystals studied with deep-level transient spectroscopy and photoluminescence. Jpn. J. Appl. Phys. **48**, 082302/1-5 (2009). https://doi.org/10.1143/jjap.48.082302
190. M. Nakamura, S. Murakami, Depth progression of dissociation reaction of the 1.014-eV photoluminescence copper center in copper-diffused silicon crystal measured by deep-level transient spectroscopy. Appl. Phys. Lett. **98**, 141909/1-3 (2011). https://doi.org/10.1063/1.3575574
191. M. Nakamura, S. Murakami, Deep-level transient spectroscopy and photoluminescence measurements of dissociation energy of the 1.014-eV copper center in copper-diffused silicon crystal. J. Appl. Phys. **111**, 073512/1-6 (2012). https://doi.org/10.1063/1.3699366
192. M. Nakamura, S. Murakami, H. Udono, Transformation reactions of copper centers in the space-charge region of a copper-diffused silicon crystal measured by deep-level transient spectroscopy. J. Appl. Phys. **112**, 063530/1-5 (2012). https://doi.org/10.1063/1.4754862
193. M. Nakamura, S. Murakami, H. Udono, Energy level(s) of the dissociation product of the 1.014-eV photoluminescence copper center in n-type silicon determined by photoluminescence and deep-level transient spectroscopy. J. Appl. Phys. **114**, 033508/1-5 (2013). https://doi.org/10.1063/1.4813878
194. K. Shirai, H. Yamaguchi, A. Yanase, H. Katayama-Yoshida, A new structure of Cu complex in Si and its photoluminescence. J. Phys. Condens. Matter. **21**, 064249/1-7 (2009). https://doi.org/10.1088/0953-8984/21/6/064249
195. A. Carvalho, D.J. Backlund, S.K. Estreicher. Four-copper complexes in Si and the Cu-related photoluminescence defect: a first-principles study. Phys. Rev. B **84**, 155322/1-8 (2011). https://doi.org/10.1103/physrevb.84.155322
196. A. Usami, Y. Tokuda, Effect of oxygen and copper on the defect cluster in neutron-irradiated p-type silicon. J. Appl. Phys. **45**, 2823–2831 (1974). https://doi.org/10.1063/1.1663686
197. S. Tamulevicius, B.G. Svensson, M.O. Aboefotoh, A. Hallén, Interaction between copper and point defects in proton-irradiated silicon. J. Appl. Phys. **71**, 4212–4216 (1992). https://doi.org/10.1063/1.350800
198. N. Yarikin, J. Weber, Copper-related deep-level centers in irradiated p-type silicon. Phys. Rev. B **83**, 125207/1-8 (2011). https://doi.org/10.1103/physrevb.83.125207
199. W. Wang, D. Yang, X. Yu, X. Ma, D. Que, Effect of point defects on copper-related deep levels in p-type Czochralski silicon. J. Appl. Phys. **102**, 073521/1-4 (2007). https://doi.org/10.1063/1.2786126
200. S. Knack, J. Weber, S.K. Estreicher, Defect reactions of copper in silicon. Electrochem. Soc. Proc. **20**, 290–298 (2002)
201. S. Knack, J. Weber, H. Lemke, H. Riemann, Copper-hydrogen complexes in silicon. Phys. Rev. B **65**, 165203/1-8 (2002). https://doi.org/10.1103/physrevb.65.165203
202. J. Kaniewski, M. Kaniewska, L. Ornoch, T. Sekiguchi, K. Sumino, Hydrogenation of copper related deep states in n-type Si containing extended defects. Mater. Sci. Forum **258–263**, 319–324 (1997). https://doi.org/10.4028/www.scientific.net/MSF.258-263.319
203. K. Kurita, T. Shingyouji, Room temperature annealing behavior of copper-related deep levels in p-type floating zone silicon wafers. Jpn. J. Appl. Phys. **40**, 1167–1171 (2001). https://doi.org/10.1143/JJAP.40.1167
204. E. Simoen, P. Clauws, M. Lamon, J. Vennik, Signature and capture cross section of copper-related hole traps in p-type high-purity germanium. Semicond. Sci. Technol. **1**, 53–57 (1986). https://doi.org/10.1088/0268-1242/1/007

205. P. Clauws, G. Huylebroeck, E. Simoen, P. Vermaercke, F. De Smet, J. Vennik, DLTS of the third acceptor level of substitutional copper in germanium. Semicond. Sci. Technol. **4**, 910–914 (1989). https://doi.org/10.1088/0268-1242/4/11/003
206. R.A. Paramonova, A.V. Rzhanov, Sov. Phys. Solid State **4**, 1335–1338 (1963)
207. Y. Kamiura, F. Hashimoto, Copper-related deep levels and their annealing kinetics in germanium. Jpn. J. App. Phys. **28**, 763–769 (1989). https://doi.org/10.1143/JJAP.28.763
208. R. Gleichmann, H. Blumtritt, J. Heydenreich, New morphological types of CuSi precipitates in silicon and their electrical effects. Phys. Status Solidi A **78**, 527–538 (1983). https://doi.org/10.1002/pssa.2210780219
209. A. Correia, D. Ballutaud, A. Boutry-Forveille, J.L. Maurice, Effects of copper and oxygen precipitation during thermal oxidation of silicon: an electron-beam-induced current study. J. Appl. Phys. **78**, 6543–6553 (1995). https://doi.org/10.1063/1.360475
210. L. Zhong, F. Shimura, Dependence of lifetime on surface concentration of copper and iron in silicon wafers. Appl. Phys. Lett. **61**, 1078–1080 (1992). https://doi.org/10.1063/1.107696
211. G.A. Adegboyega, A. Poggi, Copper impurities and their annealing behaviour in FZ silicon. Phys. Status Solidi A **143**, 373–377 (1994). https://doi.org/10.1002/pssa.2211430222
212. A.L.P. Rotondaro, T.Q. Hurd, A. Kaniava, J. Vanhellemont, E. Simoen, M.M. Heyns, C. Claeys, G. Brown, Impact of Fe and Cu contamination on the minority carrier lifetime of silicon substrates. J. Electrochem. Soc. **143**, 3014–3019 (1996). https://doi.org/10.1149/1.1837141
213. A.A. Istratov, C. Flink, H. Hieslmair, T. Heiser, E.R. Weber, Influence of interstitial copper on diffusion length and lifetime of minority carriers in p-type silicon. Appl. Phys. Lett. **71**, 2121–2123 (1997). https://doi.org/10.1063/1.119355
214. J.L. Benton, T. Boone, D.C. Jacobson, P.J. Silverman, J.M. Rosamilia, C.S. Rafferty, S. Weinzierl, B. Vu, Electrical properties of cobalt and copper contamination in processed silicon. J. Electrochem. Soc. **148**, G326–G329 (2001). https://doi.org/10.1149/1.1370970
215. R. Sachdeva, A.A. Istratov, E.R. Weber, Recombination activity of copper in silicon. Appl. Phys. Lett. **79**, 2937–2939 (2001). https://doi.org/10.1063/1.1415350
216. M.L. Polignano, D. Caputo, C. Carpanese, G. Salvà, L. Vanzetti, Low temperature drive-in of surface-deposited copper in silicon wafers. Eur. Phys. J. Appl. Phys. **27**, 435–438 (2004). https://doi.org/10.1051/epjap:2004138
217. W. Wang, D. Yang, X. Ma, Y. Zeng, D. Que, Effect of annealing atmosphere on the recombination activity of copper precipitates formed by rapid thermal process in conventional and nitrogen-doped Czochralski silicon wafers. J. Appl. Phys. **103**, 014912/1-5 (2008). https://doi.org/10.1063/1.2830859
218. A.A. Istratov, H. Hedemann, M. Seibt, P.F. Vyvenko, W. Schröter, C. Flink, T. Heiser, H. Hieslmair, E.R. Weber, Electrical and recombination properties of precipitated and interstitial copper in silicon. Electrochem. Soc. Proc. **98910**, 948–966 (1998)
219. A.A. Istratov, H. Hedemann, M. Seibt, O.F. Vyvenko, W. Schröter, T. Heiser, C. Flink, H. Hieslmair, E.R. Weber, Electrical and recombination properties of copper-silicide precipitates in silicon. J. Electrochem. Soc. **145**, 3889–3898 (1998). https://doi.org/10.1149/1.1838889
220. C. Flink, H. Feick, S.A. McHugo, A. Mohammed, W. Seifert, H. Hieslmair, T. Heiser, A.A. Istratov, E.R. Weber, Formation of copper precipitates in silicon. Physica B **273–274**, 437–440 (1999). https://doi.org/10.1016/S0921-4526(99)00499-8
221. P.S. Plekhanov, T.Y. Tan, Schottky effect model of electrical activity of metallic precipitates in silicon. Appl. Phys. Lett. **76**, 3777–3779 (2000). https://doi.org/10.1063/1.126778
222. H. Savin, M. Yli-Koski, A. Haarahiltunen, Role of copper in light induced minority-carrier lifetime degradation of silicon. Appl. Phys. Lett. **95**, 152111/1-3 (2009). https://doi.org/10.1063/1.3250161
223. J. Lindroos, M. Yli-Koski, A. Haarahiltunen, M.C. Schubert, H. Savin, Light-induced degradation in copper-contaminated gallium-doped silicon. Phys. Status Solidi RRL **7**, 262–264 (2013). https://doi.org/10.1002/pssr.201307011

224. A. Inglese, A. Focareta, F. Schindler, J. Schön, J. Lindroos, M.C. Schubert, H. Savin, Light-induced degradation in multicrystalline silicon: the role of copper. Energy Proc. **92**, 808–814 (2016). https://doi.org/10.1016/j.egypro.2016.07.073
225. P.R. Wilshaw, T.S. Fell, M.D. Coteau, EBIC contrast of defects in semiconductors. J. de Physique IV, **C6**, C6-3/C6-14 (1991)
226. M. Saritas, A.R. Peaker, Deep states associated with oxidation induced stacking faults in RTA p-type silicon before and after copper diffusion. Solid-State Electron. **38**, 1025–1034 (1995). https://doi.org/10.1016/0038-1101(95)98671-O
227. A. Broniatowski, Multicarrier trapping by copper microprecipitates in silicon. Phys. Rev. Lett. **62**, 3074–3077 (1989). https://doi.org/10.1103/PhysRevLett.62.3074
228. J.F. Hamet, R. Abelaoui, G. Nouet, G. Allais, Precipitation at grain boundaries in silicon. Mater. Sci. Eng. B **4**, 143–145 (1989). https://doi.org/10.1016/0921-S107(89)90236-6
229. A. Broniatowski, C. Haut, The electronic properties of copper-decorated twinned boundaries in silicon. Philos. Mag. Lett. **62**, 407–415 (1990). https://doi.org/10.1080/09500839008215543
230. J.F. Hamet, R. Abdelaoui, G. Nouet, Effects of copper precipitation in $\Sigma = 25$ silicon bicrystals by deep-level transient spectroscopy. J. Appl. Phys. **68**, 638–645 (1990). https://doi.org/10.1063/1.346792
231. R. Rizk, X. Portier, G. Allais, G. Nouet, Electrical and structural properties of copper and nickel precipitates in a $\Sigma = 25$ silicon bicrystal. J. Appl. Phys. **76**, 952–958 (1994). https://doi.org/10.1063/1.357773
232. Y. Ohno, K. Kutsukake, M. Deura, I. Yonenaga, Y. Shimizu, N. Ebisawa, K. Inoue, Y. Nagai, H. Yoshida, S. Takeda. Recombination activity of nickel, copper, and oxygen atoms segregating at grain boundaries in mono-like silicon crystals. Appl. Phys. Lett. **109**, 142105/1-4 (2016). https://doi.org/10.1063/1.4964440
233. X. Yu, J. Lu, G. Rozgonyi, Impact of Cu contamination on the electrical properties of a direct silicon bonded (110)/(100) interfacial grain boundary. J. Appl. Phys. **104**, 113702/1-7 (2008). https://doi.org/10.1063/1.3032655
234. M. Kittler, C. Ulhaq-Bouillet, V. Higgs, Influence of copper contamination on recombination activity of misfit dislocations in SiGe/Si epilayers: temperature dependence of activity as a marker characterizing the contamination level. J. Appl. Phys. **78**, 4573–4583 (1995). https://doi.org/10.1063/1.359802
235. J.G. Lee, S. Roy Morrison, Copper passivation of dislocations in silicon. J. Appl. Phys. **64**, 6679–6683 (1988). https://doi.org/10.1063/1.342023
236. S.K. Estreicher, Copper, lithium, and hydrogen passivation of boron in c-Si. Phys. Rev. B **41**, 5447–5450 (1990). https://doi.org/10.1103/PhysRevB.41.5447
237. M.O. Aboelfotoh, B.G. Svensson, Copper passivation of boron in silicon and boron reactivation kinetics. Phys. Rev. B **44**, 12742–12747 (1991). https://doi.org/10.1103/PhysRevB.44.12742
238. B.G. Svensson, M.O. Aboelfotoh, L.J. Lindström, Silicide formation and the generation of point defects in silicon. Phys. Rev. Lett. **66**, 3028–3031 (1991). https://doi.org/10.1103/PhysRevLett.66.3028
239. F.D. McDaniel, S.A. Datar, B.N. Guo, S.N. Renfrow, Z.Y. Zhao, J.M. Anthony, Low-level copper concentration measurements in silicon wafers using trace-element accelerator mass spectrometry. Appl. Phys. Lett. **72**, 3008–3010 (1998). https://doi.org/10.1063/1.121523
240. V. Bertagna, F. Rouelle, R. Erre, M. Chemla, Electrochemical test for silicon surface contamination by copper traces in HF, HF + Cl and HF + NH_4F dilute solutions. Semicond. Sci. Technol. **15**, 121–125 (2000). https://doi.org/10.1088/0268-1242/15/2/307
241. H. Shimizu, S. Ishiwari, Pack-extraction method combined with inductively coupled plasma mass spectroscopy to monitor metal contaminants on surfaces of silicon wafers. Semicond. Sci. Technol. **15**, 776–781 (2000). https://doi.org/10.1088/0268-1242/15/7/320
242. M.B. Shabani, Y. Shiina, F.G. Kirscht, Y. Shimanuki, Recent advanced applications of AAS and ICP-MS in the semiconductor industry. Mater. Sci. Eng. B **102**, 238–246 (2003). https://doi.org/10.1016/S0921-5107(02)00739-0

243. S.F. Koh, W.L. Lim, T.Y. Tou, Enhanced detection of copper impurity in silicon wafer by dynamic secondary ion mass spectrometry. Electrochem. Solid-State Lett. **14**, H110–H113 (2011). https://doi.org/10.1149/1.3526095

244. L. Köster, P. Blöchl, L. Fabry, Element specific diagnosis using microwave reflection photoconductive decay. Jpn. J. Appl. Phys. **34**, 932–936 (1995). https://doi.org/10.1143/JJAP.34.932

245. M. Itsumi, Y. Sato, K. Imai, N. Yabumoto, Characterization of metallic impurities in Si using a recombination-lifetime correlation method. J. Appl. Phys. **82**, 3250–3255 (1997). https://doi.org/10.1063/1.365632

246. K. Kurita, T. Shingyouji, Identification and quantification of transition metal impurities in Czochralski silicon wafers using microwave photoconductive decay lifetime measurements. Jpn. J. Appl. Phys. **37**, 5861–5865 (1998). https://doi.org/10.1143/JJAP.37.5861

247. D.A. Ramappa, W.B. Henley, Surface photovoltage analysis of copper in p-type silicon. Appl. Phys. Lett. **72**, 2298–2300 (1998). https://doi.org/10.1063/1.121341

248. M. Boehringer, J. Hauber, Improved copper detection in hydrofluoric acid by recombination lifetime measurements on dedicated silicon substrates. Appl. Phys. Lett. **80**, 527–529 (2002). https://doi.org/10.1063/1.1436272

249. M. Boehringer, J. Hauber, S. Passefort, K. Eason, In-line copper monitoring using non-contact Q-V-SPV techniques. Electrochem. Soc. Proc. **03**, 42–49 (2003)

250. M. Boehringer, J. Hauber, S. Passefort, K. Eason, In-line copper contamination monitoring using noncontact Q-VSPV techniques. J. Electrochem. Soc. **152**, G1–G6 (2005). https://doi.org/10.1149/1.1823992

251. H. Savin, M. Yli-Koski, A. Haarahiltunen, H. Talvitie, J. Sinkkonen, Contactless diagnostic tools and metallic contamination in the semiconductor industry. ECS Trans. **11**(3), 319–329 (2007). https://doi.org/10.1149/1.2778674

252. M. Itsumi, Method for determining metal contamination by combining p-type Si and n-type Si recombination lifetime measurements. Appl. Phys. Lett. **63**, 1095–1097 (1993). https://doi.org/10.1063/1.109791

253. W.B. Henley, D.A. Ramappa, L. Jastrzebski, Detection of copper contamination in silicon by surface photovoltage diffusion length measurements. Appl. Phys. Lett. **74**, 278–280 (1999). https://doi.org/10.1063/1.123280

254. D.A. Ramappa, Surface photovoltage analysis of phase transformation of copper in p-type silicon. Appl. Phys. Lett. **76**, 3756–3758 (2000). https://doi.org/10.1063/1.126797

255. M. Yli-Koski, M. Palokangas, A. Haarahiltunen, H. Väinölä, J. Storgårds, H. Holmberg, J. Sinkkonen, Detection of low-level copper contamination in p-type silicon by means of microwave photoconductive decay measurements. J. Phys. Condens. Matter **14**, 13119–13125 (2002). https://doi.org/10.1088/0953-8984/14/48/358

256. H. Väinölä, M. Yli-Koski, A. Haarahiltunen, J. Sinkkonen, Sensitive copper detection in p-type silicon using μ-PCD. Electrochem. Soc. Proc. **20**, 249–257 (2002)

257. H. Väinölä, M. Yli-Koski, A. Haarahiltunen, J. Sinkkonen, Sensitive copper detection in p-type silicon using μPCD. J. Electrochem. Soc. **150**, G790–G794 (2003). https://doi.org/10.1149/1.1624845

258. M. Yli-Koski, H. Savin, E. Saarnilehto, A. Haarahiltunen, J. Sinkkonen, G. Berenyi, T. Pavelka, Measurement of copper in p-type silicon using charge-carrier lifetime methods. Solid State Phenom. **108–109**, 643–648 (2005). https://doi.org/10.4028/www.scientific.net/SSP.108-109.643

259. H. Väinölä, E. Saarnilehto, M. Yli-Koski, A. Haarahiltunen, J. Sinkkonen, G. Berenyi, T. Pavelka, Quantitative copper measurement in oxidized p-type silicon wafers using microwave photoconductivity decay. Appl. Phys. Lett. **87**, 032109/1-3 (2005). https://doi.org/10.1063/1.1999008

260. A. Inglese, J. Lindroos, H. Savin, Accelerated light-induced degradation for detecting copper contamination in p-type silicon. Appl. Phys. Lett. **107**, 052101/1-4 (2015). https://doi.org/10.1063/1.4927838

261. A. Inglese, J. Lindroos, H. Vahlman, H. Savin, Recombination activity of light-activated copper defects in p-type silicon studied by injection- and temperature-dependent lifetime spectroscopy. J. Appl. Phys. **120**, 125703/1-8 (2016). https://doi.org/10.1063/1.4963121
262. J. Lindroos, H. Savin, Formation kinetics of copper-related light-induced degradation in crystalline silicon. J. Appl. Phys. **116**, 234901/1-5 (2014). https://doi.org/10.1063/1.4904197
263. Y. Boulfrad, J. Lindroos, M. Wagner, F. Wolny, M. Yli-Koski, H. Savin, Experimental evidence on removing copper and light-induced degradation from silicon by negative charge. Appl. Phys. Lett. **105**, 182108/1-3 (2014). https://doi.org/10.1063/1.4901533
264. Y. Boulfrad, J. Lindroos, A. Inglese, M. Yli-Koski, H. Savin, Reduction of light-induced degradation of boron-doped solar-grade Czochralski silicon by corona charging. Energy Proc. **38**, 531–535 (2013). https://doi.org/10.1016/j.egypro.2013.07.313
265. J. Lindroos, Y. Boulfrad, M. Yli-Koski, H. Savin, Preventing light-induced degradation in multicrystalline silicon. J. Appl. Phys. **115**, 154902/1-5 (2014). https://doi.org/10.1063/1.4871404
266. H. Vahlman, A. Haarahiltunen, W. Kwapil, J. Schön, A. Inglese, H. Savin, Modeling of light-induced degradation due to Cu precipitation in p-type silicon. I. General theory of precipitation under carrier injection. J. Appl. Phys. **121**, 195703/1-10 (2017). https://doi.org/10.1063/1.4983454
267. H. Vahlman, A. Haarahiltunen, W. Kwapil, J. Schön, A. Inglese, H. Savin, Modeling of light-induced degradation due to Cu precipitation in p-type silicon. II. Comparison of simulations and experiments. J. Appl. Phys. **121**, 195704/1-11 (2017). https://doi.org/10.1063/1. 4983455
268. T. Heiser, A. Mesli, Determination of the copper diffusion coefficient in silicon from transient ion-drift. Appl. Phys. A **57**, 325–328 (1993). https://doi.org/10.1007/BF00332285
269. A. Zamouche, T. Heiser, A. Mesli, Investigation of fast diffusing impurities in silicon by a transient ion drift method. Appl. Phys. Lett. **66**, 631–633 (1995). https://doi.org/10.1063/1.1.114142
270. T. Heiser, E.R. Weber, Transient ion-drift-induced capacitance signals in semiconductors. Phys. Rev. B **58**, 3893–3903 (1998). https://doi.org/10.1103/PhysRevB.58.3893
271. T. Heiser, A.A. Istratov, C. Flink, E.R. Weber, Electrical characterization of copper related defect reactions in silicon. Mater. Sci. Eng. B **58**, 149–154 (1999). https://doi.org/10.1016/S0921-5107(98)00287-6
272. T. Heiser, A. Belayachi, J.P. Schunck, Copper behavior in bulk silicon and associated characterization techniques. J. Electrochem. Soc. **150**, G831–G837 (2003). https://doi.org/10.1149/1.1627351
273. T. Heiser, A. Belayachi, J.P. Schunck, Copper behavior in bulk silicon and associated characterization techniques. Electrochem. Soc. Proc. **03**, 5–20 (2003)
274. A. Belayachi, T. Heiser, J.P. Schunck, S. Bourdais, P. Bloechl, A. Huber, A. Kempf, Optimisation of a combined transient-ion-drift/rapid thermal annealing process for copper detection in silicon. Mater. Sci. Eng. B **102**, 218–221 (2003). https://doi.org/10.1016/S0921-5107(02)00735-3
275. A. Belayachi, T. Heiser, J.P. Schunck, A. Kempf, Influence of light on interstitial copper in p-type silicon. Appl. Phys. A **80**, 201–204 (2005). https://doi.org/10.1007/s00339-004-3038-7

Chapter 6
Electrical Properties of Metals in Si and Ge

The behavior of Fe and Cu, the two dominant contaminants in semiconductor processing, was extensively discussed in Chap. 5. In this chapter the focuss is on the other transition metals, their properties and their electrical activity.

First the attention is given to some important metals commonly used in semiconductor manufacturing, implying that they are intensionally introduced in some of the process steps (e.g. as silicides, metal layers, diffusion barrier, hard masks, gate dielectric etc.). The risk that they act as a contaminant degrading the device properties has to be minimized but can in some cases not be avoided. Sections 6.1–6.3 discuss the behavior and properties of nickel, cobalt, and chromium. First their configuration in the semiconductor lattice, i.e., interstitial and/or substitutional position, interactions with dopants and other transition metals, cluster formation, precipitates and behavior during thermal anneal are analyzed. Then the electrical properties are reviewed, including the different energy levels in the band gap and the impact on electrical performance parameters. Finally the detection and identification techniques are discussed.

Sections 6.4–6.8 review the knowledge about titanium, molybdenum, palladium, platinum and gold. The other 3*d* transition metals scandium, vanadium, manganese and zinc are the subject of Sects. 6.9–6.12. The 4*d* metals including zirconium, niobium, ruthenium, rhodium, silver and cadmium are treated in Sects. 6.13–6.18. Finally the 5*d* transition metals hafnium, tantalum, tungsten, rhenium, osmium, iridium and mercury are addressed in Sects. 6.19–6.24. For some of these metals the available information in the literature is very limited but for the sake of completeness the metals are included in the list.

The behavior of these metals in silicon based on the literature up to the year 2000 was the subject of the comprehensive book by K. Graff, so that here the focus is more on later publications. In addition the behavior of these metal in SiGe and Ge is also critically reviewed.

C. Claeys and E. Simoen, *Metal Impurities in Silicon- and Germanium-Based Technologies*, Springer Series in Materials Science 270,
https://doi.org/10.1007/978-3-319-93925-4_6

6.1 Nickel in Si and Ge

6.1.1 Ni-related Point Defects and Complexes

Ni in silicon gives rise to two EPR centers: in n-type silicon a signal with orthorhombic *I* symmetry and effective spin $S = 1/2$ has been ascribed to Ni_s^- [1], while in p-type silicon and at temperatures below 4 K a line with $S = 1/2$ and monoclinic-*I* symmetry was found [2], which has more recently been assigned to Ni_s as well [3]. The corresponding structure is represented in Fig. 6.1 [3], showing that the nickel atom is bound to one of its four nearest silicon neighbors, resulting in a trigonal symmetry. This results in an amphoteric impurity with a donor level in the bottom half and an acceptor level in the top half of the band gap. However, this electrically active substitutional nickel forms only a small fraction of the total solubility in silicon, which is dominated by interstitial nickel. As will be further documented below, no electrically active level of Ni_i has been found so far, indicating that the donor level is very shallow, or is resonant with the valence band. This implies that interstitial nickel is neutral in silicon, which has important consequences for the reactions of Ni with other impurities and for its precipitation behavior.

On-line emission channeling of ^{65}Ni ion-implanted n^+, n, p and p^+ silicon has revealed different stable sites [4]. In the case of n-type Si, nickel was found in *S* sites, *BC* sites and near-*T* sites up to 400 °C. At 400 °C, the majority resides in near-*BC* sites, while at 500 °C a site change to near-*T* sites occurs. Similar results were obtained on n^+-silicon [4], whereby the site change from near-*BC* to near-*T* sites occurs at 600 °C. In p^+-silicon, a higher near-*T* fraction exists already at 400 °C, reaching a maximum at 500 °C [4]. Nickel in ideal-*S* sites is supposed to be trapped in ion-implanted-induced vacancies. In the case of interaction with V_2, ideal-*BC* sites should be preferred, which can be slightly displaced in case Ni is

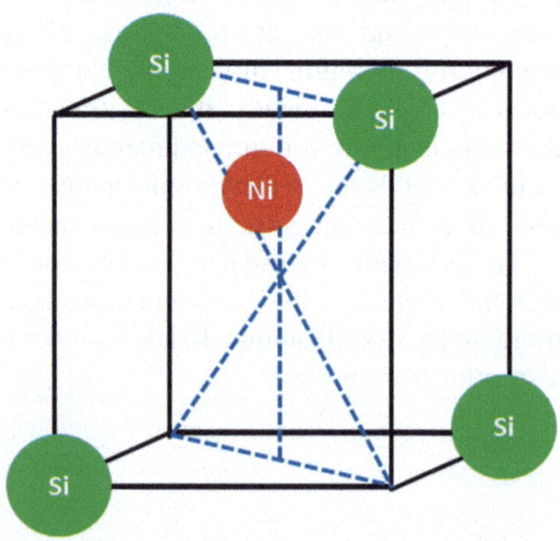

Fig. 6.1 Configuration of Ni_s^+. The nickel atom is bound to one of its four nearest Si neighbors (upper left corner), resulting in a trigonal symmetry and then shifted in a (110)-plane (indicated by dotted lines) due to the Jahn-Teller effect. After Effey-Schwickert et al. [3]

trapped in a larger multi-vacancy cluster [4]. The origin of the near-*T* site is more difficult to comprehend. The annealing behavior of the atomic displacement versus temperature suggests the existence of different complex structures. The estimated activation energies for dissociation of ^{65}Ni from near-*BC*, near-*T* and ideal-*S* sites depend on the doping type and density and are in the range of 1.6–2.8 eV [4]. It was also concluded that the ideal-*S* site (1.6–2.0 eV) is less stable for Ni, compared with Cu_s (2.9 eV) and Fe_s (3.5 eV).

Nickel is one of the few metal impurities which have been detected by EPR in Ge [5]. Substitutional nickel is paramagnetic in the singly negative charge state, whereby the ^{61}Ni isotope (abundance 1.25%) leads to hyperfine structure in the spectra, according to a spin 3/2. It has been found that Ni_s is a Jahn-Teller distorted center, whereby the atom moves out of the substitutional site into any cubic direction [5].

6.1.2 Precipitation and Co-precipitation of Ni

According to Fig. 2.1, nickel has a retrograde equilibrium solubility in silicon reducing with temperature for typical processing conditions. This implies that under slow cooling conditions close to thermodynamic equilibrium, nickel will be in super-saturation, which is the driving force for precipitation with an energy change of $k_BT\ln[C_0(T_d)] \sim 1$ eV per precipitated atom. The surface is one of the main sites for heterogeneous precipitation, giving rise to the so-called haze formation (shallow surface pits) after an RTA step of Ni contaminated silicon. In the case of fast cooling (quenching) a non-equilibrium solution of interstitial nickel can be frozen in at room temperature, which will be highly mobile and can either out-diffuse (precipitate at the surface) or become trapped by other point defects/impurities. In this section, mainly the formation of $NiSi_2$ precipitates will be discussed. As will be seen, in mc-silicon for PV applications, also co-precipitation with other TMs can occur, driven by a lowering of the system's energy.

6.1.2.1 Homogeneous and Heterogeneous Precipitation of $NiSi_2$

Early studies have reported several types of silicide phases precipitating in the bulk [6] and on extended defects [7, 8] (dislocations, twins, stacking faults) in silicon. Nevertheless, in c-Si, the preferred phase is $NiSi_2$, given the small mismatch in lattice parameter (0.4%) [6–9]. After RTA treatment, including a fast cool down to low temperature, surface precipitates give rise to shallow pits (haze formation) [10–12]. No pits are found in the bulk of the wafers, indicating the efficiency of the surface as a metal gettering site during RTA treatments. Careful TEM analysis reveals that the pits consist of $NiSi_2$ precipitates with the CaF structure (lattice parameter 0.5406 nm; 0.5248 nm for Si) and punched-out dislocation loops and stacking faults. The latter correspond with silicon interstitials emitted by the relaxation of the strained precipitates and condensed in loops or SFs. The SFs are formed at larger pits, i.e., for a higher level of Ni contamination. The occurrence of haze can be used in a standard gettering test, where back-side Ni contamination is applied and the pits at the front surface are tested, to evaluate the efficiency of gettering sites in the interior (or back-side) of the wafer [10–12].

A detailed High-Resolution TEM (HRTEM) study of the ***homogeneous precipitation*** of nickel in silicon has been carried out by the Göttingen Group [13–16]. After rapid quenching (10^3 K/s) from 1050 to 850 °C thin A-type platelets have been obtained consisting of two $NiSi_2$ {111} planes coherent with the Si{111} habit plane [13]. A-type orientation means that the platelets have the same orientation as the silicon matrix. Both precipitate-matrix interfaces are formed by Si–Si bonds, leaving Ni in a 7-fold coordination in the whole particle instead of the usual 8-fold coordination in bulk $NiSi_2$. The particle is bounded by a dislocation with a Burgers vector $\boldsymbol{b} = (a_0/4)\langle 111\rangle$ (Fig. 6.2) [15]. The growth of the platelets runs exclusively by diffusion of Ni to the dislocation and by its transformation to $NiSi_2$ within its core [13]. The diameter can change from 10 nm up to 0.9 μm by varying the diffusion temperature (750–1050 °C) and quenching rate. A considerable increase in platelet thickness was observed after additional annealing in the 500–900 °C range of samples quenched from 1050 °C, whereas the particle diameter remained the same. The driving force for this is a reduction of the interface energy [13].

Overall, the structural properties, shape and size of the $NiSi_2$ (and also $CoSi_2$) precipitates vary significantly with diffusion temperature, quenching rate, doping and the presence of other extended defects [15]. The atomic structure of $NiSi_2$ in the early stages of precipitation is a consequence of the nucleation impeded by a very large barrier and accelerated by the very fast growth (fast Ni_i diffussion). Even quenching at the fastest rate (2000 K/s) is not sufficient to keep Cu and Ni in solution at high temperatures [15]. The evolution of the precipitate shape and morphology is summarized in Fig. 6.3 [15]. The dislocation core surrounding the platelet contains the edges of two disilicide planes and offers an effective reaction channel to build the platelet from the fast interstitial impurity. The bounding

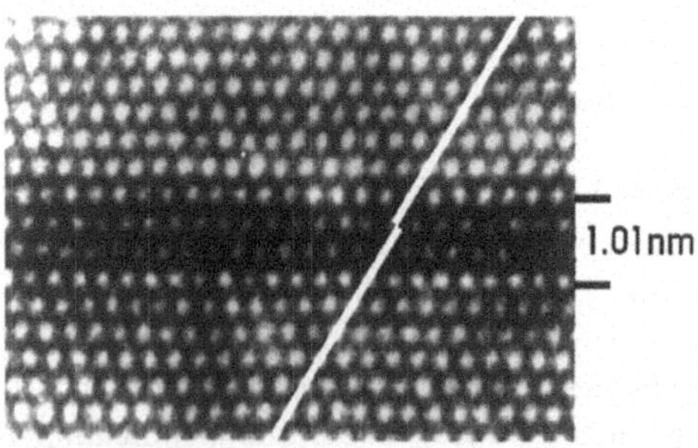

Fig. 6.2 High-resolution electron micrograph of a $NiSi_2$-platelet after quenching from high temperatures, the platelet has a thickness of two (111) layers and introduces stacking faults into silicon. The latter result from the structure of the precipitate/matrix interfaces which are built up by Si–Si bonds. Reproduced from Seibt et al. [14]. Copyright 1999 WILEY-VCH Verlag GmbH & Co

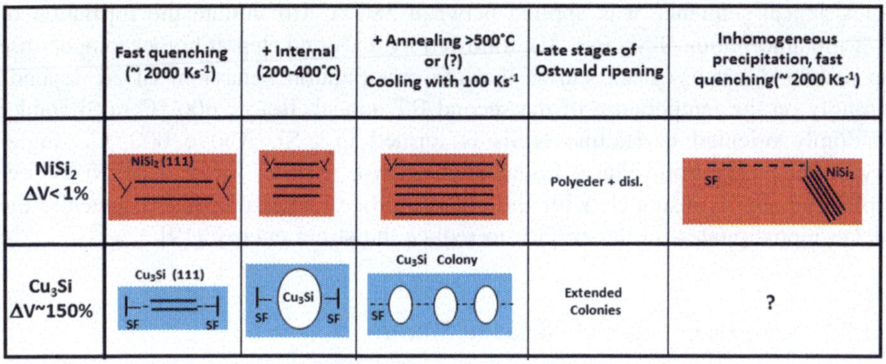

Fig. 6.3 Precipitate of $NiSi_2$ and Cu_3Si in Si obtained in a specimen saturated with Ni or Cu at diffusion temperatures T > 700 °C after fast quenching and additional annealing. Reproduced from Schröter et al. [15] with permission of Elsevier

dislocation is associated with a very large nucleation barrier. Annealing between 200 and 400 °C leads to a first transformation of the platelets into more compact ones; this occurs through what has been called an internal ripening process [15] and is associated with a reduction of the high-energy elements like the dislocation ring and its interface to the silicon. During further annealing at 500 °C or above classical Ostwald ripening in its first stage leads to platelets with increasing diameter and thickness and in the final stage to polyeders, now surrounded by dislocations to compensate the small but finite volume change linked with the $NiSi_2$ formation.

The heterogeneous precipitation of Ni on extended defects in silicon has been reviewed by Seibt et al. [17]. It has been observed that during rapid quenching from high temperature, nickel precipitates on oxygen-precipitation-induced SFs. It has been found that a type-B platelet forms at the Frank partial dislocation bounding the SF. In this case, the platelet has a twin orientation with respect to the silicon matrix, where the habit plane serves as the twinning plane. The SF plane and the habit plane of the precipitate are always different {111} planes. Precipitation of nickel at glide dislocations leads to a clearly different structure, where polyhedral $NiSi_2$ is found instead of platelets. The precipitates are bounded by {111} planes and have the shape of tetrahedra or octahedra and only occur in A-type orientation [17]. Such shapes can also be found in dislocation-free silicon after slow cooling from high temperature or as a result of precipitate coarsening (Ostwald ripening) after prolonged annealing at intermediate temperature [18]. It corresponds with a tendency of the system to reduce its total energy (surface + volume energy). In addition, the different thermal expansion coefficients of $NiSi_2$ and silicon frequently results in dislocation punching for large silicide precipitates [17].

The formation of $NiSi_2$ precipitates in nickel implanted silicon (50 keV − 2 × 10^{15} at/cm^2 above the amorphization threshold) has been investigated [19], in order to study its impact on the $NiSi_2$-induced crystallization of a-Si [20]. A two-step

RTA anneal sequence was applied between 380 °C (to initiate the formation of ~2 nm amorphous $NiSi_2$ in a-Si), followed by a second step at higher temperature up to 950 °C. It has been shown that the precipitation behaviour of Ni depends strongly on the temperature of the second RT anneal. Below 600 °C (a-Si stable) randomly oriented crystalline $NiSi_2$ is formed in a-Si. Above 600 °C, single-crystalline Si by Solid-Phase Epitaxial Regrowth (SPER) forms. The $NiSi_2$ precipitates have a mismatch with the silicon lattice. After prolonged anneals, the nickel reprecipitates on the surface to reduce the strain energy [19].

6.1.2.2 Co-precipitation of Nickel in Silicon

The advent of multi-crystalline silicon for PV applications gave the interest in the study of metal precipitates a boost, particularly in conjunction with the development of X-ray Fluorescence Spectroscopy [21–23]. Based on these studies, it has been concluded that in many cases, for energy lowering reasons, co-precipitation of several metal species occurs in silicon. This can happen according to different scenarios [23]. For example, the presence of precipitates of one metal species may aid the precipitation of another metal by providing energetically favorable nucleation sites via lattice strain or local native point defect compensation. One should distinguish between precipitates formed at low temperatures (below the eutectic point for the M-Si system—see Fig. 6.4 [23]) or at high temperatures.

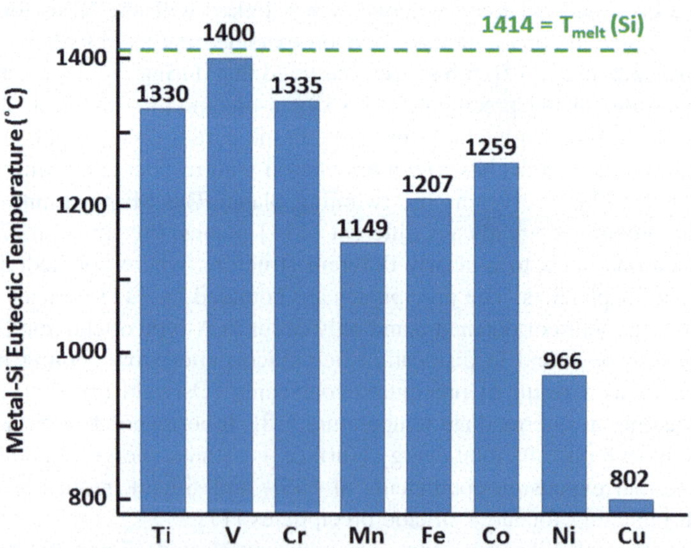

Fig. 6.4 Binary eutectic temperatures of Si and various 3*d* transition metal impurities, when cooling from the Si-rich side of the phase diagram. Note the extremely low eutectic point for Cu–Si, over 600 °C below the melting temperature of Si. Reproduced from Buonassisi et al. [23] with permission of Elsevier

Precipitates formed *at low temperature* can be explained by classical precipitation models. The chemical states of all measured precipitates are equilibrium (silicon-rich) single-metal silicides; no mixed metal silicide systems are observed in that case. Precipitate nucleation and growth is facilitated by strain minimization and the presence of defects or other metal precipitates. Precipitate or microcolony growth along preferred crystallographic planes is another manifestation of interface strain minimization [23].

Precipitates formed *at high temperature* as a result of high-temperature annealing (i.e., 1200 °C) may follow alternative routes, resulting in mixed metal-silicide systems [23]. Precipitates may be formed via:

- Liquid alloy intermediate phase
- Precipitation of several co-located dissolved impurity species into a mixed-metal silicide particle
- Via solid state diffusion of atoms between adjacent but distinct metal silicide phases.

Based on the extensive data set on mc-Si, it appears that the first mechanism is the more likely one, given the observed nano-scale compositions and large precipitate sizes [23]. For example, Ni–Fe–Cu–Si phases have been identified in mc-Si [21–23].

Co-precipitation of Cu and Ni in Cz Si wafers has also been studied, revealing the co-existence in close proximity of both Cu_3Si and $NiSi_2$ precipitates, containing small amounts of the other impurity [24, 25]. An example is shown in Fig. 6.5 [24]. It was concluded that the observed concentrations of Cu ($\sim$18–25%) and Ni ($\sim$2.3%) dissolved in the silicide precipitates may be different from the equilibrium solubilities due to the sample cooling rate and the strain in the matrix [24]. It is energetically more favorable for a Cu atom to occupy a Ni site in $NiSi_2$, as it does not require the emission of a silicon interstitial.

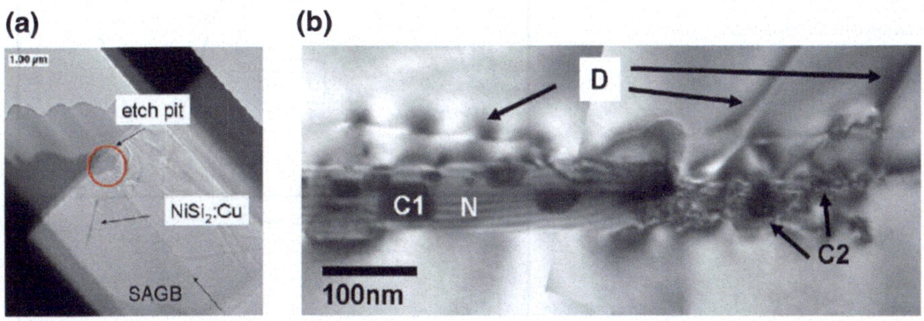

Fig. 6.5 Typical defect configuration obtained under nickel-rich conditions imaged along the Si [1$\bar{1}$0] zone axis. **a** Low magnification TEM micrograph showing the surface etch-pit (circle) used to localize the defect for FIB preparation, the grain boundary (SAGB) and a plate-shaped precipitate viewed edge-on ($NiSi_2$:Cu). **b** Bright-field image of one end of the platelet showing the main features of these defects: a nickel-rich platelet (N) containing copper-rich precipitates (C1), punched-out dislocations (D) and copper-rich precipitates attached to punched-out dislocations in the silicon matrix (C2). Reproduced from Seibt et al. [25]. Copyright 2009 WILEY-VCH Verlag GmbH & Co

6.1.3 *Electrical and Optical Activity of Ni*

From the foregoing, it is clear that interstitial nickel is electrically neutral, while only a minor fraction occupies lattice sites and has been shown to compensate the shallow-level doping density in both n- and p-type Si, pointing to an amphoteric nature [26, 27]. In addition, $NiSi_2$ precipitates are easily formed on the surface, in the interior of the wafer and on other extended defects. The question is: what is the electrical (optical) activity of these defects and how do they affect the recombination properties (lifetime) of the material. And are there other deep-level point defects associated with the presence of nickel in silicon?

6.1.3.1 Nickel-Related Point Defects and Complexes

The pioneering work of Ludwig and Woodbury [2] established the electronic structure of the nickel donor and acceptor center, although the original assignment to interstitial nickel was later corrected to the substitutional form [1, 3]. It indicates that Ni_s introduces an acceptor level ($Ni_s^{0/-}$) in the upper half of the band gap (n-type) and a donor level ($Ni_s^{+/0}$) in the lower half of the band gap (p-Si). Originally, Hall effect versus T [28] and photoconductivity measurements [29, 30] were employed to determine the levels of Ni_s in the band gap. Since the mid-seventies, the characterization of the Ni-related deep levels has been based on junction space-charge techniques and in particular DLTS [31–46]. Typical results are represented in Fig. 6.6 for n- and p-type silicon doped with Ni [42]. The main

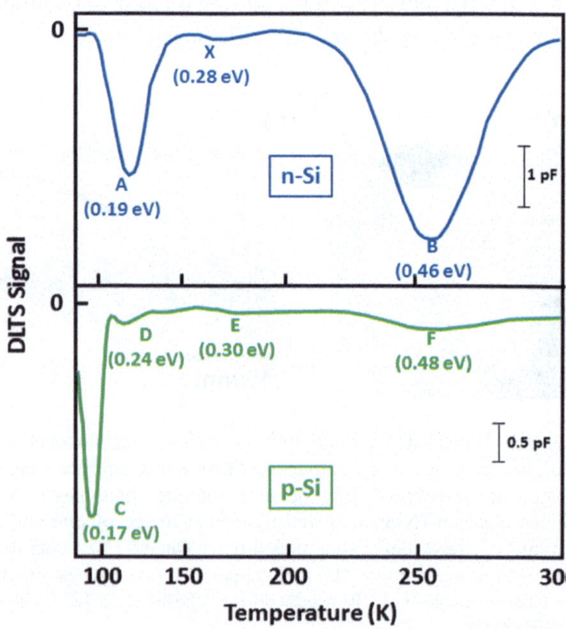

Fig. 6.6 DLT-spectra of nickel-doped n- and p-type silicon. Reproduced from Kitagawa et al. [42] by permission of Springer

Table 6.1 Activation enthalpies, apparent capture cross sections, and assignment for all defects seen in nickel-doped n- and p-type silicon. The enthalpies and capture cross sections are mean values over measurements on different samples. The errors indicated for the capture cross section are the statistical standard deviations. The statistical errors for the enthalpies are smaller than the last digit. After Scheffler et al. [47]

Label	E_a (eV)	σ_{na} (cm^2)	Assignment
E270	$E_C - 0.60$	$2.7 \pm 0.3 \times 10^{-14}$	$NiH_2(0/-)$
E230	$E_C - 0.45$	$1.4 \pm 0.1 \times 10^{-15}$	$Ni_s(0/-)$
E90	$E_C - 0.16$	$3 \pm 2 \times 10^{-15}$	CH
E90'	$E_C - 0.17$	$2.2 \pm 0.8 \times 10^{-15}$	NiH(0/−)
E45	$E_C - 0.07$	$1.2 \pm 0.7 \times 10^{-15}$	$Ni_s(-/2-)$
H280	$E_V + 0.58$	$2 \pm 1 \times 10^{-14}$	$NiH_2(0/-)$
H240	$E_V + 0.49$	$1.5 \pm 1 \times 10^{-14}$	NiH(+/0)
H190	$E_V + 0.46$	$2 \pm 2 \times 10^{-11}$	$NiH_3(0/-)$
H150	$E_V + 0.29$	$4 \pm 1 \times 10^{-15}$	Ni–H-related
H80	$E_V + 0.16$	$1.1 \pm 0.3 \times 10^{-14}$	$Ni_s(+/0)$

parameters are summarized in Table 6.1 and schematically represented in Fig. 6.7 [47]. It is clear that Ni_s has an additional (double) acceptor state close to the conduction band edge (level E45, with E standing for electron trap and the number indicating the DLTS peak temperature position in the DLT-spectra of [47]). It has been shown that the Ni_s levels are stable up to 900 °C (2 h) annealing.

As can be seen from Fig. 6.7 or Table 6.1, there are several other Ni-related deep levels present in the silicon band gap and even more have been reported in the past (see, for a summary, e.g., Fig. 1 in Ref. [42]). It has been shown that the DLT-spectrum of Ni-doped silicon first of all strongly relies on the diffusion temperature and cooling or quenching rate (sample history) [34]. Even the Ni deposition method may play a role. Finally, also the type of device (p-n junction, Schottky barrier, the presence of a silicide ohmic contact) has a strong impact on the deep levels [34, 42], which is related to the application of additional

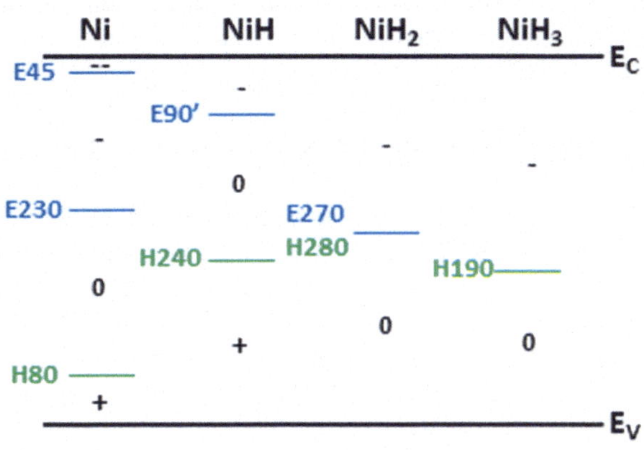

Fig. 6.7 Overview of the deep levels of nickel and the NiH complexes in the band gap of silicon with their charge states. After Scheffler et al. [47]

(low-temperature, i.e., 400 °C) annealing steps, where Ni may be removed from the depletion layer (gettering). Also the values of the majority capture cross sections vary from study to study: a temperature-independent value of $\sigma_n = 3.2 \times 10^{-17}$ cm^2 and a constant $\sigma_p = 1.8 \times 10^{-16}$ cm^2 has been reported before for Ni_s [43], which is significantly smaller than what can be found in Table 6.1. It has also been found that the capture barrier for the E45 double acceptor amounts to 33 meV, with a $\sigma_{p\infty} = 1.2 \times 10^{-14}$ cm^2 [47]. Based on the concentration profile of the Ni_s peaks, the in-diffusion [45, 46] and solubility [42] at different temperatures can be studied.

Besides the sample preparation and history, the interaction of Ni with other impurities or defects can also explain the additional deep levels found in the literature. It has been noted that Ni interacts with hydrogen [32, 47, 48], with oxygen [49] and with radiation-induced defects (vacancies and interstitials) [32, 50–53]. A recent report has clarified the different complexes formed between Ni_s and H by introducing hydrogen by a wet chemical etching ($HF:HNO_3:CH_3COOH$—3:5:3; CP4A) at room temperature [47]. The H-related deep levels, therefore, exhibit an in-diffusion profile decaying exponentially with depth from the surface (a complementary increasing profile is found for Ni_s), whereby the decay constant increases proportionally with the number of H atoms in the complex. In other words, the concentration profiles for the H_2-related deep levels vary two times faster than the H-related ones [47, 48]. The field-dependence of the levels has been studied and it was found that the H190 center exhibits a Poole-Frenkel shift, typical for a center with a Coulomb potential, i.e., it is a deep acceptor in p-type Si [47]. Also the large hole capture cross section in Table 6.1 supports this idea. Finally, in electron-irradiated n-type Si an electron trap has been found at $E_C - 0.365$ eV [51], which anneals at 300 °C [6, 53] and has been assigned to a NiVO complex, as its concentration varies complementary to the concentration of the dominant VO center.

Limited information can be found on the optical activity of Ni-doped silicon. One work has studied the Photoluminescence (PL) spectrum, demonstrating a deep donor-related PL band in p-type silicon at 812.4/811.1 meV at 11 K [54]. The PL signal disappears above 50 K by the dissociation of the localized exciton, consisting of a deeply bound hole to neutral Ni_s and a loosely bound electron. The binding energy to the trap was 11 meV and the dissociation energy turned out 14.7 meV [55]. This is similar to the case of other TMs in silicon (e.g., Cr).

6.1.3.2 Electrically Active Point Defects in Ge and SiGe

The energy level of the substitutional nickel acceptor level has been studied by DLTS in Si_xGe_{1-x} with x varying between 5 and 20% and for fully strained and fully relaxed hetero-epitaxial layers on silicon [55]. As shown in Fig. 6.8, the activation energy becomes higher for the fully strained case, while on the contrary, the acceptor level moves toward the conduction band in the strained case. Also shown is the activation energy for the Ni_s double acceptor level in Ge.

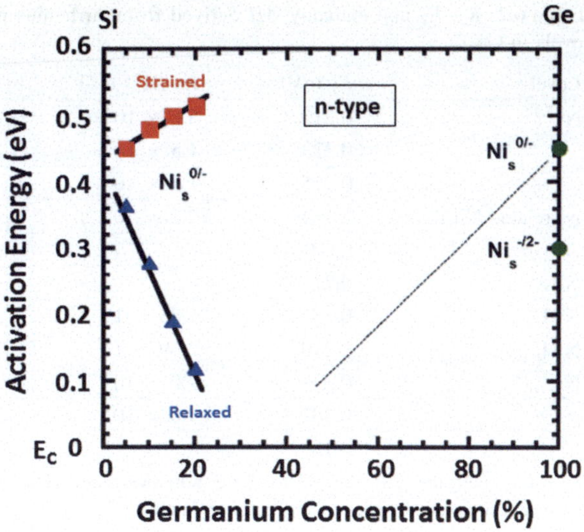

Fig. 6.8 Activation energy for the $Ni_s^{0/-}$ acceptor level in Si and Si_xGe_{1-x} in function of the strain and the Ge content, according to the data of Nauka and Kamins [55]. Also shown is the double donor level of Ni_s in Ge (see Table 6.2)

Nickel behaves very similar to copper in germanium: it is easily introduced as a thermal impurity in quenched germanium [56]. It diffuses very fast as an interstitial impurity. Similar as for the case of copper in a Ge n^+p junction, nickel impurities may be removed by drifting in a large electric field at room temperature, towards the n^+-contacts [57]. A mobility of the $E_V + 0.14$ eV level of $(2.0 \pm 1.1) \times 10^{-13}$ cm^2/Vs at 25 °C was derived, corresponding to a diffusion coefficient of $(5.1 \pm 2.7) \times 10^{-15}$ cm^2/s. However, the interstitial fraction is only a small part of the total solubility, where the electrically active substitutional configuration is the dominant one, in contrast to the case of silicon.

The deep levels associated with nickel-related point defects in germanium have been studied by DLTS. A summary of the main defects is given in Table 6.2, showing that substitutional nickel is a double acceptor, in line with the simple tetrahedral bonding model for a TM with two *s* electrons. It can accept two more electrons to complete the shell. Moreover, nickel reacts easily with Li and H, giving rise to a pair of hole traps in the bottom half of the Ge band gap. Full passivation can be achieved if two H or Li atoms are attached to Ni_s. Full details on the electrical behavior of Ni in Ge can be found in Chap. 5 of [58].

6.1.3.3 Electrical Activity of Nickel Precipitates

As discussed in Sect. 6.1.2.1, well-defined bilayer $NiSi_2$ precipitates can be formed by quenching nickel-diffused Si from high temperature [14, 15] (Fig. 6.2). The electrical activity of such precipitates has been evaluated by different techniques, mainly DLTS and EBIC [15]. Two types of DLT-spectra have been observed for $NiSi_2$ in n-type Si, illustrated by Figs. 6.9 and 6.10 [14]. In the first case a broad

Table 6.2 K_T, E_T and enthalpy ΔH derived from Arrhenius plots in DLTS for Ni-related deep levels in Ge

Level	E_T (eV)	K_T (K^{-2} s^{-1})	ΔH (eV)	References
$Ni_s^{0/-}$	0.217	5.8×10^8	–	[350]
	0.223	1.5×10^9	–	[351]
	0.215	1.5×10^8	0.207	[352]
Ni-related NiLi[a]	0.414	2.2×10^8	–	[350]
	0.40	–	–	[351]
NiLi	0.21	–	–	[353]
NiH	0.154	8.85×10^7	0.159	[352]
NiH	0.416	2.69×10^8	0.299	[353]
$Ni_s^{-/2-}$[a]	0.299	7.0×10^6	–	[350]
	0.303	7.0×10^7	–	[351]
	0.18	–	–	[213]

[a]In n-Ge, energies with respect to the conduction band. The other levels have been observed in p-Ge and are with respect to the valence band

peak is obtained in n-type Si, which widens toward lower temperatures when using a longer filling bias pulse, whereby the maximum position also shifts to lower T (Fig. 6.9). Annealing the platelets at 320 °C thickens them according to an internal ripening process and also changes the DLT-spectrum (Fig. 6.10): when changing t_p, the peak maximum position does not move but its amplitude increases proportional with $\ln(t_p)$ and the high-temperature flank of the amplitude-normalized peaks remains the same.

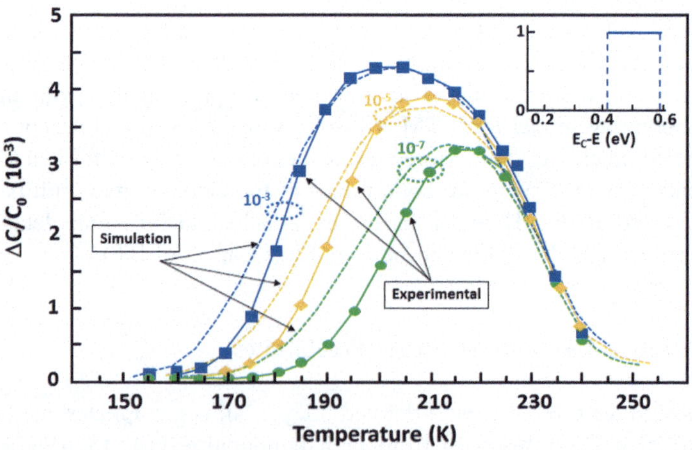

Fig. 6.9 Experimental DLT-spectra, for different pulse lengths t_p, and simulations based on a two-dimensional density-of-states function and a neutral occupation $F_p^{(N)} = 0$ for $NiSi_2$ platelets quenched from high temperature in n-type Si. Reproduced from Seibt et al. [14]. Copyright 1999 WILEY-VCH Verlag GmbH & Co

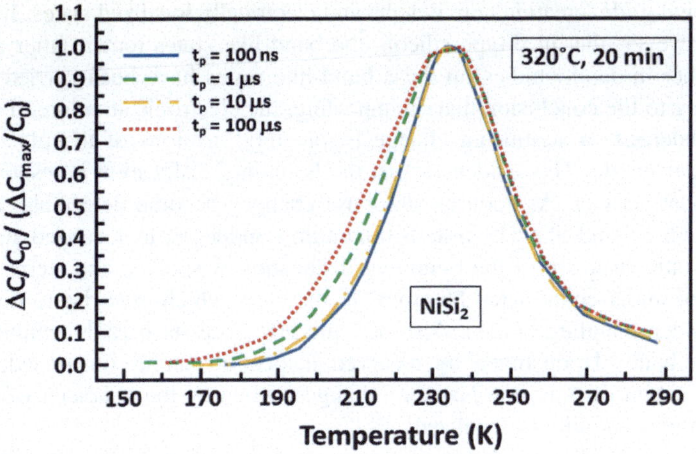

Fig. 6.10 DLT-spectrum of $NiSi_2$ platelets after internal ripening at 320 °C, for different pulse lengths t_p normalized to the line maximum. The coincidence of the high temperature sides after normalization is a fingerprint of localized states. Reproduced from Seibt et al. [14]. Copyright 1999 WILEY-VCH Verlag GmbH & Co

This different behavior has been interpreted in terms of a two-dimensional model for the density-of-state (DOS) for the extended defects, considering the internal equilibration time Γ_i, defined as the time it takes to establish electronic equilibrium between the defect levels in the band. If $\Gamma_{\rm i} \ll R_e^{-1}$, R_c^{-1} the states behave like a band and if $\Gamma_{\rm i} \gg R_e^{-1}$, R_c^{-1} they are considered as localized (R_e and R_c the emission and capture rate from a free carrier to or from the band to the defect level). Band-like states correspond with states extending along the silicide/silicon interface and along the bonding dislocation. The localized states are thought associated with steps in the interface or jogs, kinks and reconstruction defects in the dislocation core or narrow point defect clouds surrounding the platelet.

In the case of an one-dimensional defect, like, the dislocation bounding the $NiSi_2$ precipitate, the DOS is given by [16]:

$$N_p(E) = \frac{N_d}{\pi\sqrt{(E - E_p + \Delta E)(E_p - E)}} \tag{6.1}$$

with E_p and ΔE the upper edge and the width of the deep energy band, respectively and N_d the total density of active core states (cm^{-2}). For $NiSi_2$ platelets, the band of states in *n-type silicon* extends from $E_C - 0.2$ eV to $E_C - 0.4$ eV, corresponding with a capture barrier $\delta E_c = \alpha(F_p - F_p^{(N)})$ with $\alpha = 0.4$ eV and $F_p^{(N)}$ the occupation for which the defect is neutral at $\sim$0.4 [15]. With internal ripening, the platelets thicken and relax strain, whereby the bounding dislocation can leave its (111) plane by climb and glide, creating core defects and electrically localized states. It has been shown, however, that in p-type silicon, the band-like states remain after annealing [59]. Further in-depth studies on these band-like states have been carried out [60–62], leading to the conclusion that on annealing, the electronic structure of the $NiSi_2$ platelet undergoes a significant change before modifications of the platelet shape becomes detectable. This evidences that the bounding dislocation is responsible for the electrical activity. As soon as structural changes become detectable, the states change from a band-like 1D system (quantum conductor) to localized states [60]. The electronic structure of the bounding dislocation is specific and stems from the presence of under-coordinated Ni atoms in the core, which give rise to deep levels (so, no silicon dangling bonds). Also 30° and 90° glide-set partials exhibit no DBs but also no bands. Upon annealing, a curved dislocation can be constructed, whereby the coordination of only one Ni atom changes, whereby the character of the DOS changes from band-like to localized.

The electric-field dependence of the electron emission of the band-like states clearly exhibits a Poole-Frenkel (PF) lowering effect at fairly low electric fields ($<5 \times 10^4$ V/cm) [62]. This has been interpreted in terms of electric-field enhanced emission of carriers which takes into account the role of the deformation potential of the bounding dislocation loop of the precipitates. A shift of the DLTS peak on the low-T side is associated with the deformation potential of the conduction band, while the peak shift with electric field on the high-T side is due to the deformation potential of the valence band side [62]. This allows the distinction between a band of electron traps or a recombination center.

EBIC has also been performed on the $NiSi_2$ precipitates in n-type silicon [63–66]. It was observed that the minority carrier diffusion length can be approximated by $L_{diff} \sim 0.7 \times N_{EBIC}^{-1/3}$, i.e., proportional with the mean distance between the precipitates with density N_{EBIC}. In other words, in such material, the platelet precipitates form the dominant recombination path [63]. It was also noted that the image contrast was surprisingly large up to $\sim$0.4 and independent on the particle size. This could be explained by the presence of a potential barrier around the precipitates which can strongly increase the capture cross section for minority carriers and repulse the majority carriers. The presence of a potential barrier at the $NiSi_2$ precipitates was directly evidenced by electron holography [66] and amounts to 90 mV. This value is much lower than typical Schottky barrier heights which for the $NiSi_2$/n-type Si case are in the range of 0.65–0.79 eV [67, 68]. This can be explained by the screening of the barrier by the excess carriers created by the electron beam in EBIC [66]. A theoretical model for the space-charge region around a metallic platelet has been worked out by Donolato [69] to explain the EBIC contrast. No deep levels have been revealed in n- and p-type silicon for the $NiSi_2$/Si interface [70], so that there is no contribution from surface states to the electrical activity of this system.

6.1.3.4 Nickel-Decorated Extended Defects

Nickel-decorated extended defects have mainly been studied by EBIC [15, 64, 71, 72]. In the case of Ni-decorated dislocations, different types of behavior have been identified, depending on the degree of contamination [15, 64] as shown before for Fe. For high levels of contamination the recombination activity at room temperature is high (strong contrast) and reduces gradually upon cooling. For nearly 'clean' dislocations, the contrast at RT is low, indicating a poor recombination activity, while it peaks at low temperatures, which is indicative of a rather shallow level responsible for the recombination activity of dislocations in silicon along two competing paths [15]. The direct recombination through shallow core levels of the dislocations occurs directly and more efficient at low temperatures, while in the presence of impurities with a mid-gap level E_M, carriers can also tunnel from the 1D shallow states to the deep level for their recombination at higher temperatures. DLTS studies have confirmed the presence of a broad band of near mid-gap states in the case of Ni-decorated dislocations, with an amplitude which evolves according to $\ln(t_p)$ [73]. In fact, it was observed that the higher the Ni dose, the narrower the DLTS band became. This indicates that at higher contamination levels, mainly higher energy states are occupied. An additional factor which plays a role in DLTS on decorated dislocations is the fact that a high electric field can be present along the dislocation core, so that carriers trapped in the deep states can tunnel out again. This could explain why under certain circumstances, the DLTS signal of dislocations disappears or becomes negative at intermediate field strengths [73].

Also the electrical activity of grain boundaries (GBs) in silicon is dominated by metal decoration (Cu, Ni), as shown by early conductance [74] or EBIC [75] measurements. DLTS of a $NiSi_2$ decorated $\Sigma = 25$ GB in silicon as a function of annealing temperature showed marked changes [76]. For an anneal temperature $T_a = 600$ °C the electronic interface states were found to be shared between two overlapping narrow bands around $E_C - 0.63$ eV and $E_C - 0.52$ eV. For higher T_a, these bands shifted toward shallow positions and tended to reduce their width and turn into a unique trap level. This was accompanied by a lowering of the barrier height. From a structural viewpoint, the pure B-type $NiSi_2$ platelets at 800 °C became increasingly more A-type for higher T_a values, which are characterized by a lower barrier height [76]. The impact of the type of GB on the EBIC contrast has been shown by Xi et al., when studying the recombination activity of nickel-diffused mc-Si [77]. Nickel in-diffusion at 500 °C already produces strong contrast.

The impact of Ni and Au decoration on the electrical activity of misfit dislocations at the SiGe/Si interface has demonstrated that the EBIC contrast increased with increasing amounts of decoration [78–81]. At the same time, no uniform contrast along the misfits was found, but rather periodic dark spots due to the formation of $NiSi_2$ precipitates.

6.1.4 Impact of Ni on Recombination Lifetime

While Ni introduces quite a number of deep levels in silicon, from a device operation viewpoint the important parameter in many cases is the recombination (or generation) lifetime τ_r, which will depend for each deep level on the minority and majority capture cross section and on the concentration N_T. A few studies have reported the impact of nickel on the recombination lifetime of nickel-contaminated crystalline silicon [82–90]. Figure 6.11 represents the lifetime as a function of injection level Δn in p-type Cz silicon [84]. A very similar dependence is observed, irrespective of the nickel concentration, which indicates that the same Ni-related center dominates. The same conclusion was reached for n-type Si [84]. The dependence of τ on the wavelength of the excitation light source (white light versus an infrared source) indicates that surface recombination dominates in all cases. A different picture is found for n- and p-Si when comparing the dependence on the wafer resistivity. In p-Si, the recombination lifetime increases with increasing injection level Δp (Figs. 6.11 and 6.12a), while τ is rather insensitive to it for n-Si (Fig. 6.12b). Nevertheless, at high injection level ($\Delta p \gg p$ and $\Delta n \gg n$) the same lifetime value is found in both n- and p-Si. It turns out that the injection-level dependence of Ni-contaminated Si cannot be explained in terms of the classical SRH equation. This points to the fact that recombination is governed by a band of states, associated with Ni precipitates [84]. In this context, the recombination in p-Si is more representative for the deeper states of the band, while n-type material is more sensitive to the shallower levels, closer to the conduction band. It can also be derived that in p-Si, $\sigma_n > \sigma_p$ for the band of states, which dominate the carrier recombination. N-type Si is less sensitive to recombination by Ni precipitate states. Alternatively, one may hope to use the resistivity dependence of τ as a fingerprint for the presence of nickel centers. Almost no injection-level dependence is observed for 11 Ωcm n-type Si. In addition, the shape of the curves does not depend on the concentration of nickel precipitates.

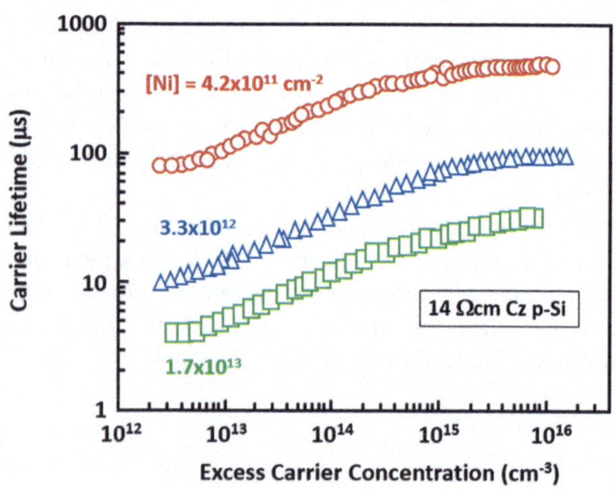

Fig. 6.11 The lifetimes due to Ni only, τ_{Ni}, after subtraction of the control data at each excess carrier density, for three 14 Ω cm p-type samples, as a function of excess carrier density. Reproduced from MacDonald [84] by permission of Springer

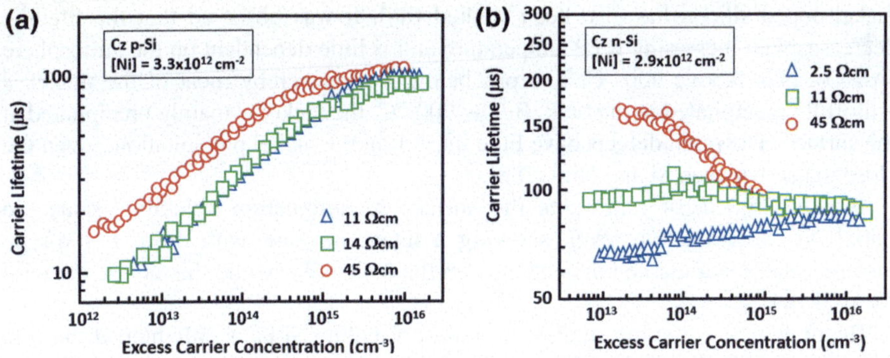

Fig. 6.12 Lifetime, τ_{Ni}, for **a** three p-type (Ni contamination 3.3×10^{12} cm^{-2}) and three **b** n-type resistivities (Ni contamination 2.9×10^{12} cm^{-2}, as a function of excess carrier density. (reproduced from MacDonald [84] by permission of Springer)

Figures 6.13 represents the Ni-related lifetime (τ_{Ni}) in function of the nickel surface concentration (contamination level controlled by dipping the wafers in a wet solution) at a fixed injection level ratio $\Delta n/n = 0.01$. It is clear that in p-type Si (Fig. 6.13a), τ_{Ni} is inversely proportional with [Ni], indicating that the density of precipitates increases with the contamination dose. When saturation is reached, the density of Ni precipitates no longer increases but the size grows [84]. This seems to happen earlier for heavily n-doped silicon (Fig. 6.13b). In comparison with Fe, Ni appears to be two decades less effective as a recombination center [84].

Overall, the recombination lifetime data is in qualitative agreement with EBIC measurements, demonstrating a relationship with the density of nickel precipitates ($L_{diff} \sim 0.7 \times N_{EBIC}^{-1/3}$). Besides classical furnace annealing, the impact on RTA-treated

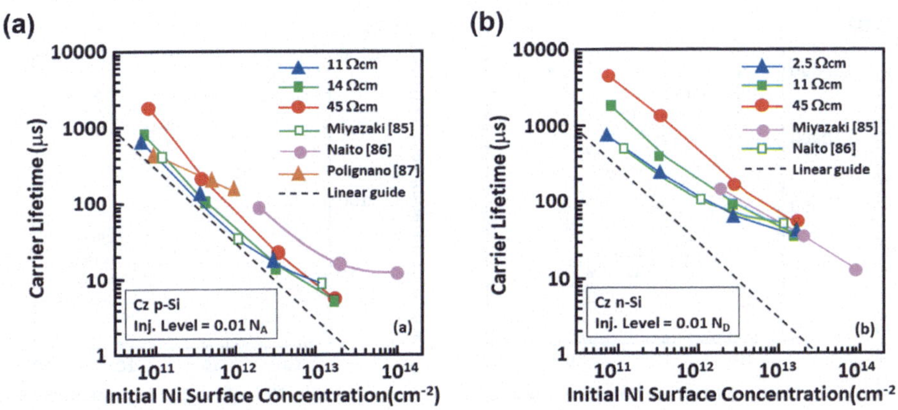

Fig. 6.13 τ_{Ni} as a function of the initial Ni surface concentration for p-type (**a**) and n-type (**b**) Si at a fixed injection level of $\eta = 0.01$. Experimental data is given from Miyazaki et al. [85], Naito and Nakashizu [86] and Polignano et al. [87]. The dashed line represents an inverse linear relationship. Reproduced from MacDonald [84] by permission of Springer

nickel-doped silicon has also been studied [88]. It was observed that the lifetime decreases with increasing RTP temperature and is little dependent on the atmosphere (Ar, O_2, N_2). Above 900 °C, τ_r drops below 1 μs, whereby most of the nickel is uniformly precipitated in the bulk. Below 900 °C, the nickel is mainly precipitated at the surface. The point defects have little impact on the nickel precipitation, given the close match between Si and $NiSi_2$ [88].

Finally, Figs. 6.14 represents the surface recombination velocity versus the initial Ni contamination level, showing a linear increase with dose. For n-type material, there is a marked dependence on the resistivity, while this is less the case for p-Si [84].

Recombination lifetime studies on multi-crystalline Si have enlightened the role of nickel in the electrical activity of grain boundaries (GBs) [91–94]. It has been shown that there exists a direct correlation between the regions of low lifetime and the presence of Fe, Ni and Cr [91]. They were found to be present in the same region, indicating the existence of preferred precipitation sites. The recombination activity of Cu_3Si and $NiSi_2$ precipitates was found to be similar, as detected by XBIC [93]. However, it was found that the recombination activity of GBs was not dominated by the presence of Ni (nickel precipitates) [93]. The release of nickel from structural defects in polycrystalline silicon has been reported by McHugo et al. [91].

Early lifetime measurements on germanium indicated that nickel acts as an efficient recombination center both in n- and p-type material [58]. At low temperatures, the acceptor level at $E_V + 0.23$ eV was found to limit the lifetime. However, as most of the early results have been obtained on dislocated material, they have to be treated with caution, as dislocations contribute to the recombination as well. A more recent microwave-probed photoconductivity study of deep levels in dislocation-free Ge wafers has identified 4 levels at room temperature [95], which agree quite well with the DLTS results reported in [96, 97].

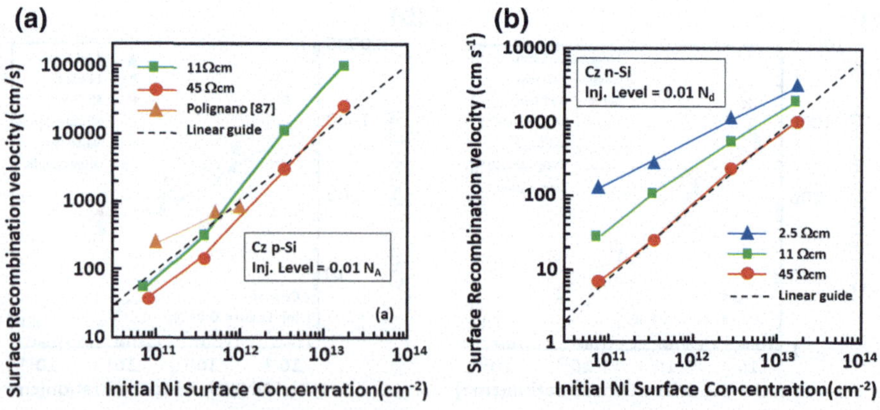

Fig. 6.14 Surface recombination velocity s as a function of initial Ni surface concentration for the **a** p-type and **b** n-type samples at a fixed low injection level $\eta = 0.01$. The dashed line represents a linear dependence. Reproduced from MacDonald [84] by permission of Springer

6.2 Cobalt in Si and Ge

6.2.1 Co-related Species in Si

One of the problems related with studying the status of Co in silicon is the fact that ESR failed to yield information—most likely because of the small concentration (solubility) of the Co_s and Co_iB species. Also more recent DLTS results (see next paragraph) give rise to controversial results. Fortunately, the decay of the ^{57}Co isotope can be used in Mössbauer spectroscopy on the 14 keV γ-irradiation of the ^{57}Fe daughter atoms [98–103]. The total amount of Co in silicon can be derived by measuring the intensity of the 122 keV γ-line, using a Ge-Li detector calibrated by a ^{57}Co standard, while different spectral lines in the Mössbauer spectrum correspond with different lattice configurations of Co (single atom, paired, clustered or precipitated), provided they contribute at least 5% to the total amount.

6.2.1.1 Atomic and Clustered Species

Overall, 6–10 different lines (species) have been identified [104, 105]. Most of the lines has been associated with Co in particles [103, 105]. However, the lines do not correspond with what is expected for pure $CoSi_2$, which has the CaF_2 lattice and a 1.2% mismatch with silicon. In such a structure, each Si atom has 4 Co nearest neighbours and each Co atom has 8 silicon ones. The observations can be explained if one considers that the response is dominated by the interface of the precipitate, where Co atoms have only 7 nearest neighbours and a dangling bond [106]. This is similar to the case of the thin $NiSi_2$ precipitates mentioned above (Sect. 6.1.2.1). It was also revealed that high B doping slows down the Co precipitation because of the pairing reaction [99, 106]. It was established that cobalt acceptor pairs anneal out between 100 and 200 °C. The dissociation of the pairs and the precipitation of the isolated interstitial Co occur simultaneously [99]. The binding energies for Co-Al and Co-Ga pairs are higher than for Co-B [99]. The presence of single-atom Co-related species has also been studied in electron-irradiated material [107], in order to investigate the interaction of Co_i (Co_s) with irradiation-induced vacancies and interstitials and several new species were identified. They were for example ascribed to a Co-divacancy (Co-V_2) and a cobalt-vacancy pair.

A critical discussion of the existing results by Utzig [104] highlighted the following facts:

- in p-type spectra, three different lines (species) with natural linewidth occur, which may correspond with different CoB species. At the same time, no DLTS lines were revealed, suggesting that they are electrically inactive.
- In n-type, Co remains part of $CoSi_2$ clusters or precipitates, also after annealing. The annealing results in a growth of the particles, so that the contribution of the surface Co atoms becomes smaller and eventually disappears. This could explain the changes in the Mössbauer spectra upon annealing and is also in line

with the annealing behaviour of $NiSi_2$ platelets (Sect. 6.1.2.1). In other words, only one Co-species is present in n-type Si, with no evidence for CoP (As, Sb) pairs, in conflict with the Göttingen results [98, 100].

- Electron emission channeling of ion-implanted ^{57}Co in silicon has revealed a dominance of the near-*BC* sites at low temperature annealing in n-type material [108]. This transforms into sites displaced from ideal-*T* towards *AB* sites, which is similar as for ^{51}Fe and ^{65}Ni in intermediately doped material. Finally, ^{57}Co is found in ideal-*S* sites only after high temperature annealing, up to 850 °C. Changing the doping from n- to p^+-type, an increase in sites near the ideal tetrahedral interstitial position has been observed by emission channeling [109]. A similar tendency was found for manganese, and iron, while this was not observed for nickel. This has been ascribed to the formation of pairs with boron, driven by Coulomb attraction. The displacement from the ideal-*T* site has been discussed in terms of the breathing mode relaxation around the boron ion [109].

6.2.1.2 Buried $CoSi_2$ Formation

For some time, there was interest in the development of buried epitaxial $CoSi_2$ layers for permeable base transistors [110–114]. This was realized by a high dose ($\sim 10^{17}$ at/cm^2) medium energy (100–500 keV) Co^+ ion implantation, followed by annealing. It was shown that there exists a critical Co dose below which only a discontinuous layer of precipitates is formed; above, a continuous epitaxial layer can be created for a sufficiently high temperature ($\sim$1000 °C). In addition, it turns out favorable to apply high-temperature ($\sim$ 350°C) implants to have some dynamic annealing of the implantation damage. This provides a defective but crystalline template for later solid-phase epitaxial regrowth. On the contrary, room or low temperature Co implantation yields a polycrystalline growth of surface layers of $CoSi_2$. The thermal history is important, whereby an intermediate stage anneal at 600 °C for 1 h is essential [113]. This provides the nucleation of a large density of $CoSi_2$ precipitates, serving as a seed for subsequent continuous layer growth, as schematically shown in Fig. 6.15 [113].

Detailed TEM investigations have shown that the growth prior to layer coalescence is exclusively due to {111} facets, which is the lowest energy Si/$CoSi_2$ interface [113]. Detailed investigations of the energetics of the $CoSi_2$–Si interface revealed that the surface tension on the {100} faces is about 43% higher than on {111} [115]. After coalescence, "filling in" of adjacent {111} facets with {100} facets may be expected to become energetically more favorable as it allows the formation of extra volume of silicide with its negative heat of formation (24.6 kcal/mole), while at the same time reducing the amount of interface area. In the case of (100) Si substrates not only first-order A and B twinning occurs, since epitaxial growth along different orientations takes place, which all have a good lattice match [114]. First-order twinning results in a B-type $CoSi_2$ orientation which is rotated through 180° around one of the ⟨111⟩ axes relative to the aligned (111) A-type

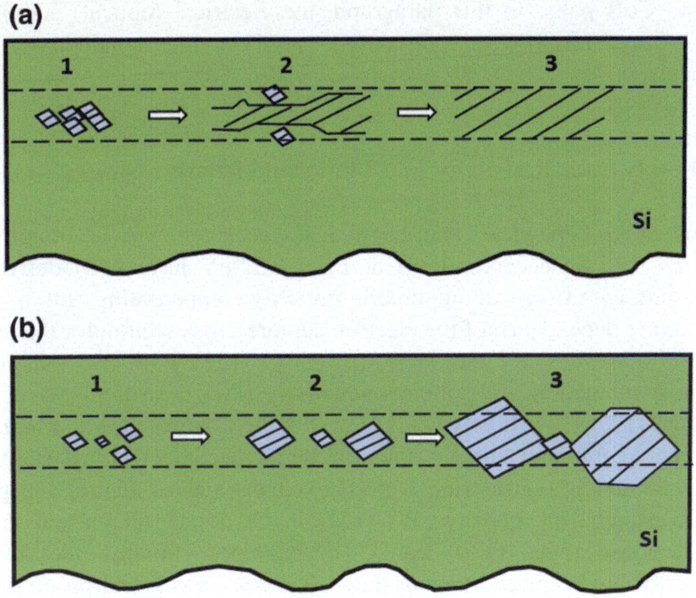

Fig. 6.15 Schematic illustration of precipitate growth processes: **a** for a supercritical dose, precipitates coalescence occurs during earlier stages of the ripening process and (100) facets grow between adjacent precipitates, establishing a continuous (100) layer. **b** For a subcritical dose, coalescence of adjoining precipitates occurs only during the later stages of the ripening process, if at all. After Hull et al. [113]

structure. Further detailed in situ TEM and modeling studies have been performed by Ruault et al. [116–122]. Finally, the endotaxial growth of hexagonal $CoSi_2$ nanoplatelets in Si(001) has been investigated by in situ grazing-incidence small-angle X-ray scattering [123].

6.2.2 Electrical and Optical Activity of Co in Si

The most simple way to evaluate the electrical impact of Co in Si is through the (Hall) resistance. Rather conflicting results have been obtained from the early days on: while some Authors found that Co compensates n-type doping density and enhances p-type [124], others came to the conclusion that Co behaves like an amphoteric impurity and compensates both n- and p-type [125, 126]. From the foregoing, it has been concluded that interstitial Co is neutral in n-type and intrinsic Si and behaves like a positively charged donor in p-type material below 900 °C [98]. There exists also general agreement on the fact that only 0.1% of the dissolved Co is electrically active at room temperature, with most of it assigned to Co_s and in p-type Si to CoB pairs. In this paragraph, the electrical (optical) activity of the different Co species in Si will be discussed, based on either conductivity or DLTS measurements.

6.2.2.1 Levels from Resistivity and Photoconductivity

From temperature-dependent Hall-coefficient measurements, a donor level at $E_V + 0.38$ eV and an acceptor level at $E_C - 0.52$ eV have been derived [127], which is in line with Co as an amphoteric impurity compensating both n- and p-Si. The temperature dependence of the electron capture cross section for the latter level was found to obey $\sigma \sim T^m$ with $m = -0.1 \pm 0.7$ [128]. Further details have been obtained in combination with photoconductivity (PC) measurements in the 1.1–2.6 μm range [127–131]. It was concluded that the Co-related donor and acceptor levels belong to different charge states of the same defect [130]. Based on the two-level scheme and considering various excited states of neutral and positively charged Co centers, the negative PC [127] in Co-doped silicon and the related transient response times [130] could be explained. Finally, the temperature dependence for hole capture by neutral cobalt donors was found proportional with T^m, with $m = -2.5 \pm 0.4$ [130]. It should, finally, be remarked that from Mössbauer studies of the ion pairing reactions, a single donor level at $E_V + 0.15$ eV was derived in p-type Si [100]. All these early interpretations have meanwhile seriously been challenged by recent DLTS.

6.2.2.2 Deep Levels from Space-Charge Transient Techniques (DLTS)

Since the seventies, space-charge-based junction techniques have overtaken conductivity (resistance) measurements to investigate Co-related deep levels in Si [115, 125, 126, 132–135]. Early DLTS results revealed the presence of two Co-related electron traps in n-type Si [125] and one hole trap in p-Si [126]. The activation energy and electron capture cross section were found to be: 0.21 eV and 5.5×10^{-15} cm^2; 0.42 eV and 1.4×10^{-15} cm^2, whereby the σ_n's were independent on temperature in the investigated range [125]. The hole trap, ascribed to a donor level was at 0.23 eV from the valence band [126]. It has also been pointed out that the deep levels introduced by Co depend heavily on the sample preparation [136]. For example, the way Co is applied to the silicon wafer has a marked influence. According to Fig. 6.16, different DLT-spectra in n- and p-type silicon can be obtained, whether the Co is deposited by evaporation from a heated Co wire (method Co) or from a Co wire which is wrapped around a heated tungsten filament (method W). The differences were explained by the occurrence of W (?) cross contamination in the latter method [136].

The impact of hydrogenation on the Co-related levels was first reported by Jost et al. [137, 138], revealing four additional H-related electron traps at 60, 90, 140 and 175 K in n-Si. Annealing for 30 min at 400 K removes the E60, E90 and

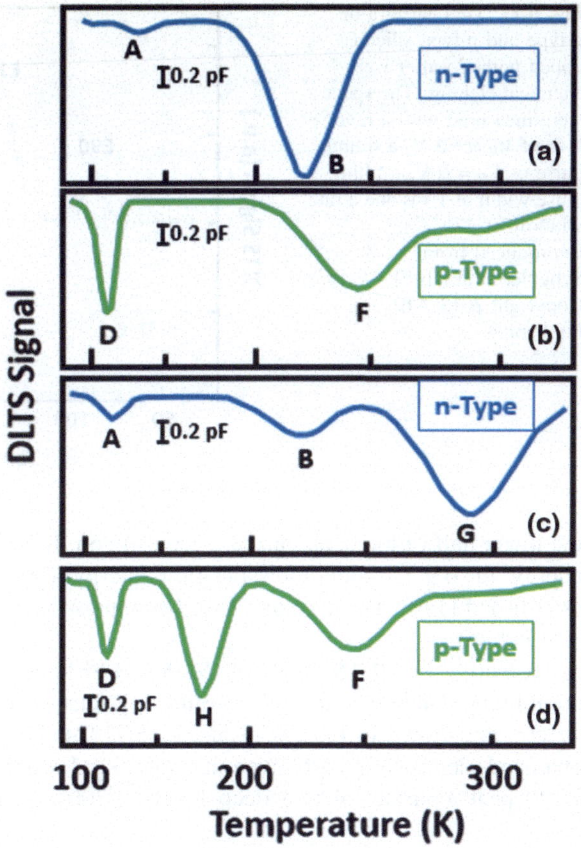

Fig. 6.16 Typical DLT-spectra for cobalt-doped silicon. Spectra **a** and **b** are for n- and p-type samples prepared by "method Co" with $t_d = 30$ min diffusion time in an evacuated sealed quartz ampoule; **c**, **d** are the corresponding spectra for samples prepared by "method W" with $t_d = 72$ h in an Ar diffusion ambient (in a quartz ampoule), respectively. The diffusion temperature $T_d = 1150$ °C. Reproduced from Nakashima et al. [136]. Copyright 1990 The Japan Society of Applied Physics

E175 levels, while the E140 trap is stable up to 470 K [137]. It should be remarked that hydrogen is readily introduced in silicon during the sample preparation for DLTS, using, e.g., mechanical polishing followed by a diluted HF dip or wet etching in HF:HNO_3 (1:2) for 30 s (so-called smoothing etch). In both cases, H can penetrate 5 μm in the material at room temperature, as derived from the doping and deep-level profiles, from C-V and DLTS, respectively [137]. In p-type Si, two additional H-related hole traps have been identified, at 50 and 150 K. For the latter trap, a thermally activated formation process was found, corresponding with an activation energy of 1.45 ± 0.03 eV and a dissociation process with activation energy of 0.68 ± 0.03 eV [138]. It was proposed that neutral Co interacts with positive hydrogen to form CoH^+ and $CoH_2{}^+$ centers.

Also the pairing reaction between Co and B has been studied in detail by DLTS [139, 140], Mathiot found several candidate electron traps for the CoB pair in P (n-Si) and B counterdoped n-type silicon [139], whereby the observed deep levels strongly depend on the applied thermal budget: furnace anneal followed by slow cooling versus RTA and fast cooling. Another study revealed no CoB donor level in

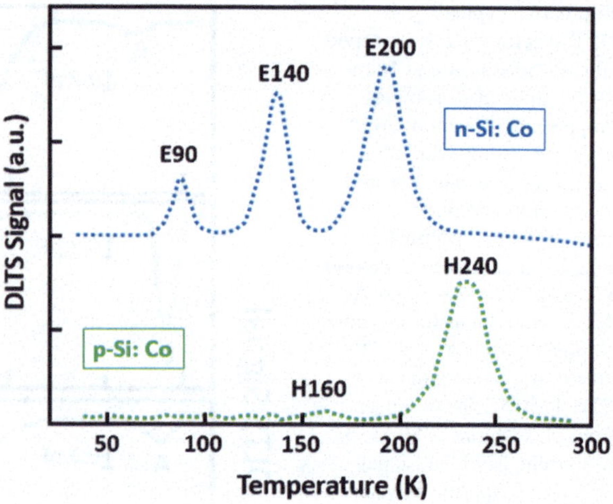

Fig. 6.17 DLT-spectra of n-type and p-type silicon doped with Co after wet chemical etching. The spectra were measured with a reverse bias of $V_R = -2$ V, a filling pulse of $V_P = 0$ V, a filling pulse width of 1 ms and a rate window of 47 s^{-1}. Reproduced from Scheffler et al. [142]. Copyright 2013 AIP Publishing

the lower half of the band gap of p-type silicon. In addition, a characteristic pairing energy of 0.9 eV and a dissociation energy (binding energy) of 0.6 eV was established [140]. These results have been questioned seriously in later DLTS work (see below).

Recently, the DLTS levels of Co-doped Si have been critically re-investigated, combining standard with high-resolution Laplace-DLTS [141–144]. The resulting spectra are shown in Figs. 6.17 and 6.18 [142]. The corresponding defect parameters and identification are given in Table 6.3. From Fig. 6.19, it is evident that the E200 peak consists of two deep levels: E200 and E200′, which show opposite

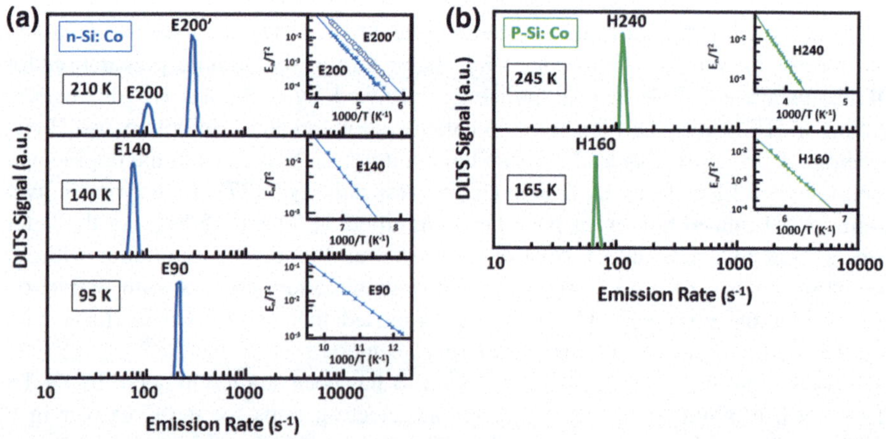

Fig. 6.18 Laplace DLT-spectra recorded for **a** E200, E140 and E90 in n-type Si and **b** H240 and H160 in p-type Si. The inset shows the Arrhenius plot obtained for each defect presented in this figure. Reproduced from Scheffler et al. [142]. Copyright 2013 AIP Publishing

Table 6.3 Activation energies, apparent capture cross section and assignment for all defects seen in Co-doped n- and p-type Si. After Scheffler et al. [142]

Label	E_a (eV)	σ_n (cm^2)	Assignment
E200	$E_C - 0.39$	8.7×10^{-16}	Co_s
E200′	$E_C - 0.35$	3.7×10^{-16}	Co_sH
E140	$E_C - 0.29$	1.1×10^{-14}	Co_sH_2
E90	$E_C - 0.16$	5.6×10^{-16}	CH
H240	$E_V + 0.46$	9.6×10^{-16}	CoB
H160	$E_V + 0.38$	1.6×10^{-14}	?

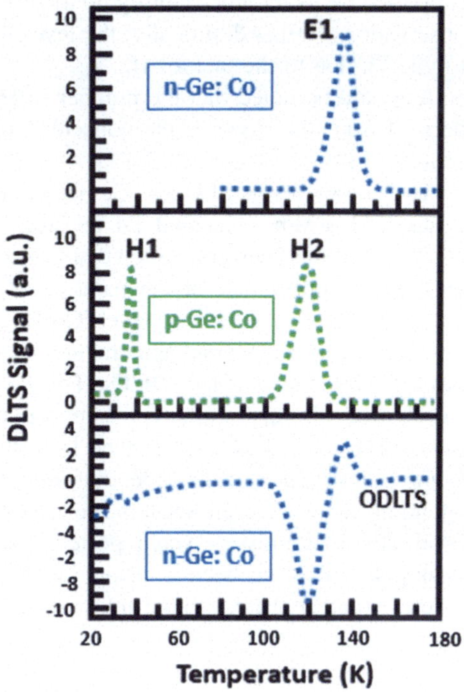

Fig. 6.19 DLTS and optical-DLTS of co-implanted germanium. Reproduced from Lauwaert et al. [145]. Copyright 2012 AIP Publishing

concentration profiles [142]. Investigating the reverse-bias dependence of the emission rate, no shift (no PF effect) has been found for E90, E140, E200 and E200′ in n-Si, indicating that they reside in the neutral charge state before capturing an electron, or in other words, they are single acceptors. On the contrary, H240 and H160 exhibit an enhancement of their emission rates when increasing the electric field F [142], suggesting the charge states of Table 6.3 (single acceptor nature). E stands for an electron trap and H for a hole trap; the number indicates the peak position in K. The thermal stability of the levels has also been investigated, showing that the hole traps disappear at 150 °C and implying that the CoB pairs are stable up to that temperature [142]. Peak E140 is more stable and anneals above 300 °C, while E90 anneals below 90 °C. E200′ and E200 disappear at 300 and 600 °C, respectively.

From Table 6.3, it is concluded that substitutional Co_s in silicon gives rise to only one acceptor level, while the donor level is too shallow to be observed. This is in line with the Mössbauer observations and in disagreement with most older DLTS results. In other words, there is no deep level in the bottom half of the band gap corresponding with Co_s [142]. The previously reported Co_i levels belong to {Co, H} related complexes, based on the strongly decaying depth profiles, when moving deeper into the material [142]. The reason why the Co_i donor level escapes DLTS observation is its unstable behavior at room temperature, as it diffuses (drifts) out of the depletion region of a junction. The identification of the E240 level as CoB is supported by its thermal stability up to 150 °C and in line with previous Mössbauer observations [99]. Additionally, the level is not found as a minority carrier peak in n-type Si, due to the lack of B.

It should be noted that the number of H atoms in a {Co, H} complex has been derived from the slope of the concentration profiles, like for the case of {Ni, H} centers [142].

The energy levels of Co in Ge have been studied by Hall effect since the early fifties [58]. From a detailed DLTS study of Co-implanted and annealed n- and p-type Ge, it has been concluded that substitutional Co introduces a double acceptor level in the upper half of the band gap (peak Co-E1 in Fig. 6.19), and a single acceptor (Co–H2) and donor level (Co-H1) in p-type material [145]. While the two acceptor levels are in good agreement with previous Hall effect data, the donor level has only been revealed in DLTS. The fact that the magnitude of the Co-H1 and Co-H2 peaks is very similar is in support of this assignment. The donor character of the Co-H1 center has been derived from the fact that at different electric fields, no Poole-Frenkel shift of the hole emission rate has been observed [145]. The corresponding deep-level parameters are summarized in Table 6.4. For the Co donor level, athermal emission was reported at temperatures below the thermal emission peak [146]. This has been associated with the photo-ionization of holes towards the valence band by back-ground infrared irradiation.

Table 6.4 Deep level parameters and assignment of the traps corresponding with substitutional Co impurities in germanium. After Lauwaert et al. [145]

Band	E_a (eV)	K_T (s^{-1} K^{-2})	σ_n (cm^2)	ΔE_σ (eV)	σ_∞ (cm^2)	ΔH (eV)	ΔS (k_B)	Level
Co-E1	0.325	3.3×10^8	1.1×10^{-13}	0.010	2.3×10^{-15}	0.315	3.9	$Co_s^{-/2-}$
Band	E_a (eV)	K_T (s^{-1} K^{-2})	σ_p (cm^2)	α_σ (cm^2 K)[a]	β_σ (cm^2)[a]	ΔH (eV)	ΔS (k_B)	Level
Co-H1	0.086	6.8×10^8	7.2×10^{-13}	8.4×10^{-13}	1.9×10^{-14}	0.086	3.1	$Co_s^{+/0}$
Co-H2	0.254	1.7×10^7	1.8×10^{-14}	7.4×10^{-13}	9.4×10^{-16}	0.261	1.4	$Co_s^{0/-}$

[a]The hole capture cross section was found to depend linearly on T according to $\sigma_p(T) = \alpha_s/T + \beta_s$

6.2.3 Impact on Lifetime

There are only few dedicated lifetime studies reported in the literature [147–149]. It was found that the generation lifetime in furnace-annealed p-type Si is quite sensitive to the Co contamination dose [147]. In the case of RTA treated samples, there is only a weak dependence, due to the fact that most of the Co remains dissolved in the interior of the wafer. The diffusion length (recombination lifetime) after FA reduces with the square root of the Co dose, as shown in Fig. 6.20 [147]. This is related to the fact that Co is only partially dissolved in the lattice, while the rest is surface segregated for a FA. There remains a constant fraction of implanted Co in solution. The surface precipitated Co gives rise to a strong increase in the surface recombination velocity at the Si/SiO_2 interface [147]. The duality between surface and bulk segregated (dissolved) Co is represented in Fig. 6.21, showing an inverse correlation between τ_g and s [147]. In the case of RTA samples, no change in s is found (no surface segregation), while the bulk diffusion length is more affected than for furnace anneal, since more Co centers remain in solution.

In principle, every recombination level has a specific impact on τ_r, through its capture cross sections and energy level(s). This is reflected in the injection level (Δn, Δp) dependence of the lifetime. This could be used to identify the responsible deep levels. Several studies on the Δn dependence have been carried out for Co-doped Si [147, 148]. Both s and the diffusion length exhibit a pronounced, specific behavior [147]. From the carrier lifetime versus injection level dependence for p-type RTA-treated Si wafers, two levels have been derived with activation energy $E_V + 0.22$ eV and $E_V + 0.42$ eV, electron capture cross sections of $0.7 - 1 \times 10^{-14}$ cm^2 and $1 - 1.2 \times 10^{-14}$ cm^2 and hole cross sections of $\sim 10^{-15}$ and 10^{-16} cm^2 [147]. Combining injection and temperature dependent lifetime

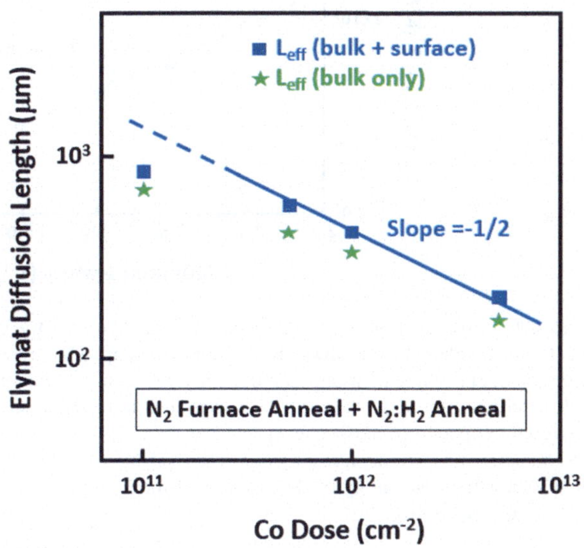

Fig. 6.20 Effective and bulk carrier diffusion length versus cobalt dose in furnace-annealed wafers (Elymat data). Reproduced from Polignano et al. [147] by permission of ECS-The Electrochemical Society

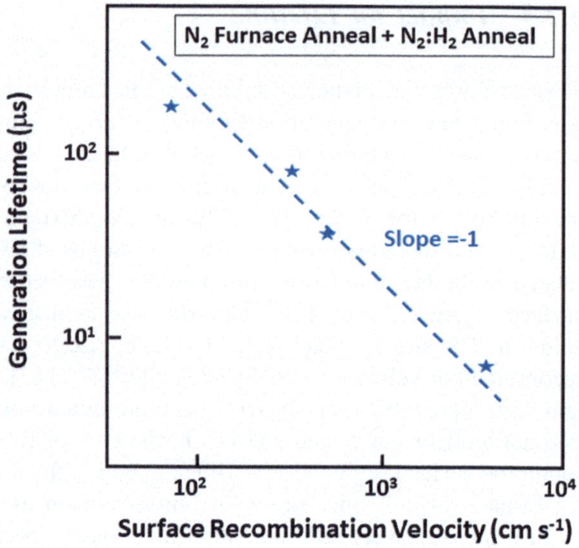

Fig. 6.21 Generation lifetime versus surface recombination velocity in Co-implanted furnace-annealed wafers. Reproduced from Polignano et al. [147] by permission of ECS-The Electrochemical Society

measurements in so-called injection-dependent lifetime spectroscopy narrows down the possible deep level parameters. Assuming that the responsible level in p-type Si is the CoB pair at $E_V + 0.41$ eV, a recombination rate, represented by Fig. 6.22 has been derived, yielding a barrier for electron capture of 52 ± 2 meV [149].

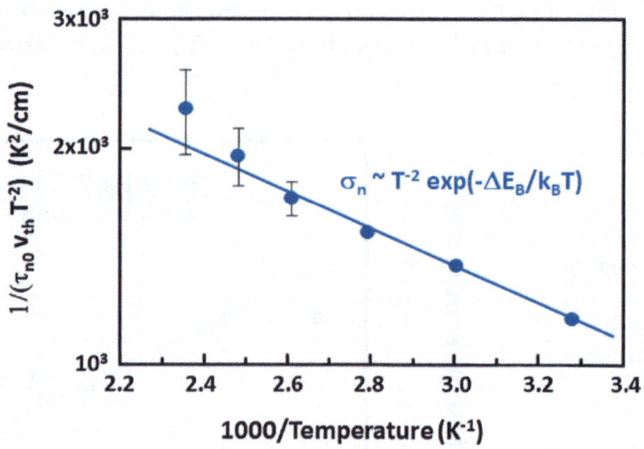

Fig. 6.22 Investigation of the temperature dependence of the capture cross section for electrons $\sigma_n(T)$, for the defect in the lower half of the band gap. The graph shows the inverse product of the electron capture time constant τ_{n0}, thermal velocity v_{th}, and a temperature-dependent factor T^{-2} in a logarithmic plot vs. the inverse of T. The data have been calculated using the $\tau_{n0}(T)$ values for different temperatures at the energy level of $E_V + 0.41$ eV. The linear correlation shown by the fit is characteristic of a capture process described by the two stage model of cascade capture and yields an activation energy of $\Delta E_B = 52 \pm 2$ meV. Reproduced from Diez et al. [149]. Copyright 2007 AIP Publishing

Since in p-type Si, the CoB pair plays a dominant role in the recombination lifetime, one can assume that similar as for FeB, it is possible to activate Co by intense light illumination or by heating to moderate temperatures (~ 100 °C) or by the combination of both. If the pair separation behaves distinctly from the FeB case, one can identify Co in this way, using the SPV method [148]. This should yield an in-line inspection method for Co contamination in a processing environment. It is known that the recovery time (Co diffusion) at room temperature is much faster for Co than for Fe. The pair separation occurs already at 75 °C for CoB instead of 200 °C for FeB. This is due to the different diffusion coefficient at RT and the different binding energy for the pairs. Cobalt can be both optically and thermally activated (very fast at 80 °C) [148]. Full photothermal dissociation occurs within 30 s at 30 °C. The degradation of L_{Diff} decreases for activation temperatures of 200 °C; this is related to the surface segregation of fast diffusing Co_i. At 350 °C no more "bulk" activation of Co is observed, as it is also segregated to the surface [144].

At the same time, it is found that the activation kinetics is opposite for the thermal and photothermal mechanism: for thermal, the CoB dissociation follows a second order kinetics, being proportional to the inverse of the annealing time t_a. For photothermal activation, no kinetics is observed, as it happens nearly instantaneously. The second order kinetics can be described as follows [148]:

$$[CoB] = \left([CoB]_0 - [CoB]_\infty\right)\frac{\tau_{diss}}{\tau_{diss} + \tau_{ass}} + [CoB]_\infty \tag{6.2}$$

with $[CoB]_0$ and $[CoB]_\infty$ the concentration of CoB pairs at time zero before the dissociation and at infinite annealing time t_a, when equilibrium between association and dissociation is reached at the annealing temperature [148]. At $T_a \geq 80$ °C, $[CoB]_\infty = 0$. We have further that τ_{diss} is the time constant for dissociation, given by:

$$\tau_{diss} = \tau_{diss0}\exp\left(\frac{E_{diss}}{k_B T_a}\right) \tag{6.3}$$

with E_{diss} the dissociation energy. Based on that, a **detection limit** for Co of 2×10^{11} at/cm^3 was derived for p-type Si after activation. Before activation, a limit of 1.5×10^{12} at/cm^3 has been derived [148]. Recombination lifetime in n-type Si is more sensitive to Co, yielding a detection limit of 3×10^9 at/cm^3 [148]. This is related to the fact that the recombination lifetime is rather dominated by the more efficient Co_s acceptor level and not by CoB pairs which are absent in n-Si.

Recently, interest has risen in so-called intermediate band (IB) solar cells, which contain a high density of deep-level impurities, forming a band of states in the band gap. Excitation of carriers to or from the band by photon absorption should increase the efficiency in the (near) IR part of the solar spectrum. One concern is of course that the deep-level impurities cause excessive carrier recombination, thereby reducing the efficiency. It is speculated, however, that the formation of such an impurity band quenches the recombination by individual atoms [150]. In order to form such a band, the Mott limit of ~5.9×10^{19} at/cm^3 needs to be exceeded in silicon, which is much higher than the solid solubility of the TMs. Such a high concentration can only be reached by strongly non-equilibrium methods like high-dose ion implantation and short-time laser annealing, to avoid out-diffusion or precipitation of the impurity atoms. This has been applied to ^{59}Co implanted silicon (60 keV, 5×10^{14} to 2×10^{16} cm^{-2}), followed by a 15 ns laser pulse of a Nd: YAG (512 nm) system at a power density of 0.7 J/cm^2 [145]. From the behavior of the resistance of the sample with T, it can be concluded that impurity band formation (the insulator-to-metal transition) occurs for a Co dose between 5×10^{14} at/cm^2 and 5×10^{15} at/cm^2. This follows from the fact that the resistivity between 100 K and 250 K is two-to-three orders of magnitude higher, whereby the resistivity of the thin Co-implanted layer amounts to 0.021 Ω cm. This is two orders below the value for the p-type Si substrate, with the highly Co-doped layer exhibiting rectifying behavior on p-type Si (junction isolation). For the highly Co-implanted samples the resistivity is independent on temperature, indicating a metallic conduction type inside the formed IB [150].

6.3 Chromium in Si and Ge

6.3.1 Configurations of Cr in Si

Chromium is a TM with intermediate diffusivity, occurring both in interstitial [151] and substitutional [152] forms in the silicon lattice, as detected by EPR. The 4*s* electrons of interstitial TMs are all transferred to the 3*d* shell. A small crystal field splits the degenerate *d* states into a 2-fold *e*-level and a threefold t_2-level, with the latter lower in energy. The spectra of interstitial Cr and the related complexes with the Cr_i atom in the nearest or second nearest neighbor position of a shallow acceptor are very similar. The acceptor ion mainly disturbs the symmetry of the cubic binding potential of the TM. CrB pairs show the same angular momentum ($S = 5/2$) as Cr_i^+ (d^5) ions. The g_L value is isotropic with a $\langle 111 \rangle$ symmetry axis. Cr_i occurs in the positive charge state, indicating that it introduces a donor level in the band gap. Detailed electron-nuclear double resonance spectroscopy of Cr_i has pointed out that at least 22% of the spin density is transferred from the impurity to the host crystal [153].

6.3.2 Electrical and Optical Activity of Cr in Si

6.3.2.1 Electrical Properties

Initially, the deep levels associated with Cr in Si have been determined by (Hall) resistivity [154–157] and photoconductivity measurements [155, 158]. A level at $E_C - 0.23$ eV (degeneracy factor 2) and at $E_V + 0.11$ eV (degeneracy factor 8) have been ascribed to the presence of Cr [156]. In another study, a donor level at $E_C - 0.226\ (\pm 0.01)$eV and $E_V + 0.128\ (\pm 0.005)$eV were reported and the corresponding degeneracy factors 1/3 and 1/4, respectively [157].

The two major Hall effect levels have been confirmed by subsequent DLTS analysis [159–162] with a typical n- and p-Si spectrum represented in Fig. 6.23 [158]. Interstitial Cr yields a donor level at $E_C - 0.23$ eV with an electron capture cross section of $\sigma_n = 2 \times 10^{-13}$ cm^2.

In Si:B, a second donor level is formed, which is associated with the CrB pair. The binding energy of the pair is 0.65 eV and the formation rate depends on the boron concentration [B], as represented in Fig. 6.24 [163]. In contrast to the case of Co and Fe, the pairs cannot be split with intense illumination. This is related to the position of the Cr_i donor in the upper half of the band gap, resulting in a neutral charge state in p-type (or intrinsic) silicon. The Fermi level cannot be lifted above the donor level, so that the interstitial Cr remains in its positive charge state and attracted by B^-; separation of the pair becomes difficult then. The dissociation energy amounts to 0.56 eV. The activation energy of the CrB level is 0.28 eV, corresponding with a hole capture cross section of 1×10^{-14} cm^2 [159]. A much lower more or less constant value of $8.4 - 9.1 \times 10^{-17}$ cm^2 at 94 K has been derived by Kunio et al. [160].

The interaction of Cr centers with hydrogen has also been studied in n-type Si, whereby H was introduced by wet chemical etching [161]. In total 4 levels were observed, the Cr_i donor and three additional levels with activation energy (NCr2) 0.28 eV, (NCr3) 0.45 eV and (NCr4) 0.54 eV [161]. The electron capture cross section for NCr2 and NCr3 is rather constant with temperature and in the range 8.2×10^{-15} cm^2 ($135 < T < 170$ K) and 3.7×10^{-15} cm^2 ($215\ \text{K} < T < 270$ K). These two levels exhibit a Poole-Frenkel shift and are thus donors (n-type Si). NCr2 and NCr4 disappear after a few days of storage at room temperature, while NCr3 grows simultaneously [161]. The hydrogen-origin is supported by the fact that the concentration profile reduces from the surface for NCr2 and NCr3, while the opposite happens for the Cr_i level NCr1. Finally, the Cr-related deep levels have

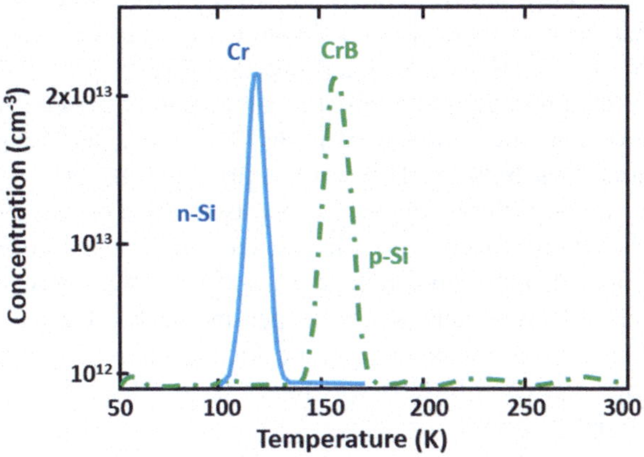

Fig. 6.23 DLTS spectra of interstitial chromium (Cr) obtained in n-type silicon and CrB pairs (CrB) measured in p-type Si:B after saturation of the pair formation. Reproduced from Conzelmann et al. [159] by permission of Springer

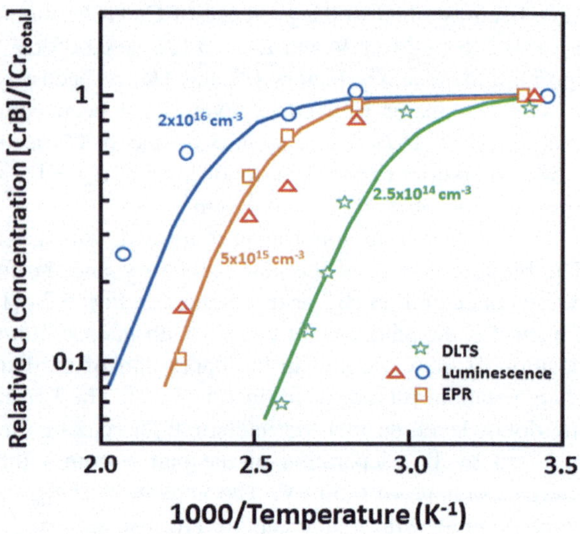

Fig. 6.24 Relative CrB pair concentration as a function of the annealing temperature in samples of various boron content. Measurement by EPR, DLTS and luminescence. The solid lines are the calculated equilibrium concentrations for a pair binding energy of 0.65 eV. Reproduced from Conzelmann et al. [163] by permission of Springer

been studied in edge-defined film-fed-growth (EFG) silicon for PV applications, showing similar levels [162]. The diffusivity in EFG Si is close to the one in c-Si:B, i.e., 5.3×10^{-17} cm^2/s at room temperature. The pair formation has also been studied, showing that the maximum concentration of [CrB] is limited by the smallest concentration of either [B] or [Cr]. The concentration saturates when [B] becomes much larger than [Cr] [162]. The minority carrier diffusion length doesn't change with time, when Cr_i transforms into CrB. This suggests that both species have similar recombination properties.

The energy levels of Cr implanted in Ge at 90 keV at a dose of 5×10^{13} at/cm^2 or 5×10^{14} at/cm^2 and annealed at 500 °C for 5 min have been studied by DLTS [164] and the corresponding spectra are reported in Fig. 6.25. One electron trap (level Cr-E1) and three hole traps (Cr-H1, Cr-H2 and Cr-H3) have been observed, with parameters reported in Table 6.5. These results are in agreement with previous Hall effect data, except for the level Cr-H1, which was revealed only by DLTS. It has been concluded that substitutional Cr in Ge gives rise to three acceptor levels (Cr-E1, Cr-H3, Cr-H2) and one donor level (Cr-H1). Similar as for the Co donor level, a non-thermal emission component has been found for the Cr-H1 and Cr-H2 levels at low temperature [146]. This effect has been exploited to firmly demonstrate that both levels belong to the same defect.

Finally, the levels of Cr in p-type $Si_{1-x}Ge_x$ have been studied for Ge concentrations up to 20% both in relaxed and strained epitaxial layers [55]. It has been found that the hole trap at $E_V + 0.270$ eV moves towards the valence band for relaxed layers and shifts deeper in the band gap for strained counterparts (see Fig. 5.5). A rather large value for the uniaxial deformation potential of this hole trap, related with CrB was found (>16) eV) [55], in line with the fact that it is a complex defect.

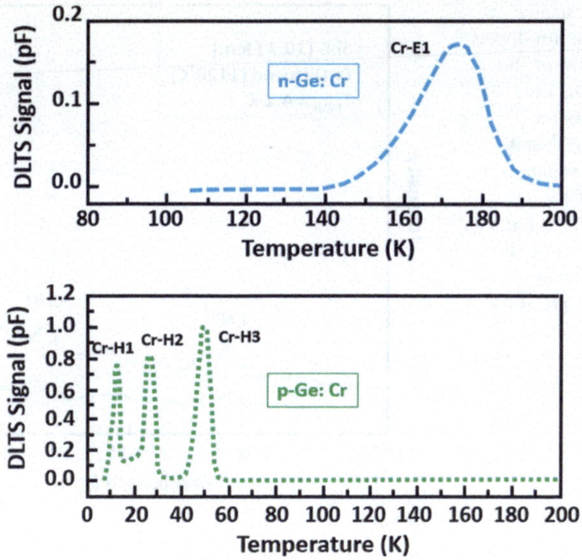

Fig. 6.25 DLTS of Cr-implanted n-type (τ_{window} = 23.5 ms) and p-type (τ_{window} = 2.1 ms) germanium. Reproduced from Lauwaert et al. [164]. Copyright 2009 AIP Publishing

6.3.2.2 Optical Properties

In p-type silicon, CrB pairs exhibit a PL spectrum, given by Fig. 6.26 [159, 163, 165] and corresponding level scheme in Fig. 6.27. The uniaxial stress-induced level splitting is consistent with an axial defect in a $\langle 111 \rangle$ crystallographic direction. It corresponds, therefore, with a trigonal center. The spectrum of Fig. 6.26 is typical for a TM-B complex. Here, it is centered at 0.84 eV—a zero-phonon line with two equidistant satellites at lower energy. This can be attributed to the excitation of one or two local modes [159]. The PL is believed to originate from the recombination of a weakly bound electron (by 20 meV), which is much lower than the spectroscopic binding energy of 326 meV, with a deeply bound hole ($\sim$0.3 eV = the deep level

Table 6.5 Deep level parameters and the assignment of the traps corresponding with Cr impurities in germanium

Band	E_{na}/E_{pa}	K_T (s^{-1} K^{-2})	σ_n/σ_p	ΔE_σ(meV) (cm^2)	σ_∞ (eV)	ΔH k_B	ΔS	Level
Cr-E1	0.364	5.2×10^7	1.8×10^{-14}	56	2.2×10^{-17}	0.310	6.7	$Cr^{2-/3-}$
Cr-H1	0.015	1.7×10^6	1.8×10^{-15}	…	…	…	…	$Cr^{+/0}$
Cr-H2	0.042 (0.048)	3.9×10^7	4.1×10^{-14}	…	…	…	…	$Cr^{0/-}$
Cr-H3	0.088 (0.095)	1.6×10^8	1.7×10^{-13}	…	…	…	…	$Cr^{-/2-}$

Values between brackets have been corrected for the Poole-Frenkel effect. After Lauwaert et al. [164]

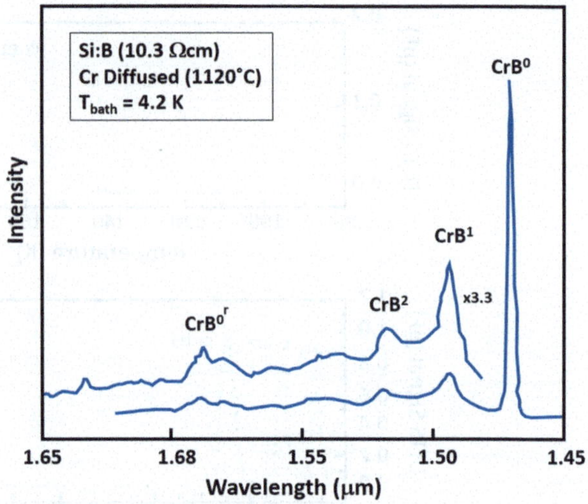

Fig. 6.26 Luminescence spectrum of chromium-doped Si:B after saturation of the CrB pair formation. Line CrB^0 is the no-phonon transition, lines CrB^1 and CrB^2 are due to excitation of local vibrations of the complex and $CrB^{0\Gamma}$ is the 0^{Γ}-phonon replica of the CrB^0 line. Reproduced from Conzelmann et al. [163] by permission of Springer

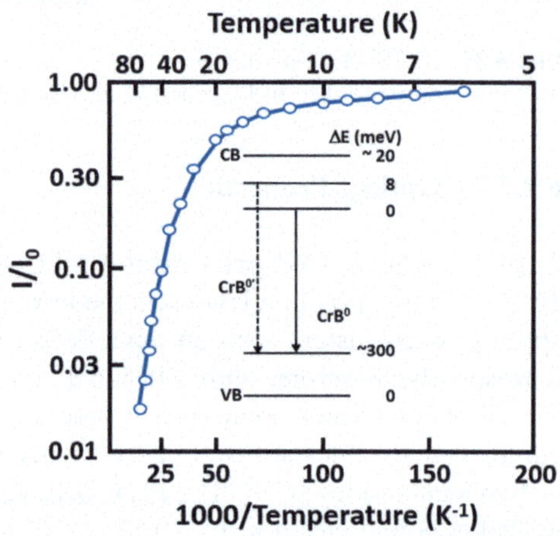

Fig. 6.27 Temperature dependence of the CrB^0-line intensity. Reproduced from Conzelmann et al. [163] by permission of Springer

activation energy). It is thus excitonic recombination at an isoelectronic impurity (CrB^--h) [159]. This picture is largely confirmed by PC measurements, yielding a negative peak at 843.8 meV, which is the position of the $(CrB)^0$ no-phonon line [166].

6.3.2.3 Impact of Cr on Lifetime in Silicon

The impact of Cr and CrB on the SRH lifetime in silicon has been studied to quite some detail [166–170]. Important is to know both the majority and minority capture cross section and their ratio σ_n/σ_p which determines to a large extent the efficiency as a recombination center in both n- and p-type Si [167]. For Cr_i in n-type Si, a ratio of 2 has been reported. The hole capture cross section for CrB pairs in p-type Si was determined from lifetime measurements at 300 K, yielding a value of $\sigma_{pCrB} = 8.4 \times 10^{-14}$ cm^2, which is quite similar to the corresponding value for Cr: $\sigma_{pCr} = 1.25 \times 10^{-13}$ cm^2 [168]. The corresponding electron capture cross sections at 300 K are $\sigma_{nCrB} = 1.5 \times 10^{-13}$ cm^2 and $\sigma_{nCr} = 2.5 \times 10^{-13}$ cm^2. This confirms the rather symmetrical hole and electron capture and the similar recombination activity of Cr and CrB at room temperature. It has also been noted that $CrSi_2$ precipitates exhibit a strongly reduced recombination activity compared to Cr_i in n-type Si [169].

Temperature and injection-dependent lifetime spectroscopy have been applied in order to more firmly identify Cr and CrB from lifetime measurements, using contactless quasi-steady-state photoconductance (QSSPC) in the range 30–170 °C [169]. For n-Si, a $\sigma_p = (4 \pm 1) \times 10^{-15}$ cm^2 was found at RT, which is T-independent and a symmetry factor of 5 ± 3, with a T-independent σ_n between 100 and 400 K [169].

For CrB in p-type silicon, an association time constant:

$$\tau_{ass} = 557\frac{T}{N_{dop}D_{Cri}} \tag{6.4}$$

has been derived from lifetime analysis [169]. On the contrary, upon the pair dissociation a recovery of the lifetime has been reported, which is a relatively fast process. It can only be accomplished thermally, not electronically (carrier injection by light or forward bias). Above 70 °C the pairs are fully dissociated. In that case, for CrB a symmetry factor of 0.4 ± 0.2 can be derived (<70 °C) [169], corresponding with a $\sigma_n = (4.5 \pm 0.5) \times 10^{-15}$ cm^2, a $\sigma_p = (1.1 \pm 0.6) \times 10^{-14}$ cm^2 and a modest temperature dependence. It was observed that the lifetime strongly depends on the doping density. For $N_{dop} = 4\times10^{15}$ cm^{-3} both species give rise to the same lifetime in p-type material [170]. For low doping, CrB yields a lower lifetime; the opposite holds for a high p-type doping. For Cr_i in p-Si, the low injection lifetime is given by [169]:

$$\tau_{SRH,Cr} = \tau_{n0} + \tau_{p0}n_1/N_{dop} \tag{6.5}$$

The high injection lifetime, on the other hand, is given by [169]

$$\tau_{SRH,Cr} = \tau_{n0} + \tau_{p0} \tag{6.6}$$

independent of the doping density.

For n-type Si with the Cr_i donor level in the upper half of the band gap, one obtains [169]:

$$\tau_{SRH,Cr} = \tau_{p0}\left(1 + n_1/N_{dop}\right) \tag{6.7}$$

This will increase strongly for an N_{dop} below $\sim 4\times10^{15}$ cm^{-3}. As a consequence, both in n- and p-Si, the recombination activity of Cr can be strongly reduced at low doping densities. On the other hand, in p-Si, the recombination lifetime of CrB is relatively independent on N_{dop} [169].

The above data has been recently critically re-assessed for partially compensated silicon, i.e., n-Si with a B counter-doping [170]. It was noted that temperatures above 230 °C are required to ensure 90% or more dissociation of the CrB pair, much higher than in the previous study (~70 °C) [169]. In p-type Si, the SRH lifetime was found less sensitive to the CrB level position, while pairing reduces (degrades) its value [170]. A stronger injection level dependence than in n-Si is observed. According to Fig. 6.28, the SRH lifetime at RT is injection-independent at a p-doping of 2.8×10^{15} cm^{-3}. Up to an $N_{dop} = 1 - 4 \times 10^{16}$ cm^{-3} CrB is more recombination active in counterdoped n-type Si and up to 10^{17} cm^{-3} in p-Si. From Fig. 6.29, one can derive that both for Cr_i and CrB, the SRH lifetime degrades with doping level [170]. The high-injection lifetime is given by:

$$\tau_{SRH,high\ inj} = \frac{1 + \sigma_n/\sigma_p}{N_T \sigma_n v_{th}} \tag{6.8}$$

and is the same for both types.

From SRH lifetime studies at low temperatures in n- and p-type Si doped with Cr, a new excitonic Auger process was proposed to explain the recombination (capture) of excess carriers [166]. In a nutshell, it is assumed that the electron of a free exciton can be captured while the hole takes away the excess energy in an Auger process. The opposite happens upon hole capture [166]. Recombination of an e-h pair requires then two subsequent exciton Auger processes.

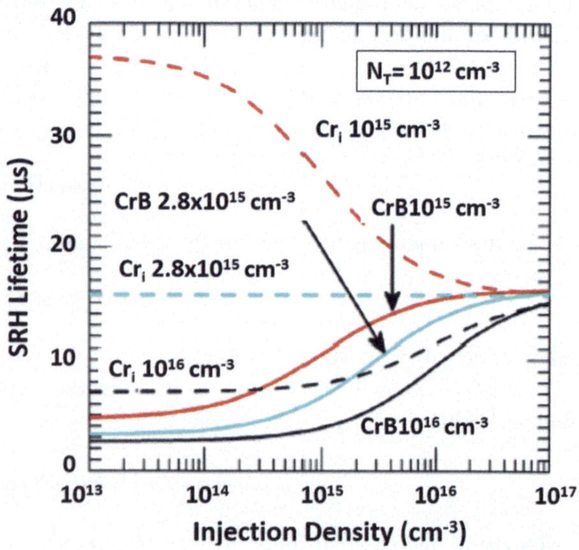

Fig. 6.28 Comparison of Cr_i and CrB SRH injection-dependent lifetime curves in p-type silicon with different doping levels. The curves are modeled with $N_T = 1 \times 10^{12}$ cm^{-3}. Reproduced from Sun et al. [170]. Copyright 2014 AIP Publishing

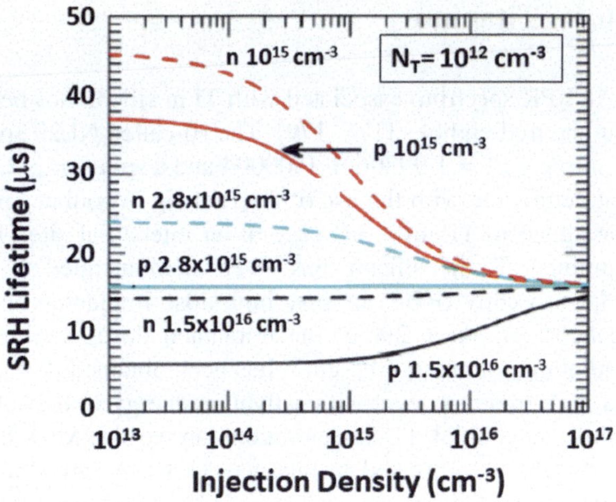

Fig. 6.29 Comparison of Cr_i SRH injection-dependent lifetime curves in p- and n-type silicon with different doping levels. The curves are modeled with $N_T = 1 \times 10^{12}$ cm^{-3}. Reproduced from Sun et al. [170]. Copyright 2014 AIP Publishing

6.3.2.4 Identification of Cr in Silicon by Lifetime Measurements

As Cr is a common contaminant in silicon processing, fast in-line detection and monitoring is of high importance. Lifetime measurements are the natural candidates for that purpose, since the Cr_i and CrB donor levels have a specific fingerprint (symmetry ratio). In addition, the pairing reaction and separation can be used on top of that for a clear identification of the presence of Cr contamination [171–174].

The detection and identification of Cr in p-type silicon can be achieved by measuring the carrier lifetime with all Cr paired (τ_{ass}) and after a dissociation anneal for 30 min at 250 °C [169]. The corresponding Cr concentration can then be derived from [169]:

$$[Cr] = C_\tau(\Delta n, N_{dop})\left[\frac{1}{\tau_{diss}} - \frac{1}{\tau_{ass}}\right] \tag{6.9}$$

with τ_{ass} and τ_{diss} the measured recombination lifetime before and after the dissociation of the CrB pairs. The calibration factor C_τ is a function of the doping density and the injection level. However, for a p-type doping density around 4×10^{15} cm^{-3}, the method fails, as $\tau_{ass} = \tau_{diss}$. For $\Delta n \ll N_{dop}$, the calibration constant approaches a saturation value, which only depends on N_{dop} [169]. Based on this idea an imaging technique for Cr mapping across a silicon wafer has been developed, based on PL imaging [175]. It should, finally, be mentioned that the SPV method has also been used to detect Cr charging effects in native and thermal oxides [176, 177].

6.4 Titanium

An EPR spectrum associated with Ti in silicon has been identified for the first time in the mid-eighties [178, 179]. The so-called NL29 spectrum is characterized by an isotropic $g_L = 1.99806 \pm 0.00004$ and a spin $S = 3/2$. On the basis of the hyperfine structure and with the use of the Ludwig-Woodbury model, this spectrum has been assigned to Ti^+ in a $3d^3$ state at an interstitial site [178]. The presence of substitutional Ti in silicon has been demonstrated by Rutherford Backscattering Spectroscopy (RBS) in very high-dose implanted and 20 ns laser-pulse annealed silicon. About 1–2% of the implanted dose, reaching far above the equilibrium solubility of about 10^{17} cm^{-3} has been obtained. It was also concluded that Ti_s acts as a deep donor level with activation energy of 0.21 ± 0.01 eV [180]. At the same time, interstitial Ti concentrations above the Mott limit of about 5×10^{19} cm^{-3} could be incorporated in this way [180]. A few studies have focused on the precipitation of Ti in silicon [181, 182]. It is known that Ti has a large affinity for oxygen and readily forms TiO_2 precipitates, which are electrically active and were shown to drastically degrade the efficiency of n^+p solar cells. Also a silicide precipitation pathway has been considered [181].

The electrical properties of Ti in silicon have been studied by different techniques like Hall effect and DLTS [181, 183–189]. Three levels have been consistently found in Ti-doped silicon: an acceptor at $E_C - 0.09$ eV (E40), and two donor levels at $E_C - 0.28$ eV (E150) and $E_V + 0.250$ eV (H180). These three levels fit nicely in the Ludwig-Woodbury framework for interstitial titanium. Recent high-resolution Laplace DLTS studies have questioned this straightforward interpretation [144, 190, 191]. The main issue is that the E40 peak does exhibit a Poole-Frenkel shift toward lower temperature with increasing electric field, while the E150 level does not, in contradiction to the supposed single donor nature [191]. In addition, after plasma hydrogenation both electron traps show a different depth profile, where the positively charged E40 donor is believed to more strongly interact with H, compared with the E150 level [191]. Both are strong arguments against the generally accepted picture that the three states belong to interstitial Ti in silicon. Also DFT calculations do not resolve this discrepancy (see Chap. 9).

The interaction of titanium with atomic hydrogen has been studied, in view of possible passivation of the deep levels [192–195]. Several Ti-H levels have been discovered in n-type Si [193–195], while none were found in p-silicon. This can be explained by the double donor nature of the H180 level, which is repulsive for positively charged hydrogen in p-type silicon. According to Table 6.6, complexes between Ti and up to three H atoms have been identified, while the metal is completely passivated in n-type Si when 4 hydrogens are bound to it. The thermal stability of all TiH levels is limited to maximum 250 °C [195].

In *germanium*, substitutional titanium gives rise to an electron trap in n-type and a hole trap in p-type material, respectively [164]. The spectra are depicted in Fig. 6.30 and the resulting deep-level parameters given in Table 6.7. The levels

Table 6.6 Activation energies, apparent capture cross sections, and assignment for all defects observed in n- and p-type silicon. After Scheffler et al. [195]

Label	E_a (eV)	σ_n or σ_p (cm^2)	Assignment
E40	$E_C - 0.08$	6×10^{-15}	Ti
E40′	$E_C - 0.07$	8×10^{-16}	TiH
E150	$E_C - 0.27$	5×10^{-16}	Ti
E170	$E_C - 0.34$	4×10^{-14}	TiH
E170′	$E_C - 0.37$	4×10^{-14}	TiH_2
E260	$E_C - 0.55$	4×10^{-15}	TiH_3
E260′	$E_C - 0.50$	3×10^{-15}	H-related
H180	$E_V + 0.29$	9×10^{-17}	Ti

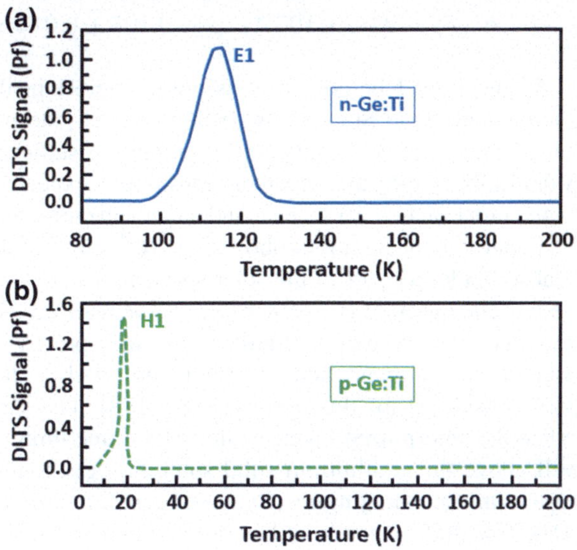

Fig. 6.30 DLTS of Ti-implanted germanium. **a** n-type, $\tau_{window} = 23.5$ ms and **b** p-type, $\tau_{window} = 2.1$ ms. Reproduced Lauwaert et al. [164]. Copyright 2009 AIP Publishing

Table 6.7 Deep level parameters and the assignment of the traps corresponding with Ti impurities in Ge. Values between brackets have been corrected for the Poole-Frenkel effect. After Lauwaert et al. [164]

Band	E_a (eV)	K_T (s^{-1} K^{-2})	σ_n or σ_p (cm^2)	ΔE_σ (meV)	σ_∞ (cm^2)	ΔH (eV)	ΔS	Level
E1	0.228	3.4×10^7	1.2×10^{-14}	27	8.7×10^{-16}	0.201	2.5	$Ti_s^{-/2-}$
H1	0.025 (0.028)	1.5×10^7	1.6×10^{-14}	…	…	…	…	$Ti_s^{0/-}$

correspond with the double and the single acceptor state of Ti_s, whereby the latter level in p-type material exhibits a Poole-Frenkel shift with the electric field, as expected for an attractive center for holes [164].

The recombination activity of Ti in Si has been probed by TDLS (PL) and IDLS (QSS-PC) [196] and by T-dependent carrier lifetime studies [197]. In p-Si, a level at $E_C - (0.24 \pm 0.03)$ eV has been identified with an asymmetry factor $k = 7650$ and another recombination center at $E_C - (0.49 \pm 0.03)$ eV and $k = 25 \pm 12$ [196]. In p-type mc-Si, it was found that the $E_V + 0.28$ eV level dominates the effective lifetime in the 140–170 °C range [197], whereby the temperature-dependence of the hole and the electron cross section was evaluated as:

$$\sigma_p(T) = 1.5 \times 10^{-16} \exp(-0.036\,\mathrm{eV}/k_B T)\left(\mathrm{cm}^2\right) \tag{6.10a}$$

and

$$\sigma_n(T) = 9.01 \times 10^{-14} \exp[(-0.107 \pm 0.01)\mathrm{eV}/k_B T]\left(\mathrm{cm}^2\right) \tag{6.10b}$$

As mentioned before, there is some interest in the development of so-called Intermediate Band Solar Cells (IBSC), whereby deep level centers form a band of states when their wave functions overlap at a sufficiently high density [180, 198]. A delocalization by an Anderson transition is expected to occur for a concentration on the order of 1×10^{20} cm^{-3} and a delocalization by a Mott transition is expected for impurity concentrations above 5×10^{19} cm^{-3} at 300 K [198]. The basic idea is to absorb a larger part of the solar spectrum by two-photon absorption to and from the IB. The question is whether this can be realized without paying the penalty of a degraded recombination lifetime by the presence of deep-level impurities. High-dose Ti implantation in combination with a short laser pulse annealing has been considered for this purpose [180, 198]. The short laser pulse should recrystallize the amorphized layer, while at the same time, activates the metal impurities well above the equilibrium solid solubility levels, above the Mott-transition concentration. Encouraging results have been demonstrated by Luque and co-workers [180, 198–200], whereby it has been demonstrated that the recombination lifetime did not suffer from recombination through individual Ti atoms, which could be present outside the IB region, deeper in the substrate [198]. It was shown that the lifetime even increased with implanted Ti dose. This proved the concept of the suppression of the non-radiative recombination lifetime by IB formation above a critical concentration [198]. A similar conclusion has been derived from the increased conductance [199] and photoconductance [200] of Ti-supersaturated silicon.

6.5 Molybdenum

As discussed in Chap. 3, Mo contamination can e.g. result from epitaxial deposition, high temperature processing and ion implantation such as BF_2. Initial DLTS studies of samples with Mo (concentrations about $4 \times 10^{17} - 1.5 \times 10^{20}$ cm^{-3}) introduced in the melt of p-Si for solar cells revealed a hole trap at $E_V + 0.31$ eV with a hole capture cross sections of 4.31 × 10^{16} cm^2 [201]. All the Mo is electrically active in the material and degrades the bulk lifetime and the solar cell performance. Although the segregation coefficient is very small (about 4.5 × 10^{-8}), a small amounts of Mo in the order of 0.1 ppb can lead to a 70% degradation of the cell performance [201].

Using a Mo spin coating and in-diffusion annealing, DLTS revealed three electron traps at $E_C - 0.27$ eV, $E_C - 0.34$ eV and $E_C - 0.58$ eV in n-Si and one hole trap at $E_V + 0.31$ eV in p-Si, as shown in Fig. 6.31. The hole trap concentration increases with the Mo concentration and is also found in Cz Si. The diffusion coefficient of Mo in Si is 2 × 10^{-10} cm^2/s at 1000 °C [202]. Based on optical-DLTS measurements it was concluded that the Mo-1, Mo-2 and Mo-3 traps are not associated with possible complex formation due to ion pairing between Mo and a doping atom [202].

Sandhu et al. [203] doped the wafers with Mo by placing Mo powder on the surface and performing a vacuum anneal at different temperatures. They observed a deep level at $E_V + 0.29$ eV and noticed that the anneal temperature plays a significant role as illustrated in Fig. 6.32. Below 650 °C no deep levels can be observed, indicating a density <10^{10} cm^{-3}. Using scanning infrared microscopy (SIRM) Mo-related precipitates having a density of 2.3 × 10^7 cm^{-3} (500 °C) to 5.8 × 10^9 cm^{-3} (800 °C) were detected in the surface region of the samples and the density is decreasing deeper in the bulk [203].

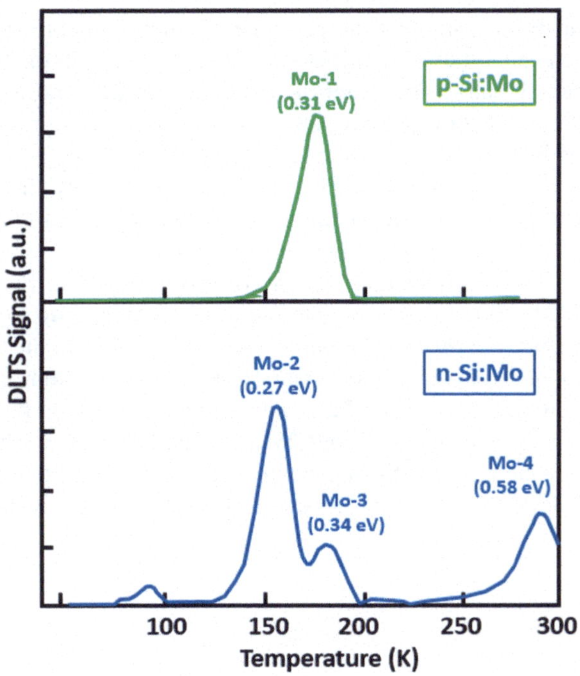

Fig. 6.31 DLTS spectra of molybdenum contaminated n- and p- type FZ Si annealed at 1000 °C in oxygen. Reproduced from Hamaguchi and Hayamizu [202]. Copyright 1991 The Japan Society of Applied Physics

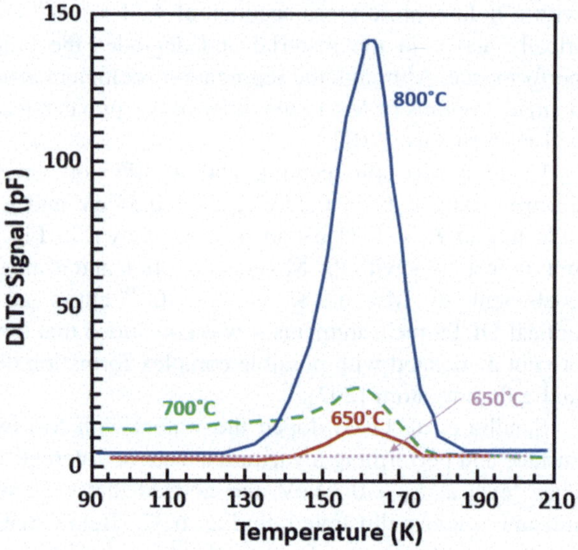

Fig. 6.32 DLTS spectra of molybdenum contaminated p-type Si annealed under vacuum at different temperatures. Reproduced from Sandhu et al. [203] with permission of Elsevier

The hole trap around $E_V + 0.30$ eV in p-type Si has also been observed in Mo implanted wafers [204, 205]. In contrast to doping in the melt, for these samples the electrically active Mo concentration is less than the implanted dose. There are also experiments performed with p-type epitaxial layers doped with Mo up to 3×10^{13} cm^{-3}, whereby the Mo doping is done by ion implantation [206]. TEM analysis points out the presence of small nano-precipitates, most likely associated with molybdenum silicides such as Mo_3Si (cubic), Mo_5Si_3 (tetragonal) or $MoSi_2$ (tetragonal). The latter is extremely stable and could be responsible for the recombination activity. DLTS reveals that in the region with nano-precipitates an additional peak is observed at $E_V + 0.47$ eV ($\sigma_p = 2.2 \times 10^{-19}$ cm^2) [206].

A recent report claims that in n-type Si only the $E_C - 0.26$ eV level is due a charge state of Mo, while the other two levels could be processing induced related to the preparation of the sample [207].

The recombination activity of Mo is shown in Fig. 6.33 and compared with Fe [204]. $POCl_3$ or HCl gettering at temperatures up to 1100 °C were insufficient to reduce the Mo concentration except very close to the surface [201] and also the use of an epitaxial p$^+$ substrate [204, 208] or a high energy B-implantation is not gettering Mo [204]. Proximity gettering methods by C and Si implantation have also been evaluated [209]. While C implantation is effective in capturing Mo, this is not the case for Si implantation. Extended defects do not seem to play a relevant role in the gettering process. Gettering will be extensively discussed in Chap. 8.

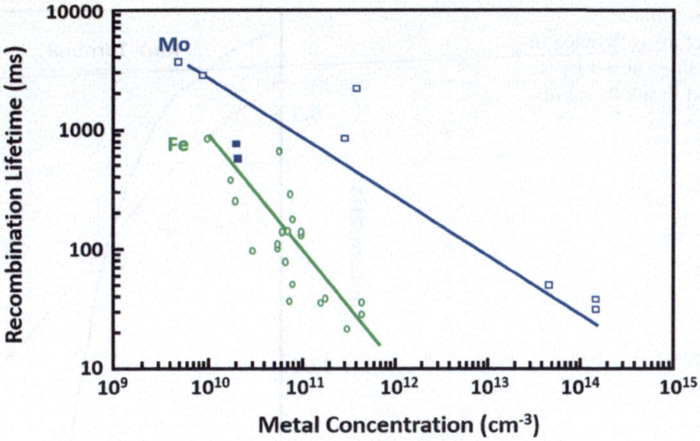

Fig. 6.33 Recombination activity of Mo in comparison with Fe. Reproduced from Benton et al. [204] by permission of ECS-The Electrochemical Society

6.6 Palladium

Palladium is a fast diffusing 4*d* transition metal with a low solubility and a tendency to form substitutional defects. This happens via interstitial defects which then transform via either a kick-out or a Frank-Turnbull mechanism. Both mechanisms have the same end result, i.e., an atom in a substitutional position but kick-out is based on the fast diffusion of interstitial atoms until they displace a lattice atom, while Frank-Turnbull occurs by the trapping of an interstitial atom by a vacancy. Analysis of the diffusion profile can differentiate between the two mechanisms as schematically shown in Fig. 6.34. Kick out gives a higher surface concentration and a shallower profile.

Substitutional palladium exhibits an acceptor state at $E_C - 0.22$ eV ($\sigma_n = 5 \times 10^{-15}$ cm^2) in p-type Si and a donor state at $E_V + 0.31$ eV ($\sigma_p = 8 \times 10^{-16}$ cm^2) and a double donor state at $E_V + 0.14$ eV ($\sigma_p = 1 \times 10^{-14}$ cm^2) in n-type Si [210]. The double donor state shows an electric field dependence of the emission rates and a thermal activation of the hole capture cross sections. These levels have been confirmed by different authors and are close to those of Pt in Si as will be discussed in the next section.

Complexes can be formed between interstitial and substitutional palladium (Pd_iPd_s) and palladium forms defects with vacancies, with dopants atoms and with other transition metals such as e.g., Fe_iPd_s complexes [211]. The latter is illustrated in Fig. 6.35, showing the appearance of an addition trapping level at $E_C - 0.32$ eV. The Pd_iPd_s defect is not electrically active. Palladium-vacancy PdV and palladium-boron PdB defect complexes have been observed by perturbed angular correlation measurements after isochronal annealing in highly doped n- and p-type silicon, respectively [212]. The fractional population of the PdV defect pair as a

Fig. 6.34 Schematic illustration of the difference in diffusion profile caused by Kick-out and Frank-Turnbull respectively

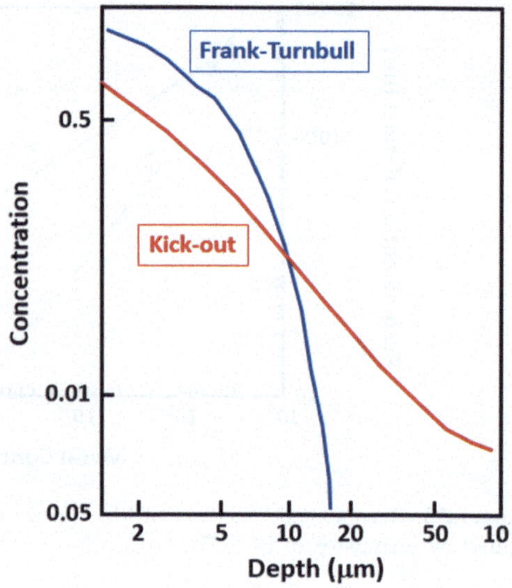

Fig. 6.35 DLTS spectrum of n-Si after Pd diffusion at 900 °C and subsequent 15 min diffusion of Fe at 1100 °C (top curve). The bottom spectrum is a reference sample diffused under the same conditions, except no Fe contamination before the second anneal step. The labels A, B, C, D refer to the same levels. A new peak due to Fe_iPd_s appears in the top spectrum. Reproduced from Czaputa [211] by permission of Springer

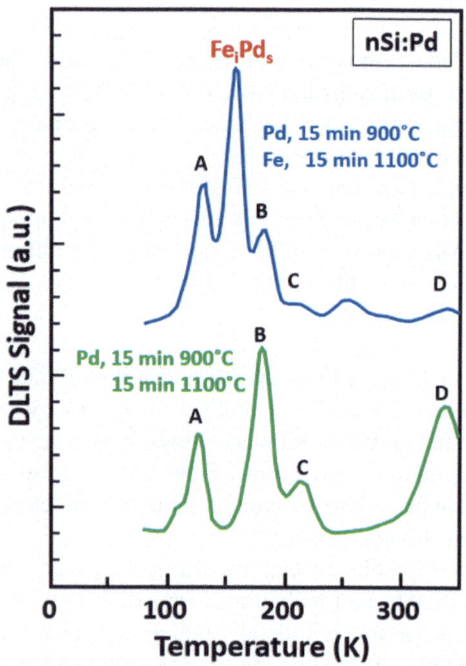

function of P concentration indicates a dominance of ionized vacancies, which is due to the Fermi-level effect. The absence of PdP pairing in n-Si suggests that it will not be possible to getter Pd via a phosphorus diffusion gettering mechanism.

After heat treatment Pd will precipitate homogeneously in so-called star-shape colonies, i.e., small precipitates surrounded by dislocations observed after a short step at the surface [213]. Similar as for Cu (see Table 2.2) about 0.5 self-interstitials are emitted per precipitated Pd atom. The precipitates have a hexagonal structure closely related to Pd_2Si particles growing on a plane parallel to the $\langle 111 \rangle$ direction and bound by an extrinsic dislocation loop. The precipitates are larger for lower cooling rates. These precipitates are also at the basis of the haze formation. For temperatures below 800 °C, PdSi is the stable silicide so that these precipitates are metastable.

Pd in Ge has been studied in the early days by DLTS and resulted in four different energy levels, i.e., $E_C - 0.23$ eV ($\sigma_n = 3 \times 10^{-17}$ cm^2), $E_C - 0.32$ eV ($\sigma_n = 4 \times 10^{-15}$ cm^2), $E_C - 0.42$ eV ($\sigma_n = 5 \times 10^{-14}$ cm^2) and $E_V + 0.22$ eV ($\sigma_p = 8 \times 10^{-20}$ cm^2) [213]. Time Dependent Perturbed Angular Correlation (TDPAC) measurements in combination with density functional calculations (DFT) point out that the PdV complex in Ge has a split-vacancy configuration, in contrast to the PdV complex in Si [214].

6.7 Platinum

Platinum is not commonly found as a contaminant. It is mainly used to control the lifetime in power devices, such as thyristors, to replace gold doping as a lifetime killer, or as a silicide for particular applications. Adding Pt to Ni also increases the thermal stability of the formed nickelsilicide. The lifetime controlling trapping levels have been identified as the $E_C - 0.26$ eV ($\sigma_n = 1 \times 10^{-14}$ cm^2) acceptor in n-type Si and $E_V + 0.32$ eV ($\sigma_p = 1 \times 10^{-15}$ cm^2) donor in p-type Si [215, 216]. Four additional levels have been observed, attributed to the possible complex formation of Pt with shallow donors or acceptors, but were not further studied in detail [217]. The fact that Pt can form different types of complexes with other elements like H, O and C is making it less suitable as a lifetime controlling element, in comparison with e.g. Au.

Pt diffuses rather fast into interstitial sites before taking a substitutional position by the kick-out mechanism. The diffusion process will be influenced by both the concentration of interstitials in the material (Cz Si versus FZ Si) and the presence of extended defects. The substitutional Pt_s can have four different charges states due to the formation of a donor and a double donor in the lower half of the band gap and an acceptor in the upper half. In the 300–500 °C temperature range, Pt_2Si and PtSi is formed. These silicides are used as long-wavelength infrared Schottky barrier detectors and focal plane arrays [218]. First, a di-silicide is formed and then transformed into mono-silicide after consumption of the Pt. This process and the continuity and roughness of the silicide layer can be controlled by using a double RTA anneal step [219]. Pt has no tendency to form precipitates so that it cannot be detected by a haze test.

The two levels mentioned above are associated to substitutional Pt_s. A third near mid-gap level ($E_C - 0.52$ eV) has also been observed [220], and is a very effective recombination center [221]. DLTS analysis can be complicated by the presence of hydrogen as several PtH complexes may be formed, i.e., besides the Pt_s levels different hydrogen-platinum complexes have been observed as illustrated in Table 6.8 [222]. The different reactions in both n- and p-type Si are schematically shown in Fig. 6.36. The H150 level is formed in both n- and p-type Si. After annealing above 600 °C all the hydrogen complexes dissociate and the same Pt_S concentration is found as before hydrogenation.

Table 6.8 Deep level parameters and the assignment of the traps observed in hydrogenated platinum-doped (by evaporation or doping in the melt) silicon. Values between brackets have been corrected for the Poole-Frenkel effect. After Sachse et al. [222]

Level	Diode type	Activation energy (eV)	σ_n or σ_p (cm^2)	Assignment
E90	*n*-type	$E_C - 0.18 \pm 0.01$	1×10^{-17}	Pt–H related
Pt(120)	*n*-type	$E_C - 0.23 \pm 0.01$	5×10^{-15}	$Pt^{-/0}$
E250	*n*-type	$E_C - 0.50 \pm 0.01$	2×10^{-16}	Pt–H related
Pt(50)	*p*-type	$E_V + 0.09 \pm 0.01$	2×10^{-17}	$Pt^{+/++}$
H150	*p*-type	$E_V + 0.30 \pm 0.02$	$9 \times X 10^{16}$	Pt–H related
Pt(160)	*p*-type	$E_V + 0.33 \pm 0.01$	8×10^{-16}	$Pt^{0/+}$
H210	*p*-type	$E_V + 0.40 \pm 0.03$	3×10^{-17}	Pt–H related

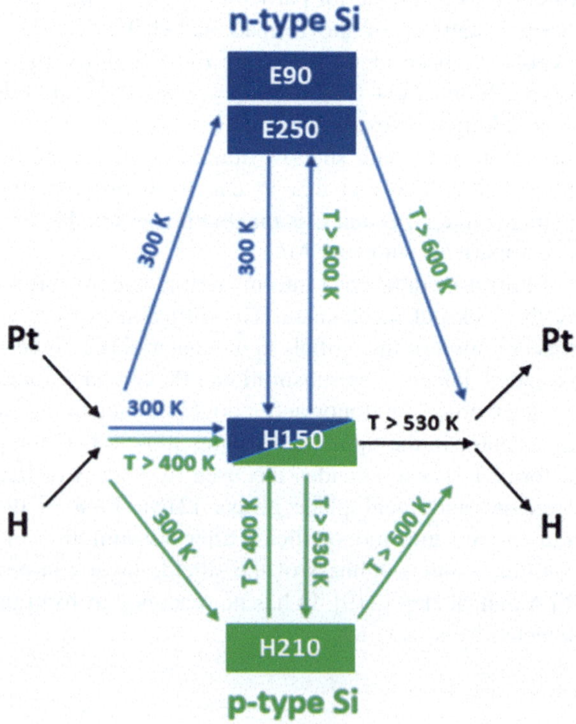

Fig. 6.36 Schematic diagram of the reactions involving Pt and H. The levels E90, E250 and H210 appear after wet-chemical etching. Level H150 is formed in *p*-type Si during annealing at $T > 400$ K, but can already be detected as compensation of phosphorus donors in *n*-type Si directly after etching. H150 dissociates at $T > 500$ K, while E90, E250, and H210 are thermally more stable. Reproduced Sachse et al. [222]. Copyright (1997) by The American Physical Society

Additional complexes, identified by EPR analysis, can be formed by the interaction of Pt with oxygen [223] or with carbon [224]. In the latter case two different defects consisting of Pt with either one or three C atoms have been observed.

The solubility is higher for Pt_i than for Pt_s and for temperatures around 1000 °C the solubility curve become retrograde. It has to be remarked that no interstitial Pt_i has been observed at room temperature.

Laplace DLTS enables to study the hyperfine structure in the spectrum of Pt-doped SiGe [225]. For Ge concentrations up to 5% analysis of the spectra revealed that Pt atoms prefer to reside at a lattice location, i.e., a substitutional Si site, in the vicinity of a substitutional Ge atom as Pt and Ge cause opposite stress effects. For higher Ge concentrations the number of features in the spectrum becomes too large. As will be discussed in the next section, a similar observation concerning the lattice position holds for Au in diluted SiGe structures.

6.8 Gold

In the early days of semiconductor technology Au played a key role to control the lifetime in bipolar, high-voltage and power devices. Au diffuses at low temperatures to interstitial sites although the interstitial solubility is low. Similar as for Pt_s, Au_s is formed by the kick-out mechanism resulting in excess Si interstitials. The self-interstitials may super-saturate in the bulk of the material, forming during further processing stacking faults [226]. The diffusion of these self-interstitials will in the absence of extended defects determine the Au diffusion profile, which results in an U-shape profile and a $\sqrt{t}$ dependence [227]. However, when extended defects are present, the diffusion is controlled by the diffusion of the interstitial Au atoms, leading to an error function and a t dependence [228].

During cooling of the samples, Au remains in the substitutional position. Although Au precipitates are formed, no haze is observed because of the low Au_s diffusivity. In contrast to Pt, no stable Au silicides are formed. The presence of extended defects may trigger the precipitate formation.

Extensive electrical characterization of the Au levels has been done since the late fifties and it is generally believed that substitutional Au is an amphoteric defect with an acceptor level at $E_C - 0.55$ eV and a donor level at $E_V + 0.35$ eV [229–232]. The donor level has a lower concentration than the acceptor level [231]. EPR studies indicate that the neutral Au center is characterized by $S = 1/2$, g_L-parallel = 2.8, g_L-perpendicular = 0 and tetragonal distortion [233], although also different g_L-values have been reported leading to a different center with an orthorhombic symmetry [234]. This will not further be discussed. Finally, important to remark that Au is easily forming donor-acceptor pairs with other transition metals which leads to a variety of different energy levels in the band gap. From a processing viewpoint, the AuFe pair is important as Fe is often observed as a contaminant.

The impact of Au contamination on the electrical activity of GBs has been studied by DLTS and EBIC in the context of mc-Si [235]. A model GB has been produced by the bonding of two slightly misaligned n-type silicon wafers, resulting in a dislocation network consisting of regular screw and edge-type dislocations. Au contamination was performed by in-diffusion of gold at 850 °C for 13 h. No Au precipitates have been observed at the GB [235], while at the same time, a strong increase of its electrical activity was derived from the strong increase of the EBIC contrast and also a pronounced reduction of the carrier concentration was observed. The latter suggest the presence of acceptor states, associated with the Au decoration of the dislocations. From DLTS, an energy level at $E_C - 0.24$ eV and $\sigma_n = 10^{-17}$ cm^2 was derived [235], The peak showed an increase of the DLTS amplitude according to $\ln(t_p)$, which is typical for a distribution of localized states at an extended defect. For the Au-contaminated case, a broad distribution of energy levels in the range $E_C - 0.33$ eV to $E_C - 0.65$ eV has been established. The DOS was in the range of $1–2 \times 10^{12}$ cm^{-2} eV^{-1}, with a charge neutrality level at $E_C - 0.70$ eV [235]. The corresponding electron capture cross section varies with energy in the band from 5.65×10^{-15} cm^2 ($E_C - 0.50$ eV) to 1.18×10^{-16} cm^2 ($E_C - 0.43$ eV). As a result of filling up the deep levels by electron trapping at the gold-contaminated GB levels, a potential barrier builds up, giving rise to a Schottky-like behavior of the current-voltage characteristic across it [235].

With the strong interest in the use of high-mobility substrates it is important to give some attention to the behavior of Au in Ge and SiGe alloys. Deep donors are easier incorporated in Si than in Ge, while the opposite is true for acceptors. Due to the lattice mismatch between Si and Ge (4%), Ge alloying to form SiGe compounds will lead to the generation of dislocations. The dislocation density increases with Ge concentration. When such SiGe crystals are doped with e.g. Au, it is possible that not all the diffused Au is measured, as a part of the Au is precipitated on the dislocations thereby reducing the concentration of interstitial and substitutional Au [236]. This can lead to a lower concentration or even the disappearance of some spectral lines. It has been reported that new peaks are revealed associated with the Au trapped in the strain field of the dislocation core [237].

Research has been performed on what is happening with the positions of the defect levels if the band gap of the material is changed, i.e., are they free to move or remain they pinned to the valence or the conduction band? Early experiments were based on using either the temperature or hydrostatic pressure to modify the band gap. A more sensitive and elegant way of band gap engineering is making use of $Si_{1-x}Ge_x$ structures, as illustrated in Fig. 6.37 showing the band gap as a function of the Ge doping [238]. Data from Braunstein et al. [239] and Weber and Alonso [240], measured at 196 and 20 K, respectively, are also shown. An increase of the germanium content from 0 to 25% will shrink the band gap with 100 meV [238]. The *X* and *L* refer to the *X*-conduction band and *L*-conduction band, respectively. The semi-empirical formula is only valid up to 85% Ge as then there is a cross-over from the Si-like *X*-conduction band to the Ge-like *L*-conduction band.

Investigation of the Au-related defect states in $Si_{1-x}Ge_x$ pointed out that both the Au donor and acceptor levels are pinned to the conduction band [238]. The deep

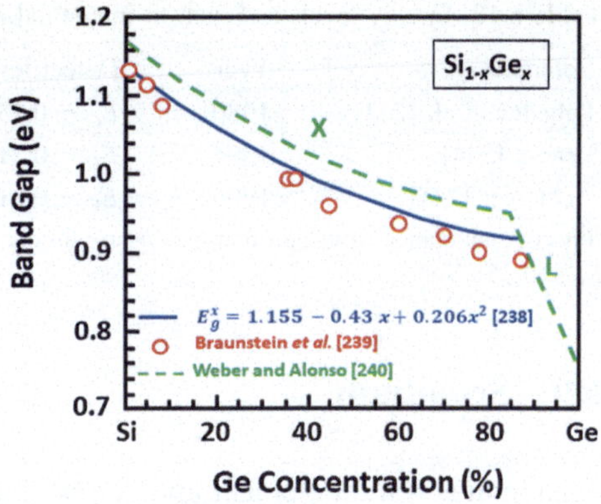

Fig. 6.37 Band gap energy of the $Si_{1-x}Ge_x$ system as function of the Ge concentration. Reproduced from Mesli et al. [238]. Copyright (1997) by The American Physical Society

acceptor level at $E_C - 0.55$ eV is a very efficient recombination-generation center and will interact simultaneously with both the conduction and the valence band. Calculations of the entropy related to the creation of electron-hole pairs will therefore have to consider the level pinning effect. For SiGe systems not only the Ge concentration impacts the band gap but also the stress in the material as the material can be either relaxed or strained. Nauka and Kamins [55] noticed an increase of the activation energy of Au-related deep levels for higher stress levels.

In SiGe the substitutional Au_s atom has in principle the choice to take position at a Si lattice site or at a Ge lattice site. A Ge atom has a larger ionic radius than a Si atom so that it is easier to kick-out Si atoms from their substitutional position. Energy considerations reveal that Au atoms prefer to reside as Si-substitutional atoms close to Ge [241]. In the whole diffusion process, one must consider the energy differences between a Si self-interstitial and a Ge interstitial site and the substitutional-site bonding of Ge and Si, respectively. The effect of the detailed atomic configuration of point defects in SiGe can be observed by Laplace DLTS as the local fluctuations influence the thermal emission rates and the fine structure [241]. This effect is called alloy splitting.

Gold contamination in Ge has been extensively studied as shown in Table 6.9. Based on DLTS investigations it is now more or less generally accepted that substitutional gold Au_s in Ge has an amphoteric character resulting in three acceptors levels at $E_C - 0.056$ eV ($Au^{2-/3-}$), $E_C - 0.215$ eV ($Au^{-/2-}$) and $E_V + 0.135$ eV ($Au^{0/-}$) and one donor level at $E_V + 0.044$ ($Au^{+/0}$) [242].

Table 6.9 Defect states in Au-doped Ge

References	Defect level (eV)								
	$E_V + E_T$					$E_C - E_T$			
Woodbury [354] 1957		0.05			0.15			0.20	
Pearton [213] 1982				0.14		0.22			
Kotina [351] 1987					0.15			0.20	
Simoen [242] 1987	0.044		0.135				0.215		0.56

Table 6.10 Energy levels assigned to interstitial scandium in Si

References	Year	Defect level (eV)		
Lebedev et al. [243]	1980[a]	$E_C - 0.15$ to -20	$E_C - 0.50$	$E_C - 0.55$
Lemke [244]	1994	$E_C - 0.21$	$E_C - 0.50$	$E_V + 0.20$
Achtziger [245]	1996	$E_C - 0.21$	$E_C - 047$	$E_C - 0.51$

[a]Photoconductance measurements. Also levels at $E_C - 0.27$, $E_C - 0.35$, $E_V + 0.45$ and $E_V + 0.35$ eV

6.9 Scandium

Scandium is a light element and can be compared with Ti. However, as it is not a common contaminant during semiconductor processing it is not much studied. Little information is available on the solubility and diffusivity in Si. In view of its place in the periodic table Sc is expected to take an interstitial position in the lattice.

The limited information available on the defect levels related to interstitial scandium is given in Table 6.10. Using photocapacitance measurements seven different levels were reported [243], but some of them may be due to possible contamination by other elements and should therefore be excluded. Two of them can be correlated with energy levels found by other authors. Using Sc doping in the Si melt of FZ crystals, Lemke was able to observe two defect levels in the upper half of the band gap, i.e., a donor level at $E_C - 0.21$ eV and a double donor at $E_C - 0.50$ eV, and a triple donor at $E_V + 0.20$ eV in the lower half of the band gap [244]. DLTS experiments, using the transmutation of ^{45}Ti to ^{45}Sc, confirmed the first two energy levels but not clearly the triple donor one [245].

No Sc precipitates have been reported. The reaction of scandium films with silicon leads to the formation of orthorhombic ScSi at a temperature of about 500 °C, while at about 900 °C ScSi reacts with Si to form $ScSi_{1.7}$ with a hexagonal structure [246].

6.10 Vanadium

Vanadium can constitute both interstitial and substitutional sites in silicon. Early investigations pointed out that interstitial vanadium lies at or close to the tetrahedral interstitial site and is characterized by two electron traps at $E_C - 0.20$ eV and at $E_C - 0.45$ eV and a hole trap at $E_V + 0.20$ eV [247]. Over the last two decades more refined DLTS measurements have been performed for both n- and p-type Si [248]. The different deep level parameters, associated with the same interstitial defect, reported in the literature are summarized in Table 6.11. The capture cross sections of the electron traps are temperature independent, in contrast to the hole trap [247]. Recent measurements give for the direct majority capture cross section of E1 = 4.3×10^{-17} cm^2 and H1 = $3.5 \times 10^{-17} \exp(-0.12/k_BT)$ cm^2 [249].

Table 6.11 Energy levels assigned to interstitial vanadium in n- and p-type Si. Data from Mullins et al. [248]

Level	References				
	Davis [355] 1980	Weber [249] 1983	Tilly [356] 1991	Sadoh [247] 1992	Mullins [252] 2016
E1	$E_C - 0.22$	$E_C - 0.16$	$E_C - 0.21$	$E_C - 0.20$	$E_C - 0.20$
E2	$E_C - 0.46$	$E_C - 0.45$	$E_C - 0.48$	$E_C - 0.45$	$E_C - 0.43$
E3	$E_V + 0.42$	$E_V + 0.30$	$E_C - 0.36$	$E_V + 0.34$	$E_V + 0.34$

Only interstitial vanadium is electrically active and the diffusion coefficient is about 1000× smaller than interstitial Fe_i [247], so that it is more difficult to getter this metal. The diffusion coefficient has been reported to be higher in Cz than in FZ Si, due to the higher diffusivity of the mobile vacancies impacting the formation of substitutional vanadium V_s defects. It is important to remark that only a small fraction of the implanted V (dose 2.5×10^{11} cm^{-2} in n-type Si and 2.6×10^{11} cm^{-2} in p-type Si, equivalent to a peak concentration of around 5×10^{15} cm^{-3}) is occupying an interstitial position and, therefore, electrically active. About 98% of the concentration is in either a substitutional position or in a precipitated form.

DFT calculations of vanadium in Si point out that the V_i lies at or close to the tetrahedral site and is characterized by a double donor level at $E_V + 0.20$ eV and $E_C - 0.52$ eV and an acceptor level at $E_C - 0.20$ eV [250], in good agreement with the experimental data given in Table 6.11. Although first principles calculations give for the substitutional V_s site an acceptor level at $E_V + 0.23$ eV [250], this has so far not been detected by DLTS [248]. Not much information is available on the precipitation behavior of vanadium. The interstitial V_i has an EPR signature characterized for the double donor $S = 3/2$ and $g_L = 1.9892$ [151, 251].

There is a strong interaction with hydrogen, as experimentally studied by an exposure to a 50 W remote plasma at room temperature for 30 min [252]. Depending on the anneal cycle used the formation of VH complexes can reduce the electrical activity of V_i with more than 20%. Annealing at 200 °C will dissociate the VH complex and restore the electrical activity of the interstitial vanadium. DFT studies predict that both V_s and the {V_s, H} and {V_s, H, H} complexes are electrically active [250] but have to be detected experimentally yet. The hydrogen complexes would be five- and sixfold coordinated, respectively.

Vanadium has also been reported to form complexes with other TMs like Au and Zn [253]. In the case of VAu two levels at $E_C - 0.20$ eV ($\sigma_n = 4.9 \times 10^{-16}$ cm^2) and $E_V + 0.42$ eV ($\sigma_p = 6.4 \times 10^{-16}$ cm^2) have been observed and the binding energy between Au and V is 1.6 eV.

Like other TM's in Si, V degrades the electrical device performance due to the generation of recombination centers. However, for high V concentrations the far infrared photoconductivity of photodetectors can be improved compared to the un-implanted samples [254]. Similar as for Ti discussed in Sect. 6.4, the non-radiative recombination lifetime is suppressed by IB formation above a critical

V concentration. The photoconductivity edge shifts to lower photon energies for lower temperatures, which is explained by the delocalization of the V associated deep levels once a critical concentration has been exceeded, thereby reducing the non-radiative recombination processes [254].

6.11 Manganese

Manganese is located in the middle of the 3*d* transition metals series and its solubility and diffusivity in silicon are relatively high and quite similar to iron in the temperature range of 300–1500 K. As it is not a common contaminant encountered in semiconductor processing, the available literature is rather limited. Mn can exist both in interstitial Mn_i and substitutional Mn_s lattice sites [151, 255]. For Mn_i a donor level at $E_C - 0.53$ eV was already observed more than 60 years ago by Carlson investigating Cz crystals doped in the melt [255]. More recent investigations give a level at $E_C - 0.43$ eV [256]. It is generally accepted that Mn_i has four charge states (Mn^-, Mn^0, Mn^+ and Mn^{++}), resulting in three energy levels, i.e., an acceptor state at $E_C - 0.12$ eV, a donor state at $E_C - 0.43$ eV and a double donor at $E_V + 0.43$ eV. For the Mn_s state a donor level at $E_V + 0.34$ eV and an acceptor level at $E_C - 0.43$ eV are reported [257].

Mn forms donor-acceptors pairs with doping elements such as boron and gallium and metals like Al, Sn and Au. A summary of the different reported levels for the MnB complex are summarized in Table 6.12 and as can be noticed there is a good agreement between the different experimental data.

It was reported that the interstitial Mn can cluster to form Mn_4 with a donor level at $E_C - 0.28$ [258]. Beside this tetragonal cluster, EPR also revealed the possibility for a trigonal $[Mn_{3i}^0 - Mn_i^x]$ cluster consisting of three interstitial Mn_i ions with $S = 3/2$ in the ground state and a fourth Mn ion with $S = 1$ [259]. Although not confirmed the fourth ion is either Mn_i^{x-} on an interstitial site or Mn_s^+ on a substitutional site, with in view of the discussion above most likely the first option. In p-type silicon another EPR spectrum of cubic symmetry was revealed, which was attributed to the Mn_{i4}^+ cluster [260]. This spectrum is highly light sensitive, with a +/0 state close to mid-gap.

Table 6.12 Energy levels assigned to the MnB complex Si

References	Year	Defect level (eV)
Carlson [255]	1956	$E_C - 0.53$
Lemke [357]	1981	$E_C - 0.55$
Czaputa et al. [258]	1983	$E_C - 0.54$
Nakashima and Hashimoto [358]	1991	$E_C - 0.50$
Roth et al. [256]	2007	$E_V + 0.55$

Slow cooling of Mn-doped Si results in precipitates, although the process has not well been studied and the role of the Mn_4 cluster is not clear. For fast cooling no haze is observed. Mn also forms silicides, i.e., MnSi after anneal in the 400–500 °C range and $MnSi_2$ above 800 °C [261, 262].

It worth mentioning that there is interest in Mn-doping, mostly achieved by ion implantation, for some specific applications. Mn implantation with a dose of 10^{14} cm^{-2} at 100 keV in combination with an 800 °C RTA step increases the resistivity of p-type Si (600 Ω cm) up to 10 kΩ cm in the top layer of the wafer due to the Mn compensation of the B concentration [263]. Such high-resistivity material is used to fabricate radio frequency devices.

Mn doping of Ge crystals in the melt has already in the fifties been studied by Hall and photoconductivity measurements and revealed two double acceptor levels at $E_C - 0.37$ eV from the conduction band and at $E_V + 0.16$ eV from the valence band, respectively [264, 265]. More recent, DLTS investigations of Ge wafers with an Mn evaporated layer which received an in-diffusion step at 700 °C and an anneal at 400 °C for 20 min, revealed three defect states [266]. These Mn related traps are shown in Fig. 6.38. The acceptor level at $E_C - 0.37$ ($\sigma_n = 2.7 \times 10^{-14}$ cm^2) and the donor level at $E_V + 0.14$ eV ($\sigma_p = 1.4 \times 10^{-13}$ cm^{-2}) correspond with the two traps mentioned above. However, the third trap at $E_V + 0.43$ ($\sigma_n = 4.7 \times 10^{14}$ cm^{-2}) has not been reported before and due to the similarity in amplitude and depth as the Mn-H2 trap it is assumed that this level can also be assigned to substitutional manganese Mn_s [266]. Furthermore, the analysis of the carrier capture cross-sections, the absence of field-assisted emission and the observation of the Mn^{2-} EPR spectrum in n-type Ge:Mn at low temperature are all compatible with Mn introducing two acceptor and one donor levels in the band gap of Ge [266].

Si:Mn and Ge:Mn can be used as ferroelectric materials enabling the fabrication of spintronic devices. The observed ferromagnetism in Mn-doped Si is associated with nanocrystalline $MnSi_{1.7}$ [267]. For Ge, the ferromagnetic operation was originally observed between 25 and 116 K [268], but $Ge_{0.94}Mn_{0.06}$ enables operation at 285 K [269]. Mn doping of Ge makes it p-type and forms secondary phases or Mn-rich nanostructures even at temperatures as low as 70 °C [267]. More work is needed to

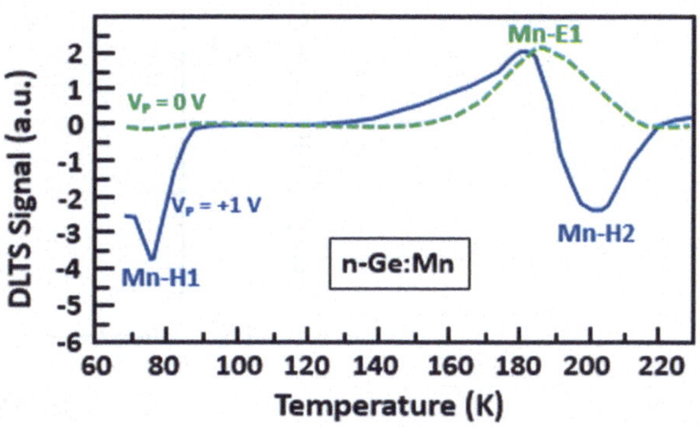

Fig. 6.38 DLTS of a n-Ge/MnGe diode, showing the 3 Mn related deep traps

determine the Mn-rich phase, although some authors refer to Mn_5Ge_3 [270, 271]. For the ferromagnetism it is important to determine the exact location of the Mn in the lattice. Electron channeling experiments in combination with ab initio calculations seems to indicate that in Ge, Mn prefers the bond-centered site, i.e., in the center of two nearest neighbor substitutional sites [272].

6.12 Zinc

Zinc can be found as a result of a polishing process depending on the slurries that are used. In addition zinc has a low evaporation temperature and a low melting temperature (419 °C). Zn is a 3*d* element that diffuses via the kick-out mechanism, as has been demonstrated by experiments in plastically deformed Si [273]. The effective diffusion coefficient in the temperature range 870 °C < *T* < 1200 °C is given by $D_{Zn_i} = 6.4 \times 10^{-5} e^{\frac{-1.85 \mp 0.07 eV}{k_B T}}$ as shown in Fig. 6.39 and compared with the coefficient of interstitial Au_i and interstitial Pt_i [274]. It should be remarked that if the metal is trapped at dislocations, the diffusion coefficient will be under-estimated. In dislocation-free Si zinc will be in super-saturation and the diffusion coefficient becomes concentration dependent.

In Si zinc forms two substitutional defects, i.e., a double acceptor near mid-gap at $E_C - 0.53$ eV and an acceptor in the lower half of the band gap at $E_V + 0.32$ eV [275]. The latter has been identified by EPR as the negative charge state of the

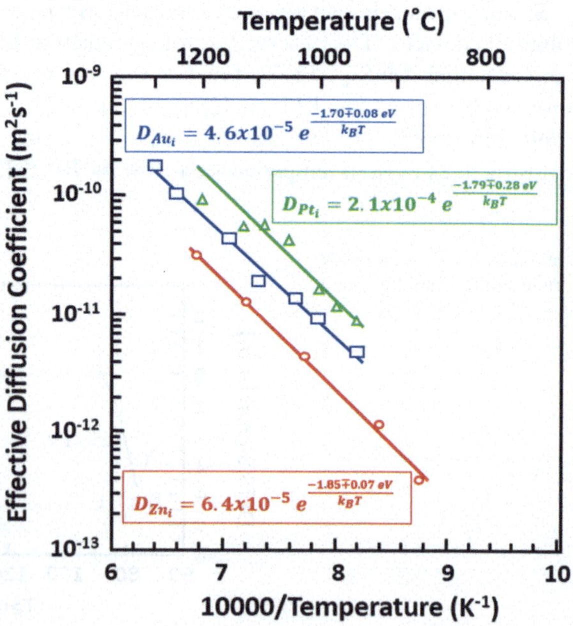

Fig. 6.39 Effective diffusion coefficient of interstitial Zn, Au and Pt in distorted Si. After Bracht and Overhof [273]. Copyright 1996 WILEY-VCH Verlag GmbH & Co

Table 6.13 Energy levels and capture cross sections assigned to the Zn complexes Si. After Lemke [277]

Defect	Defect level (eV)	Capture cross section (cm^2)
ZnCr	$E_C - 0.10$	$\sigma_n = 3 \times 10^{-15}$
ZnFe	E_C-0.47	$\sigma_n = 1 \times 10^{-16}$
ZnV	$E_V + 0.29$	$\sigma_h = 1 \times 10^{-16}$
ZnMn	$E_V + 0.18$	$\sigma_n = 2.4 \times 10^{-15}$

isolated substitutional Zn_s, characterized by $S = 3/2$ and $g_L = 1.1749 \pm 0.0005$ [276]. In contrast to the other TMs Zs does not form a donor state. Due to the acceptor nature of substitutional Zn it can react with other 3*d* TMs to form acceptor-donor pairs as indicated by the energy levels and capture cross sections given in Table 6.13 for Cr, Fe, V and Mn [277]. Some of the centers can be passivated by hydrogen rendering them electrically inactive [278].

Voss et al. [279] observed that during DLTS analysis of Zn doped $Si_{1-x}Ge_x$ samples, the position of the energy levels associated with substitutional Zn_s shifts to lower temperatures for increasing Ge concentration *x*. The capture cross sections don't depend on the composition.

The evolution of the $Si_{1-x}Ge_x$ band-edge energy and the two energy levels of the Zn acceptor states with in the band gap as a function of the composition *x* is detailed in Fig. 6.40. The kink in the profile of the conduction band edge at $x = 0.85$ is caused by the cross-over from the Si-like *X*-band E_C^X to the Ge-like *L* conduction band E_C^L. Up to about 30% Ge the first hole trap remains horizontally aligned with the SiGe band gap, giving evidence for the localized nature of this trap. The shift of the Zn energy levels can be represented by (6.11a) [280]

$$E_{Zn^{0/-}}(x) = 0.278 + 0.0043x + 0.126x^2 \quad (6.11a)$$

$$E_{Zn^{-/2-}}(x) = 0.568 + 0.301x + 0.196x^2 \quad (6.11b)$$

These equations give a good fitting of the data in Fig. 6.40. Based on ab initio calculations an alloy-broadening model has been develop in order to explain the DLTS data and enabling a refining of the formulas given in (6.11a) [281].

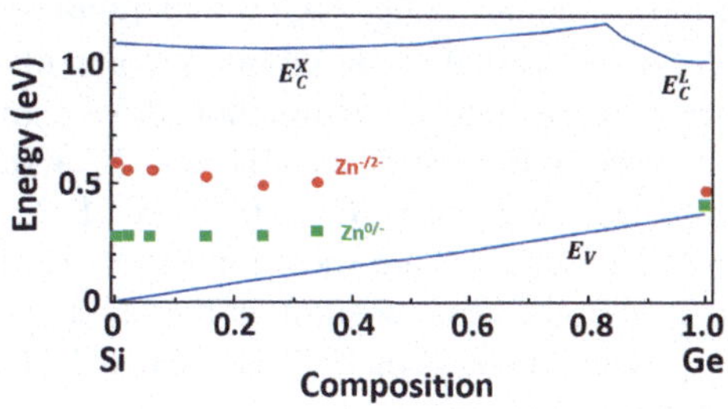

Fig. 6.40 $Si_{1-x}Ge_x$ band-edge energies and energy levels of the two Zn acceptor levels within the band gap as a function of the composition. Reproduced from Voss et al. [279]. Copyright 1991 AIP Publishing

6.13 Zirconium

Not much information is published about zirconium in Si. Lemke [282] has performed DLTS investigation of FZ Si doped in the melt with Zr and observed three energy levels; an acceptor level at $E_C - 0.13$ eV, a donor level at $E_C - 0.41$ eV and a double donor state $E_V + 0.32$ eV. These three states belong to interstitial Zr. As will be discussed in Sect. 6.18, these levels are very similar to the levels detected for Hf. Although some researchers are looking at the use of ZrO_2 [283] as a high-κ dielectric, so far, there are no reports that Zr would be a contaminant during Si processing. Experiments in a vertical furnace to study the possible cross contamination between a wafer with a ZrO_2 film and a blanket Si wafer indicated that there is no real danger for cross contamination during a 2 h anneal step at 950 °C [284].

For n-type Ge, containing a coated Zr film and annealed for 8 h at 650–750 °C in hydrogen, three electron traps have been reported, i.e., two donor levels at $E_C - 0.22$ eV and $E_C - 0.31$ eV after the diffusion process and a third donor level at $E_C - 0.37$ eV after storage at room temperature for several months [285]. A heat treatment for 2 h at 250 °C in a hydrogen plasma at a pressure of −0.5 Torr passivates the $E_C - 0.31$ eV level, has a weak effect on the $E_C - 0.22$ eV state, and no effect on the $E_C - 0.37$ eV centers. No further details have been reported on the passivation mechanism itself.

6.14 Niobium

Niobium compounds have been studied in view of their use for dielectric layers, but the information on niobium as a silicon contaminant is rather limited. The reported energy levels are an acceptor level at $E_C - 0.28$ eV ($\sigma_n = 7.5 \times 10^{-18}$ cm^2), a donor level at $E_C - 0.62$ eV ($\sigma_n > 10^{-16}$ cm^2) and a double donor state $E_V + 0.17$ eV ($\sigma_p = 3.8 \times 10^{-16}$ cm^2) [286]. These energy levels have been observed using junction space-charge techniques in Nb implanted samples [287]. More recently, Nb implantation has been systematically investigated in both n-and p-type samples, using DLTS and lifetime analysis techniques [288]. Three deep levels are revealed in niobium contaminated silicon, plus an additional level observed in high contamination dose samples only. In p-Si, two hole traps are observed at $E_V + 0.21$ eV (H1) and $E_V + 0.57$ eV (H2), while in n-Si three electron traps are observed at $E_C - 0.32$ eV (E1), $E_C - 0.54$ eV (E2), and $E_C - 0.55$ eV (E3). These levels are shown in Fig. 6.41. The E2 level at $E_C - 0.54 \pm 0.01$ eV is located at the same position in the band gap as the H2 level at $E_V + 0.57$ eV found in p-type silicon. For both H1 and H2, the trap concentration increases for a Nb dose of 10^{11} cm^{-2} to 3×10^{11} cm^{-2}, and decreases when the Nb dose is further increased to 10^{12} cm^{-2}. The peak E3 is present only at high implantation dose (10^{12} cm^{-2}) and could be related to niobium clusters.

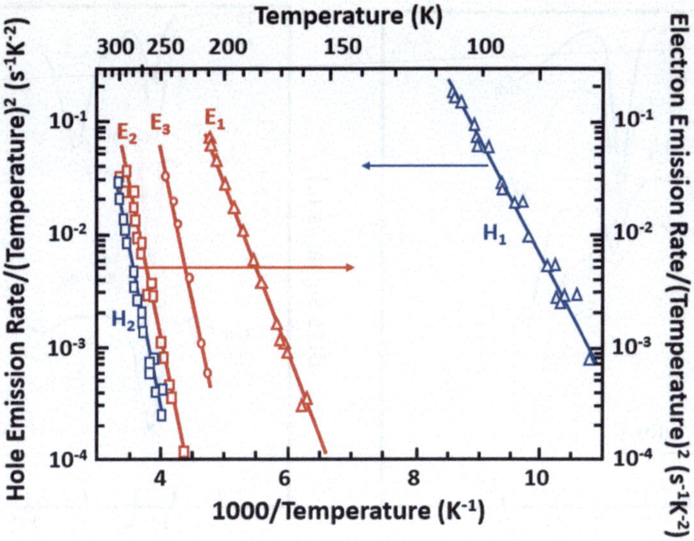

Fig. 6.41 Arrhenius plots of e_p/T^2 for the H1 and H2 levels in p-type Si:Nb and e_n/T^2 for the E1, E2 and E3 levels in n-type Si:Nb. Reproduced from Polignano et al. [288] by permission of ECS-The Electrochemical Society

Niobium impacts both the generation lifetime and the surface generation velocity and especially the trapping level near mid-gap has a strong lifetime degrading influence [288]. The deactivation of Nb at higher dose in p-type sample, i.e., a reduction of H1 and H2 for higher Nb dose, corresponds with the observed behavior of the lifetime measurement.

6.15 Ruthenium

The first report on Ru in Si was in 1991, using n- and p-Si diodes which were contaminated by plating Ru powder on the surface before an 1000 °C annealing for 1–2 h in N_2 followed by a fast quenching to room temperature [289]. For n-Si:Ru DLTS investigations revealed electron traps at $E_C - 0.25$ eV, $E_C - 0.39$ eV and $E_C - 0.61$ eV, as shown in Fig. 6.42. For increasing anneal temperature the E(0.25 eV) and E(0.39 eV) levels are transformed into a metastable E(0.58 eV) level. In the case of p-Si:Ru two energy levels H(0.34 eV) and H(0.46 eV) are observed after quenching. Annealing lowers the concentration of the last level which completely disappears after 300 °C anneal. This leads to the conclusion that Ru doping of Si results in two dominant levels, i.e., an electron trap E(0.58 eV) and a hole trap H(0.34 eV), which have been associated with a complex of Ru atoms with vacancies and substitutional Ru_s, respectively [289]. Ru_s has a donor charge state. Electron irradiation reduces the electron trap and increases the hole trap density.

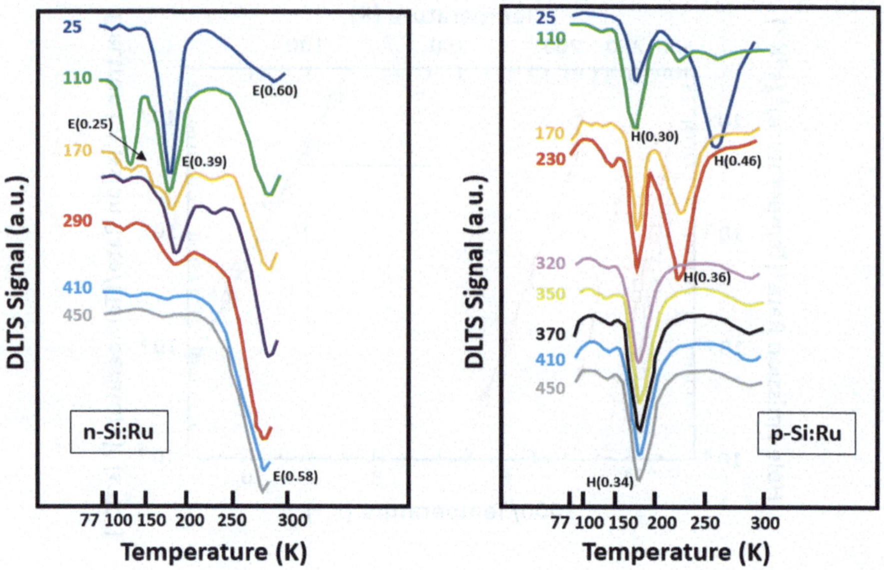

Fig. 6.42 DLTS spectra of isochronal annealed n-Si:Ru and p-Si:Ru samples. Reproduced from Zhou et al. [289]. Copyright 1991 AIP Publishing

An experiment using Ru doping of the melt of FZ crystals resulted in two different defect levels at $E_C - 0.14$ eV ($\sigma_p = 1.1 \times 10^{-16}$ cm^2) and E_V +0.26 eV ($\sigma_p = 9.2 \times 10^{-16}$ cm^2) [244]. These levels are, however, close to those ($E_C - 0.184$ eV and $E_V + 0.265$ eV) observed in Ru ion implanted (2×10^{14} cm^{-2}, 300 keV) FZ Si and electrically activated at 1200 °C for 30 s [290]. The origin of these levels has not been identified but may be attributed to the same amphoteric defect. The hole capture cross section is about 10^{-15} cm^2 and has a small capture enthalpy of around 1 meV, while the enthalpy of the electron capture is −8 meV [290].

Ru thin films on Si will during a temperature treatment react leading to the formation of an isomorphous silicide phase Ru_2Si_3 [291]. Ru_2Si_3 forms by a Si diffusion controlled mechanism from approximately 450–525 °C with an activation energy of 1.8 eV. The Ru silicides can also be fabricated using other techniques such as electroplating, ion implantation, epitaxial growth, etc.

Ru has a high melting point (2334 °C), a good wettability with Cu and can improve the filling properties of Cu electroplating so that it is often used as a diffusion barrier in Cu metallization systems. The effectiveness of a pure Ru barrier depends on the thickness and decreases from 450 to 300 °C if the thickness changes from 15 to 5 nm [292]. Therefore, much attention is given to Ru-based alloys such as Ru–Ta or Ru–MgO–Ta [293–295]. Using a Ru–Ta/Ru–Ta(N) structure by adding some N to the bottom Ta film further improves the filling properties for scaled down technologies and enhances the reliability performance [295].

For very narrow metal lines in scaled down technologies, the use of Cu causes some problems due to the increasing surface and grain boundary scattering effects, the strong increase of the Cu resistivity and issues related to electromigration in Cu. This has triggered the strong interest in the use of CVD Ru on a barrier as interconnect metal [296].

6.16 Rhodium

Similar to Co, Ni, Cu and Pd, haze formation is also observed for rhodium [286]. This implies that Rh has a high diffusivity and high solubility, although no data is reported in the literature. For rhodium at higher temperatures silicides are formed, i.e., RhSi up to 425 °C, Rh_4Si_5 at 850 °C and Rh_3Si_4 at 925 °C [261, 297]. A variety of analytical techniques, including XPS/UPS, XRD, SIMS, SEM and AFM, have been used to characterize the different silicides [298].

Quenching Rh-doped Si results in electrically active defects. Substitutional Rh_s gives an acceptor level at $E_C - 0.32$ eV and a donor level at $E_C - 0.52$ eV [261]. However, the complex Rh_iRh_s is electrically inactive [299]. Hydrogenation experiments, using a mixture of HF–HNO_3–CH_3COOH and a 1 h diffusion at 400 °C or 600 °C, resulted in new trap levels as shown in Table 6.14 [300]. In n-type samples the levels EI150 at $E_C - 0.33$ eV, E90 at $E_C - 0.16$ eV and E70 at $E_C - 0.14$ eV are observed. Levels EI150 and E270 belong to the substitutional rhodium donor and acceptor. However, there is also a newly formed EII150 level due to a hydrogen-rhodium complex. The level E70 is assigned to a hydrogen-rhodium complex. In p-type samples two levels were detected: H280 at $E_V + 0.50$ eV and H200 at $E_V + 0.37$ eV, both assigned to RhH complexes.

Lemke and Zulehner [301] investigated the impact of the concentration of vacancies and interstitials in the material on the formation of substitutional Rh_s in FZ crystals doped with Rh during the melt. First interstitial Rh_i is formed which is then trapped during the cooling by a vacancy to transform the defect in

Table 6.14 Energy levels, capture cross sections and possible identification of deep levels observed in hydrogenated rhodium-doped p-Si. After Knack et al. [300]

Level	Defect level (eV)	Capture cross section (cm^2)	Identification
E70	$E_C - 0.14$	$\sigma_n = 2.4 \times 10^{-14}$	RhH_1
E90	$E_C - 0.16$	$\sigma_n = 4.9 \times 10^{-16}$	CH donor
EI150	$E_C - 0.33$	$\sigma_n = 1.1 \times 10^{-14}$	Rh-acceptor
EII150	$E_C - 0.33$	$\sigma_p = 2.2 \times 10^{-17}$	RhH_2?
E270	$E_C - 0.54$	$\sigma_p = 3.8 \times 10^{-15}$	Rh-donor
H55	$E_V + 0.10$	$\sigma_p = 6.5 \times 10^{-15}$	?
H200	$E_V + 0.37$	$\sigma_p = 6.2 \times 10^{-16}$	RhH_2
H280	$E_V + 0.50$	$\sigma_p = 1.4 \times 10^{-17}$	RhH_1

substitutional Rh_s. This occurs in a temperature range $T < T_c$, where T_c depends mainly on the binding energy of the metal atom on the lattice site. The work of Voronkov [302] pointed out that the parameter P_g, given by

$$P_g = V_g/G_g \tag{6.12}$$

with V_g the growth velocity and G_g the temperature gradient at the melt/solid interface is crucial to determine the growth regime of the crystal. The value of the ratio in comparison to a critical value $P_{gcrit} = 2.23 \times 10^{-5}$ cm^2/Ks determines whether the crystal growth results in either a vacancy-rich ($P_g > P_{gcrit}$) or an interstitial-rich regime ($P_g < P_{gcrit}$). This implies that by changing the growth parameters the concentration of point defects in the material can be influenced. Experimental data indicated that in the interstitial-dominated growth regime a lower concentration of Rh defects is found, compared to the situation in the vacancy-rich regime [301]. Further calculations revealed that for Rh $T_c = 1410$ K resulting in a capture coefficient of interstitial Rh_i by vacancies >3.5×10^{-14} cm^3/s. A similar analysis for Pt_s leads to a $T_c = 1530$ K and a capture coefficient >3.9×10^{-14} cm^3/s. The developed model can be used to calculate the axial and radial distribution of the Rh_s defects [301]. Figure 6.43 gives a comparison between the experimental and theoretical axial distribution of Rh_s defects for n-type Si with a doping concentration of 3×10^{13} cm^{-3}. First there is a steep decrease of the concentration before reaching a flat zone. The steep increase at the crystal end is caused by the fast cooling process. The other way around, analysis of Rh-doped Si can be used to determine the vacancy concentration in the material [301, 303].

In the early days of semiconductors there was interest in the rhodium-germanium system and different structures such as Rh_2Ge, Rh_5Ge_3 and RhGe have been studied [304]. The main interest was triggered by the superconductivity of Rh_5Ge. RhSi/Si Schottky junctions which can be used as photodetectors and the $RhSi_{1-y}Ge_y/Si_{1-x}Ge_x$

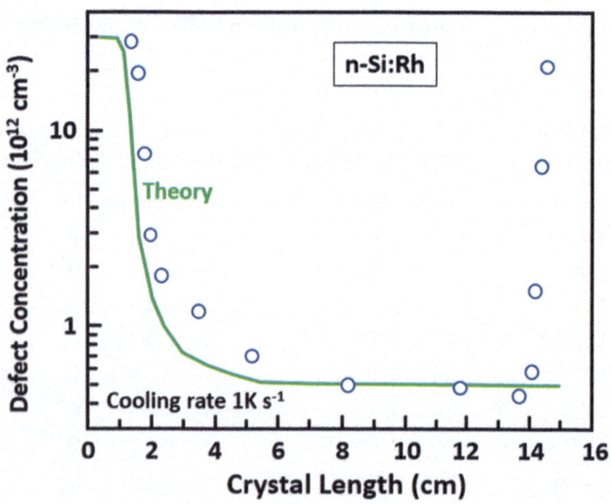

Fig. 6.43 Comparison between the theoretical and experimental axial distribution of Rh_s defects in n-type FZ-Si with a doping concentration of 3×10^{13} cm^{-3}. Reproduced from Lemke and Zulehner [301] with permission of Elsevier

Schottky barrier are a part of a SiGe heterostructure bipolar phototransistor enabling a monolithic high-speed photodetector [305]. This type of devices is used for system-on-chip imaging and communication applications.

6.17 Silver

Although silver has a very good conductivity, it was not selected for metallization processes of integrated circuits as it cannot be manufactured to the same purity level as aluminum. The real interest in Ag metallization came with the breakthrough of PV devices or solar cells where cost issues strongly lowered the otherwise stringent contamination specifications. However, Ag is also known as a contaminant during semiconductor processing. In addition, Ag can penetrate easily through passivating oxide layers on electronic devices under normal operating conditions, resulting in e.g. an increase of the interface state density or a degradation of the solar cell parameters [306]. Ag is a relatively fast diffuser, with the same diffusivity as Au_i

$$D_{Ag_i}(\mathrm{Si}) = 6 \times 10^{-5} \exp(-1.15\,\mathrm{eV}/k_B T)\ \mathrm{m^2 s^{-1}} \tag{6.13}$$

and forming interstitial defects Ag_i in dislocation-free Si, but substitutional defects in distorted Si [307]. Although Ag_i and Au_i have nearly equal diffusion coefficients the involved diffusion mechanism is different, i.e., Ag diffuses via the dissociative or Frank-Turnbull mechanism, whereas the Au diffusion occurs predominantly via the kick-out mechanism. In Si with dislocations, substitutional Ag_s may arise from Ag_i–Ag_s transitions.

DLTS analysis reveals that Ag in Si leads to two deep levels, an acceptor around $E_C - 0.54$ eV ($\sigma_n = 8 \times 10^{-17}$ cm^2) and a donor level around $E_V + 0.34$ eV ($\sigma_p = 8.7 \times 10^{-16}$ cm^2) associated with substitutional silver Ag_s [308, 362]. Using emission channeling experiments the near-substitutional location of implanted Si has been confirmed [309]. After implantation only about 30% of the Ag is in the substitutional position and that increases to maximum 80% after a 200–300 °C anneal step.

A hydrogenation treatment will result in the observation of additional deep levels as indicated in Table 6.15 [310]. The interaction of Ag with one hydrogen atom introduces three deep levels. These defects are stable up to 300–350 °C. Also an

Table 6.15 Energy levels, capture cross sections and possible identification of deep levels observed in hydrogenated silver-doped p-Si. After Yarykin et al. [310]

Level	Defect level (eV)	Capture cross section (cm^2)	Identification
E50	$E_C - 0.09$	$\sigma_n = 2 \times 10^{-18}$	−/2− AgH_1
E240	$E_C - 0.45$	$\sigma_n = 4 \times 10^{-16}$	0/− AgH_1
H200	$E_V + 0.28$	$\sigma_p = 2 \times 10^{-15}$	+/0 AgH_1
H145	$E_V + 0.38$	$\sigma_p = 7 \times 10^{-16}$	0/− AgH_2

electrically not active complex with three hydrogen atoms has been observed, which anneals out around 450 °C [310]. The transformation in hydrogen-rich complexes proceeds via the successive addition of H atoms although not all the steps can be detected directly [311]. Some complexes dissociate at higher temperatures. In the case of Ag the complex with most H atoms is the most stable one and dissociates directly into the isolated metal defect and hydrogen.

Silver-doped Si has also been studied by EPR and in n-type Si three spectra Si-NL45, Si-NL46 and Si-NL47 have been identified, all having an effective $S = ½$ and a trigonal symmetry, although an analysis with higher spin results into true spin values $S = 5/2$ and $S = 7/2$ for Si-NL45 and Si-NL46, respectively [312]. These centers could correspond with the negatively and positively charged silver pair. Si-NL47 is also Ag-related but not further identified.

There is not much information available on the precipitation of Ag in Si and Ag doesn't form haze. Experiments with Ag films on (111)Si showed that after 30 min anneal at 650 °C and Schimmel etch after removal of the film, triangular etch pits appear which are characteristic for crystallographic defects [313]. The etch pit density, which is only visible for anneal temperatures above 450 °C, increases linearly with increasing anneal time. It is assumed that the Ag-correlated pits are either caused by the solubility decrease for decreasing temperature or a homogeneous precipitation process. More detailed research is needed for clarification.

Silver in Ge has extensively been studied for a long time. Concerning the diffusion coefficient of Ag in Ge there exists some discrepancy but recent data gives for the temperature range 1050–1190 K the equation [314]

$$D_{Ag_i}(Ge) = 1.62 \times 10^{-8} \exp(-0.45\,\mathrm{eV}/k_B T)\ \mathrm{m^2\ s^{-1}} \tag{6.14}$$

Similar as for Si the Ag solubility in Ge shows a retrograde behavior with a maximum near 1160 K. The electrical activity of the Ag impurities have been studied first by Hall measurements and later by DLTS. The deep levels observed by Huylebroeck et al. [315] are in good agreement with the majority of the older data and their observations are given in Table 6.16. The concentrations of H1 and H2 are similar within the experimental error and are therefore likely related to the same defect. A similar observation holds for E1 and E2. All four levels are associated with amphoteric substitutional silver Ag_s in different charge states. This makes the behavior of Ag in Ge very similar to Ag in Si [316].

Table 6.16 Energy levels, capture cross sections of deep levels observed in hydrogenated silver-doped Ge. After Huylebroeck et al. [315]

Level	Defect level (eV)	Capture cross section (cm^2)	Identification
E1	$E_C - 0.144$	$\sigma_n = 1.1 \times 10^{-17}$ (80 K)	2−/3− Acceptor
E2	$E_C - 0.302$	$\sigma_n = 1.6 \times 10^{-17}$ (192 K)	−/2− Acceptor
H1	$E_V + 0.035$	$\sigma_p > 10^{-15}$ (18 K)	+/0 Donor
H2	$E_V + 0.116$	$\sigma_p > 10^{-15}$ (62 K)	0/− Acceptor

6.18 Cadmium

Cadmium is not a common contaminant and its properties have not well been studied. There is only information on the deep energy levels. Lang et al. [317] used radioactive ^{111}In implanted samples, whereby indium ions decrease with a half-life time of 2.83 day by a decay into stable cadmium ions, and observed a double acceptor level at $E_C - 0.45$ eV and a single acceptor level at $E_V + 0.145$ eV associated with substitutional Cd_s. These levels are in agreement with those reported earlier by Hall measurements [318, 319]. The double acceptor can react with other TMs, such as e.g. with Fe to form a Fe_iCd_s pair with the Fe atom in the center axis in the $\langle 111 \rangle$ direction [320]. The deep level corresponding with this center may be the observed state at $E_V + 0.20$ eV [317]. Photoluminescence studies of Cz Si resulted in the observation of three hole trap PL centers at 1026, 983 and 935 meV, which are most-likely associated with an interaction between Cd and oxygen [321]. The most dominant center is the one at 983 meV.

The substitutional position of Cd in Si has been confirmed by EPR measurement on Cd implanted Si, analyzing the temperature dependence of the hyperfine interactions [363].

6.19 Hafnium

In the last decade the interest in hafnium has strongly increased by the commercial use of Hf-based gate dielectrics (HfO_2, $HfSi_xO_y$, HfO_yN_z) for scaled down CMOS technologies. The first publication on Hf contamination in Si is from Lemke [282] who detected in Hf-doped in the melt during FZ crystal growth three traps, i.e., an acceptor at $E_C - 0.10$ eV ($\sigma_n > 2 \times 10^{-14}$ cm^2), a donor trap at $E_C - 0.40$ eV ($\sigma_n > 2 \times 10^{-14}$ cm^2) and a double donor at $E_V + 0.32$ eV ($\sigma_p > 5 \times 10^{-18}$ cm^2) associated with interstitial defects.

Photoluminescence investigations of slowly cooled Si:Hf samples annealed at 1000 °C revealed that besides more than 25 sharp lines in the spectrum, there were three clear PL emission lines, i.e., an 896.6 meV PL emission line related to a deep Hf defect at $E_C - 0.22$ eV, an 845 meV line related to a defect at $E_C - 0.27$ eV associated with an internal transition and an 934.8 meV line associated with an exciton complex [322]. It is interesting to remark that for samples co-implanted with oxygen, the intensities of the PL lines increase with increasing oxygen concentration for an oxygen implant dose higher than 10^{13} cm^{-2} and a fast cooling. This is illustrated in Fig. 6.44. No strong impact is observed for slow cooling of the sample.

A detailed DLTS study of Hf implanted n- and p-type Cz Si was performed by Sachdeva et al. [323] and they observed after fast quenching from 1000 °C. Six defects levels in the upper half band gap and four defect levels in the bottom half of the band gap, as summarized in Table 6.17. The capture cross sections are determined from an Arrhenius plot. For H143 the capture cross section is temperature

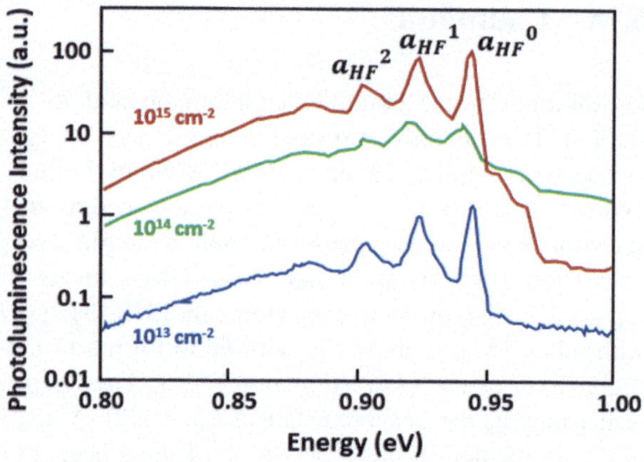

Fig. 6.44 PLspectra of Hf-implanted Si, co-implanted with oxygen at a dose of 10^{13}, 10^{14} and 10^{15} cm^{-2} and annealed at 650 °C for 20 min, 1000 °C for 3 h and fast quenched. Reproduced from Sachdeva et al. [322]. Copyright (2005) by The American Physical Society

Table 6.17 Energy levels, capture cross sections (Arrhenius plot) and defect density of deep levels observed in Hf implanted Si after slow or fast quenching from 1000 °C

Level	Defect level (eV)		Capture cross section (cm^2)	Defect density (cm^{-3})
	Slow	Fast		
E55		$E_C - 0.09$	$\sigma_n = 4.4 \times 10^{-15}$	3.3×10^{14}
E70	$E_C - 0.117$		$\sigma_n = 2.94 \times 10^{-11}$	4.6×10^{13}
E81		$E_C - 0.15$	$\sigma_n = 4.28 \times 10^{-14}$	5.92×10^{14}
E84	$E_C - 0.20$		$\sigma_n = 1.61 \times 10^{-11}$	
E92		$E_C - 0.17$	$\sigma_n = 4.14 \times 10^{-13}$	5.8×10^{14}
E109	$E_C - 0.22$		$\sigma_n = 1.08 \times 10^{-13}$	3.4×10^{13}
E127		$E_C - 0.22$	$\sigma_n = 6.8 \times 10^{-15}$	5.5×10^{14}
E146	$E_C - 0.27$		$\sigma_n = 1.53 \times 10^{-14}$	1.08×10^{14}
E146		$E_C - 0.29$	$\sigma_n = 1.21 \times 10^{-14}$	4.7×10^{14}
E178		$E_C - 0.39$	$\sigma_n = 5.05 \times 10^{-13}$	4.8×10^{14}
E202	$E_C - 0.4$		$\sigma_n = 3.26 \times 10^{-14}$	6.59×10^{13}
E232	$E_C - 0.44$		$\sigma_n = 2.25 \times 10^{-14}$	5.26×10^{13}
E232		$E_C - 0.49$	$\sigma_n = 2.8 \times 10^{-14}$	4.17×10^{14}
H64	$E_V + 0.13$		$\sigma_p = 1.79 \times 10^{-13}$	8×10^{13}
H81	$E_V + 0.15$		$\sigma_p = 1.49 \times 10^{-14}$	2.5×10^{13}
H109	$E_V + 0.25$		$\sigma_p = 1.39 \times 10^{-16}$	1.3×10^{13}
H132		$E_V + 0.26$	$\sigma_p = 4.15 \times 10^{-14}$	2.5×10^{12}
H143	$E_V + 0.24$		$\sigma_p = 1.89 \times 10^{-15} \exp(-0.04\ eV/k_B T)$	3.4×10^{13}
H191		$E_V + 0.43$	$\sigma_p = 3.2 \times 10^{-13}$	1.1×10^{13}
H249		$E_V + 0.50$	$\sigma_p = 3.08 \times 10^{-14}$	1.5×10^{13}

dependent. After fast quenching other defect levels are observed and the trap density is a factor of five higher [323]. Some levels are in agreement with the above reported levels by Lemke [282]. It should be kept in mind that ion implantation (compared to doping in the melt) leads to a concentration higher than the solid solubility so that clustering and/or precipitation may occur. In addition, there may be an influence of the used phosphorus/boron doping of the samples and the possible in-diffusion of hydrogen during sample preparation. As mentioned in the previous paragraph there can also be an impact of the oxygen concentration in the Cz samples leading to the formation of HfO pairs.

Only limited information about the diffusion properties of Hf in Si is reported in the literature. It is very unlikely that below 800 °C Hf diffuses into Si [364]. Experiments at higher temperatures and long anneal times give a diffusion coefficient between 2×10^{-16} cm^2/s at 1100 °C and 1×10^{-15} cm^2/s at 1250 °C [323]. Diffusion seems to occur via a two-component mechanism (fast via interstitial Hf_i, slow via substitutional Hf_s) and the estimated solubility of interstitial and substitutional Hf is 1.1×10^{16} cm^{-3} and 1.4×10^{19} cm^{-3}, respectively [323]. The interstitial Hf_i might be associated with the $E_C - 0.27$ eV level, while substitutional Hf_s could be associated with a trap at $E_V + 0.43$ eV.

Investigations concerning the effects of Hf implanted (1×10^{11} to 1×10^{13} cm^{-2}) into p-type Si substrates on the properties of n^+-polycrystalline-Si/SiO_2/Si capacitors and MOSFETs indicated that there is no influence on the flat-band voltage, the calculated substrate doping density, the electron channel mobility or the leakage current [324].

6.20 Tantalum

Tantalum is nowadays standardly used to form diffusion barriers in VLSI metallization systems preventing Cu to penetrate into the silicon [325]. Ta is a slow diffuser in Si and Ge. Lemke [244] studied the defects states of Ta in silicon and observed three levels associated with interstitial Ta_i, i.e., an acceptor at $E_C - 0.22$ eV, a donor at $E_C - 0.58$ eV and a double donor at $E_V + 0.24$ eV. The first level was also detected earlier by transient capacitance measurements on Schottky barrier diodes in n-type Si [326].

After deposition of a Ta film on Si, the Ta–Si intermixing starts at 550 °C, while nucleation of hexagonal $TaSi_2$ silicide initiates around 620 °C [327]. The formation of a second Ta silicide layer is only detected for temperatures around 900 °C. For the Ta-silicidation process the diffusion of Si into the Ta film is the dominant reaction [328].

6.21 Tungsten

Tungsten is one of the common metals in semiconductor manufacturing and standardly used in the back-end of line processing so that the electrical properties of this refractory metal in Si are well documented. Photoconductivity measurements on W-contaminated Si have been performed since the early sixties and up to 5 different trap levels were reported, later more levels were detected by DLTS. Table 6.18 summarizes the different levels that have been reported in the literature. Main problem with the identification of these levels is that the preparation of all the sample were based on different techniques, enhancing the possibility to also trap other transition metals and possibly hydrogen. The trap at $E_V + 0.36$ eV is not further identified but could be related with W at the surface and it seems to be influenced by the reactive ion etching step (RIE) which could point to an interaction between W and etching induced defects [205, 329] It is now generally accepted that W doping of Si results in a hole level at $E_V + 0.24$ eV associated with substitutional W_s. Another trap around mid-gap has recently been associated with some form of W precipitate or cluster, occurring very close to the implanted surface of the wafer, i.e., within 0.1 μm [330]. In the same study, it was shown that the hole trap at $E_V + 0.409$ eV, with a fixed hole capture cross section of 3–4 × 10^{-16} cm^2 exhibits an athermal emission component in the range of 210–230 K, which could be related with impact ionization of the defect center [330].

The solubility of W in Si is between 7 × 10^{19} and 1 × 10^{20} cm^{-3} for temperatures between 837 and 934 °C [331]. The diffusion coefficient is concentration dependent and defined by both an interstitial and a substitutional component [332, 333]. An enhanced diffusion is observed for high concentrations. The interstitial component is given by [331]

$$D_{W_i}(\mathrm{Si}) = 5.59 \times 10^{-5} \exp(-1.35\,\mathrm{eV}/k_B T)\ \mathrm{cm}^2\ \mathrm{s}^{-1} \tag{6.15}$$

As W also reacts with self-interstitials to form WI pairs, these also contribute to the diffusion process but with a 5 times lower diffusivity than W_i. The W_i formation is promoted by the injection of self-interstitials during e.g. thermal oxidation or the formation of oxygen precipitates in Cz Si. Simulations indicated that W atoms occupy substitutional sites and the W_s profile is nearly identical to the total W profile, while the maximum concentrations of W_i, W_s and WI are located near the surface [331].

6.22 Rhenium–Osmium

Rhenium, a refractory metal with a melting temperature at 3190 °C, belongs to one of the transition metals which are very difficult to incorporate in Si because of the very low segregation coefficient and low diffusion coefficient. According to Lemke [244] the segregation coefficient is only 5 × 10^{-9}. Using doping in the melt of FZ crystals he could detect two defect states, i.e., an acceptor level at $E_C - 0.07$ eV ($\sigma_p = 8.7 \times 10^{-16}$ cm^2) and a donor level at $E_C - 0.35$ eV ($\sigma_p = 5.1 \times 10^{-16}$ cm^2).

Similar as rhenium, osmium is also a 5*d* transition metal with a very low segregation coefficient (1.5 × 10^{-8}) making it also difficult to incorporate in Si and resulting in rather low DLTS signals. For this metal Lemke [244] observed two

Table 6.18 Energy levels in W-doped Si

References	Defect level (eV)													
	$E_V + E_T$								$E_C - E_T$					
Zibuts [359] 1964		0.31		0.34						0.37	0.30			0.22
Busta [332] 1981												0.28	0.25	
Fujisaki [360] 1988								0.41						0.22
Boughaba [333] 1991	0.22		0.33						0.59					
Ando [361] 1991								0.41						
Lemke [25] 1994						0.39								
Polignano [205] 2015 Simoen [330] 2016					0.36		0.40	0.41						0.22

deep levels, i.e., an acceptor level at $E_C - 0.22$ eV ($\sigma_p = 4.6 \times 10^{-17}$ cm^2) in the upper half of the band gap and a donor level at $E_V + 0.30$ eV ($\sigma_p = 8 \times 10^{-16}$ cm^2) in the lower half of the band gap, both assigned to interstitial Os_i.

Simular as for Ru discussed in Sect. 6.14, Os deposited on Si will after a thermal anneal react with Si to form silicides such as Os_2Si_3 and $OsSi_2$ [291]. Ru_2Si_3 forms by a diffusion controlled mechanism from approximately 450–525 °C with an activation energy of 1.8 eV. Os_2Si_3 is formed by a nucleation controlled process with also an activation energy of 1.8 eV at temperatures around 750 °C. Os silicides can also be formed using other fabrication techniques such as electroplating, ion implantation, epitaxial growth, etc.

Doping Si with Os has been reported to enable the fabrication of photodetectors with a low reverse voltage and a broad spectral range [334]. Osmium, like ruthenium, can also be used as diffusion barrier and contact filling for Cu damascene metallization technologies [335].

6.23 Iridium

The diffusion coefficient of Ir in Si has first been studied in the seventies based on Ir diffusion from the vapor phase [336], and only thirty years later additional information based on RBS analysis of Ir implanted samples became available, pointing out that the diffusion coefficient is strongly dependent on the Ir concentration. This is illustrated in Fig. 6.45 [337]. The difference can be explained by assuming that for higher concentrations the substitutional diffusion component plays a dominant role, while for low concentrations it is more a mixed vacancy-interstitial diffusion mechanism. The compressive lattice stress can be the cause for the suppression of the interstitial component for high concentrations [337].

For iridium-doped in the melt during FZ Si crystal growth DLTS revealed an acceptor level at $E_C - 0.24$ eV ($\sigma_p = 9.1 \times 10^{-15}$ cm^2) and a donor level at $E_C - 0.62$ eV ($\sigma_n = 7.2 \times 10^{-16}$ cm^2) and both levels are assigned to substitutional Ir_s [338]. The used ambient during crystal growth has an impact on the defect distribution as shown in Fig. 6.46. For crystals grown under vacuum the defect distribution is more or less constant, while there is a lower and more fluctuating distribution for crystals grown in an Ar ambient. Only at the beginning and the end of the crystal there is a comparable concentration of substitutional defects, which depends on the cooling rate of the crystal. The lower defect concentration for crystals grown in an Ar ambient is due to the formation of defect complexes with substitutional Ir_s. In p-type Si this results in an additional DLTS level at $E_V + 0.14$ eV. The defect complex, which has not further been identified but could be related to Ir_iIr_s, dissociates after an 600 °C anneal for 20 min, so that the same distribution as for the vacuum-grown crystal is obtained [338].

The two Ir_s defect levels have also been observed in p^+pnn^{++} power diodes fabricated by Ir diffusion in the 820–940 °C temperature range [339]. However, for long diffusions at high temperature (940 °C) an additional defect level at

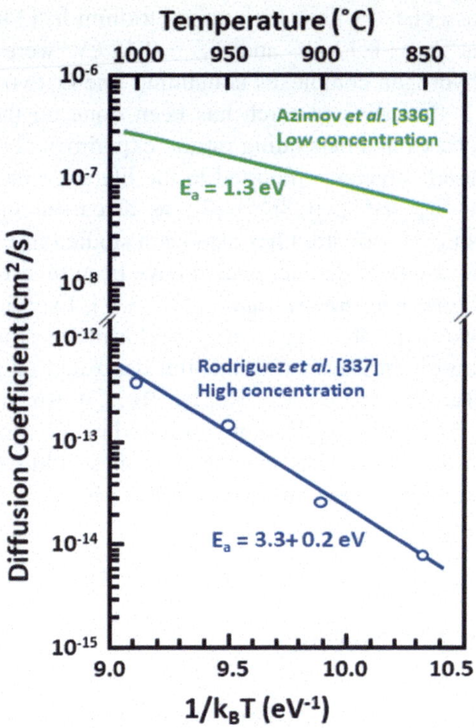

Fig. 6.45 Diffusion coefficient of iridium in Si for low and high Ir concentrations, respectively. Reproduced from Rodriguez et al. [337] with permission of Elsevier

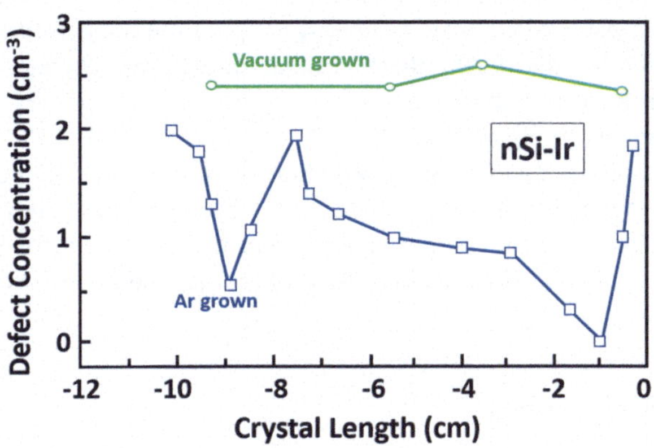

Fig. 6.46 Axial distribution of substitutional Ir_s defects in Ir-doped in the melt FZ Si grown under vacuum or in an Ar ambient, respectively. Reproduced from Lemke [338] by permission of EC-The Electrochemical Society

$E_C - 0.16$ eV is observed which may be associated with the interstitial iridium donor Ir_i. The substitutional defect level around $E_C - 0.24$ eV turns out to be an effective recombination center degrading the carrier lifetime for operating temperatures below 120 °C.

DLTS investigations of the electrical properties of stable Ir isotopes in Si revealed two levels in close agreement with the two mentioned above and associated with substitutional iridium Ir_s [340]. In addition, two other defects states at $E_C - 0.17$ eV and $E_C - 0.45$ eV were observed and identified as iridium-hydrogen complexes containing one or two hydrogen atoms, respectively.

Extensive research has been done on the formation of the different Ir-silicide phases and depending on the experimental conditions and the anneal temperatures used, structures reported in the literature include IrSi, $IrSi_{1.75}$, $IrSi_2$, $IrSi_{\sim 3}$, Ir_2Si_3, Ir_3Si_4, Ir_3Si_5, Ir_4Si_5, Ir_4Si_7 as discussed in [341–343] and references therein. Ir silicide formation has also been studied in SiGe structures [344]. In the case of Ge, a variety of distinct phases have been investigated such as IrGe, Ir_3Ge_7, Ir_4Ge_5 in a broad temperature range [345–347]. Especially the work of Gaudet et al. [345] is worth mentioning as they performed a structural characterization of 20 different transition metals (30 nm film deposited on Ge(001)) including Ti, Zr, Hf, V, Nb, Ta, Cr, Mo, W, Mn, Re, Fe, Ru, Co, Rh, Ir, Ni, Pd, Pt, and Cu.

From an application point of view it is worthwhile mentioning that the interest in iridium films has increased as these films can be used as electrodes in dynamic random access memories (DRAMs) and ferroelectric random access memories (FeRAMs) [348].

6.24 Mercury

Mercury is very commonly used in semiconductor technology as mercury probes have and are ample used for non-destructive characterization of semiconductor devices and wafers such as e.g. for capacitance-voltage measurements and lifetime measurements. Due to its low evaporation temperature (48 °C) it will not diffuse in the Si bulk during thermal treatments [286]. Therefore, in the literature there is no information available concerning diffusion properties, solubility and possible defect levels of Hg in Si.

Jones and Corbett [349] studied the impact of Hg contamination on the Si surface on the electrical performance of Schottky diodes. They noticed a higher capacitance-voltage barrier height and a higher leakge current for Schottky diodes contaminated with mercury residues. The problem can be solved by an appropriate cleaning step.

References

1. N.T. Son, A.B. van Oosten, C.A.J. Ammerlaan, Electron paramagnetic resonance of nickel in silicon—II. Hyperfine and quadrupole interactions. Solid State Commun. **80**, 439–445 (1991). https://doi.org/10.1016/0038-1098(91)90723-9
2. G.W. Ludwig, H.H. Woodbury, Electron spin resonance in semiconductors. Solid State Phys. **13**, 223–304 (1962). https://doi.org/10.1016/s0081-1947(08)60458-0
3. B. Effey-Schwickert, M. Wiegand, H. Vollmer, R. Labusch, Nickel in silicon studied by electron paramagnetic resonance. Appl. Phys. A **77**, 711–716 (2003). https://doi.org/10.1007/s00339-002-1891-9

4. D.J. Silva, U. Wahl, J.G. Correia, L.M.C. Pereira, L.M. Amorim, M.R. da Silva, E. Bosne, J. P. Araújo, Lattice location and thermal stability of implanted nickel in silicon studied by on-line emission channeling. J. Appl. Phys. **115**, 023504/1–9 (2014). https://doi.org/10.1063/1.4861142
5. G.W. Ludwig, H.H. Woodbury, Electron spin resonance in nickel-doped germanium. Phys. Rev. **113**, 1014–1018 (1959). https://doi.org/10.1103/PhysRev.113.1014
6. P.D. Augustus, Prismatic punching of dislocations from $NiSi_2$ precipitates in epitaxial silicon. Electrochem. Soc. Proc. **83**(9), 414–422 (1983)
7. D.M. Vanderwalker, The nucleation of Ni silicides on dislocations and twins in silicon. Phys. Status Solidi A **86**, 507–516 (1984). https://doi.org/10.1002/pssa.2210860207
8. J. Chung, H.J. Möller, Cobalt- and nickel-disilicide formation at twin boundaries in silicon. Phys. Status Solidi A **138**, 473–482 (1993). https://doi.org/10.1002/pssa.2211380215
9. R.C. Cammarata, C.V. Thompson, K.N. Tu, $NiSi_2$ precipitation in nickel-implanted silicon films. Appl. Phys. Lett. **51**, 1106–1108 (1987). https://doi.org/10.1063/1.99003
10. R.R. Kola, G.A. Rozgonyi, J. Li, W.B. Rogers, T.Y. Tan, K.E. Bean, K. Lindberg, Transition metal silicide precipitation in silicon induced by rapid thermal processing and free-surface gettering. Appl. Phys. Lett. **55**, 2108–2110 (1989). https://doi.org/10.1063/1.102342
11. S. Sadamitsu, M. Sano, M. Hourai, S. Sumita, N. Fujino, T. Shiraiwa, TEM observation of defects induced by Ni contamination on a Si(100) surface. Jpn. J. Appl. Phys. **28**, L333–L336 (1989). https://doi.org/10.1143/JJAP.28.L133
12. M. Hourai, K. Murakami, T. Shigematsu, N. Fujino, T. Shiraiwa, Behavior of defects induced by metallic impurities on Si(100) surfaces. Jpn. J. Appl. Phys. **28**, 2413–2420 (1989). https://doi.org/10.1143/JJAP.28.2413
13. M. Seibt, W. Schröter, Precipitation behaviour of nickel in silicon. Philos. Mag. A **59**, 337–352 (1989). https://doi.org/10.1080/01418618908205063
14. M. Seibt, H. Hedemann, A.A. Istratov, F. Riedel, A. Sattler, W. Schröter, Structural and electrical properties of metal silicide precipitates in silicon. Phys. Status Solidi A **171**, 301–310 (1999). https://doi.org/10.1002/(SICI)1521-396X(199901)171:13.3.CO;2-G
15. W. Schröter, V. Kveder, M. Seibt, H. Ewe, H. Hedemann, F. Riedel, A. Sattler, Atomic structure and electronic states of nickel and copper silicides in silicon. Mater. Sci. Eng. B **72**, 80–86 (2000). https://doi.org/10.1016/S0921-5107(99)00499-7
16. M. Seibt, R. Khalil, V. Kveder, W. Schröter, Electronic states at dislocations and metal silicide precipitates in crystalline silicon and their role in solar cell materials. Appl. Phys. A **96**, 235–253 (2009). https://doi.org/10.1007/s00339-008-5027-8
17. M. Seibt, V. Kveder, W. Schröter, O. Voß, Structural and electrical properties of metal impurities at dislocations in silicon. Phys. Status Solidi A **202**, 911–920 (2005). https://doi.org/10.1002/pssa.200460515
18. M. Seibt, K. Graff, Characterization of haze-forming precipitates in silicon. J. Appl. Phys. **63**, 4444–4450 (1988). https://doi.org/10.1063/1.340164
19. C.-J. Choi, S.-Y. Chang, Y.-W. Ok, T.-Y. Seong, H. Gan, G.Z. Pan, K.N. Tu, Formation of nickel disilicide using nickel implantation and rapid thermal annealing. J. Electron. Mater. **32**, 1072–1078 (2003). https://doi.org/10.1007/s11664-003-0090-6
20. A.Yu. Kuznetsov, B.G. Svensson, O. Nur, L. Hultman, Nickel distribution in crystalline and amorphous silicon during solid phase epitaxy of amorphous silicon. J. Appl. Phys. **84**, 6644–6649 (1998). https://doi.org/10.1063/1.369039
21. M. Heuer, T. Buonassisi, M.A. Marcus, A.A. Istratov, M.D. Pickett, T. Shibata, E.R. Weber, Complex intermetallic phase in multicrystalline silicon doped with transition metals. Phys. Rev. B **73**, 235204/1–5 (2006). https://doi.org/10.1103/physrevb.73.235204
22. M. Heuer, T. Buonassisi, A.A. Istratov, M.D. Pickett, M.A. Marcus, A.M. Minor, E.R. Weber, Transition metal interaction and Ni–Fe–Cu–Si phases in silicon. J. Appl. Phys. **101**, 123510/1–6 (2007). https://doi.org/10.1063/1.2748346

23. T. Buonassisi, M. Heuer, A.A. Istratov, M.D. Pickett, M.A. Marcus, B. Lai, Z. Cai, S.M. Heald, E.R. Weber, Transition metal co-precipitation mechanism in silicon. Acta Mater. **55**, 6119–6126 (2007). https://doi.org/10.1016/j.actamat.2007.07.030
24. C. Rudolf, P. Saring, L. Stolze, M. Seibt, Co-precipitation of copper and nickel in crystalline silicon. Mater. Sci. Eng. B **159–160**, 365–368 (2009). https://doi.org/10.1016/j.mseb.2008.10.015
25. M. Seibt, D. Abdelbarey, V. Kveder, C. Rudolf, P. Saring, L. Stolze, O. Voß, Structure, chemistry and electrical properties of extended defects in crystalline silicon for photovoltaics. Phys. Status Solidi C **6**, 1847–1855 (2009). https://doi.org/10.1002/pssc.200881470
26. M. Yoshida, K. Furusho, Behavior of nickel as an impurity in silicon. Jpn. J. Appl. Phys. **3**, 521–529 (1964). https://doi.org/10.1143/JJAP.3.521
27. M. Yoshida, K. Saito, Dissociative diffusion of nickel in silicon and self-diffusion of silicon. Jpn. J. Appl. Phys. **6**, 573–581 (1967). https://doi.org/10.1143/JJAP.6.573
28. W.B. Chua, K. Rose, Electrical properties of high-resistivity nickel-doped silicon. J. Appl. Phys. **41**, 2644–2647 (1970). https://doi.org/10.1063/1.1659275
29. A.A. Lebedev, A.T. Mamadalimov, Thermal and field quenching of the photoconductivity in Ni-doped silicon. Sov. Phys. Semicond. **6**, 96–98 (1972)
30. S.A. Azimov, N.A. Sultanov, L. Islamov, R.N. Nagmatov, Infrared quenching of the photoconductivity of nickel-doped silicon. Sov. Phys. Semicond. **7**, 1227–1228 (1974)
31. G.P. Chiavarotti, M. Conti, A. Messina, Characterisation of properties of nickel in silicon using thermally stimulated capacitance method. Solid-State Electron. **20**, 907–909 (1977). https://doi.org/10.1016/0038-1101(77)90012-0
32. S.J. Pearton, A.J. Tavendale, The electrical properties of deep copper- and nickel-related centers in silicon. J. Appl. Phys. **54**, 1375–1379 (1983). https://doi.org/10.1063/1.332160
33. M. Jaraiz, S. Dueñas, J. Vicente, L. Bailón, J. Barbolla, Electron thermal emission rates of nickel centers in silicon. Solid-State Electron. **29**, 883–884 (1986). https://doi.org/10.1016/0038-1101(86)90008-0
34. H. Indusekhar, V. Kumar, Electrical properties of nickel-related deep levels in silicon. J. Appl. Phys. **61**, 1449–1455 (1987). https://doi.org/10.1063/1.338075
35. H. Lemke, Dotierungseigenschaften von Nickel in Silizium. Phys. Status Solidi A **99**, 205–213 (1987). https://doi.org/10.1002/pssa.2210990124
36. H. Kitagawa, H. Nakashima, Nickel-related deep levels in silicon studied by combined Hall effect and DLTS measurement. Phys. Status Solidi A **99**, K49–K52 (1987). https://doi.org/10.1002/pssa.2210990150
37. H. Kitagawa, H. Nakashima, Nickel-related donor level in silicon. Phys. Status Solidi A **102**, K23–K27 (1987). https://doi.org/10.1002/pssa.2211020152
38. M. Gong, Z.-P. You, The deep levels in nickel-doped silicon. Phys. Status Solidi A **111**, K49–K52 (1989). https://doi.org/10.1002/pssa.2211110152
39. H. Kitagawa, H. Nakashima, Amphoteric property of electrically active nickel in silicon. Jpn. J. Appl. Phys. **28**, 305–310 (1989). https://doi.org/10.1143/JJAP.28.305
40. H. Kitagawa, S. Tanaka, Electrically active nickel in silicon studied by DLTS in several kinds of silicon diodes. Phys. Status Solidi A **120**, K67–K70 (1990). https://doi.org/10.1002/pssa.2211200141
41. J. Bartoš, L. Tesař, Some electrical and optical properties of nickel-related deep levels in silicon. Phys. Status Solidi A **122**, 607–616 (1990). https://doi.org/10.1002/pssa.2211220222
42. H. Kitagawa, S. Tanaka, H. Nakashima, M. Yoshida, Electrical properties of nickel in silicon. J. Electron. Mater. **20**, 441–447 (1991). https://doi.org/10.1007/BF02657824
43. S. Tanaka, K. Matsushita, H. Kitagawa, Majority-carrier capture cross section of amphoteric nickel center in silicon studied by isothermal capacitance transient spectroscopy. Jpn. J. Appl. Phys. **35**, 4624–4625 (1996). https://doi.org/10.1143/JJAP.35.4624

44. A.A. Istratov, E.R. Weber, Electrical properties and recombination activity of copper, nickel and cobalt in silicon. Appl. Phys. A **66**, 123–136 (1998). https://doi.org/10.1007/s003390050649
45. S. Tanaka, H. Kitagawa, Distribution of electrically active nickel atoms in silicon crystals measured by means of deep level transient spectroscopy. Physica B **401–402**, 115–118 (2007). https://doi.org/10.1016/j.physb.2007.08.125
46. M. Nakamura, S. Murakami, H. Udono, Depth profiles of the nickel donor center in p-type silicon diffused with dilute nickel measured by deep-level transient spectroscopy. Jpn. J. Appl. Phys. **53**, 091301/1–5 (2014). https://doi.org/10.7567/jjap.53.091301
47. L. Scheffler, Vl. Kolkovsky, J. Weber, Electrical levels in nickel doped silicon. J. Appl. Phys. **116**, 173704/1–8 (2014). https://doi.org/10.1063/1.4901003
48. M. Shiraishi, J.-U. Sachse, H. Lemke, J. Weber, DLTS analysis of nickel-hydrogen complex defects in silicon. Mater. Sci. Eng. B **58**, 130–133 (1999). https://doi.org/10.1016/S0921-5107(98)00280-3
49. M.K. Bakhadyrkhanov, S. Zaĭnobidinov, A.T. Teshabaev, M.A. Khodzhaeva, Influence of heat treatment on the interaction of nickel atoms with oxygen in silicon. Sov. Phys. Semicond. **10**, 593–594 (1976)
50. M.K. Bakhadyrkhanov, S. Zaĭnobidinov, Influence of neutron irradiation on electrical properties of silicon doped with nickel, cobalt, or manganese. Sov. Phys. Semicond. **11**, 1205–1206 (1977)
51. V.P. Markevich, A.R. Peaker, I.F. Medvedeva, V.E. Gusakov, L.I. Murin, B.G. Svensson, Interactions of Cu and Ni impurities with vacancy-related point defects in Czochralski-grown silicon crystals. ECS Trans. **18**(1), 1013–1018 (2009). https://doi.org/10.1149/1.3096565
52. N. Yarykin, J. Weber, Evidence for room-temperature in-diffusion of nickel into silicon. Appl. Phys. Lett. **109**, 102101/1–3 (2016). https://doi.org/10.1063/1.4962394
53. N. Yarykin, J. Weber, Nickel in silicon: Room-temperature in-diffusion and interaction with radiation defects. Phys. Status Solidi C **14**, 1700005/1–3 (2017). https://doi.org/10.1002/pssc.201700005
54. M.H. Nazare, M.F. Thomaz, Photoluminescence study of nickel doped silicon. Mater. Sci. Forum **38–41**, 433–438 (1989). www.scientific.net/MSF.38-41.433
55. K. Nauka, T.I. Kamins, Deep state defects in strained and relaxed epitaxial $Si_{1-x}Ge_x$ on Si introduced by 3*d* transition metal and 5*d* noble metal impurities. Physica B **273–274**, 603–607 (1999). https://doi.org/10.1016/S0921-4526(99)00584-0
56. S.H. Segers, J. Lauwaert, P. Clauws, E. Simoen, J. Vanhellemont, F. Callens, H. Vrielinck, Direct estimation of capture cross sections in the presence of slow capture: application to the identification of quenched-in deep-level defects in Ge. Semicond. Sci. Technol. **29**, 125007/1–9 (2014). https://doi.org/10.1088/0268-1242/29/12/125007
57. S.J. Pearton, A.J. Tavendale, The mobility of a nickel-related center in reverse-biased germanium n^+p diodes. Solid State Electron. **26**, 1019–1021 (1983). https://doi.org/10.1016/0038-1101(83)90079-5
58. C. Claeys, E. Simoen (eds.), *Germanium-Based Technologies—From Materials to Devices* (Chap. 5, Elsevier, 2007)
59. M.V. Trushin, O.F. Vyvenko, Electrical levels of nanoscale $NiSi_2$ precipitates in silicon band gap. Phys. Status Solidi C **4**, 3056–3060 (2007). https://doi.org/10.1002/pssc.200775453
60. F. Riedel, W. Schröter, Electrical and structural properties of nanoscale $NiSi_2$ precipitates in silicon. Phys. Rev. B **62**, 7150–7156 (2000). https://doi.org/10.1103/PhysRevB.62.7150
61. W. Schröter, H. Hedemann, V. Kveder, F. Riedel, Measurements of energy spectra of extended defects. J. Phys. Condens. Matter. **14**, 13047–13059 (2002). https://doi.org/10.1088/0953-8984/14/48/350
62. H. Hedemann, W. Schröter, Influence of electric field-enhanced emission on deep level transient spectra of bandlike extended defects: $NiSi_2$-precipitates in silicon. Solid State Phenom. **57–58**, 293–298 (1997). www.scientific.net/SSP.57-58.293

63. M. Kittler, J. Lärz, W. Seifert, M. Seibt, W. Schröter, Recombination properties of structurally well defined $NiSi_2$ precipitates in silicon. Appl. Phys. Lett. **58**, 911–913 (1991). https://doi.org/10.1063/1.104474
64. M. Kittler, W. Seifert, Z.J. Radzimski, Two classes of recombination behavior as studied by the technique of the electron beam induced current: $NiSi_2$ particles and misfit dislocations in Ni contaminated n-type silicon. Appl. Phys. Lett. **62**, 2513–2515 (1993). https://doi.org/10.1063/1.109632
65. M. Kittler, W. Seifert, Analysis of recombination activity of $NiSi_2$ platelets in Si. Phys. Status Solidi A **150**, 463–470 (1995). https://doi.org/10.1002/pssa.2211500140
66. P. Formanek, M. Kittler, Direct evidence of internal Schottky barriers at $NiSi_2$ precipitates in silicon by electron holography. J. Appl. Phys. **97**, 063707/1–5 (2005). https://doi.org/10.1063/1.1863432
67. R.T. Tung, Schottky-barrier formation at single-crystal metal-semiconductor interfaces. Phys. Rev. Lett. **52**, 461–464 (1984). https://doi.org/10.1103/PhysRevLett.52.461
68. R.T. Tung, K.K. Ng, J.M. Gibson, A.F.J. Levi, Schottky-barrier heights of single-crystal $NiSi_2$ on Si(111): the effect of a surface p-n junction. Phys. Rev. B **33**, 7077–7090 (1986). https://doi.org/10.1103/PhysRevB.33.7077
69. C. Donolato, The space-charge region around a metallic platelet in a semiconductor. Semicond. Sci. Technol. **8**, 45–49 (1993). https://doi.org/10.1088/0268-1242/8/1/007
70. A. Chantre, A.F.J. Levi, R.T. Tung, W.C. Dautremont-Smith, M. Anzlowar, States at epitaxial $NiSi_2$/Ni heterojunctions studied by deep-level transient spectroscopy and hydrogenation. Phys. Rev. B **34**, 4415–4418 (1986). https://doi.org/10.1103/PhysRevB.34.4415
71. T.S. Fell, P.R. Wilshaw, M.D. de Coteau, EBIC investigations of dislocations and their interaction with impurities in silicon. Phys. Status Solidi A **138**, 695–704 (1993). https://doi.org/10.1002/pssa.2211380241
72. P.R. Wilshaw, T.S. Fell, Electron beam induced current investigations of transition metal impurities at extended defects in silicon. J. Electrochem. Soc. **142**, 4298–4304 (1995). https://doi.org/10.1149/1.2048501
73. V. Kveder, W. Schröter, M. Seibt, A. Sattler, Electrical activity of dislocations in Si decorated by Ni. Solid State Phenom. **82–84**, 361–366 (2002). https://doi.org/10.4028/www.scientific.net/SSP.82-84.361
74. F.J. Stützler, H.J. Queisser, Charged defect states at silicon grain boundaries. J. Appl. Phys. **60**, 3910–3915 (1986). https://doi.org/10.1063/1.337513
75. J.L. Maurice, C. Colliex, Fast diffusers Cu and Ni as the origin of electrical activity in a silicon grain boundary. Appl. Phys. Lett. **55**, 241–243 (1989). https://doi.org/10.1063/1.101919
76. R. Rizk, A. Ihlal, X. Portier, Evolution of electrical activity and structure of nickel precipitates with the treatment temperature of a $\Sigma = 25$ silicon bicrystal. J. Appl. Phys. **77**, 1875–1880 (1995). https://doi.org/10.1063/1.358888
77. Z. Xi, D. Yang, J. Chen, T. Sekiguchi, Recombination behavior of nickel cast multicrystalline silicon. Mater. Sci. Semicond. Process. **9**, 304–307 (2006). https://doi.org/10.1016/j.mssp.2006.01.052
78. Z.J. Radzimski, T.Q. Zhou, A.B. Buczkowski, G.A. Rozgonyi, Electrical activity of dislocations: prospects for practical utilization. Appl. Phys. A **53**, 189–193 (1991). https://doi.org/10.1007/BF00324249
79. Z.J. Radzimski, T.Q. Zhou, A. Buczkowski, G.A. Rozgonyi, D. Finn, L.G. Hellwig, J.A. Ross, Recombination at clean and decorated misfit dislocations. Appl. Phys. Lett. **60**, 1096–1098 (1992). https://doi.org/10.1063/1.106455
80. M. Kittler, W. Seifert, V. Higgs, Recombination activity of misfit dislocations in silicon. Phys. Status Solidi A **137**, 327–335 (1993). https://doi.org/10.1002/pssa.2211370206
81. M. Kittler, W. Seifert, Two types of electron-beam-induced current behaviour of misfit dislocations in Si(Ge): experimental observations and modelling. Mater. Sci. Eng. B **24**, 78–81 (1994). https://doi.org/10.1016/0921-5107(94)90302-6

82. D. Walz, J.-P. Joly, R. Falster, G. Kamarinos, Characterization of nickel contamination in Float Zone and Czochralski silicon wafers by using electrolytic metal tracer or microwave photoconductivity decay measurement. Jpn. J. Appl. Phys. **34**, 4091–4095 (1995). https://doi.org/10.1143/JJAP.34.4091
83. J. Sakuma, Y. Okui, H. Miyazawa, F. Inoue, M. Miyajima, Influence of metal contamination on minority carrier diffusion length and oxide charge. Mat. Res. Soc. Proc. **510**, 245–252 (1998). https://doi.org/10.1557/PROC-510-245
84. D. MacDonald, Impact of nickel contamination on carrier recombination in n- and p-type crystalline silicon wafers. Appl. Phys. A **81**, 1619–1625 (2005). https://doi.org/10.1007/s00339-005-3371-5
85. M. Miyazaki, Influence of metal impurities on lifetime. ASTM STP **1340**, 294–304 (1998)
86. S. Naito, T. Nakashizu, Electric degradation and defect formation of silicon due to Cu, Fe and Ni contamination. Proc. Mat. Res. Soc. **262**, 641 (1992). https://doi.org/10.1557/proc-262-641
87. M.L. Polignano, F. Cazzaniga, A. Sabbadini, G. Queirolo, A. Cacciato, A. Di Bartolo, Comparison among lifetime techniques for the detection of transition metal contamination. Mat. Sci. Eng. B **42**, 157–163 (1996). https://doi.org/10.1016/S0921-5107(96)01698-4
88. W. Wang, Z. Xi, D. Yang, D. Que, Recombination activity of nickel in Czochralski silicon during rapid thermal process. Mater. Sci. Semicond. Process. **9**, 296–299 (2006). https://doi.org/10.1016/j.mssp.2006.01.057
89. M.L. Polignano, D. Codegoni, S. Grasso, A. Riva, F. Sammiceli, D. Caputo, V. Privitera, ECS Trans. **16**(6), 195–206 (2008). https://doi.org/10.1149/1.2980303
90. Y. Yoon, B. Paudyal, K. Kim, Y.-W. Ok, P. Kulshreshtha, S. Johnston, G. Rozgonyi, Effect of nickel contamination on high carrier lifetime n-type crystalline silicon. J Appl Phys **111**, 033702/1–5 (2012). https://doi.org/10.1063/1.3680880
91. S.A. McHugo, A.C. Thompson, I. Périchaud, S. Martinuzzi, Direct correlation of transition metal impurities and minority carrier recombination in multicrystalline silicon. Appl. Phys. Lett. **72**, 3482–3484 (1998). https://doi.org/10.1063/1.121673
92. A.A. Istratov, T. Buonassisi, R.J. McDonald, A.R. Smith, R. Schindler, J.A. Rand, J.P. Kalejs, E.R. Weber, Metal content of multicrystalline silicon for solar cells and its impact on minority carrier diffusion length. J. Appl. Phys. **94**, 6552–6559 (2003). https://doi.org/10.1063/1.1618912
93. T. Buonassisi, A.A. Istratov, M.D. Pickett, M.A. Marcus, G. Hahn, S. Riepe, J. Isenberg, W. Warta, G. Willeke, T.F. Ciszek, E.R. Weber, Quantifying the effect of metal-rich precipitates on minority carrier diffusion length in multicrystalline silicon using synchrotron-based spectrally resolved X-ray beam-induced current. Appl. Phys. Lett. **87**, 044101/1–3 (2005). https://doi.org/10.1063/1.1997274
94. T. Kojima, T. Tachibana, N. Kojima, Y. Ohshita, K. Arafune, A. Ogura, M. Yamaguchi, Nickel distribution and recombination activity in as-grown and annealed multicrystalline silicon. Jpn. J. Appl. Phys. **53**, 04ER20/1–6 (2014). https://doi.org/10.7567/jjap.53.04er20
95. E. Gaubas, A. Uleckas, R. Grigonis, V. Sirutkaitis, J. Vanhellemont, Microwave probed photoconductivity spectroscopy of deep levels in Ni doped Ge. Appl. Phys. Lett. **92**, 222102/1–3 (2008). https://doi.org/10.1063/1.2939096
96. S. Forment, J. Vanhellemont, P. Clauws, J. Van Steenbergen, S. Sioncke, M. Meuris, E. Simoen, A. Theuwis, A deep level transient spectroscopy study of transition metals in n- and p-type germanium. Mater. Sci. Semicond. Process. **9**, 559–563 (2006). https://doi.org/10.1016/j.mssp.2006.08.048
97. P. Clauws, E. Simoen, Metals in germanium. Mater. Sci. Semicond. Process. **9**, 546–553 (2006). https://doi.org/10.1016/j.mssp.2006.08.047
98. D. Gilles, W. Schröter, W. Bergholz, Impact of the electronic structure on the solubility and diffusion of 3*d* transition elements in silicon. Phys. Rev. B **41**, 5770–5782 (1990). https://doi.org/10.1103/PhysRevB.41.5770

99. W. Bergholz, On the diffusion of Co in Si and its applicability to the Si intrinsic defect problem. J. Phys. D Appl. Phys. **14**, 1099–1113 (1981). https://doi.org/10.1088/0022-3727/14/6/018
100. D. Gilles, W. Schröter, Solubility, diffusion and ion-pairing of cobalt in extrinsic silicon at 700 °C. Mater. Sci. Forum **10–12**, 169–174 (1986). https://doi.org/10.4028/www.scientific.net/MSF.10-12.169
101. H. Kitagawa, K. Hashimoto, Solid solubility of cobalt in silicon. Jpn. J. Appl. Phys. **16**, 857–858 (1977). https://doi.org/10.1143/JJAP.16.857
102. W. Bergholz, W. Schröter, Precipitation of cobalt in silicon studied by Mössbauer spectroscopy. Phys. Status Solidi A **49**, 489–498 (1978). https://doi.org/10.1002/pssa.2210490209
103. E. Scheibe, W. Schröter, Investigations in cobalt doped silicon by DLTS and Mössbauer effect. Phys. B **116**, 318–322 (1983). https://doi.org/10.1016/0378-4363(83)90267-X
104. J. Utzig, Properties of Co in FZ and CZ silicon studied by Mössbauer spectroscopy. J. Appl. Phys. **64**, 3629–3633 (1988). https://doi.org/10.1063/1.341400
105. W. Bergholz, S. Damgaard, J.W. Petersen, G. Weyer, A Mössbauer-spectroscopy study of the annealing of supersaturated solutions of ^{57}Co in silicon. Phys. Status Solidi A **75**, 289–300 (1983). https://doi.org/10.1002/pssa.2210750133
106. M. Pasemann, W. Bergholz, W. Schröter, Mössbauer spectroscopical and electron microscopical investigations of the behaviour of cobalt in silicon. Phys. Status Solidi A **81**, 273–280 (1984). https://doi.org/10.1002/pssa.2210810130
107. W. Bergholz, S. Damgaard, J.W. Petersen, G. Weyer, Reactions of cobalt in silicon with electron-irradiation-induced defects. Phys. Status Solidi A **81**, 239–246 (1984). https://doi.org/10.1002/pssa.2210810125
108. D.J. Silva, U. Wahl, J.G. Correia, L.M.C. Pereira, L.M. Amorim, M.R. da Silva, J.P. Araújo, Origin of the lattice sites occupied by implanted Co in Si. Semicond. Sci. Technol. **29**, 125006/1–5 (2014). https://doi.org/10.1088/0268-1242/29/12/125006
109. D.J. Silva, U. Wahl, J.G. Correia, V. Augustyns, T.A.L. Lima, A. Costa, E. Bosne, M.R. da Silva, J.P. Araújo, L.M.C. Pereira, Nucl. Instrum. Methods Phys. Res. B **371**, 59–62 (2016). https://doi.org/10.1016/j.nimb.2015.09.051
110. A.E. White, K.T. Short, R.C. Dynes, J.P. Garno, J.M. Gibson, Mesotaxy: single-crystal growth of buried $CoSi_2$ layers. Appl. Phys. Lett. **50**, 95–97 (1987). https://doi.org/10.1063/1.97830
111. A.H. van Ommen, J.J.M. Ottenheim, A.M.L. Theunissen, A.G. Mouwen, Synthesis of heteroepitaxial Si/$CoSi_2$/Si structures by Co implantation into Si. Appl. Phys. Lett. **53**, 669–671 (1988). https://doi.org/10.1063/1.100641
112. C.W.T. Bulle-Lieuwma, A.H. van Ommen, L.J. van Ijzendoorn, Microstructure of heteroepitaxial Si/$CoSi_2$/Si formed by Co implantation into (100) and (111) Si. Appl. Phys. Lett. **54**, 244–246 (1989). https://doi.org/10.1063/1.101446
113. R. Hull, A.E. White, K.T. Short, J.M. Bonar, Formation of continuous $CoSi_2$ layers by high Co dose implantation into Si(100). J. Appl. Phys. **68**, 1629–1634 (1990). https://doi.org/10.1063/1.346643
114. A.H. van Ommen, C.W.T. Bulle-Lieuwma, J.J.M. Ottenheim, A.M.L. Theunissen, Ion beam synthesis of heteroepitaxial Si/$CoSi_2$/Si structures. J. Appl. Phys. **67**, 1767–1778 (1990). https://doi.org/10.1063/1.345602
115. D.P. Adams, S.M. Yalisove, D.J. Eaglesham, Interfacial and surface energetics of $CoSi_2$. J. Appl. Phys. **76**, 5190–5194 (1994). https://doi.org/10.1063/1.357237
116. M.-O. Ruault, F. Fortuna, H. Bernas, O. Kaïtasov, In situ study of in-beam cobalt silicide growth in silicon. Nucl. Instrum. Methods Phys. Res. B **84**, 135–138 (1994). https://doi.org/10.1016/0168-583X(94)95741-X
117. M. Palard, M.-O. Ruault, O. Kaïtasov, H. Bernas, K.H. Heinig, Irradiation induced growth of $CoSi_2$ precipitates in Si at 650 °C: An in situ study. Nucl. Instrum. Methods Phys. Res. B **120**, 212–215 (1996). https://doi.org/10.1016/s0168-583x(96)00511-3

118. A.E. Volkov, M.-O. Ruault, H. Bernas, V.A. Borodin, Ion beam synthesis of $CoSi_2$: influence of surface kinetics on nucleation. Nucl. Instrum. Methods Phys. Res. B **178**, 327–330 (2001). https://doi.org/10.1016/S0168-583X(01)00514-6
119. M.-O. Ruault, F. Fortuna, V.A. Borodin, M.G. Ganchenkova, M.A. Kirk, Nucleation and growth of cobalt disilicide precipitates during in situ transmission electron microscopy implantation. J. Appl. Phys. **104**, 033527/1–21 (2008). https://doi.org/10.1063/1.2964098
120. F. Fortuna, V.A. Borodin, M.-O. Ruault, E. Oliviero, M.A. Kirk, Synergetic effects of dual-beam implantation on the microstructural development in silicon. Phys. Rev. B **84**, 144118/1–16 (2011). https://doi.org/10.1103/physrevb.84.144118
121. F. Fortuna, M.-A. Nguyen, M.-O. Ruault, M.A. Kirk, V.A. Borodin, M.G. Ganchenkova, Selective nucleation induced by defect nanostructures: a way to control cobalt disilicide precipitation during ion implantation. J. Appl. Phys. **112**, 123504/1–15 (2012). https://doi.org/10.1063/1.4769213
122. F. Fortuna, M.-O. Ruault, V.A. Borodin, M.G. Ganchenkova, O. Kaïtasov, Sensitivity of $CoSi_2$ precipitation in silicon to extra-low dopant concentrations. I. Experiment. J. Appl. Phys. **117**: 045703/1–8 (2015). https://doi.org/10.1063/1.4906402
123. D. da Silva Costa, C. Huck-Iriart, G. Kellermann, L.J. Giovanetti, A.F. Craievich, F.G. Requejo, In situ study of the endotaxial growth of hexagonal $CoSi_2$ nanoplatelets in Si(001). Appl. Phys. Lett. **107**, 223101/1–5 (2015). https://doi.org/10.1063/1.4936377
124. M.K. Bakhadyrkhanov, B.I. Boltaks, G.S. Kulikov, Diffusion, solubility, and electrical properties of cobalt in silicon. Sov. Phys. Solid State **12**, 144–149 (1970)
125. H. Nakashima, H. Tomokage, H. Kitagawa, K. Hashimoto, Energy level and solid solubility of cobalt in silicon by in-depth profile measurement. Jpn. J. Appl. Phys. **23**, 776–777 (1984). https://doi.org/10.1143/JJAP.23.776
126. H. Kitagawa, H. Nakashima, K. Hashimoto, Energy levels and solubility of electrically active cobalt in silicon studied by combined Hall and DLTS measurements. Jpn. J. Appl. Phys. **24**, 373–374 (1985). https://doi.org/10.1143/JJAP.24.373
127. J.S. Moore, M.C.P. Chang, C.M. Penchina, Energy levels in cobalt compensated silicon. J. Appl. Phys. **41**, 5282–5285 (1970). https://doi.org/10.1063/1.1658663
128. C.M. Penchina, J.S. Moore, Cobalt acceptor state in silicon: temperature dependence of the energy level and capture cross section. Phys. Rev. B **9**, 5217–5221 (1974). https://doi.org/10.1103/PhysRevB.9.5217
129. C.M. Penchina, J.S. Moore, N. Holonyak Jr., Energy levels and negative photoconductivity in cobalt-doped silicon. Phys. Rev. **143**, 634–636 (1966). https://doi.org/10.1103/PhysRev.143.634
130. M.C.P. Chang, C.M. Penchina, J.S. Moore, Excited impurity states and transient photoconductivity in cobalt-doped silicon. Phys. Rev. B **4**, 1229–1234 (1971). https://doi.org/10.1103/PhysRevB.4.1229
131. D.C. Wong, C.M. Penchina, Deep donor state of cobalt in silicon. Appl. Phys. Lett. **28**, 142–144 (1976). https://doi.org/10.1063/1.88668
132. H. Suwaki, K. Hashimoto, H. Nakashima, K. Hashimoto, Annealing of supersaturated cobalt in silicon. Jpn. J. Appl. Phys. **25**, 1952–1953 (1986). https://doi.org/10.1143/JJAP.25.1952
133. L.D. Yau, W.W. Chan, C.T. Sah, Thermal emission rates and activation energies of electrons and holes at cobalt centers in silicon. Phys. Status Solidi A **14**, 655–662 (1972). https://doi.org/10.1002/pssa.2210140234
134. A.O. Evwaraye, Impurity states in cobalt-doped silicon. J. Electron. Mater. **7**, 383–401 (1978). https://doi.org/10.1007/BF02655644
135. H. Lemke, Eigenschaften der Energieniveaus von Kobalt in Silizium. Phys. Status Solidi A **85**, K133–K136 (1984). https://doi.org/10.1002/pssa.2210850251
136. H. Nakashima, Y. Tsumori, T. Miyagawa, K. Hashimoto, Deep impurity levels of cobalt in silicon. Jpn. J. Appl. Phys. **29**, 1395–1398 (1990). https://doi.org/10.1143/JJAP.29.1395
137. W. Jost, J. Weber, H. Lemke, Hydrogen-induced defects in cobalt-doped n-type silicon. Semicond. Sci. Technol. **11**, 22–26 (1996). https://doi.org/10.1088/0268-1242/11/1/007

138. W. Jost, J. Weber, H. Lemke, Hydrogen-cobalt complexes in p-type silicon. Semicond. Sci. Technol. **11**, 525–530 (1996). https://doi.org/10.1088/0268-1242/11/4/011
139. D. Mathiot, Cobalt related levels in P and (P + B) doped n-type silicon: Possible observation of the (CoB) pair. J. Appl. Phys. **65**, 1554–1558 (1989). https://doi.org/10.1063/1.342972
140. H. Lemke, K. Irmscher, Proof of interstitial cobalt defects in silicon Float Zone crystals doped during crystal growth. ECS Trans. **3**(4), 299–310 (2006). https://doi.org/10.1149/1.2355765
141. L. Scheffler, V. Kolkovsky, J.J. Weber, A re-examination of cobalt-related defects in n- and p-type silicon. Phys. Status Solidi A **209**, 1913–1916 (2012). https://doi.org/10.1002/pssa201200140
142. L. Scheffler, V. Kolkovsky, J. Weber, Isolated substitutional cobalt and Co-related complexes in silicon. J. Appl. Phys. **113**, 183714/1–7 (2013). https://doi.org/10.1063/1.4804321
143. J. Weber, L. Scheffler, V. Kolkovsky, N. Yarykin, New results on the electrical activity of 3*d*-transition metal impurities in silicon. Solid-State Phenom. **205–206**, 245–254 (2014). http://www.scientific.net/SSP.205-206.245
144. Vl. Kolkovsky, L. Scheffler, J. Weber, Transition metals (Ti and Co) in silicon and their complexes with hydrogen: a Laplace DLTS study. Physica B **439**, 24–28 (2014). https://doi.org/10.1016/j.physb.2013.11.005
145. J. Lauwaert, J. Vanhellemont, E. Simoen, H. Vrielinck, P. Clauws, Electronic properties of iron and cobalt impurity centres in germanium. J. Appl. Phys. **111**, 113713/1–7 (2012). https://doi.org/10.1063/1.4729037
146. S.H. Segers, J. Lauwaert, P. Clauws, E. Simoen, J. Vanhellemont, F. Callens, H. Vrielinck, Temperature-independent slow carrier emission from deep-level defects in p-type germanium. J. Phys. D: Appl. Phys. **46**, 425101/1–7 (2013). stacks.iop.org/JPhysD/46/425101
147. M.L. Polignano, A. Guissani, D. Caputo, C. Clementi, G. Pavia, F. Priolo, Detection of metal segregation at the oxide-silicon interface. J. Electrochem. Soc. **149**, G429–G439 (2002). https://doi.org/10.1149/1.1483101
148. N. Pic, A. Danel, M.-L. Polignano, G. Salvà, M. Sardo, S. Rey, Influence of cobalt contamination in the measurement of diffusion length of silicon wafers. Solid State Phenomena **95–96**, 373–380 (2004). www.scientific.net/SSP.95-96.373
149. S. Diez, S. Rein, T. Roth, S.W. Glunz, Cobalt related defect levels in silicon analyzed by temperature- and injection-dependent lifetime spectroscopy. J. Appl. Phys. **101**, 033710/1–6 (2007). https://doi.org/10.1063/1.2433743
150. Y. Zhou, F. Liu, X. Song, The insulator-to-metal transition of Co hyperdoped crystalline silicon. J. Appl. Phys. **113**, 103702/1–5 (2013). https://doi.org/10.1063/1.4794818
151. H.H. Woodbury, G.W. Ludwig, Spin resonance of transition metals in silicon. Phys. Rev. **117**, 102–108 (1960). https://doi.org/10.1103/PhysRev.117.102
152. H.H. Woodbury, G.W. Ludwig, Vacancy interactions in silicon. Phys. Rev. Lett. **5**, 96–97 (1960). https://doi.org/10.1103/PhysRevLett.5.96
153. R. van Kemp, E.G. Sieverts, C.A.J. Ammerlaan, Electron-nuclear double resonance of interstitial chromium in silicon. Phys. Rev. B **36**, 3528–3541 (1987). https://doi.org/10.1103/PhysRevB.36.3528
154. H. Feichtinger, R. Czaputa, Energy levels and solubility of interstitial chromium in silicon. Appl. Phys. Lett. **39**, 706–708 (1981). https://doi.org/10.1063/1.92856
155. A.A. Lebedev, N.A. Sultanov, Some properties of chromium-doped silicon. Sov. Phys. Semicond. **4**, 1900–1902 (1971)
156. A.A. Zolotukhon, L.S. Milevskii, Effect of the Fermi energy on the decomposition kinetics of a solid solution of chromium in silicon. Sov. Phys. – Solid State 13: 1598–1603 (1972)
157. T. Kunio Jr., T. Nishino, E. Ohta, M. Sakata, Energy levels and degeneracy ratios for chromium in silicon. Solid-State Electron. **24**, 1087–1091 (1981). https://doi.org/10.1016/0038-1101(81)90174-X
158. A.A. Zolutukhin, L.S. Milevskiĭ, Photoconductivity of silicon containing site and interstitial chromium impurities. Sov. Phys. Semicond. **10**, 914–915 (1976)

159. H. Conzelmann, K. Graff, E.R. Weber, Chromium and chromium-boron pairs in silicon. Appl. Phys. A **30**, 169–175 (1983). https://doi.org/10.1007/BF00620536
160. T. Kunio Jr., T. Yamazaki, E. Ohta, M. Sakata, Defect levels in chromium-doped silicon. Solid-State Electron. **26**, 155–160 (1983). https://doi.org/10.1016/0038-1101(83)90117-X
161. T. Sadoh, M. Watanabe, H. Nakashima, T. Tsurushima, Deep levels of chromium-hydrogen complexes in silicon. J. Appl. Phys. **75**, 3978–3981 (1994). https://doi.org/10.1063/1.356018
162. S.H. Park, D.K. Schroder, Deep-level impurities in edge-defined film-fed-growth silicon. J. Appl. Phys. **78**, 801–810 (1995). https://doi.org/10.1063/1.360338
163. H. Conzelmann, Photoluminescence of transition metal complexes in silicon. Appl. Phys. A **42**, 1–18 (1987). https://doi.org/10.1007/BF00618154
164. J. Lauwaert, J. Van Gheluwe, J. Vanhellemont, E. Simoen, P. Clauws, Electronic properties of titanium and chromium impurity centers in germanium. J. Appl. Phys. **105**, 073707/1–7 (2009). https://doi.org/10.1063/1.3082124
165. A. Schlette, R. Kienle, A. Dörnen, W. Kürner, K. Thonke, Photoconductivity study of CrB and Cr_i in silicon. Mat. Res. Soc. Symp. Proc. **183**, 239–244 (1990). https://doi.org/10.1557/PROC-163-239
166. A. Hangleiter, Nonradiative recombination via deep impurity levels in silicon: experiment. Phys. Rev. B **35**, 9149–9161 (1987). https://doi.org/10.1103/PhysRevB.35.9149
167. D. Macdonald, L.J. Geerligs, Recombination activity of interstitial iron and other transition metal point defects in p- and n-type crystalline silicon. Appl. Phys. Lett. **85**, 4061–4063 (2004). https://doi.org/10.1063/1.1812833
168. S. Dubois, O. Palais, P.J. Ribeyron, Determination at 300 K of the hole capture cross section of chromium-boron pairs in p-type silicon. Appl. Phys. Lett. **89**, 232112/1–3 (2006). https://doi.org/10.1063/1.2402261
169. J. Schmidt, R. Krain, K. Bothe, G. Pensl, S. Beljakowa, Recombination activity of interstitial chromium and chromium-boron pairs in silicon. J. Appl. Phys. **102**, 123701/1–10 (2007). https://doi.org/10.1063/1.2822452
170. C. Sun, F.E. Rougieux, D. Macdonald, Reassessment of the recombination parameters of chromium in n- and p-type crystalline silicon and chromium-boron pairs in p-type crystalline silicon. J. Appl. Phys. **115**, 214907/1–9 (2014). https://doi.org/10.1063/1.4881497
171. K. Mishra, Identification of Cr in p-type silicon using the minority carrier lifetime measurement by the surface photovoltage method. Appl. Phys. Lett. **68**, 3281–3283 (1996). https://doi.org/10.1063/1.116574
172. M.L. Polignano, F. Cazzaniga, A. Sabbadini, F. Zanderigo, F. Priolo, Metal contamination monitoring and gettering. Mater. Sci. Semicond. Process. **1**, 119–130 (1998). https://doi.org/10.1016/S1369-8001(98)00015-8
173. M.L. Polignano, E. Bellandi, D. Lodi, F. Pipia, A. Sabbadini, F. Zanderigo, G. Queirolo, F. Priolo, Quantitative evaluation of bulk-diffused metal contamination by lifetime techniques. Mater. Sci. Eng. B **55**, 21–33 (1998). https://doi.org/10.1016/S0921-5107(98)00192-5
174. K. Kurita, T. Shingyouji, Identification and quantification of transition metal impurities in Czochralski silicon wafers using microwave photoconductive decay lifetime measurements. Jpn. J. Appl. Phys. **37**, 5861–5865 (1998). https://doi.org/10.1143/JJAP.37.5861
175. H. Habenicht, M.C. Schubert, W. Warta, Imaging of chromium point defects in p-type silicon. J. Appl. Phys. **108**, 034909/1–7 (2010). https://doi.org/10.1063/1.3459892
176. H. Shimizu, E. Omori, M. Ikeda, Anomalous behavior of Schottky barrier-type surface photovoltages in chromium-contaminated n-type silicon wafers exposed to air. Jpn. J. Appl. Phys. **45**, 4982–4984 (2006). https://doi.org/10.1143/JJAP.45.4982
177. H. Shimizu, S. Shimada, M. Ikeda, Negative oxide charge in thermally oxidized Cr-contaminated n-type silicon wafers. Jpn. J. Appl. Phys. **49**, 038001/1–2 (2010). https://doi.org/10.1143/jjap.49.038001

178. D.A. van Wezep, R. van Kemp, E.G. Sieverts, C.A.J. Ammerlaan, Electron-nuclear double resonance of titanium in silicon: ^{29}Si ENDOR. Phys. Rev. B **32**, 7129–7138 (1985). https://doi.org/10.1103/PhysRevB.32.7129
179. D.A. van Wezep, C.A.J. Ammerlaan, Electron-nuclear double resonance of titanium in silicon: ^{47}Ti and ^{49}Ti ENDOR. Phys. Rev. B **37**, 7268–7275 (1988). https://doi.org/10.1103/PhysRevB.37.7268
180. J. Olea, M. Toledano-Luque, D. Pastor, G. González-Díaz, I. Mártil, Titanium doped silicon layers with very high concentration. J. Appl. Phys. **104**, 016105/1–3 (2008). https://doi.org/10.1063/1.294258
181. A.M. Salama, L.J. Cheng, The effects of titanium impurities in n^{+}/p silicon solar cells. J. Electrochem. Soc. **127**, 1164–1167 (1980). https://doi.org/10.1149/1.2129839
182. W. Wijaranakula, S.S. Kim, Precipitation of 3*d* transition-metal silicides in Czochralski silicon crystals. J. Appl. Phys. **76**, 6017–6019 (1994). https://doi.org/10.1063/1.358355
183. W. Fahrner, A. Goetzberger, Determination of deep energy levels in Si by MOS techniques. Appl. Phys. Lett. **21**, 329–331 (1972). https://doi.org/10.1063/1.1654399
184. J.-W. Chen, A.G. Milnes, A. Rohatgi, Titanium in silicon as a deep level impurity. Solid-State Electron. **22**, 801–808 (1979). https://doi.org/10.1016/0038-1101(79)90130-8
185. J.R. Morante, J.E. Carceller, P. Cartujo, J. Barbolla, Thermal emission rates and capture cross-section of majority carriers at titanium levels in silicon. Solid-State Electron. **26**, 1–6 (1983). https://doi.org/10.1016/0038-1101(83)90153-3
186. A.C. Wang, C.T. Sah, Complete electrical characterization of recombination properties of titanium in silicon. J. Appl. Phys. **56**, 1021–1031 (1984). https://doi.org/10.1063/1.334095
187. D. Mathiot, S. Hocine, Titanium-related deep levels in silicon: a reexamination. J. Appl. Phys. **66**, 5862–5867 (1989). https://doi.org/10.1063/1.343608
188. L. Tilly, H.G. Grimmeiss, H. Petterson, K. Schmalz, K. Tittelbach, H. Kerkow, Electrical and optical properties of titanium-related centers in silicon. Phys. Rev. B **43**, 9171–9177 (1991). https://doi.org/10.1103/PhysRevB.43.9171
189. V.P. Markevich, S. Leonard, A.R. Peaker, B. Hamilton, A.G. Marinopoulos, J. Coutinho, Titanium in silicon: lattice positions and electronic properties. Appl. Phys. Lett. **104**, 152105/1–4 (2014). https://doi.org/10.1063/1.487102
190. Kolkovsky Vl, L. Scheffler, J. Weber, A re-examination of the interstitial Ti levels in Si. Phys. Status Solidi C **9**, 1996–1999 (2012). https://doi.org/10.1002/pssc.201200141
191. L. Scheffler, V. Kolkovsky, J. Weber, Isolated Ti in Si: deep level transient spectroscopy, minority carrier transient spectroscopy, and high-resolution Laplace deep level transient spectroscopy studies. J. Appl. Phys. **117**, 045713/1–7 (2015). https://doi.org/10.1063/1.4906855
192. R. Singh, S.J. Fonash, A. Rohatgi, Interaction of low-energy implanted atomic H with slow and fast diffusing metallic impurities in Si. Appl. Phys. Lett. **49**, 800–802 (1986). https://doi.org/10.1063/1.97551
193. W. Jost, J. Weber, Titanium-hydrogen defects in silicon. Phys. Rev. B **54**, R11038–R11041 (1996). https://doi.org/10.1103/PhysRevB.54.R11038
194. S. Leonard, V.P. Markevich, A.R. Peaker, B. Hamilton, Passivation of titanium by hydrogen in silicon. Appl Phys Lett **103**, 132103/1–4 (2013). https://doi.org/10.1063/1.4822329
195. L. Scheffler, Vl. Kolkovsky, J. Weber, Identification of titanium-hydrogen complexes with up to four hydrogen atoms in silicon. J. Appl. Phys. **117**, 085707/1–7 (2015). https://doi.org/10.1063/1.4913524
196. T. Roth, M. Rüdiger, W. Warta, S.W. Glunz, Electronic properties of titanium in boron-doped silicon analyzed by temperature-dependent photoluminescence and injection-dependent photoconductance decay lifetime spectroscopy. J. Appl. Phys. **104**, 075410/1–6 (2008). https://doi.org/10.1063/1.2996252
197. B.B. Paudyal, K.R. McIntosh, D.H. Macdonald, Temperature dependent carrier lifetime studies on Ti-doped multicrystalline silicon. J. Appl. Phys. **105**, 124510/1–5 (2009). https://doi.org/10.1063/1.3139286

198. E. Antolín, A. Martí, J. Olea, D. Pastor, G. Gonzaléz-Díaz, I. Mártil, A. Luque, Lifetime recovery in ultrahighly titanium-doped silicon for the implementation of an intermediate band material. Appl. Phys. Lett. **94**, 042115/1–3 (2009). https://doi.org/10.1063/1.3077202
199. E. García-Hemme, R. García-Hernansanz, J. Olea, D. Pastor, A. del Prado, I. Mártil, G. Gónzalez-Díaz, Sub-bandgap spectral photo-response analysis of Ti supersaturated Si. Appl. Phys. Lett. **101**, 192101/1–5 (2012). https://doi.org/10.1063/1.4766171
200. J. Olea, D. Pastor, A. del Prado, E. García-Hemme, R. García-Hernansanz, I. Martíl, G. Gónzalez-Díaz, Ruling out the impact of defects on the below band gap photoconductivity of Ti supersaturated Si. J. Appl. Phys. **114**, 053110/1–7 (2013). https://doi.org/10.1063/1.4817254
201. A. Rohatgi, R.H. Hopkins, J.R. Davis, R.B. Cambell, H.C. Mollenkoph, The impact of molybdenum on silicon and silicon solar cell performance. Solid-State Electron. **23**, 1185–1190 (1980). https://doi.org/10.1016/0038-1101(80)90032-5
202. T. Hamaguchi, Y. Hayamizu, Deep levels associated with molybdenum in silicon. Jpn. J. Appl. Phys. **30**, L1837–L1839 (1991). https://doi.org/10.1143/JJAP.30.L1837
203. A. Sandhu, T. Ogikubo, H. Goto, V. Csapo, T. Pavelka, Investigation of deep levels and precipitates related to molybdenum in silicon by DLTS and scanning infrared microscopy. J. Cryst. Growth **210**, 116–121 (2000). https://doi.org/10.1016/S0022-0248(99)00663-6
204. J.L. Benton, D.C. Jacobson, B. Jackson, J.A. Johnson, T. Boone, D.J. Eaglesham, E.A. Stevie, J. Becerro, Behavior of molybdenum in silicon evaluated for integrated circuit processing. J. Electrochem. Soc. **146**, 1929–1933 (1999). https://doi.org/10.1149/1.1391868
205. M.L. Polignano, I. Mica, F. Barbarossa, A. Galbiati, S. Grasso, V. Soncini, Contamination by slow diffusers in ion implantation processes: The examples of molybdenum and tungsten. Nucl. Instr. Methods Phys. Res. B **356**, 164–171 (2015). https://doi.org/10.1016/jn.nimb.2015.04.069
206. S. Leonard, V.P. Markevich, A.R. Peaker, B. Hamilton, K. Yousseff, G. Rozgonyi, Molybdenum nano-precipitates in silicon: a TEM and DLTS study. Phys. Status Solidi B **251**, 2201–2204 (2014). https://doi.org/10.1002/pssb.201400065
207. S.M. Cox, Detection of a molybdenum acceptor level in n-type silicon. Electrochem. Soc. J. Solid-State Technol. **3**, P397–P402 (2014). https://doi.org/10.1149/2.0081412jss
208. M. Aoki, T. Itakura, N. Sasaki, Mo contamination in p/p^+ epitaxial silicon wafers. Jpn. J. Appl. Phys. **34**, 712–714 (1995). https://doi.org/10.1143/JJAP.34.712
209. F. Russo, G. Moccia, G. Nardone, R. Alfonsetti, G. Polsinelli, A. D'Angelo, A. Patacchiola, M. Liverani, P. Pianezza, T. Lippa, M. Carlini, M.L. Polignano, I. Mica, E. Cazzini, M. Ceresoli, D. Codegoni, Proximity gettering of slow diffuser contaminants in CMOS imager sensors. Solid-State Electron. **91**, 91–99 (2014). https://doi.org/10.1016/j.sse.2013.10.011
210. J.U. Sachse, W. Jost, J. Weber, H. Lemke, Trivalent behavior of palladium in silicon. Appl. Phys. Lett. **71**, 1379–1381 (1997). https://doi.org/10.1063/1.119899
211. R. Czaputa, Transition metal impurities in silicon: New defect reactions. Appl. Phys. A **49**, 431–436 (1989). https://doi.org/10.1007/BF00615028
212. R. Dogra, A.K. Sharma, A.P. Byrne, M.C. Ridgway, Evolution of palladium related defects in silicon. AIP Conf. Proc. **1447**, 1033–1034 (2012). https://doi.org/10.1063/1.4710358
213. S.J. Pearton, Deep metal-related centers in germanium. Solid-State Electron. **25**, 305–311 (1982). https://doi.org/10.1016/0038-1101(82)90139-3
214. A.A. Abiona, W. Kemp, H. Timmers, Pd-vacancy complex in Ge: TDPAC and ab initio study. AIP Conf. Proc. **1583**, 105–108 (2014). https://doi.org/10.1063/1.4865614
215. M.D. Miller, H. Schade, C.J. Nuesse, Lifetime-controlling recombination centers in platinum-diffused silicon. J. Appl. Phys. **47**, 2569–2578 (1976). https://doi.org/10.1063/1.322976
216. S.D. Brotherton, P. Bradley, J. Bicknell, Electrical properties of platinum in silicon. J. Appl. Phys. **50**, 3396–3403 (1979). https://doi.org/10.1063/1.326331
217. A.O. Evwaraye, E. Sun, Electrical properties of platinum in silicon determined by deep-level transient spectroscopy. J. Appl. Phys. **47**, 3172–3176 (1976). https://doi.org/10.1063/1.323112

218. M. Kimata, M. Ueno, H. Yagi, T. Shiraishi, M. Kawai, K. Endo, Y. Kosasayama, T. Sone, T. Ozeki, N. Tsubouchi, PtSi Schottky-barrier infrared focal plane arrays. Opto-Electron. Rev. **6**, 1–10 (1998)
219. R.A. Donaton, S. Jin, H. Bender, M. Zagrebnov, K. Baert, K. Maex, A. Vantomme, G. Langouche, Formation of ultra-thin PtSi layers with a 2-step silicidation process. Microelectron. Eng. **37**, 507–514 (1979). https://doi.org/10.1016/S0167-9317(97)00153-6
220. Y.K. Kwon, T. Ishikawa, H. Kuwano, Properties of platinum-associated deep levels in silicon. J. Appl. Phys. **61**, 1055–1058 (1987). https://doi.org/10.1063/1.338197
221. A.A. Gill, N. Baber, M.Z. Iqbal, Role of the mid-gap level as the dominant recombination center in platinum-doped silicon. J. Appl. Phys. **67**, 1130–1132 (1990). https://doi.org/10.1063/1.345783
222. J.U. Sachse, E.Ö. Sveinbjörnsson, W. Jost, J. Weber, H. Lemke, Electrical properties of platinum-hydrogen complexes in silicon. Phys. Rev. B **55**, 16176–16185 (1997). https://doi.org/10.1103/PhysRevB.55.16176
223. U. Juda, O. Scheerer, M. Höhne, H. Riemann, H.J. Schilling, J. Donecker, A. Gerhardt, Oxygen-related 1-platinum defects in silicon: An electron paramagnetic resonance study. J. Appl. Phys. **80**, 3435–3444 (1996). https://doi.org/10.1063/1.363212
224. O. Scheerer, M. Höhne, U. Juda, H. Riemann, Carbon-related platinum defects in silicon: An electron paramagnetic resonance study of high spin states. J. Appl. Phys. **82**, 3456–3461 (1997). https://doi.org/10.1063/1.365661
225. L. Dobaczewski, K. Bonde Nielsen, A. Nylandsted Larsen, J. Lundsgaard Hansen, B.K. Gościński, A.R. Peaker, Lattice sitting of platinum atoms in diluted SiGe alloys. Acta Phys. Polonica **94**, 297–299 (1998)
226. M. Morooka, M. Takahashi, F. Hashimoto, Stacking fault induced by gold diffusion in silicon. Jpn. J. Appl. Phys. **31**, 2327–2332 (1992). https://doi.org/10.1143/JJAP.31.2327
227. W. Lerch, N.A. Stolwijk, Diffusion of gold in silicon during rapid thermal annealing: Effectiveness of the surface as a sink for self-interstitials. J. Appl. Phys. **83**, 1312–1320 (1998). https://doi.org/10.1063/1.366831
228. M. Morooka, Limiting process for gold in-diffusion in silicon with and without extended defects. Jpn. J. Appl. Phys. **35**, 2537–2543 (1996). https://doi.org/10.1143/JJAP.35.2537
229. C.B. Collins, R.O. Carlson, C.J. Gallagher, Properties of gold-doped silicon. Phys. Rev. **105**, 1168–1173 (1957). https://doi.org/10.1103/PhysRev.105.1168
230. A.F. Tasch, C.T. Sah, Recombination-generation and optical properties of gold acceptor in silicon. Phys. Rev. B **1**, 800–808 (1970). https://doi.org/10.1103/PhysRevB.1.800
231. S. Braun, H.G. Grimmeiss, Optical properties of gold acceptor and donor levels in silicon. J. Appl. Phys. **45**, 2658–2665 (1974). https://doi.org/10.1063/1.1663646
232. K.S.R.K. Rao, V. Kumar, S.K. Premachandran, K.P. Raghunath, Relationship of the gold related donor and acceptor levels in silicon. J. Appl. Phys. **69**, 2714–2716 (1991). https://doi.org/10.1063/1.348622
233. G.D. Watkins, M. Kleverman, A. Thilderkvist, H.G. Grimmeiss, Structure of gold in silicon. Phys. Rev. Lett. **67**, 1149–1152 (1991). https://doi.org/10.1103/PhysRevLett.67.1149
234. N.T. Son, T. Gregorkiewicz, C.A.J. Ammerlaan, Paramagnetic state of the isolated gold impurity in silicon. Phys. Rev. Lett. **69**, 3185–3188 (1992). https://doi.org/10.1103/PhysRevLett.69.3185
235. X. Yu, X. Li, R. Fan, D. Yang, M. Kittler, M. Reiche, M. Seibt, G. Rozgonyi, Effect of Au contamination on the electrical characteristics of a "model" small-angle grain boundary in n-type direct silicon bonded wafer. J. Appl. Phys. **108**, 053719/1–6 (2010). https://doi.org/10.1063/1.3471817
236. G. Kissinger, H.G. Grimmeis, Difficulties in doping SiGe alloys with transition metal point defects. Phys. Status Solidi A **145**, K5–K9 (1994). https://doi.org/10.1002/pssa.2211450122
237. O. Voss, V.V. Kveder, M. Seibt, Electrical properties of gold in dislocated silicon. Phys. Status Solidi A **204**, 2185–2189 (2007). https://doi.org/10.1002/pssa.200675434

238. A. Mesli, P. Kringhoj, A. Nylandsted Larsen, Pinning behavior of gold-related levels in Si using $Si_{1-x}Ge_x$ alloy layers. Phys. Rev. B **56**, 13202–13217 (1997). https://doi.org/10.1103/PhysRevB.56.1320
239. R. Braunstein, A.R. Moore, F. Herman, Intrinsic optical absorption in germanium-silicon alloys. Phys. Rev. **109**, 695–710 (1958). https://doi.org/10.1103/PhysRev.109.695
240. J. Weber, M.I. Alonso, Near-band-gap photoluminescence of Si-Ge alloys. Phys. Rev. B **40**, 5683–5693 (1989). https://doi.org/10.1103/PhysRev.109.695
241. L. Dobaczewski, K. Goscinski, K. Bonde Nielsen, A. Nylandsted Larsen, J. Lundsgaard Hansen, A.R. Peaker, Alloy splitting of gold and platinum levels in SiGe. Phys. Rev. Lett. **83**, 4582–4585 (1999). https://doi.org/10.1103/PhysRevLett.83.4582
242. E. Simoen, P. Clauws, G. Huylebroeck, J. Vennik, DLTS of gold impurities in germanium. Semicond. Sci. Technol. **2**, 507–512 (1987). https://doi.org/10.1088/0268-1242/2/8/006
243. A.A. Lebedev, N.A. Sultanov, P. Yuspov, Photoelectric properties of yttrium doped silicon. Sov. Phys. Semicond. **14**, 342–343 (1980)
244. H. Lemke, Characterization of transition metal-doped silicon crystals prepared by float zone technique. Proc. Electrochem. Soc. **94**(10), 695–710 (1994)
245. N. Achtziger, Scandium correlated deep levels in silicon. J. Appl. Phys. **80**, 6286–6292 (1996). https://doi.org/10.1063/1.363648
246. O. Thomas, C.S. Petersson, F.M. d'Heurle, The reaction of scandium thin films with silicon: diffusion, nucleation, resistivities. Appl. Surf. Sci. **53**, 138–146 (1991). https://doi.org/10.1016/0169-4332(91)90254-H
247. T. Sadoh, H. Nakashima, T. Tsurushima, Deep levels of vanadium and vanadium-hydrogen complex in silicon. J. Appl. Phys. **72**, 520–524 (1992). https://doi.org/10.1063/1.352353
248. J. Mullins, V.P. Markevich, M.P. Halsall, A.R. Peaker, Vanadium in silicon: lattice positions and electronic properties. Appl. Phys. Lett. **110**, 142105 (2017). https://doi.org/10.1063/1.4979697
249. E.R. Weber, Transition metals in silicon. Appl. Phys. A **30**, 1–22 (1983). https://doi.org/10.1007/BF00617708
250. D.J. Backlund, T.M. Gibbons, S.K. Estreicher, Vanadium interactions in crystalline silicon. Phys. Rev. B **94**, 195210/1-6 (2016). https://doi.org/10.1103/PhysRevB.94.195210
251. J.J. van Kooten, D. van Kootwijk, C.A.J. Ammerlaan, Self-ENDOR of vanadium in silicon. J. Phys. C: Solid State Phys. **20**, 841–847 (1987). https://doi.org/10.1088/0022-3719/20/6/007
252. J. Mullins, V.P. Markevich, M.P. Halsall, A.R. Peaker, Interactions of hydrogen with vanadium in crystalline silicon. Phys. Status Solidi A **213**, 2838–2843 (2016). https://doi.org/10.1002/pssa.201600493
253. H. Lemke, Eigenschaften einiger Storstellenkomplexe von Gold in Silizium. Phys. Status Solidi A **75**, 473–482 (1983). https://doi.org/10.1002/pssa.2210750217
254. E. García-Hemme, R. García-Hernansanz, J. Olea, D. Pastor, A. del Prado, I. Mártil, G. Gónzalez-Díaz, Far infrared photoconductivity in a silicon-based material: vanadium supersaturated silicon. Appl. Phys. Lett. **103**, 032101–1/5 (2013). https://doi.org/10.1063/1.4813823
255. R.O. Carlson, Properties of silicon doped with manganese. Phys. Rev. **104**, 937–941 (1956). https://doi.org/10.1103/PhysRev.104.9372
256. T. Roth, P. Rosenits, S. Diez, S.W. Glunz, D. Macdonald, S. Beljakowa, G. Pensl, Electronic properties and dopant pairing behavior of manganese in boron-doped silicon. J. Appl. Phys. **102**, 103716/1–8 (2007). https://doi.org/10.1063/1.2812698
257. H. Lemke, Eigenschaften von substitutionell gelöstem Mangan in Silizium. Phys. Status Solidi A **83**, 637–643 (1984). https://doi.org/10.1002/pssa.2210830228
258. R. Czaputa, H. Feichtinger, J. Oswald, Energy levels of interstitial manganese in silicon. Solid State Commun. **47**, 223–226 (1983). https://doi.org/10.1016/0038-1098(83)90549-5
259. J. Kreissl, W. Gehlhoff, H. Vollmer, Trigonal manganese cluster in silicon: an electron-paramagnetic study. Phys. Rev. **49**, 10307–10317 (1994). https://doi.org/10.1103/PhysRevB.49.10307

260. J. Wedekind, H. Vollmer, R. Labusch, Tetrahedral Mn_{i4} cluster in silicon. Phys. Rev. B **63**, 233202/1–4 (2001). https://doi.org/10.1103/physrevb.63.233202
261. M.A. Nicolet, S.S. Lau, Formation and characterization of transition-metal silicides. VLSI Electron. **6**, 330–459 (1983). https://doi.org/10.1016/B978-0-12-234106-9.50011-8
262. K.H. Kim, J.D. Lee, J.J. Lee, S.W. Han, J.S. Kang, Formation and characterization of manganese silicide on Si(111) and Si(100) substrates. J. Korean Phys. Soc. **51**, 1032–1037 (2007)
263. K. Mallik, C. de Groot, P. Ashburn, Enhancement of resistivity of Czochralski silicon by deep level manganese doping. Appl. Phys. Lett. **89**, 112122/1–3 (2006). https://doi.org/10.1063/1.2349836
264. H.H. Woodbury, W.W. Tyler, Properties of germanium doped with manganese. Phys. Rev. **100**, 659–661 (1955). https://doi.org/10.1103/PhysRev.100.659
265. R. Newman, H.H. Woodbury, W.W. Tyler, Photoconductivity in manganese doped germanium. Phys. Rev. **102**, 613–617 (1956). https://doi.org/10.1103/PhysRev.102.613
266. J. Lauwaert, S.H. Segers, F. Moens, K. Opsomer, P. Clauws, F. Callens, E. Simoen, H. Vrielinck, Electronic properties of manganese impurities in germanium. J. Phys. D: Appl. Phys. **48**, 175101/1–5 (2015). https://doi.org/10.1088/0022-3727/48/17/17510
267. S. Zhou, H. Schmidt, Mn-doped Ge and Si: a review of the experimental status. Materials **3**, 5054–5092 (2010). https://doi.org/10.3390/ma3125054
268. Y.D. Park, A.T. Hanbicki, S.C. Erwin, C.S. Hellberg, J.M. Sullivan, J.E. Mattson, T.F. Ambrose, A.G. Wilson Spanos, B.T. Jonker, A group-IV ferromagnetic semiconductor: Mn_xGe_{1-x}. Science **295**, 651–654 (2002). https://doi.org/10.1126/science.1066348
269. S. Cho, S. Choi, S.C. Hong, Y. Kim, J.B. Ketterson, B.J. Kim, Y.C. Kim, J.H. Jung, Ferromagnetism in Mn-doped Ge. Phys. Rev. B **66**, 033303/1–3 (2002). https://doi.org/10.1103/physrevb.66.033303
270. L. Ottaviano, M. Passacantando, S. Picozzi, A. Continenza, R. Gunnella, A. Verna, G. Bihlmayer, G. Impellizzeri, F. Priolo, Phase separation and dilution in implanted Mn_xGe_{1-x} alloys. Appl. Phys. Lett. **88**, 061907/1–3 (2006). https://doi.org/10.1063/1.2171485
271. E. Biegger, L. Staheli, M. Fonin, U. Rudiger, Y.S. Dedkov, Intrinsic ferromagnetism versus phase segregation in Mn-doped Ge. J. Appl. Phys. **101**, 103912/1 (2007). https://doi.org/10.1063/1.2718276
272. S. Decoster, S. Cottenier, U. Wahl, J.G. Correia, L.M.C. Pereira, C. Lacasta, M.R. Da Silva, A. Vantomme, Diluted manganese on the bond-centered site in germanium. Appl. Phys. Lett. **97**, 1914/1–3 (2010). https://doi.org/10.1063/1.3501123
273. H. Bracht, H. Overhof, Kinetics of interstitial-substitutional exchange of Zn, Pt, and Au in Si: experimental results and theoretical calculations. Phys. Status Solidi A **158**, 47–55 (1996). https://doi.org/10.1002/pssa.2211580107
274. N.A. Stolwijk, D. Grünebaum, M. Perret, M. Brohl, Zinc and sulphur in silicon: Experimental evidence for kick-out diffusion behavior. Mat. Sci. Forum **38–41**, 701–706 (1989). www.scientific.net/MSF.38-41.70
275. S. Weiss, R. Beckmann, R. Kassing, The electrical properties of zinc in silicon. Appl. Phys. A **50**, 151–156 (1990). https://doi.org/10.1007/BF003
276. W. Gehlhoff, A. Näser, H. Bracht, EPR proof of the negatively charged acceptor state Zn^- in silicon. Phys. B **273–274**, 264–267 (1999). https://doi.org/10.1016/S0921-4526(99)00481-0
277. H. Lemke, Eigenschaften einiger Storstellenkomplexe von Zink in Silizium. Phys. Status Solidi A **71**, 172–187 (1982). https://doi.org/10.1002/pssa.2210720118
278. P. Stolz, G. Pensl, D. Grunebaum, N. Stolwijk, Hydrogen passivation and thermal reactivation of zinc double acceptors in silicon. Mat. Sci. Eng. B **4**, 31–34 (1989). https://doi.org/10.1016/0921-5107(89)90211-0
279. S. Voss, H. Bracht, N.A. Stolwijk, P. Kringhøj, A. Nylandsted Larsen, Energy levels of Zn in $Si_{1-x}Ge_x$ alloys. Appl. Phys. Lett. **73**, 2331–2333 (1998). https://doi.org/10.1063/1.121813

280. A.K. Ramdas, S. Rodriguez, Spectroscopy of the solid-state analogues of the hydrogen atom: donors and acceptors in semiconductors. Rep. Progr. Phys. **44**, 1297–1387 (1981). https://doi.org/10.1088/0034-4885/44/12/002
281. S. Voss, N.A. Stolwijk, H. Bracht, A. Nylandsted Larsen, H. Overhof, Substitutional Zn in SiGe: deep-level transient spectroscopy and electron density calculations. Phys. Rev. B **68**, 035208/1–9 (2003). https://doi.org/10.1103/physrevb.68.035208
282. H. Lemke, Properties of single crystals doped with zirconium or hafnium. Phys. Status Solidi A **122**, 617–630 (1990). https://doi.org/10.1002/pssa.2211220223
283. P. Raghu, N. Rana, C. Yim, E. Shero, F. Shadman, Adsorption of moisture and organic contaminants on hafnium oxide, zirconium oxide, and silicon oxide gate dielectrics. J. Electrochem. Soc. **150**, F186–1943 (2003). https://doi.org/10.1149/1.1605747
284. B. Vermeire, K. Delbridge, V. Pandit, H.G. Parks, K. Ramkumar, S. Geha, S. Raghavan, J. Jeon, The effect of hafnium and zirconium contamination on MOS processing. Proc. Adv. Semicond. Manuf. **2002**, 299–303 (2002). https://doi.org/10.1109/ASMC.2002.1001622
285. S.J. Pearton, A.J. Tavendale, Hydrogen passivation of deep-metal related donor centers in germanium. J. Appl. Phys. **54**, 820–823 (1983). https://doi.org/10.1063/1.3320
286. D. Graff, Metal impurities in Silicon-Device Fabrication (Springer, Heidelberg) (2013). https://doi.org/10.1007/978-3-642-57121-3
287. H. Petersson, H.G. Grimmeiss, L. Tilly, K. Schmalz, H. Kerkow, Electrical and optical characterization of niobium-related centers in silicon. Semicond. Sci. Technol. **8**, 1247–1252 (1993). https://doi.org/10.1088/0268-1242/8/7/011
288. M.L. Polignano, D. Codegoni, G. Borionetti, F. Bonoli, J. Brivio, S. Greco, A. Marino, P. Monge. I. Patoprsta, V. Privitera, C. Riva, Niobium contamination in silicon. Electochem Soc Trans **33**(11), 133–144 (2010). https://doi.org/0.1149/1.3485688
289. J. Zhou, J. Wu, L. Lu, Z. Han, Physical behavior of ruthenium in silicon. Appl. Phys. Lett. **69**, 2746–2748 (1991). https://doi.org/10.1063/1.348632
290. H. Pettersson, H.G. Grimmeiss, K. Schmalz, A. Knecht, R. Passier, Electrical and optical properties of ruthenium-related defects in silicon. J. Appl. Phys. **77**, 2495–2500 (1995). https://doi.org/10.1063/1.358778
291. C.S. Petersson, J.E.E. Baglin, J.J. Dempsey, F.M. d'Heurle, S.J. LaPlaca, Silicides of ruthenium and osmium: thin film reactions, diffusion, nucleation and stability. J. Appl. Phys. **53**, 4866 (1982). https://doi.org/10.1063/1.331319
292. R. Chan, T.N. Arunagiri, Y. Zhang, O. Chyan, R.M. Wallace, M.J. Kim, T.Q. Hurd, Diffusion studies of copper on ruthenium thin film: a plateable copper diffusion barrier. Electrochem. Solid-State Lett. **7**, G154–G157 (2004). https://doi.org/10.1149/1.1757113
293. S. Armini, Z. El-Mekki, M. Nagar, A. Radisic, W. Ruythooren, P.M. Vereecken, Wafer scale copper direct plating on thin PVD RuTa layers: a route to enable filling 30 nm features and below? Electrochem. Soc. Trans. **58**(17), 3–15 (2014). https://doi.org/10.1149/05817.0003ecst
294. S.H. Hsieh, W.J. Chen, C.M. Chien, Structural stability of diffusion barriers in Cu/MgO/Ta/Si. Nanomater. **5**, 1840–1852 (2015). https://doi.org/10.3390/nano5041840
295. N. Torazawa, S. Hirao, S. Kanayama, H. Korogi, S. Matsumoto, The development of Cu filling and reliability performance with Ru-Ta alloy barrier for Cu interconnects. J. Electrochem. Soc. **163**, E173–E178 (2016). https://doi.org/10.1149/2.0901606jes
296. S. Dey, K.H. Yu, S. Consiglio, K. Tapily, T. Hakamata, C.S. Wajda, G.J. Leusink, J. Jordan-Sweet, C. Lavoie, D. Muir, B. Moreno, A.C. Diebold, Atomic layer deposited ultrathin metal nitrided barrier layers for ruthenium interconnect applications. J. Vac. Sci. Technol. A **35**, 03E109/1–8 (2017). https://doi.org/10.1116/1.4979709
297. S. Petersson, R. Anderson, J. Baglin, J. Dempsey, W. Hammer, F. d'Heurle, S. LaPaca, Thin film formation of rhodium silicides. J. Appl. Phys. **51**, 373–382 (1980). https://doi.org/10.1063/1.327381
298. L. Marot, R. Schoch, R. Steiner, V. Thommen, D. Mathys, E. Meyer, Rhodium and silicon system: II. Rhodium silicide formation. Nanotechnology **21**, 365707 (2010). https://doi.org/10.1088/0957-4484/21/36/365707

299. H. Lemke, Substitutional transition metal defects in silicon grown-in by the float zone technique. Mat. Sci. Forum **196–201**, 683–688 (1995). www.scientific.net/MSF.196-201.683
300. S. Knack, J. Weber, H. Lemke, Hydrogen-rhodium complexes in silicon. Mat. Sci. Eng. B **58**, 141–145 (1999). https://doi.org/10.1016/S0291-5107(98)00284-0
301. H. Lemke, W. Zulehner, Impact of vacancies and self-interstitials on the formation of transition metal defect in float-zone silicon crystals. Physica B **273–274**, 398-403 (1999). https://doi.org/10.1016/s0921-4526(99)00490-1
302. V. Voronkov, The mechanism of swirl defects formation in silicon. J. Cryst. Growth **69**, 625–643 (1982). https://doi.org/10.1016/0022-0248(82)90386-4
303. H. Rieman, A. Ludge, K. Schwerd, Defects in FZ silicon and their impact on X-ray PIN sensor parameters. Electrochem. Soc. Proc. **17**, 509–514 (2000)
304. S. Geller, The rhodium-germanium system. I. The crystal structures of Rh_2Ge, Rh_5Ge_3 and RhGe. Acta Cryst. **8**, 15–21 (1955). https://doi.org/10.1107/S0365110X55000030
305. M.P. Lepselter, A.T. Fiory, N.M. Ravindra, Platinum and Rhodium Silicide-Germanium optoelectronics. J. Electron. Mater. **37**, 403–416 (2008). https://doi.org/10.1007/s11664-007-0216-3
306. S. Kar, R. Varghese, On the mechanism of degradation in Si/SiO_x/Ag metal oxide semiconductor solar cells. J. Appl. Phys. **53** (1998). https://doi.org/10.1063/1.331228
307. F. Rollert, N.A. Stolwijk, H. Mehrer, Solubility, diffusion and thermodynamic properties of silver in Silicon. J. Phys. D: Appl. Phys. **20**, 1148–1155 (1987). https://doi.org/10.1088/0022-3727/20/9/010/meta
308. N. Baber, H.G. Grimmeiss, M. Kleverman, P. Omling, M. Zafer Iqbal, Characterization of silver-related deep levels in silicon. J. Appl. Phys. **62**, 2853–2857 (1987). https://doi.org/10.1063/1.339425
309. U. Wahl, J.G. Correia, A. Vantomme, Location of implanted Ag in Si. Nucl. Instr. Methods Phys. B **190**, 543–546 (2002). https://doi.org/10.1016/S0168-583X(01)01191-0
310. N. Yarykin, J.U. Sachse, H. Lemke, J. Weber, Silver-hydrogen interactions in silicon. Phys. Rev. B **59**, 5551–5560 (1999). https://doi.org/10.1016/S0921-5107(98)00282-7
311. J.U. Sachse, E.Ö. Sveinbjörnsson, N. Yarykin, J. Weber, Similarities in the electrical properties of transition metal-hydrogen complexes in silicon. Mat. Sci. Forum B **58**, 134–140 (1999). https://doi.org/10.1016/S0921-5107(98)00282-7
312. N.T. Son, C.A. Ammerlaan, T. Gregorkiewicz, Magnetic resonance spectroscopy in silver-doped silicon. J. Appl. Phys. **73**, 1797–1801 (1993). https://doi.org/10.1063/1.353188
313. L. Chen, Y. Zeng, P. Nyucanh, T.L. Alford, Silver diffusion and defect formation in Si (111) substrate at elevated temperatures. Mater. Chem. Phys. **76**, 224–227 (2002). https://doi.org/10.1016/S0254-0584(01)00529-6
314. H. Bracht, N.A. Stolwijk, H. Mehrer, Diffusion and solubility of copper, silver and gold in germanium. Phys. Rev. B **43**, 14465–14477 (1999). https://doi.org/10.1103/PhysRevB.43.14465
315. G. Huylebroeck, P. Clauws, E. Simoen, E. Rotsaert, J. Vennik, DLTS of silver in germanium: evidence for an amphoteric impurity. Semicond. Sci. Technol. **4**, 529–535 (1989). https://doi.org/10.1088/0268-1242/4/7/006
316. A. Fazzio, M.J. Caldas, A. Zunger, Electronic structure of copper, silver and gold impurities in silicon. Phys. Rev. B **32**, 934–954 (1985). https://doi.org/10.1103/PhysRevB.32.934
317. N. Lang, G. Pensl, M. Gebhard, N. Achtziger, M. Uhrmacher, Deep level transient spectroscopy on radioactive impurities: Demonstration for Si:^{111}In*. Appl. Phys. A **53**, 85–101 (1991). https://doi.org/10.1007/BF0032386
318. M.A. Gulamova, N.A. Karimova, M.V. Gafurova, Cadmium levels in silicon. Sov. Phys. Semicond. **5**, 687–689 (1971). https://doi.org/10.1002/pssa.2210660168
319. S.S. Dyunaidov, N.A. Urmanov, M.V. Gafurova, The cadmium levels in silicon. Phys. Status Solidi A **66**, K79–K81 (1981). https://doi.org/10.1002/pssa.2210660168
320. A. Näser, W. Gehlhoff, H. Overhof, Identification of cadmium-related centers in silicon. Physica B **273–274**, 279–282 (1999). https://doi.org/10.1016/S0921-4526(99)00504-9

321. E. McGlynn, M.O. Henry, K.G. McGuigan, M.C. doCarma, Photoluminescence study of cadmium-related defects in oxygen-rich silicon. Phys. Rev. B **54**, 14494–14503 (1996). https://doi.org/10.1103/PhysRevB.54.14494
322. R. Sachdeva, A.A. Istratov, P.N.K. Deenapanray, E.R. Weber, Photoluminescence band of Hf origin in hafnium-implanted silicon in the energy range 700 meV to 950 meV. Phys. Rev. B **71**, 195208/1–7 (2005). https://doi.org/10.1103/physrevb.71.195208
323. R. Sachdeva, A.A. Istratov, P.N.K. Deenapanray, E.R. Weber, Electrical properties and diffusion behavior of hafnium in single crystal silicon. Appl. Phys. A **84**, 351–367 (2006). https://doi.org/10.1007/s00339-006-3637-6
324. C.S. Kang, K. Onishi, L. Kang, J.C. Lee, Effects of Hf contamination on the properties of silicon oxide metal-oxide-semiconductor devices. Appl. Phys. Lett. **81**, 518–520 (2002). https://doi.org/10.1063/1.1532755
325. E. Kolawa, J.S. Chen, J.A. Reid, P.J. Pokela, M.A. Nicolet, Tantalum-based diffusion barrier in Si/Cu VLSI metallizations. J. Appl. Phys. **70**, 1369–1373 (1991). https://doi.org/10.1063/1.349594
326. K. Miyata, C.T. Sah, Thermal emission rates and activation energies of electrons at tantalum centers in silicon. Solid-State Electron. **19**, 611–613 (1976). https://doi.org/10.1016/0038-1101(76)90059-9
327. R. Hübner, M. Hecker, N. Mattern, V. Hoffmann, K. Wetzig, H.J. Engelmann, E. Zschech, Comparison of the annealing behavior of thin Ta films deposited onto Si and SiO_2 substrates. Analyt. Bioanalytic. Chem. **379**, 568–575 (2004). https://doi.org/10.1007/s00216-004-2602-5
328. J. Pelleg, N. Goldshleger, Silicide formation in the Ta/Ti/Si system by reaction of codeposited Ta and Ti with Si (100) and (111) substrates. J. Appl. Phys. **85**, 1531–1539 (1999). https://doi.org/10.1063/1.369283
329. M.L. Polignano, F. Barbarossa, A. Galbiati, D. Magni, I. Mica, Tungsten contamination in ion implantation. Nucl. Instr. Methods Phys. Res. B **377**, 99–104 (2016). https://doi.org/10.1016/j.nimb.2016.04.026
330. E. Simoen, K. Saga, H. Vrielinck, J. Lauwaert, Deep levels in W-doped Czochralski silicon. ECS J. Solid State Sci. Technol. **5**, P3001–P3007 (2016). https://doi.org/10.1149/2.00111604jss
331. A. De Luca, A. Portavoce, M. Texier, C. Grosjean, N. Burle, V. Oison, B. Pichaud, Tungsten in silicon. J. Appl. Phys. **115**, 013501/1 (2014).https://doi.org/10.1063/1.4859455
332. H.H. Busta, H.A. Waggener, Precipitation-induced currents and generation-recombination currents in intentionally contaminated silicon p^+n junctions. J. Electrochem. Soc. **124**, 1414–1429 (1977). https://doi.org/10.1149/1.213366
333. S. Boughaba, D. Mathiot, Deep level transient spectroscopy characterization of tungsten-related deep levels in silicon. J. Appl. Phys. **69**, 278–283 (1991). https://doi.org/10.1063/1.347708
334. M.S. Yunusov, R.A. Muminov, G. Nurkuziev, N. Gapparov, A. Kholboev, Photodetectors based on osmium-doped silicon. Semiconductors **33**, 1318–1319 (1999). https://doi.org/10.1134/1.1187916
335. D. Josell, C. Witt, T.P. Moffat, Osmium barriers for direct copper electrodeposition in damascene processing. Electrochem. Solid-State Lett. **9**, C41–C43 (2006). https://doi.org/10.1149/1.2149214
336. S.A. Azimov, B.V. Umarov, M.S. Yumusov, Investigation of the diffusion and solubility of iridium in silicon. Sov. Phys. Semicond. **10**, 842–843 (1976)
337. A. Rodríguez, C. González, T. Rodríguez, A. Kling, M.F. da Silva, J.C. Soares, RBS characterization of the iridium diffusion in silicon. Nucl. Instr. Methods Phys. Res. B **161–163**, 663–667 (2000). https://doi.org/10.1016/S0168-583X(99)00922-2
338. H. Lemke, Substitutional transition metal defects in silicon grown-in by the float zone technique. Electrochem. Soc. Proc. **96**(13), 272–286 (1996)
339. V. Benda, M. Cernik, D. Stepkova, Deep energy levels in power diodes introduced by iridium diffusion. Microelectron. J. **29**, 695–699 (1998). https://doi.org/10.1016/S0026-2692(98)00035-4

340. J. Bollmann, S. Knack, J. Weber, Iridium-related deep levels in n-type silicon. Phys. Status Solidi B **222**, 251–260 (2000). https://doi.org/10.1002/1521-3951(200011)222: 1 < 251:: AID-PSSB251 > 3.0.CO;2-9
341. S. Petersson, J.A. Reimer, M.H. Brodsky, D.K. Campbell, F. d'Heurle, B. Karlsson, P.S. Tove, $IrSi_{1.75}$ a new semiconductor compound. J. Appl. Phys. **53**, 3342–3343 (1982). https://doi.org/10.1063/1.330999
342. M. Wittmer, P. Oelhafen, K.N. Tu, Chemical reaction and Schottky-barrier formation at the Ir/Si interface. Phys. Rev. B **35**, 9073–9084 (1987). https://doi.org/10.1103/PhysRevB.35.9073
343. C.E. Allevato, C.B. Vining, Phase diagram and electrical behavior of silicon-rich iridium silicide compounds. J. Alloys Compounds **200**, 99–105 (1993). https://doi.org/10.1016/0925-8388(93)90478-6
344. G. Curello, R. Gwilliam, M. Harry, R.J. Wilson, B.J. Sealy, T. Rodriguez, F.J. Jimenez-Leube, Iridium silicide formation on high dose Ge^+ implanted Si layers. Proc. Mat. Res. Soc. **402**, 411 (2001). https://doi.org/10.1557/PROC-402-411
345. S. Gaudet, C. Detavernier, A.J. Kellock, P. Desjardins, C. Lavoie, Thin film reaction of transition metals with germanium. J. Vac. Sci. Technol., A **24**, 474–485 (2006). https://doi.org/10.1116/1.2191861
346. A. Habanyama, C.M. Comrie, Study of iridium/germanium interaction in a lateral diffusion couple. Thin Solid Films **516**, 5137–5143 (2008). https://doi.org/10.1016/j.tsf.2008.01.005
347. C.H. Mullet, B.H. Stenger, A.M. Durand, J.A. Morad, Y. Sato, E.C. Poppenheimer, S. Chiang, Growth and phase transformations of Ir on Ge(111). Surf. Sci. **666**, 96–103 (2017). https://doi.org/10.1016/j.susc.2017.08.021
348. S. Shirvaliloo, H. Kangarloo, Produciton of iridium metal thin films for application as electrodes in DRAMs and FRAMs. J. Electr. Electron Eng. **3**, 35–38 (2015). https://doi.org/10.11648/j.jeee.s.2015030201
349. P.L. Jones, J.W. Corbett, Investigation of the electrical degradation of silicon Schottky contacts due to mercury contamination. Appl. Phys. Lett. **55**, 2331–2333 (1989). https://doi.org/10.1063/1.102052
350. G. Huylebroeck, P. Clauws, E. Simoen, J. Vennik, DLTS of nickel impurities in germanium. Solid State Commun. **82**, 367–369 (1992). https://doi.org/10.1016/0038-1098(92)90370-O
351. I.M. Kotina, V.V. Kuryatkov, S.R. Novikov, T.I. Pirozhkova, Capacitance spectroscopy of deep Cu, Au, Ag, and Ni centers in germanium. Sov. Phys. Semicond. **21**, 635–637 (1987)
352. F.X. Zach, H.G. Grimmeiss, E.E. Haller, Nickel related deep levels in germanium. Mat. Sci. Forum. **83–87**, 245–250 (1992). https://doi.org/10.4028/www.scientific.net/MSF.83-87.245
353. A. Giese, H. Bracht, N.A. Stolwijk, H. Mehrer, Diffusion of nickel and zinc in germanium. Defect Diff. Forum **143–147**, 1059–1066 (1997)
354. H.H. Woodbury, W.W. Tyler, Triple acceptors in germanium. Phys. Rev. **105**, 84–92 (1957). https://doi.org/10.1103/PhysRev.105.84
355. J.R. Davis, A. Rohatgi, R.H. Hopkins, P.D. Blais, P. Rai-Choudhury, R. McCornick, H.C. Mollenkorf, Impurities in silicon solar cells. IEEE Trans. Electron. Dev. **27**, 677–687 (1980). https://doi.org/10.1109/T-ED.1980.19922
356. T. Tilly, H.G. Grimmeiss, H. Pettersson, K. Schmalz, K. Tittelbach, H. Kerkow, Electrical and optical properties of vanadium-related centers in silicon. Phys. Rev. B **44**, 12809–12814 (1991). https://doi.org/10.1103/PhysRevB.44.12809
357. H. Lemke, Eigenschaften repulsiver Zentren in Silizium. Phys. Status Solidi A **66**, 641–649 (1981). https://doi.org/10.1002/pssa.2210660216
358. H. Nakashima, K. Hashimoto, Deep impurity levels and diffusion coefficient of manganese in silicon. J. Appl. Phys. **69**, 1440–1445 (1991). https://doi.org/10.1063/1.347285
359. Y.A. Zibuts, L.G. Paritskii, S.M. Ryvkin, Some properties of silicon containing mercury, tungsten, molybdenum and platinum impurities. Sov. Phys. Solid State **5**, 2416–2419 (1964)
360. Y. Fujisaki, T. Ando, H. Kozuka, Y. Takano, Characterization of tungsten related deep levels in bulk silicon crystals. J. Appl. Phys. **63**, 2304–2306 (1988). https://doi.org/10.1063/1.341045

361. T. Ando, S. Isomae, C. Munakata, T. Ane, Deep level transient spectroscopy on p-type silicon crystals containing tungsten impurities. J. Appl. Phys. **70**, 5401–5403 (1991). https://doi.org/10.1063/1.350196
362. H. Lemke, Dotierungseigenschaften von Silber in Silizium. Phys. Status Solidi A **94**, K55–K59 (1986). https://doi.org/10.1002/pssa.221094017
363. A.F. Pasquevich, R. Vianden, Temperature-dependence of the hyperfine interactions of ^{111}Cd in silicon. Phys. Rev. B **35**, 1560–1565 (1987). https://doi.org/10.1103/PhysRevB.37.10858
364. M. Quevedo-Lopez, M. El-Bouanani, S. Addepalli, J.L. Duggan, B.E. Gnade, R.M. Wallace, M.R. Visokay, M. Douglas, L. Colombo, Hafnium interdiffusion studies from hafnium silicate into silicon. Appl. Phys. Lett. **79**, 4191–4194 (2001). https://doi.org/10.1063/1.1425466

Chapter 7
Impact of Metals on Silicon Devices and Circuits

It has been clearly established in the previous Chapters that even a small concentration of metallic contamination can dominate the generation and recombination properties of group IV semiconductor materials. It has also been shown that metals can degrade the generation properties of simple devices, like a MOS capacitor (Sect. 4.3.8). Deep level centers present in the depletion region of a device will impact on the generation lifetime and, hence, on the leakage current. At the same time, one can benefit from this sensitivity of certain types of devices to metal-related effects and design test methodologies relying on simple structures close to the sensitive areas. The advantage is that these structures undergo the same processing steps as the final products, so that the obtained information can be more relevant to assess the defect origin of circuit failure and yield. In fact, quite often, devices are sensitive to levels of contamination which can hardly be detected by other established techniques, like DLTS. Ultimately for extremely scaled devices—e.g., a pixel of an image sensor—a single metal atom can be detected through its contribution to the dark current or the low-frequency noise.

This Chapter overviews the possible impact of metal impurities on the operation, reliability and yield of devices and circuits. In a first part, the presence of metal-related centers in a MOS capacitor will be described. Both dissolved atoms and precipitates can cause deleterious effects on the MOS characteristics, giving rise to excess gate leakage, hysteresis and instability and in the worst case, early voltage breakdown (failure). A second part will have a look on p-n junctions and Schottky barriers. Again, excess leakage current can be generated in reverse operation and soft breakdown characteristics may be obtained. In forward operation, excess recombination mediated by metal-related deep levels gives rise to an ideality factor between 1 and 2. A third part considers the impact on transistors and circuits, including CCDs and CMOS Imager Sensors (CIS).

As a final remark, it should be mentioned that the recombination properties of specific metals, like, substitutional Au and Pt have been beneficially used in the past for certain device applications. The corresponding recombination centers can assist in the fast recombination of excess carriers when switching a silicon power device

C. Claeys and E. Simoen, *Metal Impurities in Silicon- and Germanium-Based Technologies*, Springer Series in Materials Science 270,
https://doi.org/10.1007/978-3-319-93925-4_7

from the on-state to the off-state (see, e.g., [1]). In the meantime, better results have been obtained by introducing H- or He-implantation-induced deep levels, as the location of the recombination centers can be controlled better than by metal in-diffusion. Therefore, these examples will not be considered in this Chapter.

7.1 MOS Capacitors

Historically speaking, one of the first issues related to metals in SiO_2 was the presence of mobile alkali ions (Na, K, Ca, …) originating among others from the high purity water [2–4]. As these ions are mobile in the electrical field existing in the gate dielectric, they can respond to a gate voltage sweep and give rise to hysteresis in the capacitance, Basically, the presence of these mobile charges changes the surface potential, causing a drift with gate bias [2]. This issue is nowadays well under control and does not affect markedly the characteristics of thin gate oxides, employed in state-of-the-art transistors, However, according to Chap. 3 , there is still the possibility of metal contamination during device processing (surface cleaning, ion implantation, RTP, …) which can become incorporated in a grown or deposited oxide (or nitride) or segregates at the interface with the substrate. In addition, some process steps rely on the application of specific metals (source/drain and gate salicidation; copper metallization (CMOS); 3D integration using copper Through-Silicon-Vias (TSVs); Ag and Cu metallization in solar cells) in the process flow. This means that there is a certain risk of in-diffusion of these metals in the substrate or other parts of the active devices, where they can be harmful.

MOS devices can be affected in different ways by metal contamination. Firstly, when metals are present at the semiconductor/dielectric interface, the surface recombination and generation velocity will increase, as a consequence of the higher density of interface states that is created by individual atoms or precipitates. The presence of precipitates at the surface will also affect the oxidation, resulting in a locally thinner dielectric, a higher gate leakage current (I_G) by direct or Fowler-Nordheim (FN) tunneling and a lower breakdown voltage V_{BD}. Metals can segregate in the oxide, creating local trap centers, mobile or fixed charges, resulting in a hysteresis of the C-V characteristics and a shift in the flat-band voltage V_{FB}. Finally, in extreme cases they can nucleate Oxidation-Induced Stacking Faults (OISF) and at the same time activate these extended defects [5]. All these effects have an impact on the performance and the reliability of a MOS capacitor, which is an essential part of a MOSFET. In this section, the effect of different TMs (Fe, Ni, …) on the behavior of MOS capacitors will be discussed. As an example impurity, copper will be treated separately, given its relevance for today's Back-End-Of-Line (BEOL) metallization both in microelectronics and photovoltaics.

7.1.1 *Impact of Metal Contamination on MOS Capacitors*

In order to understand the likelihood of metal contamination in gate dielectrics it is important to have information on their diffusion, precipitation and segregation behavior in SiO_2 and other dielectrics commonly used in device processing. An important remark is that since these are amorphous materials, diffusion can generally not be described in terms of the mechanisms established for crystalline materials in Sect. 2.2. First, a summary of the available literature data will be given, followed by a description of the behavior of iron and nickel in SiO_2 and other dielectric layers.

7.1.1.1 Diffusion, Precipitation and Segregation of Metals in Dielectric Layers

Table 7.1 gives a summary of the available data for the diffusion coefficient of some TMs in different types of insulating materials. D is described by an Arrhenius law like (2.2), with D_0 the pre-exponential factor from the intercept and H the activation enthalpy derived from the experimental slope. It can be noted that the latter parameter decreases with the size of the metal species, which suggests that different types of point defects can be involved in the diffusion, i.e., oxygen vacancies or metal interstitials. Sodium clearly stands out as the fastest diffuser in SiO_2. Data for copper will be further discussed in the following section.

With respect to the diffusion mechanism of iron in SiO_2, it has been concluded that the positive oxygen vacancies cause an ambipolar type of diffusion of negatively charged Fe, by Coulomb attraction [6]. The ambipolar diffusion is governed by the slowest of the two ions, most likely V_O^+ [7]. Iron has a relatively small diffusivity through SiO_2 so that the question raises whether it can be used as a masking agent against it. It was shown that this is the case for a 120 nm thick thermally grown layer (dry oxide at 850 °C) up to about 900 °C [6]. Beyond that, the D_{Fe} becomes too high to effectively block the in-diffusion [6]. The diffusion of Fe in the buried oxide of bonded and etched-back Silicon-On-Insulator (BESOI) and Separation by Implanted Oxygen (SIMOX) wafers has also been investigated

Table 7.1 Typical diffusivities and activation enthalpies of different metal impurities in SiO_2 and other dielectrics (after Ramappa and Henley [6])

Metal	Oxide	D_0 (cm²/s)	H (eV)	D (@ T °C) (cm²/s)	References
Au	SiO_2	1.52×10^{-7}	2.14	4.6×10^{-16} (1000)	[235]
Cu	SiO_2		1.82	1.2×10^{-11} (450)	[30]
Ni	SiO_2	7.2×10^{-10}	1.61	2.5×10^{-16} (1000)	[16]
Na	SiO_2	6.9	1.3	5.2×10^{-5} (1000)	[236]
Fe	Al_2O_3	9.2×10^{-8}	1.17		[237]
Fe	SiO_2	4×10^{-8}	1.51	3.8×10^{-14} (1000)	[6]

[7]. It was found that a diffusivity on the order of 10^{-15} to 10^{-14} cm^2/s was obtained for BESOI wafers in the temperature range of 900–1050 °C. The corresponding activation energy was 2.8 eV [7]. The diffusivity at 1000 °C is sufficiently high for enabling Fe gettering by the residual oxygen ion implantation damage, present below the buried oxide in SIMOX wafers [8]. Numerical modeling of the diffusion of Fe in SiO_2 has also been performed [9], based on a Monte Carlo methodology of the random walk diffusion model with a mean free path which is dependent on the concentration of oxygen atoms in excess. It was, namely, observed that the diffusion of Fe is reduced in an oxidizing ambient at 1000 °C [9], whereby oxygen acts as a trap or diffusion barrier for iron.

It was also observed that Fe has the tendency to aggregate in the oxide near the interface. The underlying segregation coefficient k_{Fe} is shown in Fig. 7.1 and corresponds with [6]:

$$k_{Fe} = 3.76 \exp\left(\frac{-1.96\ (\text{eV})}{k_B T}\right) \tag{7.1}$$

This segregation hinders the in-diffusion of iron in the bulk silicon and may lead to a supersaturation and precipitation in the oxide. This can cause gate oxide breakdown, reducing the Gate Oxide Integrity (GOI) [6]. A ratio of 1.3 between the iron concentration in SiO_2 and silicon has been derived from instrumental NAA [10].

Iron pile-up at the Si–SiO_2 interface was also observed after in-diffusion from the backside of the wafer [11]. It was furthermore found that the iron entered the oxide region near the interface, possibly to reduce SiO_2. The formation of

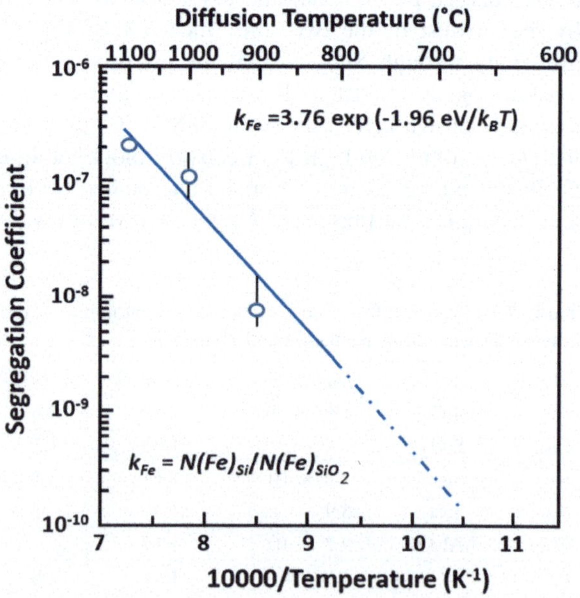

Fig. 7.1 Temperature dependence of iron impurity segregation coefficient at the SiO_2–Si interface. k_{Fe} is defined as the ratio of the Fe concentration in the silicon wafer adjacent to the interface $N(Fe)_{Si}$ to the Fe concentration in the SiO_2 close to the interface $N(Fe)_{SiO_2}$. Reproduced from Ramappa and Henley [6] by permission of ECS—The Electrochemical Society

crystalline α-$FeSiO_2$ and β-$FeSi_2$ in SiO_2, after hot iron implantation at 300, 600 and 800 °C has been investigated by TEM [12]. In the 800 °C implantation case, most of the Fe segregated either just on the SiO_2 surface or at the Si/SO_2 interface, the latter in line with previous observations [6]. The precipitation of Fe in silicon with surface or bulk contamination has been investigated after dry thermal oxidation at 900 °C [13]. The precipitating phase is β-$FeSi_2$, independent of the quench rate or the surface Fe concentration. Due to the lattice mismatch with silicon, the formation of β-$FeSi_2$ is accompanied by the creation of Fe vacancies and the absorption of silicon interstitials. The oxidation thus promotes the formation of the equilibrium β phase, as this injects interstitials in the substrate [13]. Most likely, the nucleation barrier is reduced but not necessarily the growth kinetics, as the size of the particles stays the same (diameter ~ 10 nm), irrespective of the Fe concentration. The interface precipitate density is proportional to the surface Fe concentration. These precipitates correlate with the presence of sharp pyramidal asperities at the interface, resulting in a local thinning of the oxide. This could significantly contribute to the local early breakdown of the oxide [13]. Finally, it has been noted that thermal oxidation is initially enhanced for Fe-contaminated n-type Si(100) wafers [14]. After sufficiently long oxidation, an Fe_2O_3 layer is formed which blocks the in-diffusion of oxygen and retards the oxidation process.

A combined SIMS, TEM and Atom Probe Tomography (APT) study of silicon implanted with Fe, followed by thermal oxidation at 900 or 1100 °C has led to a detailed insight in the formation of the pyramidal defects, as sketched in Fig. 7.2 [15]. In a first stage, an iron silicide precipitate is formed below the Si/SiO_2 interface. In stage 2, both the precipitate and the oxidation proceeds, while the

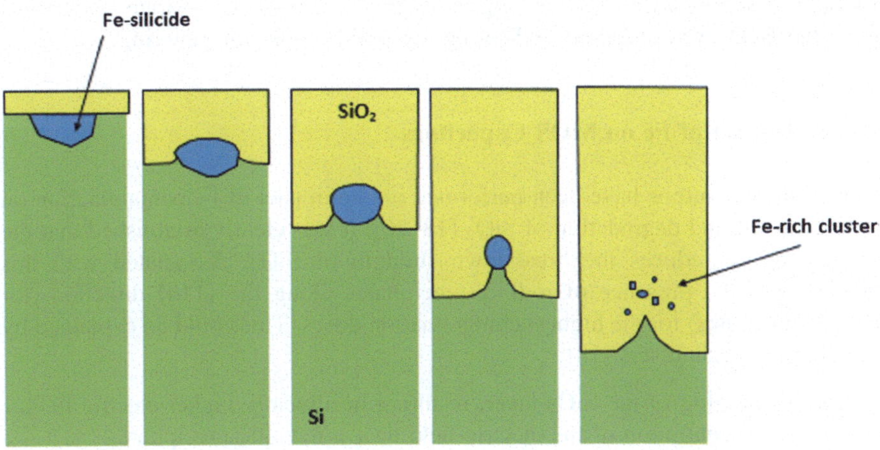

Fig. 7.2 Schematic representation of the formation of a pyramidal defect during thermal oxidation of an Fe-contaminated Si wafer. The different phases are described in the text (after De Luca et al. [15])

oxide front moves forward. In a third stage, the morphology of the precipitate changes, with a hindering effect for the silicon matrix. In phase 4, Fe enrichment of the particle occurs and formation of a silicon jag below the precipitate at the Si/SiO_2 interface. Finally, the full oxidation of the iron precipitate leaves Fe-rich sub-nanometric clusters in the oxide layer on top of a pyramidal defect (stage 5).

In the case of Fe ion-implanted SiO_2, it has been found that vacuum annealing with an electron beam forms iron nanoparticles at the surface of the SiO_2 layer [9]. In that process, Si atoms are substituted by iron in local tetrahedral coordination, giving rise to small nuclei above a certain concentration threshold. Under e-beam annealing enhanced desorption of the SiO_2 surface induces protruding Fe nanoparticles from the surface [9]. On the contrary, fast diffusion under vacuum annealing and standard furnace annealing results in the loss of Fe in the near surface region in favor of segregation at the SiO_2/Si interface.

For nickel in SiO_2, it was concluded that the same type of point defect was responsible for its diffusion, irrespective of the type of layer, i.e., thermally grown or deposited at high temperature [16]. Moreover, in the same work the diffusion coefficient of Ni has been determined in Chemical Vapor Deposited Si_3N_4 at 1320 K (SiH_4 and NH_3) yielding a value:

$$D_{Ni} = 1.8 \times 10^{-6} \exp\left(\frac{-2.11 \pm 0.05\ (\mathrm{eV})}{k_B T}\right) \tag{7.2}$$

At the same time, it was observed that the diffusivity strongly depended on the stoichiometry of the nitride layer and on the hydrogen content [16]. In the case of SIMOX wafers with a BOX thickness up to 320 nm, it was concluded that both Cu and Ni can diffuse through it down to 600 °C (30–60 min annealing) [17]. Pronounced segregation of these impurities in the damaged substrate region just below the BOX was observed, indicating a possible route for gettering.

7.1.1.2 Impact of Fe on MOS Capacitors

Detailed investigations have been performed on the impact of Fe contamination on the breakdown and degradation of SiO_2 [18–20]. It was clearly established that the presence of Fe reduces the breakdown strength of SiO_2. Associated with this behavior was the presence of rodlike precipitates along the $\langle 110 \rangle$ direction (for (100) Si substrates) for the highest contamination doses. This could be explained by a number of mechanisms [18]:

- A local thinning of the SiO_2 layer, resulting in a locally higher electric field,
- A local concentration of the electric field near a metallic precipitate,
- A reduction of the local barrier height for tunneling when a $FeSi_2$ precipitate is present.

Further studies revealed that there appears to be a critical Fe concentration for the onset of oxide breakdown, which increases with smaller t_{ox} [19]. Based on

TCAD, a model was derived expressing the local breakdown field in terms of the defect size, according to:

$$F_{max}^{loc} = \frac{V_G}{t_{ox}}(1 + \alpha_{pr} d_{pr}) \tag{7.3}$$

In (7.3), d_{pr} is the size of the $FeSi_2$ precipitate and α_{pr} is a constant equal to 7×10^6 cm^{-1}. The simulations indicated that the local breakdown field is higher than can be expected from oxide thinning alone; the enhancement is mainly due to the presence of a conductive phase at the interface and resulting in an increase in F_{ox} [19].

An expression has been proposed for the critical Fe concentration for oxide breakdown with a thickness in the range of 30–5 nm [20]:

$$[Fe]_{crit} = 1.53 \times 10^{11} 10^{0.12 t_{ox}} \tag{7.4}$$

When studying ultrathin oxides, a much weaker impact has been observed [20]. While 5 nm oxides are still sensitive to an Fe density of 4×10^{10} cm^{-3}, much less effect is seen for 3 and 4 nm oxides.

7.1.1.3 Impact of Nickel on MOS Capacitors

Nickel has a tendency to precipitate at the wafer surface in case of slow enough cooling, giving rise to so-called haze. These surface $NiSi_2$ precipitates are electrically active and will increase the surface recombination velocity. This also has a detrimental impact on the operation of MOS-based devices [21, 22]. It has been found that Ni impurities introduce weak spots in thermally grown gate oxide (SiO_2) [21]. As a consequence, the gate leakage current is locally higher and the breakdown voltage is reduced at these sites. A correlation has been found between the presence of Ni impurities, oxide degradation and increased surface roughness. The surface roughness directly affects the gate leakage current, as it leads to a higher density of interface states. The latter increases with the surface step density [21]. For thermal oxidation after Ni contamination by ion implantation, it was found that nickel precipitates penetrate both the silicon substrate and the oxide. The model for the oxide degradation is then related to the local thinning of the SiO_2 and/or the local electric field enhancement by the presence of a conductive phase. The density of precipitates and the associated thermal oxide degradation increase with the Ni implantation dose [21].

The surface segregation of nickel incorporated during crystal pulling of p-type Cz silicon is more pronounced for a boron doping density in the range of 5.5–7.4×10^{14} cm^{-3} than for a doping density of 1.1–1.7×10^{15} cm^{-3} [23]. This is evaluated by the impact on the GOI of 20 nm thermal oxide, grown at 900 °C. A possible explanation relies on the formation of Ni-B pairs, reducing both the surface segregation of Ni and the formation of nickel-related GOI defects at the Si/SiO_2 interface for higher doped wafers.

7.1.2 *Impact of Copper on MOS Capacitors*

Copper may have a detrimental effect on the integrity and the reliability of gate oxides. The presence of Cu^+ ions results in positive charge which affects the flat-band voltage of an MOS capacitor and the drift of copper ions in the gate oxide field results in drift and hysteresis in the *C-V* characteristics. The presence of copper may also yield a higher leakage current, a lowering of the breakdown voltage and early failures, resulting in a reduced GOI. Early reports indicated that Cu precipitates near the Si/SiO_2 interface were responsible for early breakdown [5, 24–29]. This triggered immediately more fundamental studies on the transport properties of copper through SiO_2 and on its impact on oxidation. In this part, the effect of copper contamination before (pre-) or after gate oxidation, during typical BEOL metallization on the properties of MOS capacitors will be described. Understanding of the observed phenomena is based on a deeper insight in the fundamental mechanisms. In the first two paragraphs of this part, the basic properties of Cu in SiO_2 and in low-κ dielectrics will be summarized. Next, the impact of Cu on the performance of MOS capacitors will be given, followed by a description of the effect on the properties of so-called Interlayer Dielectrics (ILD) between two metallization levels.

7.1.2.1 Basic Properties of Copper in SiO_2

The basic properties of Cu in SiO_2 have been studied to some detail [25, 30]. It has been established that the transport of Cu in SiO_2 is accelerated under a positive field, indicating that it diffuses as Cu^+, with an activation energy of 1.82 eV at a field of +1 MV/cm [30]. Based on the experimental behavior, a model has been established for the copper diffusion profile in SiO_2 during bias-temperature stress (BTS) [31]. The basic idea is to combine the copper flux from thermal diffusion, becoming important at low fields, with the drift in the oxide field. This results in the following expression [31]:

$$J = -D_{Cu-ox} \cosh \frac{qb_{Cu}F}{k_B T} \frac{\partial [Cu]}{\partial x} + \frac{2D_{Cu-ox}}{b} \sinh\left(\frac{qb_{Cu}F}{2k_B T}\right)[Cu] \tag{7.5}$$

where $D_{Cu\text{-}ox}$ is the diffusivity of Cu in SiO_2, F is the electric field and b_{Cu} is the jump distance of a Cu ion (=0.27 nm) and x is the depth in SiO_2 from the Cu/SiO_2 interface (gate). For $F \ll 2k_BT/qb_{Cu}$, (7.5) reduces to the well-known 1-D Fick's law for diffusion:

$$J = -D_{Cu-ox} \frac{\partial [Cu]}{\partial x} + \frac{D_{Cu-ox} q F}{k_B T}[Cu] \tag{7.6}$$

Combined with Fick's second law ($\partial[\text{Cu}]/\partial t = -\partial J/\partial x$) the diffusion of Cu can be divided in two regimes: a constant field regime and a variable field regime [31]. This is related to the fact that Cu^+ changes the potential distribution in the oxide and, hence, the local electric field F. Based on this, a good agreement between calculated and measured copper profiles in SiO_2 has been obtained.

The behavior of copper contamination during thermal oxidation has been studied as well [32–36], showing a tendency for copper to accumulate at the Si/SiO_2 interface. This is due to its high diffusivity and low solubility in SiO_2 at high temperatures. Moreover, when present in the silicon substrate, copper will also be gettered at or near the interface upon cooling because of the reduced solubility [32]. The heat of formation of CuO_2 is 1.74 eV/atom, which is much lower than for SiO_2. During oxidation, oxygen interacts preferably with Si forming SiO_2 instead of CuO_2. Copper also does not form a network in SiO_2 [32], because the solubility is equal to the vapor pressure of Cu and, hence, is low. Upon oxidation at high temperatures (1000 °C) copper is driven out of the forming SiO_2 and segregates at the Si/SiO_2 interface. It has been shown that the in-diffusion of copper from the surface during oxidation is orders of magnitude slower than expected from its interstitial diffusion coefficient. This is interpreted in terms of a parallel diffusion of interstitial and substitutional copper, whereby the slow Cu_s component gives rise to the copper-enriched near surface region [33].

In another study, the diffusion of copper from the backside of the wafer towards the Si/SiO_2 interface was investigated at typical back-end of line processing temperatures ($T \leq 400$ °C) [35]. The degradation of the gate oxide was relatively small, resulting in a V_T shift of 1–15 mV at 450 °C for 5 h. It was also found that the effect of copper greatly depends on the presence of a backside SiO_2 layer [35]. The diffusion behavior at 400 °C annealing varies strongly with the back-surface contamination level and depends on the frontside SiO_2 thickness. It was observed that most of the copper atoms at the contaminated surface were unable to diffuse into the silicon but remained on the back surface [35]. The impact of Cu on the backside does not depend on its concentration when it exceeds 10^{12} cm^{-2} at 400 ° C, which is determined by the solubility at that temperature. At high $[Cu]_{back}$, most copper precipitates, forming a self-gettering layer and, thus, reducing the flux of Cu_i^+ in-diffusing in the substrate. However, the presence of a backside SiO_2 layer increases the Cu in-diffusion and $[Cu]_{surf}$ at the front side. This is related to the fact that the copper solubility in SiO_2 is much higher than in Si at 400 °C, so that a concentration gradient builds up without precipitation [35].

In addition, the Cu diffusion to the surface also depends on the front oxide thickness. For a $t_{ox} \leq 3$ nm no difference with a bare silicon surface has been observed [35]: the highest copper concentration is found on top of the SiO_2, with a strong gradient towards the substrate. On the other hand, for a $t_{ox} \sim 6$ nm, no copper was found on the surface of the front oxide even after 4 h annealing. During subsequent room temperature storage, the copper in the thin oxide redistributes to homogenize the copper profile. In simple terms, the front SiO_2 acts like a switch which determines whether copper will pass (≤ 3 nm) or not (~ 6 nm). This behavior depends on the electrical activity of the SiO_2 layer: at 400 °C, electrons generated by thermal energy can tunnel through a 3 nm oxide, thereby charging the SiO_2 surface negatively. This attracts Cu_i^+ ions and facilitates its diffusion through the thin oxide up to the surface [35].

It has been shown that MOS capacitors with a Cu gate exhibit some specific leakage mechanisms under 300 °C and 1 MV/cm BTS [36]. The presence of copper in the oxide results in an increased conductivity, which allows the leakage of minority carriers, generated at the inversion layer. Recently, a method has been proposed to mimic the unintentional metal contamination by a corona discharge. The metals are driven in from the top of the gate oxide by the plasma mimic (PM) and a 170 °C thermal activation (corona—temperature stress) [37]. Applying this method, it was concluded that Co has a different degradation trend compared with Ni, Fe and Cu [37]. Moreover, trace metals applied post-oxidation lower the barrier for tunneling whatever the contamination method. The data suggest that Cu, Ni and Fe must reach the Si/SiO_2 interface to degrade the tunneling (70 nm thermal oxide). Due to the large electronegativity of the TMs, they do not react with SiO_2. In the case of Co, CoO is formed, which may lead to a degradation of the oxide surface [32].

It has been observed that the presence of Cu_3Si catalyzes the oxidation of silicon at room temperature [38–40]. This may cause a local thickening of SiO_2 and has an impact on the local GOI and leakage current.

7.1.2.2 Diffusion of Cu in Low-κ Dielectrics

In order to understand the degradation of ILD, diffusion studies of copper in several types of layers have been performed in the past [41–48]. The copper diffusion coefficient has been determined either by chemical techniques (SIMS; radiotracer method; …) or using electrical parameters (recombination lifetime; flat band voltage V_{FB} of a Cu/ILD/silicon capacitor). Some of the results have been summarized in Table 7.2. The flat-band voltage method is related to the charge Q_{Cu} associated with Cu ions in the dielectric and is also related with the increase in the low-field leakage current density J_G through the layer [48], according to:

$$\Delta V_{FB[Cu^+]} = \frac{Q_{Cu}(t)}{C_{ox}} \sim \int_0^t J_G(\tau)\, d\tau \tag{7.7}$$

In addition, the charge to breakdown is defined by [48]:

$$Q_{BD} \sim \int_0^{TTF} J_G(\tau)\, d\tau \tag{7.8}$$

Table 7.2 Diffusion data of copper in ILD

Material	D_{Cu} (cm^2/s)	H (eV)	F (MV/cm)	Method	References
4% P-Si glass	5.3×10^{-11}	0.5	n.a.	Radiotracer	[41]
SiN:16 at.% H	4×10^{-6}	1.1	n.a.	Radiotracer	[41]
SiO_2 (thermal)	–	1.2	n.a.	AAS/SIMS	[42]
PSG/SiO_2	–	1.6	n.a.	AAS/SIMS	[42]
PE-SiO	–	1.30	2 (low)	TTF[a]	[43]
PE-SiN	–	1.85	2 (low)	TTF[b]	[44]
PECVD oxynitride	–	1.39	+1.0	ΔV_{FB}^{c}	[45]
PECVD oxide	–	1.13	+1.0	ΔV_{FB}	[45]
Methyl-doped SiO_2	–	0.76	+1.5	ΔV_{FB}	[48]
Thermal SiO_2	2.22×10^{-3}	1.54	+1.0		[49]
PECVD oxynitride	3.09×10^{-5}	1.34			[49]
PECVD SiO_2	2.59×10^{-5}	1.18			[49]
Methyl-doped SiO_2	6.06×10^{-9}	0.71			[49]
Porous SiCOH	6.01×10-9	0.71			[50]

[a]Time to failure of MOS structure with Cu gate
[b]Strongly field dependent
[c]Flat-band voltage shift of a MOS capacitor with Cu gate

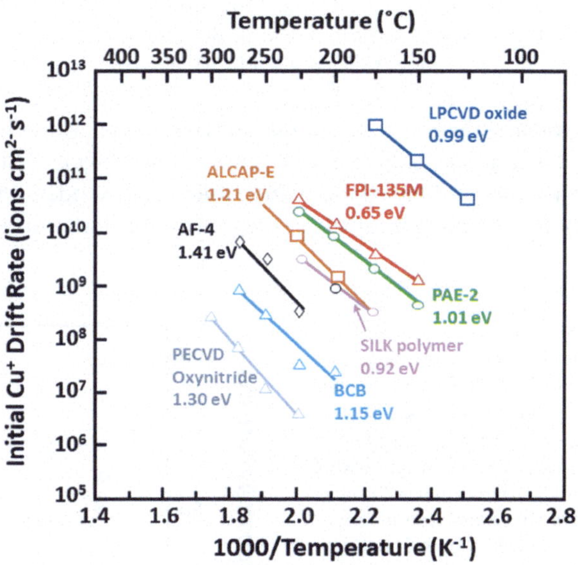

Fig. 7.3 Arrhenius plot of initial Cu^+ drift rates in various dielectrics. Drift rates were extracted from I_G-V_G measurements at an electric field of 0.8 MV/cm, except for PECVD oxynitride which was extracted at a field of 1.0 MV/cm. Reproduced from Loke et al. [46]. Copyright 1999 IEEE

For example, Fig. 7.3 represents the Arrhenius plots for a number of low-κ polymer ILD derived from BTS measurements, showing E_a values in the range of 0.85–1.41 eV [46]. In general, SiN-based dielectrics form a good diffusion barrier for Cu, compared with SiO_2 [44].

Based on the model of (7.6) and (7.7) for Cu transport in a dielectric layer, copper diffusivities have been extracted from *C-V* measurements, using the negative V_{FB} shift [49]. The activation energy for Cu diffusion in a dielectric is composed of the sum of the network strain energy generated by the enlargement of the network due to the diffusing ions and by the electrostatic energy between Cu^+ and the dielectric lattice. The activation energy should have a large value for dielectric materials with a large shear modulus, a dense network structure and a low dielectric constant [49]. This probably explains the ordering for D_{Cu}: SiN > SiON > SiO_2 > SiOCH. In the cases of porous dielectrics, other mechanisms due to structural porosity can occur and give rise to a non-Arrhenius type of behavior [49].

A model has been established for the diffusion of Cu through porous low-κ dielectrics, accounting for the amount of porosity and its shape/distribution [50]. The model reproduces the reduction of the copper drift with increasing porosity, as there are less vertical paths for unhindered transport. The highest transport time was found for laterally elongated pores, while vertical ones do not yield an efficient obstruction for copper transport [50].

7.1.2.3 Impact of Cu on MOS Capacitors

As noted before, the presence of copper contamination on the silicon surface will lead to the formation of Cu_3Si precipitates, which degrades the GOI [25]. As shown by Fig. 7.4, a drastic increase in the gate oxide failures occurs around $>5\times10^{12}$ at/ cm^2 for pre-annealed wafers (1150 °C for 1 h in N_2—well drive-in). The higher breakdown rate is explained by assuming that when a precipitate is present before oxidation, it penetrates in the growing oxide, as the local oxidation rate is smaller. This results in a thinner and, hence, weaker oxide [25]. While the results of Fig. 7.5 have been obtained on rather thick gate oxides, the results on thinner oxides (<10 nm) are not always straightforward and sometimes contradictory. In some

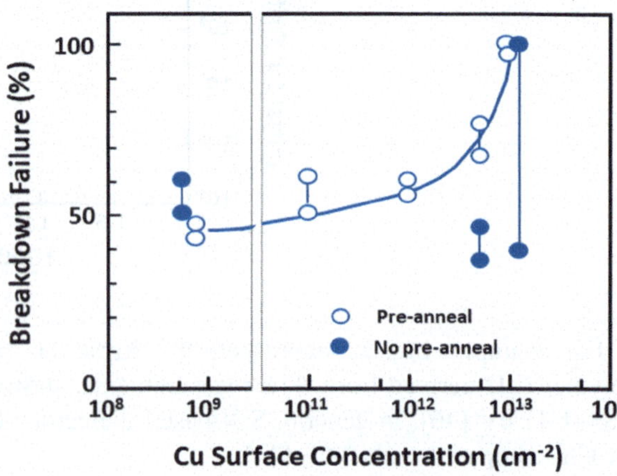

Fig. 7.4 Dependence of breakdown failure of 25 nm thermal SiO_2 on surface metal contamination. Open circles and filled circles are samples after pre-annealing at 1150 °C for 1 h in N_2 and without pre-annealing, respectively. Reproduced from Hiramoto et al. [25]. Copyright 1989 The Japan Society of Applied Physics

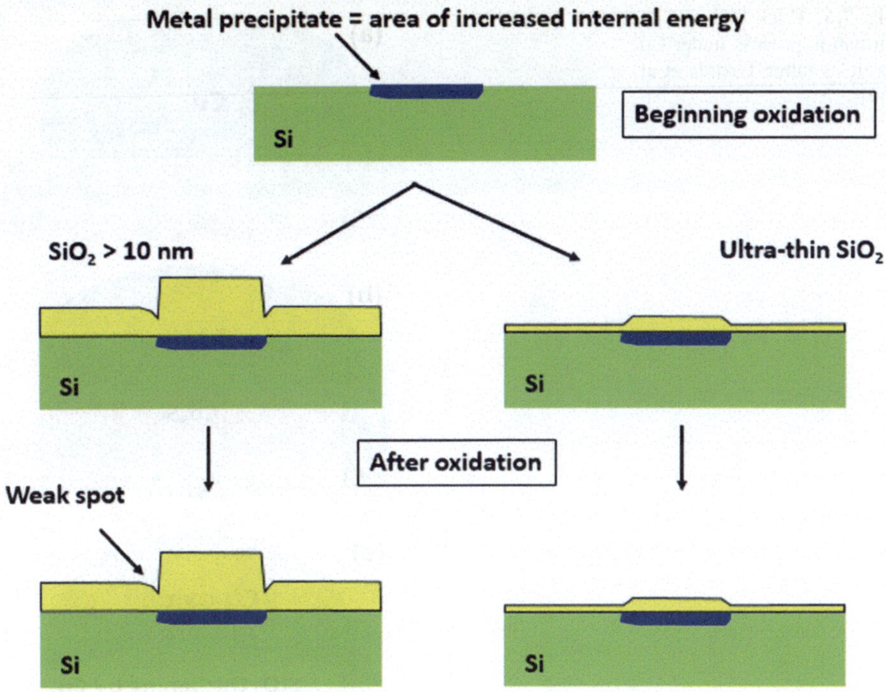

Fig. 7.5 Model to explain the extraordinary robustness of ultrathin gate oxides against metallic contamination (after Hölzl et al. [52])

cases, negligible effect of Cu contamination was reported [51, 52] while in other cases, a strong degradation was observed [53–57], which becomes more pronounced for thinner oxides.

The difference between the two groups of studies is that in the first case, the Cu contamination is applied on the front surface before the oxidation, while in the second case, backside copper and low-temperature heat treatments were applied, to mimic the effect of BEOL contamination. For example, it has been shown that thin gate oxides are more robust against *pre-oxidation metal contamination* of the *front surface*. This has been explained by the mechanism sketched in Fig. 7.5 [52]. At the same time, the surface roughness prior to gate oxidation impacts on the copper contamination effect (12 nm oxides on p-Si) [57]. This has been ascribed to the increase of the effective interface area by increased roughness, while the copper contamination results in a reduction of the effective oxide thickness.

Whether the oxide thinning hypothesis by the presence of pre-existing Cu_3Si precipitates is right or not has been investigated by a detailed Conductive Atomic Force Microscopy (C-AFM) investigation [58–60]. Surprisingly, it was concluded that the local leakage current underneath a Cu_3Si precipitate is smaller than in the adjacent SiO_2 regions [58–60], for frontside Cu contamination prior to a 600 °C thermal oxidation of p-type Si (1.3 nm). This is also assumed for the model of

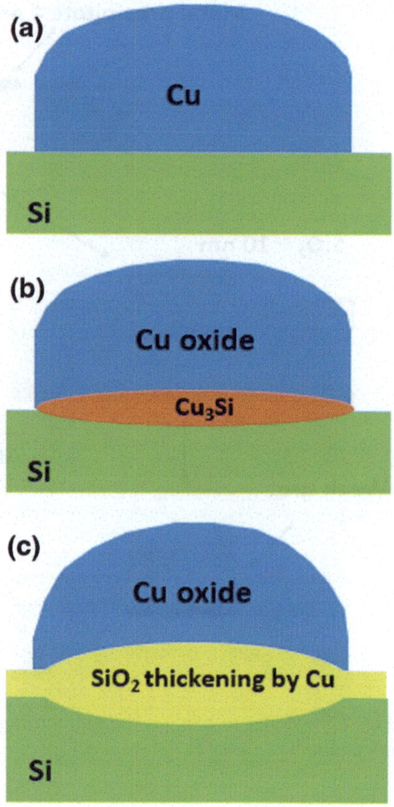

Fig. 7.6 Thick SiO_2 formation process under Cu particles (after Tokuda et al. [60])

Fig. 7.5. The lower leakage current is explained by the catalytic oxidation mechanism [38–40, 61, 62], illustrated in Fig. 7.6 and resulting in ***a local thickening*** of the gate oxide under Cu_3Si. On the other hand, ***a high density of copper on top of the*** SiO_2 ***surface induces dielectric breakdown***, while the presence of small amounts of copper results in a high leakage current at low electric fields [58–60].

As shown in Fig. 7.7 for ***backside Cu contamination*** in the range of 300–600 °C (***n-type*** Si), the [Cu] threshold for degradation of a pre-existing gate oxide reduces for a lower t_{ox} [53]. Two mechanisms may be responsible for this behavior: Cu precipitates in or near the interface can create a short or a geometric singularity, which enhances the local electric field and reduces the breakdown voltage [53]. It appears that thin oxides on n-type Si are more sensitive to Fe than to Cu contamination [32]. At the same time, thin gate oxides on ***p-type*** substrates turn out to be more robust against BEOL Cu contamination [54, 55]: while 5 nm SiO_2 exhibits no reduction of V_{BD} after a 400 °C anneal for 30 min in N_2 [54], a much poorer charge-to-breakdown and a higher stress-induced leakage current are observed for 3 nm thermal oxides [55]. This is explained by assuming a lowering of the direct tunneling barrier and increased interface-state-assisted tunneling by the presence of copper at the interface and in the oxide.

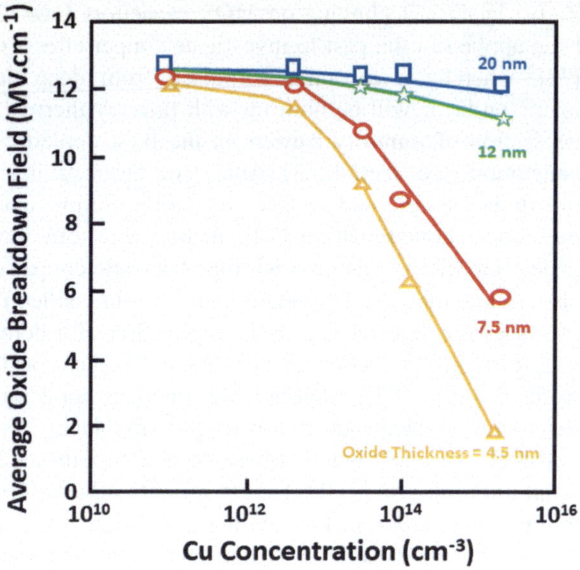

Fig. 7.7 Average oxide breakdown electric field as a function of copper concentration for various oxide thicknesses. Reproduced from Ramappa and Henley [53] by permission of ECS—The Electrochemical Society)

The difference between n- and p-type Si can be explained by considering the higher tendency for copper precipitation with a Fermi level closer to the conduction band. It has also been reported that the presence of Cu at the interface increases the D_{it} and the positive oxide charge density for 5 nm SiO_2 on p-type Si [54]. In addition, the pre-tunneling gate leakage current increases with backside [Cu], probably by the creation of additional neutral traps in the gate oxide after a 400 °C anneal [54]. Whether backside copper contamination degrades the front oxide depends largely on the presence of a backside SiO_2 layer and on the thickness of the front oxide [56], as also discussed previously [35]. The presence of an oxide layer at the back enhances the copper in-diffusion and its deleterious effects on the gate oxide, which, furthermore worsen for thinner gate oxides.

Compared with thermal SiO_2 on p-type Cz Si, oxynitride gate dielectrics with the same physical thickness of 3–5 nm show a higher robustness against BEOL copper contamination [63]. The behavior of metal contamination at different isolation structures has been studied as well [64]. In the case of LOCOS (Local Oxidation of Silicon), a thick isolation oxide is grown, which is characterized by stress concentrations near the so-called bird's beak. Similar stress peaks exist near the corners of Shallow Trench Isolation (STI), which are larger than in the case of LOCOS. This attracts copper to the active device area [64].

Surprisingly, little studies have been devoted so far to the effect of copper contamination on MOS capacitors with a high-κ dielectric. Liao et al. [65] have investigated the impact on ~4.2 nm Al_2O_3, showing good copper contamination resistance, which appears to be similar as for oxynitride, with high nitrogen content.

In a series of studies, the impact of copper contamination in thinned silicon wafers for 3D integration, based on Cu TSVs has been investigated by the *C-t* or

Zerbst analysis technique on MOS capacitors [66–72]. This Zerbst technique has been applied in the past to investigate copper effects on traditional MOS structures [72], whereby a capacitor is pulsed from deep depletion into inversion. The inversion layer will be built up with time by thermal diffusion and by the thermal generation of minority carriers in the bulk depletion region and by surface state generation (see Sect. 4.3.8). Analyzing the resulting capacitance transient enables to extract the bulk and surface generation lifetime and can be used as a measure of the copper contamination [72]. In this way, one can clearly show that the bulk generation lifetime reduces with the backside copper contamination level [66]. It is also found that the degradation of the thin silicon film (copper concentration) increases for reduced t_{film} [67]. Application of a denuded zone (internal gettering) suppresses the reduction of τ_g in the active layer at the top surface of the thinned wafer (50 μm) [67]. Alternatively, one can apply different types of back surface treatments to create an extrinsic gettering layer. According to Fig. 7.8, the dry polish (DP) sample behaves best, compared with a CMP treatment of the backside or an ultra-poligrind (UPG) stress-relief treatment [69]. One can discern in Fig. 7.8 that no degradation of the lifetime in the film by copper in-diffusion occurs below 10 min at 300 °C anneal. A damaged zone of around 100 nm is required at the backside of a thinned silicon wafer to withstand a thermal budget of 30 min at 300 °C [71]. In the case of a CMP treatment, most of the damaged layer is removed, resulting in a smooth back surface but with little robustness against copper diffusion.

When considering a copper TSV, one of the issues is the scalloping of the sidewalls by repeated etching and passivation. When the scalloping roughness is high, a conformal deposition of the dielectric liner and the barrier liner becomes more difficult [68]. A poor coverage of the barrier layer may induce the diffusion of Cu atoms from the Cu TSV into the silicon substrate, during a post-annealing process. The impact of scalloping on the barrier properties of a Ta barrier has been studied by the MOS *C-t* technique. The favorable barrier properties of Ta are related

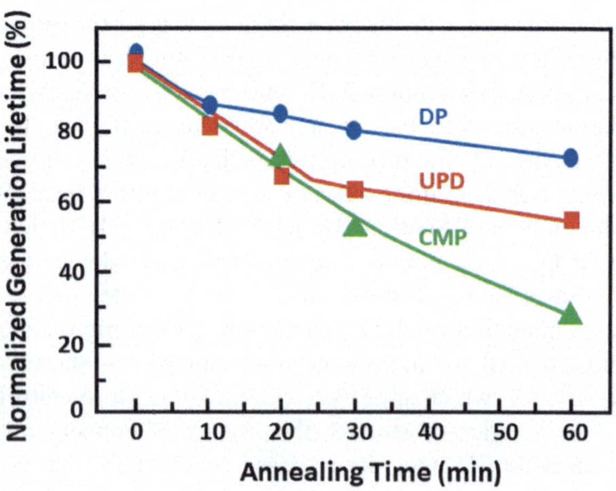

Fig. 7.8 Normalized generation lifetime τ_g of the MOS capacitors formed on thinned wafers (100 μm) with UPG, DP, and CMP versus annealing time. Reproduced from Lee et al. [69] by permission of ECS—The Electrochemical Society

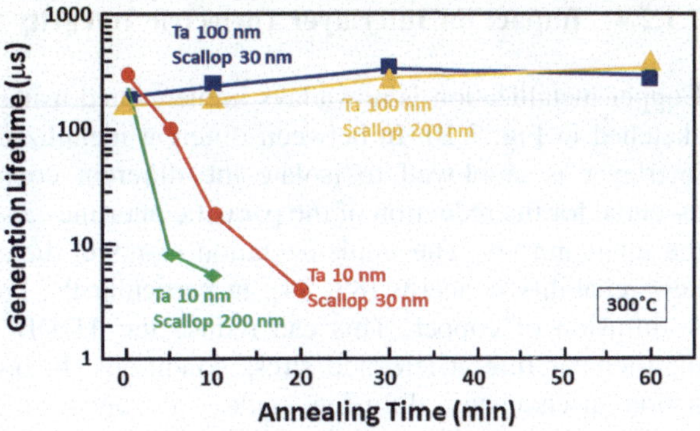

Fig. 7.9 Generation lifetime (τ_g) of minority carriers measured from *C-t* analysis versus the annealing time at 300 °C. Reproduced from Bea et al. [68]. Copyright 2011 IEEE

to its high melting point, giving rise to a high activation energy for copper diffusion. It has been shown that a 10 nm Ta barrier is effective against copper diffusion at 300 °C up to 90 min annealing for a planar structure. This is different in a Cu TSV, with roughness of 200 or 30 nm, as illustrated in Fig. 7.9 [68]. This implies that either a thicker barrier or a smoother profile has to be targeted for a Cu TSV. Another technique to evaluate copper diffusion issues in Cu TSVs is the use of infrared-optical beam irradiation, which measures the resistance change, associated with enhanced leakage current [73]. As an alternative to the MOS *C-t* test, it has been proposed to employ a gated diode on the n^+/n-well region, including the via [74]. Besides *C-t* analysis, one can perform gate diode current measurements and Charge Pumping (CP) in order to assess the impact of copper diffusion on the density of interface states at the interface.

When testing the impact of Cu ions on the TSV properties, the polarity of the applied gate bias plays an important role, as was concluded by studying the Time-to-Dielectric-Breakdown (TDDB) [75]. A negative field drives Cu^+ back again to the via. This results in a shorter TDDB for positive bias drifted TSVs at 75 °C and 7 MV/cm field strength. However, the opposite trend was observed at higher stress temperature (175 °C) and 5 MV/cm field [75]. This indicates that different breakdown mechanisms may be active, related to the interaction between copper ions and electrons. Cu acts as an active trap to electrons, reducing electron leakage transport. Cu may also serve as a scattering center, reducing the number of impact-generated defects by hot carriers. This leads to the conclusion that the breakdown mechanism of the dielectric layer containing copper depends on temperature and bias (polarity), through the impact on the role of copper. Normally, TDDB is reduced by the presence of copper. However, under severe stress conditions the opposite can occur.

Finally, MOS capacitors have been used as a test vehicle to validate the stability of different barrier layer materials in copper metallization [71, 76–80]. Nowadays, there is also growing interest in using ultrathin 2D materials, like graphene [81] and hexagonal boron nitride and MoS_2 [82] as ultimate barrier materials.

7.1.2.4 Impact on Interlayer Dielectric Integrity

Copper metallization is nowadays implemented using the damascene process flow sketched in Fig. 7.10. In between different metallization levels, a low-κ dielectric interlayer is employed to isolate the different copper lines. A low κ value is essential for the reduction of the parasitic interline capacitance and the delay time of the interconnects. The implementation of novel dielectric materials can introduce new reliability concerns [83, 84], in particular the breakdown associated with the in-diffusion of copper. This can reduce the TDDB under high-electric-field and high-temperature accelerated stress conditions. In order to understand the degradation mechanisms, diffusion studies of copper in low-κ dielectrics have been performed, generally described in terms of an Arrhenius model, whereby the activation enthalpy H depends on the electric field F. Dedicated degradation models have been proposed for the case of copper ILD, which may be different from the traditional F or $1/F$ models for TDDB in SiO_2.

The Time-To-Failure (TTF) in ILD is usually performed under accelerated stress conditions, i.e., at a larger electric field than under normal operation conditions. Prediction of the device lifetime should happen then by extrapolation to low

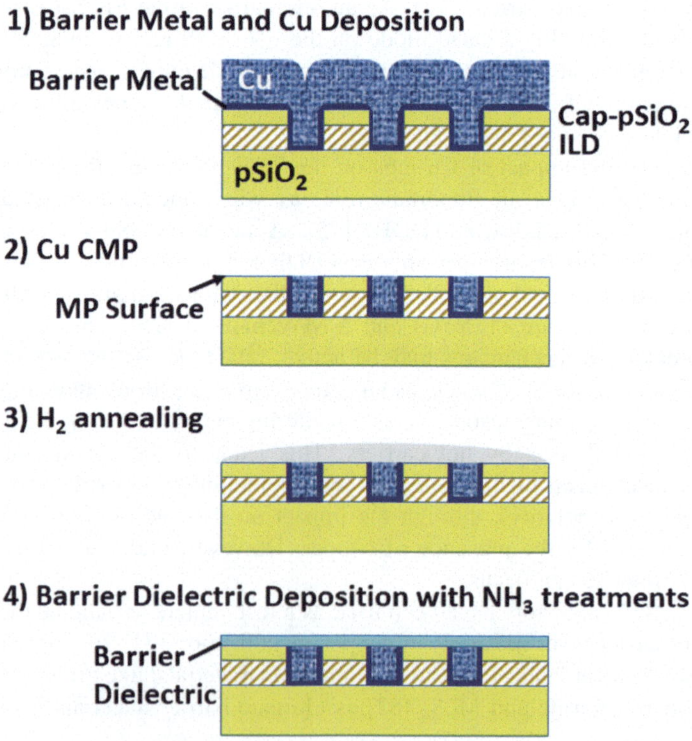

Fig. 7.10 Process flow for Cu damascene wiring (after Noguchi [84])

F values, implying that the exact experimental F-dependence of the TTF is of vital importance. While traditionally the TTF for SiO_2 is explained by an F or a 1/F model, this may no longer apply for low-κ ILD, suffering from copper diffusion. In that case, several different experimental behavior has been derived and various models advanced, given by (7.9)–(7.14) [85].

$$F \text{ model: TTF} \sim \exp(-\gamma F) \tag{7.9}$$

$$1/F \text{ model: TTF} \sim \exp(G/F) \tag{7.10}$$

$$\text{Power-law model: TTF} \sim F^m \tag{7.11}$$

$$\left(1/F+\sqrt{F}\right)\text{model: TTF} \sim \exp\left(-\delta\sqrt{F}+a/F\right) \tag{7.12}$$

$$\sqrt{F} \text{ model: TTF} \sim \exp\left(-2\beta_s\sqrt{F}\right) \tag{7.13}$$

$$F^2 \text{ model: TTF} \sim \exp\left[\left(E_a-\eta F^2\right)/\mathrm{kT}\right]f(C_e,T,F) \tag{7.14}$$

The different models are associated with different physical breakdown mechanisms. A thermo-mechanical model for SiO_2 bond breakage was introduced to explain the F dependence. The 1/F dependence for gate oxide breakdown is based on high-energy-hole induced damage. Similarly, the 1/F term in the $1/F+\sqrt{F}$ model for interconnects comes from high-energy electron-induced damage [85]. The BD mechanism for the power law model is related to hydrogen-induced defects. The $\sqrt{F}$ model involves Cu^+ generation at the anode side through leakage current. It has also been found that the intrinsic bond breakage in the presence of copper should rather follow a F^2 dependence than proportional with F, because of an internal field enhancement due to the presence of Cu^+ drift.

Clearly, this matter is far from settled, as can be judged from the different experimental behavior reported in the literature and the different models (physical mechanisms) invoked (see, e.g., [86–96]). In addition, testing conditions are also important: it has been demonstrated that under ac gate bias stress the TTF of the ILD can be larger than for standard DC stressing [97, 98]. This is related to the fact that during the negative stress period, the MIS or damascene structure partially recovers, since the diffusive and drift flux of copper is opposite, with the first one directed towards the dielectric and the second one towards the copper electrode [97]. During the positive stress cycle, both fluxes work in the same way towards the dielectric, yielding a higher degradation. It was found that the TTF increases with frequency in the range 0.001–0.003 Hz range and then saturates. For alternating stress (duty cycle 50%) above 0.1 Hz on the other hand, the TTF becomes smaller [98]. This is related to the fact that Cu^+ cannot be recovered during the negative $V_{Gstress}$ cycle and more copper is driven in the low-κ layer from the copper electrodes at higher ac stress frequencies [98].

7.2 Impact on p-n Junction Devices and Schottky Barriers

The first report on the impact of TM precipitates on the forward and reverse current of Si p-n junctions dates from 1960 [99], revealing the occurrence of so-called soft breakdown. Since then, many reports have been published demonstrating the negative impact of metal centers in the depletion region of a junction. In this part, first the effects of metal contamination on the *I-V* characteristics of diodes will be described. Next, the case of silicided n^+p and p^+n junctions will be discussed, as this nicely illustrates the use of a simple device for the assessment of metal contamination effects in semiconductor processing. An area where metal contamination is even more crucial is PV. Especially in the case of mc-Si, containing a certain level of as-grown metal contamination, the question arises which is a tolerable level to achieve reasonable efficiencies. Finally, the impact of contamination on Schottky barrier structures will be discussed.

7.2.1 Metal Contamination in p-n Junctions

In the early days of semiconductor processing, massive metal contamination took place, leading to supersaturation and precipitation of metals. Several reports illustrate this [99–103]. In addition, it was found that other extended defects, present in a junction, showed pronounced electrical activity (i.e., leakage current generation) when decorated by metal precipitates [104]. Since then, large progress has been made in cleaning and gettering technologies, so that the expected contamination levels are usually well below the solubility limits at typical processing temperatures. Nevertheless, there is always a probability for metal contamination during processing (Chap. 3), especially when novel materials and process modules are being introduced. Therefore, a good knowledge of the expected effects of metal contamination on for example p-n junctions is relevant. Quite some studies have been devoted to the presence of copper, which will be treated first. Next, the impact of other metals on p-n junction characteristics will be summarized.

7.2.1.1 Impact of Copper

Several groups have investigated the correlation between junction leakage and the surface copper, nickel and iron contamination [105–107]. According to Fig. 7.11, a significant impact is observed from about 2×10^{12} cm^{-2} for n^+p diodes [105]. A stronger effect is observed for Cu, compared with Fe and Ni. This was explained by the fact that Cu is gettered by the n^+-layer, forming extended defects (dislocations) which cause leakage current when they penetrate the depletion region. A deep-trap model has been developed to explain the measured trend for copper-contaminated p-n junctions [106]. Experimentally, an activation energy of

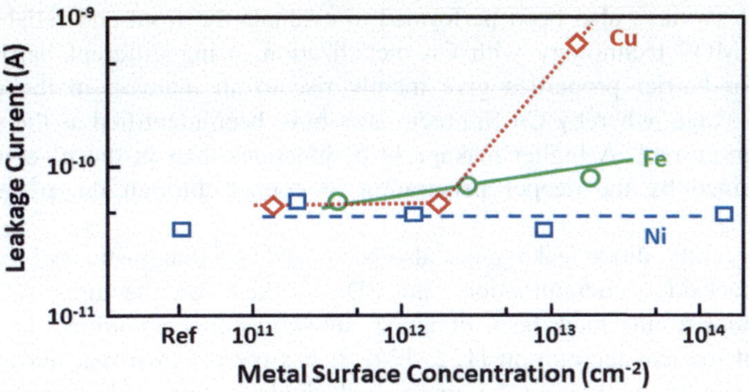

Fig. 7.11 The relationship between leakage current and surface metal concentration before heat treatment (junction area of $n^{+}p$ diode is 1 mm^2). Reproduced from Miyazaki et al. [105]. Copyright 1991 The Japan Society of Applied Physics

0.775 eV was derived for the reverse current [106], while the generation lifetime has been derived from Zerbst analysis on a MOS capacitor. Based on this, a methodology was developed to monitor the effects of contamination during a pre-gate oxidation clean [106].

In the case of active source/drain junctions in CMOS transistors the Perimeter (P) to Area (A) ratio becomes significant, so that the corresponding peripheral leakage current starts to dominate. This has been illustrated by investigating the effect of surface copper contamination on p-n junctions with different gate overlap structures [107]. It was concluded that copper is preferentially found at the field overlap region, where the copper is gettered by the mechanical stress which is present there. This mainly increases the peripheral leakage current component [107].

A copper bulk concentration of around 1×10^{13} cm^{-3} has been quoted as detrimental for junction leakage [108], which corresponds with the solubility limit at 425 °C, a typical back-end processing temperature.

Thermal stability investigations of Cu/NiSi contacted $p^{+}n$ junctions have revealed that at 375 °C anneal, a significant increase of the reverse current occurs, by the penetration of copper through the ~30 nm thick NiSi layer [109]. After annealing at 425 °C, a Cu_3Si phase was formed in the junction.

The leakage current of p-n junctions has also been used as a test vehicle to explore the reliability of barrier layers against copper diffusion [108, 110]. Advanced metallization technology brings copper in close contact with the silicon substrate, in some cases with only less than a 5 nm diffusion barrier layer between the two. The detection of the penetration of copper through a barrier is challenging. In the case of copper, it is the analysis of the electrical mode of failure that is of particular importance for the copper metallization. Many reports can be found, therefore, where the barrier failure is monitored by the impact of copper penetration on the leakage current of an adjacent junction [111–120].

Diode tests have also been performed to evaluate the front-end reliability of a 130 nm CMOS technology with Cu metallization, using different barrier layers [121]. Poor barrier properties give mainly rise to an increase in the p^+/n-well junction leakage, whereby Cu_3Si precipitates have been identified as the source of the leakage current. A higher leakage in p^+-junctions than in the n^+-counterparts was explained by the deeper penetration of copper through the p^+-layer than through n^+.

More recently, diode leakage has also been used as a diagnostic tool to evaluate copper backside contamination in 3D stacked architectures [122–124]. Controversial results have been obtained, showing either no impact [122] or an increase of the leakage current [122, 124]. It has been shown that under realistic operation conditions, mainly the source-bulk diode is degraded, according to the mechanism depicted in Fig. 7.12 [123]. Due to the direction of the electric field in the channel ($V_{DS} > 0$ V), copper ions most easily penetrate into the n^+-source region. It was also reported that the presence of a backside SiO_2 layer drastically enhances the degradation [123]. This is related to the fact that SiO_2 is a poor diffusion barrier for copper, while without SiO_2 present at the backside, a Cu_3Si self-gettering layer is formed, which prevents copper from diffusing in above the solubility limit.

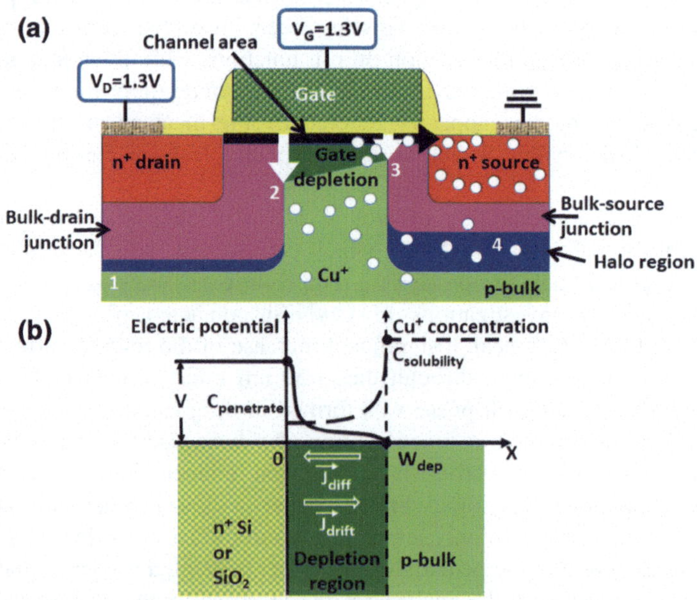

Fig. 7.12 **a** A schematic cross-section of an n-MOSFET and the Cu^+ migration path to the n^+-source region in the field-temperature stressing. **b** The one-dimensional distribution of the electric potential and the Cu^+ concentration at the depletion region (after Yeon et al. [123])

While there is little information of metal contamination on ***Ge-based devices***, there is one case where a clear correlation has been found between the performance and the concentration of as-grown copper-related defects, namely radiation detectors for γ-ray spectroscopy [125, 126]. Such a detector is basically a large n^+p junction fabricated in a high-purity germanium (HPGe) crystal with a diameter of several cm. The doping density is in the range of a few 10^{10} cm^{-3} (p-type) in order to enable full depletion of the detector at a reasonable operation bias. Owing to the low band gap of Ge, such detectors are operated at 77 K (at room temperature, the material is intrinsic with a carrier concentration in the range of 1×10^{13} cm^{-3}). It has been established that in order to have a sufficiently high spectral resolution, defined by the FWHM of a 1.33 MeV ^{60}Co γ-ray, the so-called "active copper" concentration at 77 K should be below 4.5×10^9 cm^{-3}, resulting in a resolution of <2 keV [126]. For higher copper concentrations in the fully depleted HPGe, hole trapping, resulting from the presence of copper-related deep levels with activation energy larger than 70 meV from the valence band edge will create a broadening of the collected charge signal and a too high FWHM.

7.2.1.2 Impact of Other TMs

As shown in Fig. 7.12, ***iron*** has a smaller impact on n^+p junction leakage than for example copper. Slight increases in the saturation current (forward intercept current) was reported in [127]. This could be related with the fact that in p-type silicon, Fe is generally paired with B, resulting in rather shallow defect levels (Chap. 5). This gives rise to inefficient carrier generation (low leakage current), while the recombination lifetime (saturation current) will be affected.

While interstitial ***nickel*** is electrically neutral, both substitutional and precipitated Ni introduce deep levels in the band gap (Chap. 6) and will, therefore, impact on the electrical performance of devices fabricated in a silicon wafer. From the early days on, it has been shown that the presence of TM-related precipitates in the depletion region of a silicon diode will create a strong increase of the reverse current, giving rise to so-called soft current-voltage (*I-V*) characteristics [99, 102]. On the other hand, it has even been considered to use nickel-doped silicon for special types of impact-ionization diodes [128, 129].

Early on, it has been demonstrated that the presence of compensating ***Co***-related deep levels in p-n junctions [130] or silicon resistors [131, 132] may lead to oscillations in the (forward) *I-V* characteristics with a frequency in the range of a few kHz up to 2 MHz [132]. This behavior is associated with the impact ionization of occupied deep Co centers and arises from a repeated process of trapping and re-ionization of conduction electrons on deep levels. These oscillations are very sensitive to light exposure and occur at a certain threshold field in the range of 3 kV/cm [132]. This threshold is lowered significantly under white light illumination. While originally, there was some interest in utilizing this phenomenon for certain applications, the idea was later abandoned.

7.2.2 *Silicidation-Induced Metal Contamination*

Self-aligned silicidation of source/drain contacts is up to now standard practice in CMOS technologies. Originally, Ti was used, later replaced by cobalt and, finally, nickel. This is related to the increasing resistivity for narrow lines, which is better in the order NiSi < $CoSi_2$ < $TiSi_2$, the better film integrity for thin layers and the lower stress level in the silicides. In fact, in recent bulk FinFET technologies, Ti-silicidation has been re-introduced. Silicidation normally consists of deposition of the metal layer, followed typically by a two-step RTA process. The first step at lower temperature (e.g. 625–675 °C for Ti) serves to react the metal with the silicon substrate, forming a silicide. In between the two steps, the unreacted metal is removed by wet etching. The second step at higher T (800 °C for Ti-silicidation) forms the final low-resistive $TiSi_2$ phase. It is clear that during the RTP steps metal atoms will diffuse in the silicon substrate and potentially form a risk for high leakage current, if they can penetrate the depletion region. In this section, this issue will be discussed for $TiSi_2$, $CoSi_2$ and NiSi, respectively.

7.2.2.1 Ti-Silicidation

Analysis of the leakage current of Ti-silicided diodes revealed that, especially for p^+n diodes with a thin p^+-region, higher values were obtained [133] and that the thermal budget should be kept below 900 °C in order to preserve good characteristics. Several possible mechanisms have been proposed over the years [133–136]. Originally, it was believed that the silicide penetrated the shallow junction, particularly at the device edges [133], resulting in a $TiSi_2$/n-Si Schottky barrier contact. Later work suggested that the soft reverse I-V could be explained by Poole-Frenkel barrier lowering of a trap potential [135], suggesting the in-diffusion of Ti in the depletion region. The activation energy of good diodes showed high values ($\sim$0.8 eV) in between mid-gap SRH and diffusion current (E_g) [136]. This points to carrier generation by a deep level which is close(r) to one of the band edges. DLTS revealed the presence of a hole trap at $E_V + 0.3$ eV [136]. However, this does not coincide with the $E_V + 0.26$ eV Ti-related double-donor level (see Chap. 6), but rather corresponds with the dominant metal-contamination-related thermal defect associated with RTP (see Sect. 3.7). As it is far from mid-gap, it is not an efficient generation center and rather large concentrations can be tolerated. The leaky p^+n diodes, on the other hand, revealed at sufficiently large V_R athermal currents that hardly change with an increase in the temperature. This suggests a field-assisted (tunneling) generation mechanism, which could originate from $TiSi_2$ protrusions through the p^+-layer into the depletion region. Excess carriers are then generated by local breakdown in the high electric field existing in the metal/n-Si Schottky barrier [136]. Such effect was not observed for the n^+p junctions, where the metallurgical junction was usually deeper.

7.2.2.2 Co-silicidation

For $CoSi_2$ concerns mainly arose with respect to the leakage current of n^+p junctions [137–142] (see also Sect. 3.8.2) and how this was affected by the process conditions. In some cases, it was found that the origin of the excess leakage current was not uniformly distributed over the wafer and was stemming from the local penetration of $CoSi_2$ spikes [139–142]. These give rise to a local short of the n^+p junction. It was also found that smaller area junctions have a much higher yield, due to the lower probability for having one (or more) spikes [139, 140]. A detailed study indicated that the increase in the leakage current mainly happens for the first RTA temperature at 450 °C where Co_2Si changes to CoSi. At that point CoSi spikes have been observed in cross-sectional TEM with a length of 20–100 nm, sufficient to cross the junction [139, 140]. For a 1st RTA temperature of 550 °C where CoSi transforms into the desired low-resistive $CoSi_2$ phase, the spikes become roundish and their density is much reduced. Keeping the 1st RTA step at 550 °C and varying the 2nd RTA step (to convert the formed silicide into the $CoSi_2$ phase) between 750 and 900 °C (30 s) indicates that 800–850 °C is the optimal temperature range for low-leakage, high-yielding junctions [139, 140]. At 900 °C, the in-diffusion of Co creates a strong but uniform increase of the leakage current [139, 140]. From an analysis of the electric field dependence of the leakage current, it has been concluded that local Band-to-Band-Tunneling (BTBT) could explain the strong increase of I_R by the presence of a $CoSi_2$ spike [141]. Other work suggested an increase of the reverse current by the Poole-Frenkel effect [142], although the derived local electric fields are extremely high [141]. At the same time, the forward current of Co-spiked diodes was also degraded, showing a much higher I_F for large-perimeter diodes [142], which suggests that $CoSi_2$ penetration preferentially happens at the periphery of the diodes, where the n^+p junctions is thinnest. This gives rise to a local $CoSi_2$/p-Si Schottky barrier with a higher forward current [142].

The inherent (uniform) leakage mechanism by the in-diffusion of Co in shallow n^+p junctions has been studied in detail by Tsuchiaki et al. [143, 144]. A dedicated diode structure was used, which eliminates sidewall isolation effects on the reverse current. Solid-state source P-doping has been employed to avoid the impact of ion-implantation-related end-of-range (EOR) damage. According to Fig. 7.13, substantial leakage increase is found for Co-silicided n^+p junctions compared to non-silicided counterparts. The effect increases exponentially the shallower the junction is [143]. From the statistical distribution of the leakage current, it can be concluded that the source of the leakage is uniform [143] and thermally activated with an E_a close to mid-gap (Fig. 7.14). In other words, the excess current is associated with Co-related GR centers, obeying the SRH mechanism. Post annealing at 800 °C indicates the presence of a fast in-diffusing species (Co_i...).

Backside SIMS, avoiding the knock-on-atom effect, shows a good agreement between the Co chemical profile and the leakage profile as shown in Fig. 7.15 [143]. This points to a mechanism whereby Co bursts into silicon during the 2nd RTA step (815 °C, 15 s) used for the transformation to $CoSi_2$. From the figure, a carrier generation rate per Co atom of $2.5 \times 10^4\ s^{-1}$ can be derived, corresponding

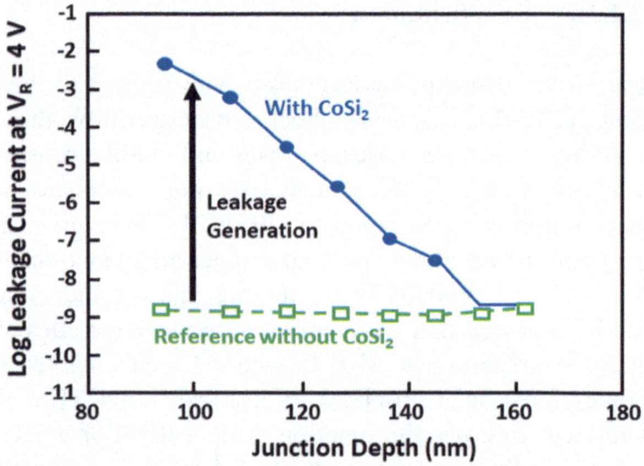

Fig. 7.13 Leakage levels (I_R at $V_R = 4$ V) as a function of the junction depth. Closed symbols and the solid line: junctions with $CoSi_2$ formation. Open symbols and dotted line: junction without $CoSi_2$. Reproduced from Tsuchiaki et al. [143]. Copyright 2002 The Japan Society of Applied Physics

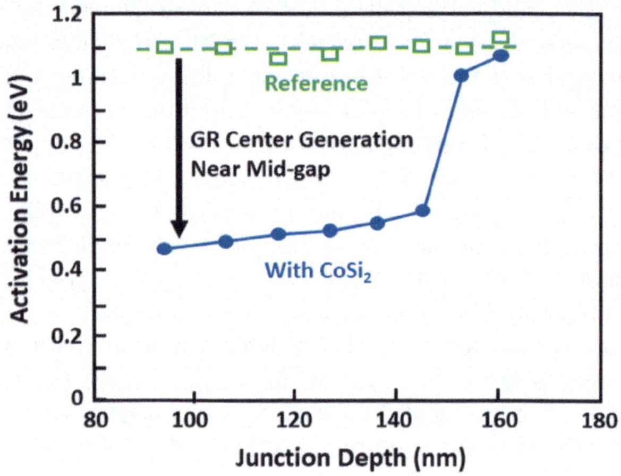

Fig. 7.14 Activation energy plotted as a function of the junction depth x_j. Excess leakage currents due to $CoSi_2$ formation have activation energies around mid-gap, indicating GR center formation by silicidation. Activation energies for reference devices are around the Si energy gap, indicating its diffusion-type origin. Reproduced from Tsuchiaki et al. [143]. Copyright 2002 The Japan Society of Applied Physics

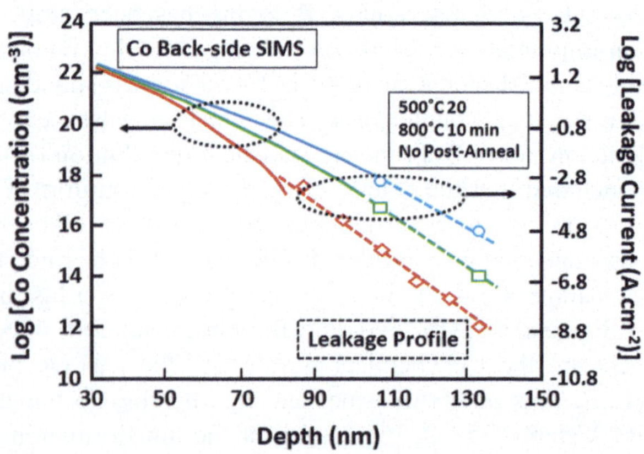

Fig. 7.15 Comparison between Co SIMS profile (solid lines, left axis) and leakage-depth profile (open symbol and dotted line, right axis) after various annealing steps. Connecting the profiles allows a rough estimate of effective leakage per Co atom. Reproduced from Tsuchiaki et al. [143]. Copyright 2002 The Japan Society of Applied Physics

with an effective capture cross section of 2.4×10^{-13} cm^2. A schematic of the intrinsic Co-related leakage mechanism is shown in Fig. 7.16 and compared with the well-known spike-related mechanism. Unlike the spike case, uniformly distributed Co-related GR centers are created in the depletion region of the junction. However, since their effective generation rate per atom is rather large (large cross section), it is assumed that Co occurs in a clustered form. This also follows from the extracted diffusion length per center, which is only 15 nm at 800 °C—much too low for fast diffusing Co_i [143]. Also the presence of Co in the substrate at concentrations over 10^{17} cm^{-3}—way above the solubility limit at 800 °C—indicates that the responsible leakage centers are clusters rather than individual atoms. In addition, no levels were found in DLTS.

The impact of several technological parameters on this leakage mechanism has been investigated [143, 144]. It was first of all concluded that the 2nd RTA temperature has no marked impact. It indicates that the burst of the Co-related centers already happens quite fast in the $CoSi_2$ transformation process, at the early stage of the temperature ramping [143]. The presence of a TiN or a Ti/TiN cap also does not make a difference [143]. Next, the impact of a pre-silicidation ion implantation (PSII) has been investigated. In order to separate chemical from physical (damage) effects, As- was compared with Ge-implantation [144]. Maximum energy of 50 keV and a dose up to 10^{14} at/cm^2 for both species with similar mass (damage) was considered. The depth of the damaged region should be completely consumed by the $CoSi_2$ process, with a 2nd RTA temperature of 815 °C in N_2 for 30 s, transforming CoSi into $CoSi_2$. The impact of the amorphization effect (physical or lattice damage) becomes pronounced above a dose of 5×10^{13} at/cm^2 in Fig. 7.17 [144]. Increasing the lattice damage with dose yields an exponential reduction of I_R

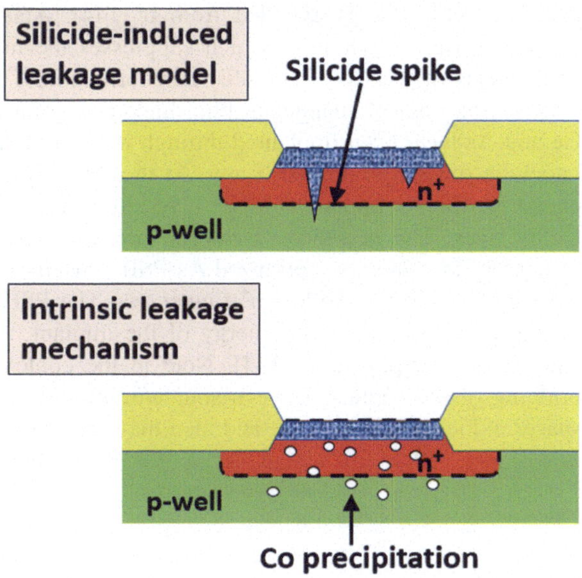

Fig. 7.16 Schematic illustration of the intrinsic leakage mechanism by $CoSi_2$ formation, in comparison with the spike-induced leakage model. Unlike the spike model above, Co in-diffusion and resultant Co clustering is revealed as the intrinsic leakage mode at least in the damage-free environment (after Tsuchiaki et al. [143])

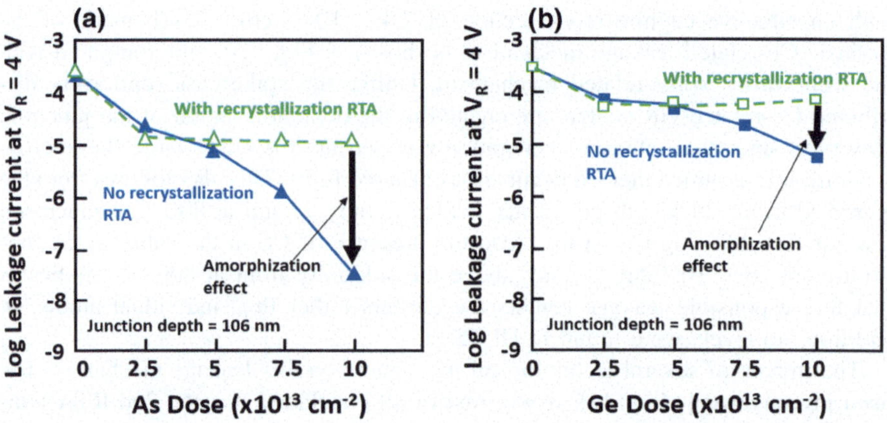

Fig. 7.17 Leakage levels (at $V_R = -4$ V, junction depth $x_j = 106$ nm) plotted as a function of PSII dose for **a** As and **b** Ge. Implantation is performed through a 20-nm-thick oxide with PSII energy of 50 keV. Leakage levels of samples with additional recrystallization RTA applied prior to silicidation are also plotted (dotted lines). Reproduced from Tsuchiaki et al. [144]. Copyright 2003 The Japan Society of Applied Physics

at −4 V. In addition, the effect of As is largely superior to Ge, indicating a significant chemical effect [144]. By As-PSII a reduction of the reverse current by four orders of magnitude can be obtained in Co-silicided n^+p junctions [144]. In the frame of the above model, it indicates that As is able to stop the massive in-diffusion of Co (diffusion barrier). This is supported by the SIMS profiles of Fig. 7.18 [144].

Further analysis has revealed that As is retained near the original CoSi/Si interface, while Ge is expelled from it [144]. Arsenic may have formed some compound (alloy) with Co, which suppresses the massive Co in-diffusion at the early stages of the CoSi → $CoSi_2$ transformation. TEM inspection did not reveal clear morphological changes in film thickness, grain structure… [144]. Since both Ge and As have been implanted through a 20-nm-thick screen oxide, the effect of knock-on oxygen is limited if any. At the same time, the oxide screening is beneficial for the As implantation, as it results in an As and oxygen peak inside the $CoSi_2$ layer. The presence of oxygen probably facilitates the Co-As compound formation. In summary, optimized As-PSII is useful in reducing the Co-silicidation intrinsic leakage for shallow n^+p junctions. The PSII optimization involves maximizing the dose and/or the energy of the implant, in order to obtain a sufficient amount of amorphization [144]. Note in the context of Co-silicidation, that the bridging of the contact between source/drain and gate, by up-diffusion along the spacer oxide can be an issue and also may require process optimization [145].

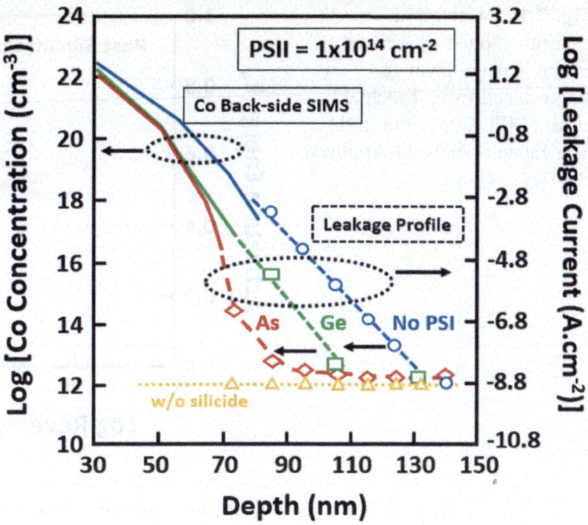

Fig. 7.18 Comparison between Co backside SIMS profiles and leakage-depth profiles of no-PSII, As-PSII, and Ge-PSII (both 50 keV through oxide) samples. Good matches are obtained between the SIMS and leakage-depth profiles. As-PSII allows x_j reduction about 50 nm compared with the no-PSII sample. Reproduced from Tsuchiaki et al. [144]. Copyright 2003 The Japan Society of Applied Physics

7.2.2.3 Ni-Silicidation

Besides inadverted contamination [105] Ni can also be introduced in a silicon depletion region by the nickel-silicidation of a p^+n (n^+p) junction [146–148]. In the case of contamination from a chemical solution, it has been found that the impact of Ni on the leakage current of n^+p junctions is negligible compared with similar Fe or Cu doses [105]. This was explained by the fact that Ni was gettered in the n^+-region without forming a dislocation network.

Tsuchiaki et al. have carried out a systematic study of the impact of nickel in-diffusion during a post-silicidation annealing at low temperatures (400–500 °C range, under N_2 up to 90 min) on the generation current of silicided n^+p junctions, using a dedicated diode test structure, eliminating peripheral leakage current effects associated with the oxide isolation [147, 148]. At the same time, a range of junction depths was investigated, in order to assess the scaling potential of the nickel silicidation. It was found that the reverse current I_R at a fixed reverse voltage V_R increases consistently upon annealing and more for shallower junctions [147]. The statistical leakage distribution was tight, eliminating in this case spiking of the silicide as a possible cause of the degradation, as was found earlier for Ti- and Co-silicidation. In other words, there is an uniform distribution of leakage generators. In addition, the *I-V* curves remain qualitatively the same, only shifting up with a scaling factor, which indicates that the same degradation mechanism is active for all annealing conditions and junction depths studied [147]. This is supported by the single activation energy of the leakage current at fixed V_R found in Fig. 7.19, showing a simple activation type of dependence, with an E_a around mid-gap [147]. However, closer inspection reveals that E_a depends on the depth of the junction and on the annealing time. This indicates progressive in-diffusion of GR centers and their structural transformation with annealing time.

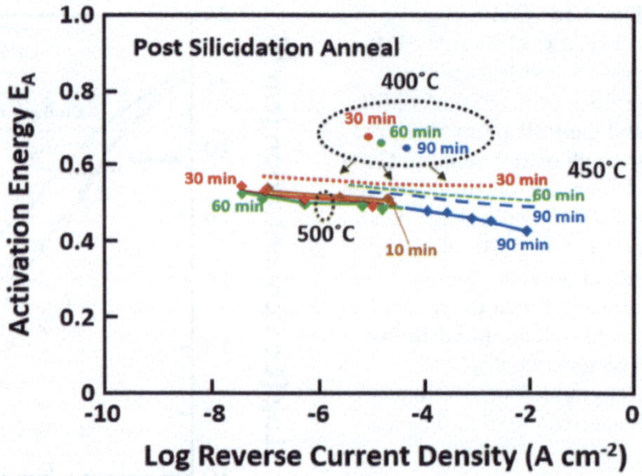

Fig. 7.19 Activation energies plotted as a function of the leakage level I_R. Reproduced from Tsuchiaki et al. [147]. Copyright 2004 The Japan Society of Applied Physics

From a comparison of the leakage current versus depth and the back-side SIMS profile of Ni (Fig. 7.20), it can be concluded that the leakage current is a sensitive indicator of the Ni in-diffusion (profile). An effective capture cross section of $\sim 2.0 \times 10^{-14}$ cm^2 per in-diffused Ni atom has been derived from Fig. 7.20 [147]. Given this rather large value, it has been concluded that the leakage centers are Ni clusters (precipitates). At the same time, the effective diffusion constant of the GR

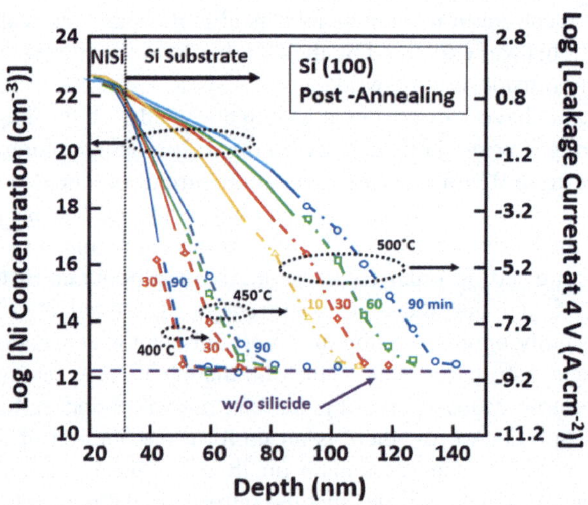

Fig. 7.20 Ni atomic concentration in the Si substrate (left axis) after each annealing time at 400, 450, and 500 °C and the corresponding leakage depth profiles. The depth is defined as the distance from the surface layer (i.e., inclusive of the 30-nm thick NiSi layer). Encroachment of the depletion layer into the n$^+$-region is taken into account. The excellent matching obtained between the Ni depth profiles and the leakage depth profiles irrespective of the annealing conditions provides direct evidence of Ni involvement in GR center formation. Reproduced from Tsuchiaki et al. [147]. Copyright 2004 The Japan Society of Applied Physics

centers in Fig. 7.21 is much smaller than for interstitial nickel and corresponds with a much higher migration enthalpy of 2.0 eV, compared with <0.47 eV for Ni_i [147].

These data has been interpreted in terms of the model depicted in Fig. 7.22. Initially, Ni undergoes a fast migration phase, which is manifested as a large offset in the temporal evolution of the leakage depth at 500 °C. According to Fig. 7.23, an initial burst of Ni_i occurs at higher annealing temperature, which is released from the $NiSi_2$ [147]. In fact, Ni_i is already introduced during silicidation, resulting in a lower bound for the junction depth of 20 nm, in order to maintain the integrity of the junction. At the same time, post-silicidation annealing temperatures should be kept below about 450 °C for shallow junctions [147].

Similar studies have been performed on Si(110) substrates in order to investigate the orientation dependence of the Ni in-diffusion and precipitation/leakage generation [148]. In principle, the preferential orientation of the nickel precipitates should be different in both cases and may give rise to an orientation-dependent leakage current. This is supported by comparing Figs. 7.20 and 7.24 [148]. A clear difference is found for the leakage profiles at 500 °C annealing. In Si(110) the profiles are virtually immobile with annealing time (while this is not the case at 400 °C or 450 °C), whereas a strong variation is found for the Si(100) case. This points towards a different leakage mechanism or a different type of centers in Si(110) [148]. Nevertheless, for both substrate types, a similar effective capture cross

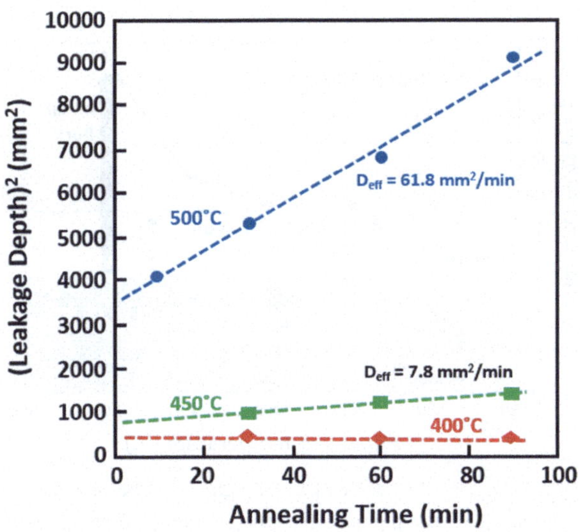

Fig. 7.21 Square of leakage depth (i.e., depth at $I_R = 10^{-7}$ A/cm^2, measured from the bottom of the NiSi layer) as a function of the annealing time. The linear dependence obtained for both 450 and 500 °C annealing indicates a steady diffusion as the governing migration kinetics of Ni. Although the obtained diffusion coefficients are much smaller than the atomic values. Also, the intersections with the *Y*-axis are far from the origin, particularly for 500 °C annealing (i.e., there is an initial Ni burst). Reproduced from Tsuchiaki et al. [147]. Copyright 2004 The Japan Society of Applied Physics

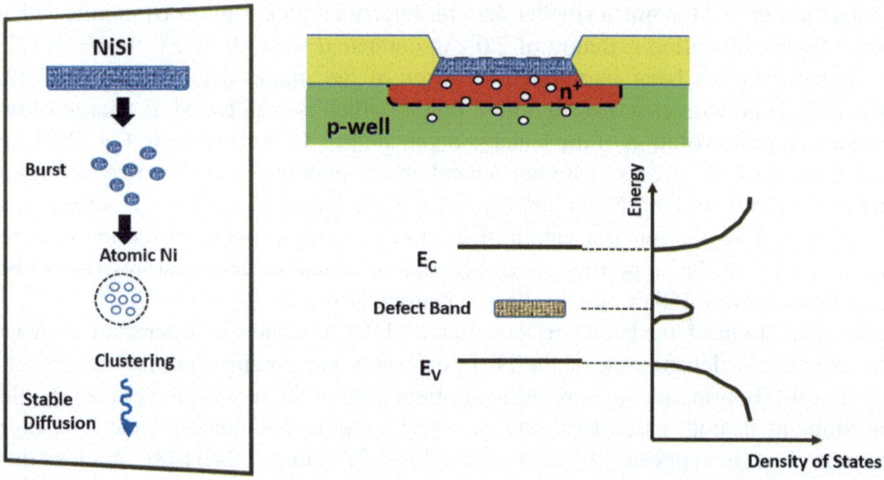

Fig. 7.22 The generation mechanism of thermally induced leakage currents for NiSi junctions. Ni is originally released from the NiSi layer in atomic form. An ultra-fast atomic diffusion temporarily ensues. Rapid coalescence of the migrating Ni marks the onset of the subsequent slower but stable diffusion. The thermally induced GR centers in the NiSi junction are the result of the clustering of Ni. Continuous structural evolution of the complex defects at an elevated temperature leads to the formation of a defect-band in the Si bandgap (after Tsuchiaki et al. [147])

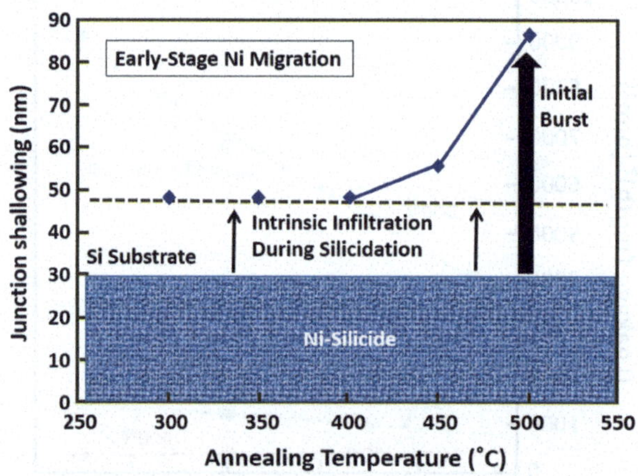

Fig. 7.23 Practical limitation of junction shallowing (inclusive of NiSi thickness) as a function of the highest process temperature. The initial Ni burst is superimposed on the intrinsic Ni infiltration during silicidation. Clearly, thin NiSi films are inherently sensitive to the process temperatures and the thermal instability of the NiSi films poses a strong constraint on junction scaling. Reproduced from Tsuchiaki et al. [147], Copyright 2004 The Japan Society of Applied Physics

section of 2×10^{-14} cm^2 is derived and the temporal evolution is similar: a burst of small precursors of Ni clusters from the NiSi layer, leading to a fast coalescence and followed by a slower but constant in-diffusion [148]. It is believed that similar clusters are formed in both orientations. However, the immobility in the case of Si (110) reflects a different crystal structure or symmetry of the clusters with respect to the junction plane—it suggests a preferential orientation for the case of Si(110) of the nickel precipitates, which should be most pronounced at the junction periphery [148]. Because of the specific test structure used, the peripheral leakage current is coming from $NiSi_2$ sidewalls and not from oxide isolation sidewalls. In other words, both the peripheral and area leakage have the same origin which is reflected in the similar shape of the corresponding *I-V* characteristics [148]. The difference in the case of Si(110) could indicate an additional diffusion channel of Ni (and Ni clusters) different from the (110) junction plane, which is not present in the case of Si(100) [148]. These observations may have a strong impact on FinFET type of devices as well, which usually include different planes for the channel orientation.

One way to improve the thermal stability of Ni-silicide (NiSi, low resistive phase) is by adding a small amount, e.g., 5% of Pt into the Ni [149–151]. This increases the nucleation temperature of the high-temperature and high-resistive di-silicide phase to 900 °C and the agglomeration T to 750 °C and above. A detailed study by DLTS and TEM revealed that at 600 °C silicidation extended defects could be found aligning along {311} planes (5–10 nm long) near the silicide/silicon interface [149, 150] for Ni(Pt). On the other hand, no extended defects were found for samples silicided at 700 and 800 °C [149]. The

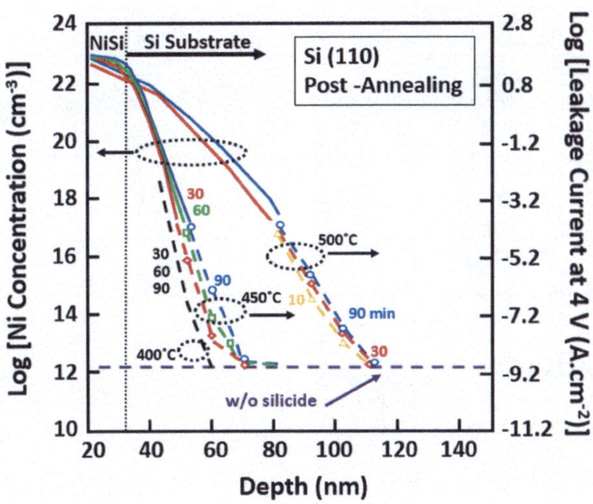

Fig. 7.24 Areal leakage levels (I_R at a reverse bias $V_R = 4$ V) of Si(110) substrates plotted as a function of the junction depth for each post-annealing condition (right axis) and corresponding Ni atomic concentration profiles in the substrate (left axis). The leakage-depth profiles and correlating Ni back-side SIMS depth profiles remain virtually immobile throughout the annealing at 500 °C. Reproduced from Tsuchiaki et al. [148]. Copyright 2007 The Japan Society of Applied Physics

DLT-spectrum of the 600 °C silicided sample showed a broad peak between 140 and 250 K, possibly the superposition of different individual peaks. At higher temperatures, two electron traps (E2 = $E_C - 0.42$ eV; E3 = $E_C - 0.50$ eV) and one hole trap (H1 = $E_V + 0.22$ eV) have been observed [149]. H1 and E2 are Ni-related traps (Chap. 6) and E3 is a Pt-related deep level. The deep levels exhibit a decaying concentration profile within 2 μm from the NiSi/n-Si interface. It was concluded that the degradation of the p^+n junctions was due to the formation of local NiSi/n-Si Schottky barriers with a height of 0.65 eV, corresponding approximately with the observed activation energy of the reverse current [149, 151]. However, no evidence was found for the formation of such spikes in Ni(Pt) films, so that the addition of Pt not only stabilizes the nickel silicide film but also improves the integrity of the junctions. Optimal performance was found for a silicidation temperature of 700 °C, using Ni(Pt) [149, 151].

When integrating self-aligned nickel silicidation in a full CMOS process flow, another issue became obvious, as illustrated by Fig. 7.25 [152, 153]. A lateral encroachment of the silicide is observed in the SEM cross section of an nMOSFET, which is not observed for p-channel transistors [152]. In the case of (110) silicon substrates, the issue is even more pronounced [153]. This phenomenon leads to an anomalous gate edge leakage current, which is two orders of magnitude higher for n^+p compared with p^+n junctions. The origin of the encroachment is the enhanced lateral nickel diffusion, which is thought to occur during the $NiSi_2 \rightarrow NiSi$ phase

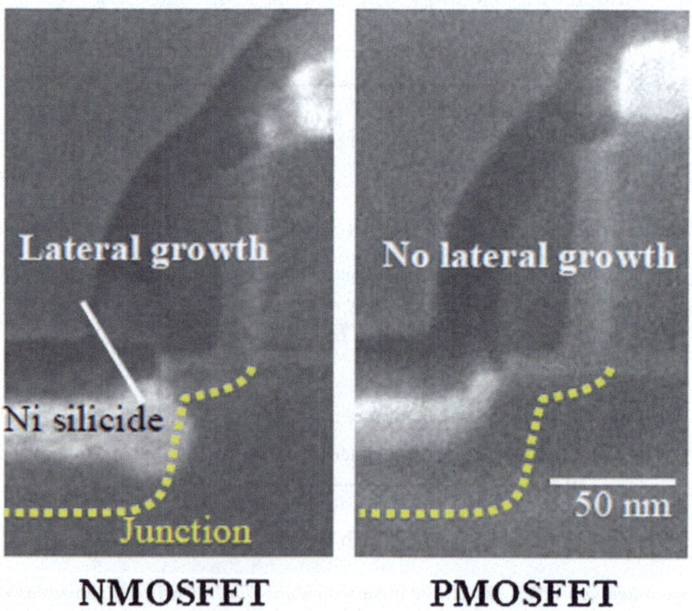

Fig. 7.25 Cross-sectional SEM image of Ni silicide in nMOS and pMOS. At the gate edge of the nMOS, the lateral growth is observed and penetrates into the junction (after Yamaguchi et al. [152])

transformation and is assisted by residual ion-implantation damage at the gate sidewall [154, 155]. Detailed TEM investigation pointed out the presence of ion-implantation-induced SFs as the root cause [153]. It is more easily observed in dense structures like SRAM arrays (45 nm node devices and beyond) and scales with the As-dose employed for the source/drain implantation [155]. Also the stress related with the STI regions can contribute to the formation of the crystalline defects [153]. The encroachment can be mitigated for example by the application of a pre-amorphization prior to nickel film deposition by a Si [152] or N_2 ion implantation [153]. Also the optimization of the gate sidewall etching helps in reducing the anomalous leakage.

Few reports have discussed defect-related issues in nickel germano-silicidation: it has been shown that improved leakage current characteristics of p^+(SiGe)/n(Si) junctions could be obtained by the application of a thin silicon cap on top of the SiGe [156]. A reduction of the leakage current, composed of SRH (low V_R) and BTBT (high V_R) by one decade can be obtained in this way. This is associated with the higher thermal stability of the layer, whereby agglomeration is suppressed. The presence of Ge in the silicide enhances the agglomeration tendency. Leakage current can be caused by agglomeration through the fact that extended defects are formed and in addition, current crowding occurs due to the concentration of the electric field lines at the germano-silicide grains. The former gives rise to enhanced SRH current and the latter to pronounced BTBT [156]. Also in the development of germanium nMOSFETs, enhanced off-state leakage current has been associated with the in-diffusion of nickel during nickel-germanidation [157]. Performing a 150 °C so-called hot source-drain P implant lowers the ion-implantation damage by dynamic annealing and reduces the leakage current.

7.2.3 Metal Contamination in Silicon Solar Cells

The impact of metal doping on the performance of solar cells has been systematically investigated by the Westinghouse Group around 1980 [158–160]. In that study, a series of p-type Czochralski silicon crystals have been pulled, doped with different concentrations of metals and solar cells were fabricated with a p-type base. The resulting normalized efficiency as a function of concentration has been plotted in Fig. 7.26. It is clear that there is a wide difference between different metals, with the more common, fast diffusing TMs (Fe, Ni, Cu, Co, Cr, …) having a relatively lower impact than slow diffusing elements like W, Mo and Ti. A similar study was repeated later by Pizzini et al. for solar-grade polycrystalline silicon [161]. In order to understand these trends more clearly, one should first of all establish how the solar cell parameters are affected by deep-level recombination centers. Therefore, in a first part, the main solar cell parameters will be defined and it will be pointed out how they depend on the material quality, i.e., the recombination lifetime. Next, the impact of specific TMs on the solar cell performance will be discussed.

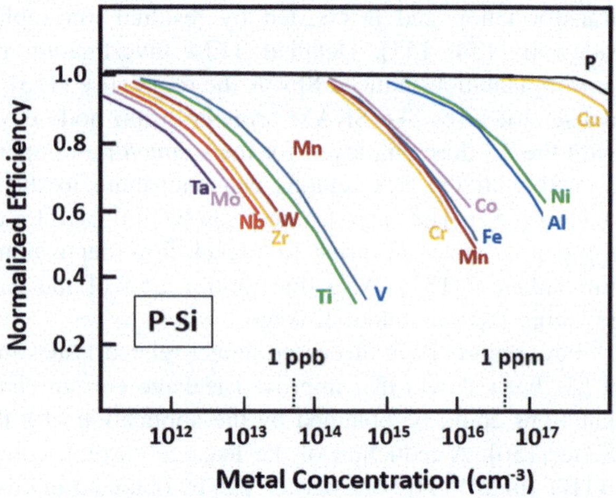

Fig. 7.26 The effect of metallic contaminants on solar cell efficiency. The impurity concentration is that in the wafer prior to cell processing. Reproduced from Davis Jr. et al. [159]. Copyright 1980 IEEE

7.2.3.1 Impact Defects on Solar Cell Parameters

The efficiency η_{sc} of mc-Si solar cells is typically 2–4% smaller than for a crystalline Si counterpart. This is not only related to the presence of GBs and intragrain extended defects, but also to the metal-related recombination centers present in the as-grown substrates. As shown before, this is both related to the deep levels associated with individual metal atoms, complexes or precipitates and to the activation of extended defects by metal decoration [162–164]. The basic parameter which is affected by metals is the bulk (or base) recombination lifetime τ_r. For low to intermediate injection (or illumination) levels Δn, the τ_r will be governed by SRH recombination at the surface or in the bulk [165]. From mid to high Δn recombination in the n$^+$-emitter will take over, while for very high excess carrier densities, Auger recombination will dominate. The corresponding effective recombination lifetime can be written as [165]:

$$\frac{1}{\tau_r} = \frac{1}{\tau_{SRH}} + \frac{1}{\tau_{emit}} \tag{7.15}$$

The emitter recombination lifetime is connected to the emitter saturation current density J_{0e} through [165]:

$$\frac{1}{\tau_{emit}} = \frac{J_{0e}(N_a + \Delta n)}{qn_i^2 t_{sc}} \tag{7.16}$$

In (7.16), N_a is the base p-type doping density and t_{sc} is the thickness of the cell. It is assumed that the transport of the carriers from the bulk to the emitter is not the rate limiting factor, so that one can consider that a close-to-uniform carrier density profile Δn exists. This is the case for not too high values of J_{0e} or when the contribution of the emitter to the total recombination is small [165].

The excess carrier density Δn determines the implied open circuit voltage V_{oc} under illumination through:

$$V_{oc} = \frac{k_B T}{q} \ln\left(\frac{\Delta n(N_a + \Delta n)}{n_i^2}\right) \tag{7.17}$$

assuming Δn does not vary in terms of spatial density profile in the base [165]. For efficient solar cells, high values of V_{oc} are desired. Conversely, the presence of defects will reduce Δn through carrier recombination and, hence, will reduce V_{oc}. Another important parameter is the short-circuit current I_{sc}, which is proportional to the illumination power. Both V_{oc} and I_{sc} determine then the Fill Factor (*FF*) of a solar cell. In order to optimize the *FF* and thus the η_{sc}, high values of both V_{oc} and I_{sc} are desired. The presence of electrically active defects in mc-Si can significantly reduce these parameters [166].

Metal precipitates can also be the origin of (pre-) breakdown of a mc-Si solar cell [167–169]. This has been investigated by monitoring the local light emission, using a CCD camera in an electroluminescence set-up, where the reverse diode bias is increased slowly. From the three different types of BD behavior observed, one could definitively be assigned to the recombination activity of metal-related defects [168]. It corresponds with a "soft" *I-V* characteristic, i.e., $I_R \sim V_R^n$, with $n > 1$. The BD is interpreted in terms of the Schottky barrier which is formed at the metal precipitate/p-mc Si interface, locally enhancing the electric field. This has been further investigated for deliberately TM-contaminated ingots, showing the results of Fig. 7.27 [168]: it is evident that the fraction of surface area which shows BD-related light emission increases with the amount of metal contamination.

Metals can also be the origin of shunts in solar cells [170], by locally reducing the built-in potential barrier of a p-n junction. These shunts form a parallel conducting path and drain power from other regions of the solar cell, with better

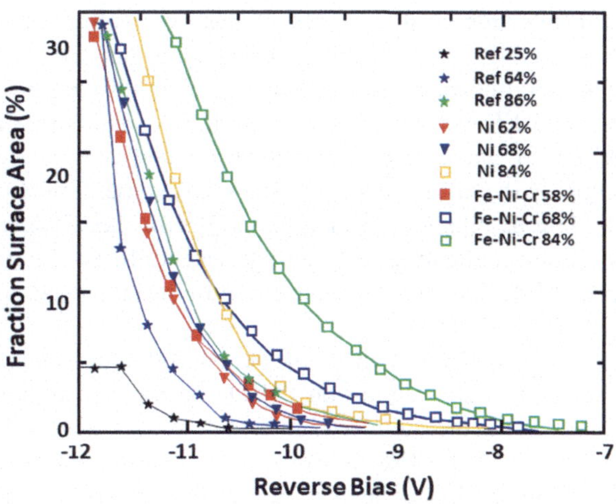

Fig. 7.27 Fraction of the surface area which shows light emission due to breakdown. This graph shows measurements made on solar cells which were taken from three ingots at different ingot heights (denoted by the percentage: 0% = bottom; 100% = top). Reproduced from Kwapil et al. [168]). Copyright 2009 AIP Publishing

junction properties. This significantly reduces the current and voltage outputs and thus the *FF*. A shunt can also be viewed as a highly localized defect in a p-n junction, which is ohmic in nature and somehow shorts the junction. However, many shunt sites exhibit a rectifying *I-V* characteristic and not a linear one. Therefore, one can define in a more general way a shunt as a site of increased local current density under forward bias. One way to investigate the presence of shunts is by lock-in thermography, which detects hot spots, corresponding with local high power dissipation [170]. Combining LBIC, XBIC and μXRF a clear correlation between such hot spots and the presence of metals (Cu, Fe, Ti or Ag—contact metal) has been established in mc-Si solar cells [170].

7.2.3.2 Impact Specific TMs on Solar Cells

The calculated effect of ***iron*** contamination on mc-Si solar cells has been investigated experimentally [171–174] and by 1D device modeling [164, 165, 175, 176]. Early studies on c-Si solar cells indicated that iron affects the bulk lifetime above a concentration of 1×10^{15} cm^{-3} [171]. Later on, it has been clearly evidenced that gettering plays a crucial role in the effect of iron contamination on solar cell performance [173, 174]. Comparing n$^+$-emitter (P-doped) with p$^+$-emitter (B-doped) cells, it was concluded that the J_{0e} was not impacted in the former case, while it increased for B-doped emitters [173]. This is surprising as PDG is much more efficient to remove Fe from the base region and collect it in the emitter than B-gettering. The presence of Fe in the n$^+$-region should increase J_{0e} through enhanced carrier recombination. This can only be explained by considering the much smaller hole capture cross section than the σ_n of Fe in p$^+$-silicon, provided that Auger recombination is not dominating [173]. For the considered solar cell processing, a reduction of the V_{oc} of p$^+$-emitters from 655 to 630 mV was reported, for Fe-contaminated solar cells [173].

Besides gettering steps implemented in a solar cell process, the activity of Fe can be reduced by hydrogen passivation. H can be released for example from the Si_3N_4 anti-reflection coating layer. It has been observed that ion-implanted Fe (dose 1×10^{14}–1×10^{15} at/cm^2), followed by an anneal at 850 °C has only a minor effect on the solar cell performance with a p-type mc-Si base [174]: the efficiency, V_{oc} and *FF* were unaltered, while there was a slight reduction of the J_{sc}. Also J_{0e} remained unchanged. Both gettering and plasma hydrogenation were shown to improve the diffusion length. The improvement by H-passivation was also visible in the Internal Quantum Efficiency (IQE) for the whole wavelength range from 350 to 1000 nm, indicating that hydrogen has diffused through the whole thickness of the cell [174]. This may be resulting from the enhanced diffusion of hydrogen along the GB and dislocations.

The calculated impact of Fe on different solar cell parameters is shown in Fig. 7.28 [165], relying on PC1D (Personal Computer One Dimensional) simulations, using measured parameters for the Fe-related recombination centers. It is observed that for [Fe] > 10^{10} cm^{-3} a strong reduction of the *FF* occurs, due to the injection-level dependence of this parameter around one-sun maximum power

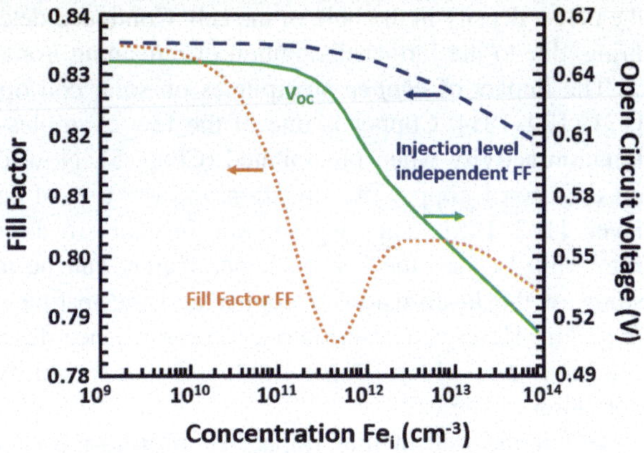

Fig. 7.28 Effect of Fe_i concentration on V_{oc} and fill factor as modeled by PC1D for a 1.5 Ω cm cell of thickness 0.03 cm and with $J_{0e} = 3 \times 10^{-13}$ A/cm^2. Also shown is the fill factor curve for a recombination center that has an injection-level independent bulk lifetime. Reproduced from Macdonald and Cuevas [165]. Copyright © 2000 WILEY-VCH Verlag GmbH & Co

conditions. The V_{oc} starts to be degraded for an Fe concentration around 5×10^{11} cm^{-3}. The detrimental impact of Fe_i on the *FF* is primarily due to the large asymmetry between the capture cross sections [165].

Further PC1D modeling has shown that the impact of Fe on J_{sc} becomes higher for a higher p-type base doping density [175]. However, in practice, I_{sc}, V_{oc} and *FF* are hardly affected in real solar cells, due to the efficient PDG and H-passivation [175]. The calculated impact of gettering and H-passivation on the solar cell efficiency in p-type FZ Si is demonstrated in Fig. 7.29 [175].

The evolution of Fe-related defects during mc-Si n^+p solar cell processing has been systematically studied by DLTS [177] and lifetime analysis [178]. It has been shown that Fe-related defects in and near the n^+p junction volume are formed directly after phosphorus diffusion due to the inflow of iron atoms from the dissolving iron silicide precipitates. Partial dissolution or gettering of these defects is shown to occur during the anti-reflection coating deposition, resulting in an increase in

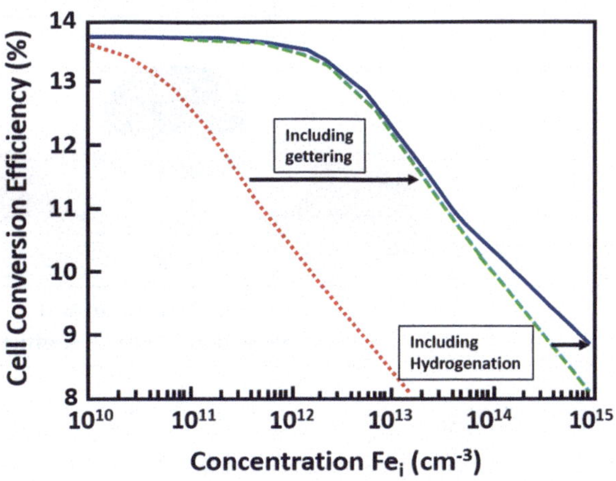

Fig. 7.29 Computed solar cell conversion efficiency as a function of Fe_i concentration, integrating gettering and hydrogenation effects in FZ-Si doped to 10^{16} cm^{-3}. Reproduced from Dubois et al. [175]. Copyright 2006 AIP Publishing

the defect density in the bulk of the cell. Continued deterioration occurs during contact firing, due to the further dissolution of remaining iron-silicide precipitates [175].

The impact of ***copper*** precipitates on solar cell operation has also been studied [171, 179–181]. Copper is one of the few examples which show stronger recombination activity when precipitated (Chap. 5). Nowadays, interest focusses mainly on the use of copper metallization and the risk of copper in-diffusion in the base layer [182–185]. The degradation mechanism is represented schematically in Fig. 7.30. In fact, the solar cell parameters can be used as sensitive tools for the study of the in-diffusion of copper and the testing of potential diffusion barriers (e.g., Ni). Hereby, a criterion of 5% performance degradation can be handled, i.e., a 5% lower *FF*, which is also employed as a reliability criterion for 20–25 years of operation of the cell.

Given the detrimental impact of ***titanium*** on solar cell efficiency shown in Fig. 7.27, some detailed studies were carried out in the past [171, 172, 186, 187]. In p-type silicon, a hole trap at $E_V + 0.26$ eV has been detected and found responsible for the degradation of the solar cell parameters [187]. The formation of TiO_2 precipitates has also been considered as a possible origin of degradation [186].

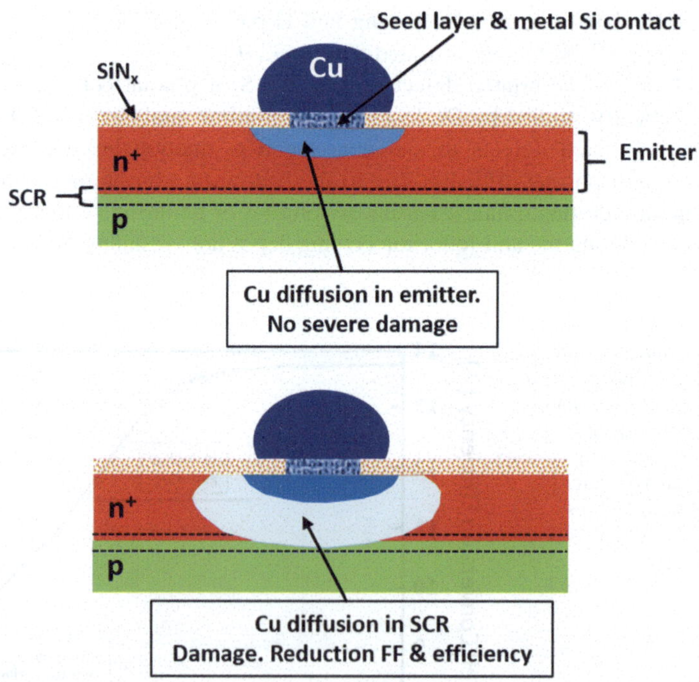

Fig. 7.30 Damaging mechanism of copper diffusion in silicon solar cells. Reduction of *FF* and thus of cell efficiency occurs when copper enters the space charge region. Reproduced from Bartsch et al. [182] by permission of ECS—The Electrochemical Society

The efficiency of ***gold-diffused*** solar cells fabricated on p-type FZ, Cz and mc-Si was compared in [188]. Au is a prototype 5*d* TM impurity with preferential substitutional incorporation. According to Chap. 6, it introduces two defect states in the band gap at $E_V + 0.347$ eV (donor) and $E_C - 0.554$ eV (acceptor). At 300 K, the donor level is dominating the recombination. A hole capture cross section of 2.7×10^{-15} cm^2 has been reported [188], while a rather similar electron capture cross section was derived at room temperature ($\sigma_n = 1.9 \times 10^{-15}$ cm^2). This means that the asymmetry is rather low ($\sim$2.4) and the recombination lifetime is weakly injection-level dependent. It has also been found that in mc-Si, the diffusion and concentration of Au is enhanced, due to the presence of dislocations, acting as a sink for silicon interstitials, which are created during the diffusion of Au through the kick-out mechanism [188].

It was observed that the solar cell efficiency was strongly reduced by the presence of Au for all substrate types [188], while this was not the case for the *FF*, due to the limited Δn-dependence of τ_r. The impact of gettering was relatively poor, which is expected for substitutional TMs [188]. In mc-Si, a difference has been noted between highly dislocated zones and relatively defect-free areas: in the first case, Au was released from the precipitates present at the extended defects, during solar cell processing, thus reducing the efficiency. The effect of H-passivation was much more pronounced for mc-Si than for crystalline FZ and Cz cells [188]. This originates from the much deeper H diffusion along the dislocations in the mc-Si case. As a consequence, also the IQE was improved for all wavelengths, while in the case of FZ and Cz substrates, some increase occurs at shorter wavelengths (900 nm), indicating that H only penetrated some 50 μm in the material.

As a slow diffusor in silicon, molybdenum has a strong degrading impact on the solar cell performance (Fig. 7.26). This has been studied in more detail by Rohatgi et al. [189], showing that the onset of the efficiency degradation starts at a level of about 5×10^{11} cm^{-3} in p-type Cz material. The dominant effect is the reduction of the bulk recombination lifetime, while the reverse current characteristics are less degraded. The major cause of the poor performance in Si:Mo is the drop in cell short circuit current, due to the lower τ_r [189].

A specific potential-induced degradation (PID) effect has been reported in both mc- and crystalline thin-film Si solar cells, fabricated on soda lime glass [190, 191]. This gives rise to a parallel shunt resistance, which has been ascribed to the presence of sodium-decorated stacking faults. The conductivity of the SFs is thought to arise under the influence of sodium ion movement through the high electric field across the SiN_x anti-reflective coating, resulting in PID [191].

7.2.3.3 Acceptable Metal Levels in Mc-Si Solar Cells

In a similar effort as for Cz silicon in the past [158–160], the impact of controlled amounts of metal doping on the performance of mc-Si was investigated, showing that adding 8 ppm weight of Cr, 11 ppm wt of Fe, 0.1 ppm wt of Ti, 13 ppm wt of Ni and 8 ppm wt of Cu are all equivalent in terms of their effect on solar cell

Table 7.3 Limiting concentrations of metal impurities in the feedstock material corresponding with a certain maximum solar cell degradation

Metal	C_L [192] (ppma)	C_L [193] (ppma)
Ti	0.022	0.1
Cr	0.026	8
Fe	12.5	11
Cu	4.6	8
Ni	–	13

performance, giving a degradation of the efficiency of about 2%. The question rises then: what is an acceptable level of contaminant concentration for mc-Si solar cells? This issue has been tackled by a number of research groups [192, 193] based on different criteria. In the first case, an effective lifetime of 20 μs was considered, which should correspond with an efficiency of about 15% [192]. The contribution of both dissolved and precipitated TMs were included in the lifetime calculation. The resulting concentrations in the feedstock material are summarized in Table 7.3 for the most prominent TMs [193]. A second more detailed study defines the limiting concentration C_L of impurities in the molten silicon charge that produces a $J_{sc} \times V_{oc}$ product with about 2% degradation with respect to a reference, uncontaminated ingot [193]. The higher is the C_L value, the lower is the impact on the solar cell performance. These are also given in Table 7.3.

It is clear that also for mc-Si, Ti is one of the most harmful impurities for solar cell performance, which is, moreover, hard to getter. Much higher concentrations can be tolerated for Fe, Cu and Ni, which are fast diffusers and more easy to getter. The results for Cr appear to be scattered, although it is expected to behave more similar to the fast diffusers, like in c-Si (Fig. 7.26).

7.2.4 Impact on Schottky Barriers

The *I-V* characteristics of Schottky barriers can be affected in different ways when metal contamination is present. First of all, if metal-related GR centers or precipitates exist in the depletion region, an excess leakage current will be generated [194–197].

The presence of metal contamination can also impact the thermionic emission current by changing the Schottky barrier height. Based on that principle, it has been suggested that the use of Schottky barriers—i.e., metal-silicon contacts—offers a more sensitive tool to monitor the penetration of copper through a barrier layer [198, 199]. In that case, the test structure consists of a Cu/barrier layer/Si layer stack. When using TiN as a barrier, a height of 0.6–0.7 eV is expected on p-type Si [199]. When Cu or Cu_3Si is present at the surface, the barrier reduces to 0.4–0.54 eV or 0.48 eV, respectively. This will increase the current through the SB and reduces the built-in voltage as measured by a reverse-bias *C-V* curve. In fact, even small patches of copper at the surface can be detected as the current flows through the region with the smallest barrier [199], while the *C-V* characteristic is sensitive to the spatial average SB height. According to the data of Fig. 7.31 only 20 nm TiN is robust up to 500 °C [199]. This is related to the fact that up to 20 nm, the layer is amorphous-like, while for thicker layers a columnar growth starts, with

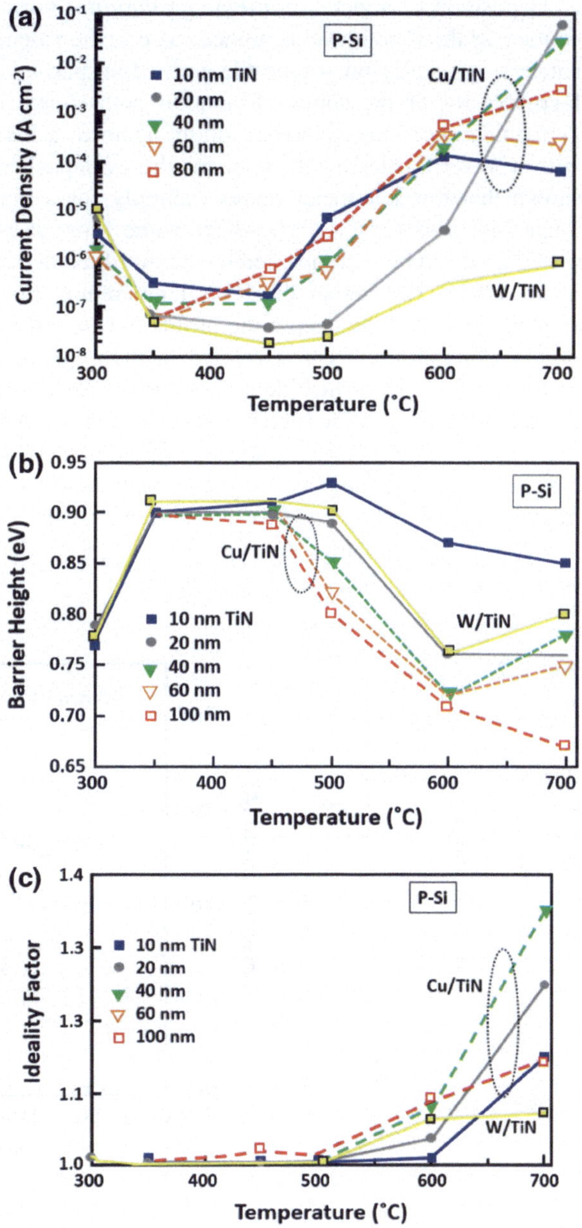

Fig. 7.31 Current densities at 5 V (**a**), barrier height (**b**) and ideality factor (**c**) of Cu/TiN/p-Si Schottky diodes versus annealing temperature for different barrier thicknesses. A graph for a W/TiN/p-Si sample is also shown. Reproduced from Ahrens et al. [199] with permission of Elsevier

intergranular voids, which form preferential diffusion paths for copper. As can be seen, the SB height drops upon copper penetration and the forward current ideality factor increases. However, the latter parameter is less sensitive. Barrier failure is earlier detected by leakage current measurements [199].

Alternatively, the TID technique (Sect. 5.2.3.2) based on a SB has been proposed for the electrical detection of copper penetration [200].

Deposition of a metal on ***n-type germanium*** normally gives rise to a rectifying contact, as the Fermi level is pinned close to the valence band. This results in a high barrier (~0.6 eV) on n-type Ge and a low barrier on p-type material, which is favorable for ohmic contact formation. Subsequent thermal annealing allows for germanidation to take place, resulting again in a SB contact on n-Ge [201, 202], with a lower barrier height than for the as-deposited contacts [202]. It has been shown that this parameter reduces slightly for a higher RTA temperature in the range of 300–500 °C [201]. At the same time, as shown in Sect. 3.8.1 the germanidation can be accompanied by the in-diffusion of the corresponding metal, e.g., nickel [201]. The higher the RTA temperature, the higher the in-diffused metal concentration and the deeper the profile extends in the material [201]. As a result, excess generation current is created in the depletion region, which enhances the reverse *I-V* of the germanide/n-Ge contact. An example is given in Fig. 7.32 for Pt-germanide [202]. The higher reverse current for the as-deposited SB is related to the presence of Pt-sputtering-damage in the depletion region, which anneals out at 300 °C [202]. This implies that the germanidation temperature has to be optimized in order to control the reverse SB current. At the same time, the in-diffusion of metals during germanidation allows to prepare reference samples for the DLTS study of metal-related deep levels in n-type Ge [203].

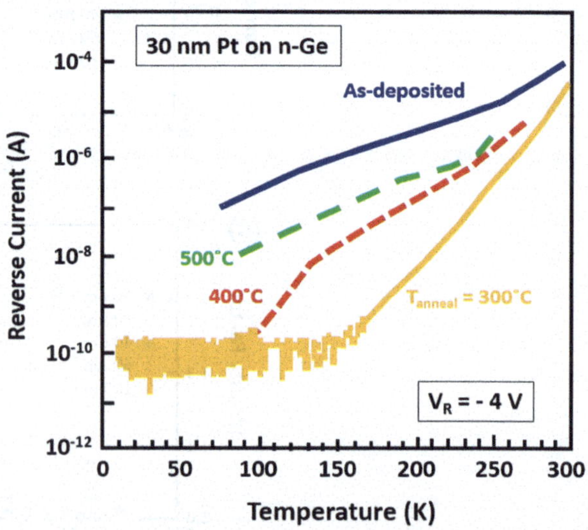

Fig. 7.32 *I-T* curves at $V_R = -4$ V for Pt-germanide and an as-deposited Pt Schottky barrier on n-type Ge. Reproduced from Simoen et al. [202] by permission of ECS—The Electrochemical Society

7.3 Impact on Transistors and on Circuit Operation and Yield

First the impact of metal contamination on transistor parameters (V_T, D_{it}, …) is discussed. In a second section, the effect on circuit operation is described. In particular, DRAM and image sensors are quite prone to contamination, giving rise to an increase in the dark current. As is shown, due to scaling of the pixel size, CMOS image sensors are sensitive to a single metal atom in the depletion region. In fact, this allows a new form of spectroscopy with unprecedented sensitivity. Finally, also the yield of CMOS and bipolar processing is affected by metal contamination. Scaling lowers the tolerable levels, explaining the efforts spent over the years in the development of ultra-clean processing and gettering treatments (Chap. 8).

7.3.1 Impact on Transistors

Several studies have investigated the impact of mainly Cu contamination on the DC parameters [34, 35, 56, 204–209] and reliability [121, 210] of MOSFETs. Usually, a small degradation of the threshold voltage (V_T) and a slight increase of the subthreshold slope (*SS*) with copper contamination is reported. In both cases, the increase of the D_{it} may account for the observations. In principle, also the positive oxide charge associated with copper ions yields a reduction of the V_T for n- and an increase for p-channel transistors. Associated with the increase in D_{it}, a reduction of the mobility by increased Coulomb scattering has been noted [79]. A strong degradation of the short-channel effect on V_T has also been reported due to impurity diffusion [56]. In some cases, however, little or no degradation of the transistor parameters has been observed from copper backside contamination [206, 209]. This was explained by the absence of pronounced copper in-diffusion due to the self-gettering effect by Cu_3Si formed at the back surface.

It has been found that p-channel transistors are less affected by the Cu-induced increase in the density of interface states than n-MOSFETs [208], suggesting a different degradation mechanism. For n-channel transistors, a linear correlation between ΔD_{it} and ΔV_T was found, while this trend was less clear for pMOSFETs [208].

In the case of 3D integration with a Cu TSV no transistor degradation was reported related to possible barrier failure, liner breakdown or Cu diffusion, when respecting the Keep Out Zone (KOZ) [211]. This was achieved for a minimum Ta barrier of 3 nm, a silicon oxide liner with a minimum thickness of 200 nm and TSVs with a depth of 50 μm and a diameter of 5 μm [211]. The gate stack consisted of 1.8 nm HfO_2 and a 5 nm TiN metal electrode with a total EOT of 0.9 nm. As a test vehicle for the study of Cu in-diffusion, the use of the low-frequency noise of a MOSFET has also been proposed, as this parameter is sensitive to trapping centers in the gate oxide (border traps) [54]. An increase in the LF noise power spectral density has been considered as evidence for the in-diffusion of copper after annealing at 500 °C. Along the same lines, GR noise spectroscopy has been employed for the study of metal impurities in bipolar junction transistors [212].

7.3.2 Impact on Circuits

DRAM circuits are extremely sensitive to excess leakage currents, which change the charge stored in the capacitor in a too short time. This is evaluated through the retention time, which should be sufficiently long to yield an acceptable refresh time and power consumption. DRAMs can also be stacked on top of other circuits, using a copper TSV scheme, where the impact of copper contamination should be carefully investigated. It was indeed shown that Cu contamination in a *thinned DRAM* chip for 3D integration may not affect the on-current of an nMOSFET, while the retention time of the memory is drastically reduced by enhanced carrier generation by Cu centers [51, 203–215]. This implies that *a proper choice of the electrical test vehicle* (parameter) *is mandatory.* For example, the *C-t* analysis of a MOS capacitor is a valuable tool to access the generation lifetime, which should be well correlated with the DRAM retention time. It has been shown that for a cell at 20 μm distance from a TSV with a diameter of 10 μm and a 50 nm Ta barrier no degradation is found after annealing at 250 °C [214]. Some degradation starts at 300 °C, indicating excess generation by Cu in-diffusion. This could be improved by replacing Ta through a 20 nm Mn barrier in the TSV. This moves up the temperature to 500 °C where degradation is found after 10 min [213]. Thinning the DRAM chip further from 100 to 50 μm enhances the effect of back-side copper contamination on the retention time [215].

Schmid et al. [216] used a 1D SRH model to calculate the leakage current due to surface contamination in order to estimate the maximum tolerable contamination level for DRAM memory types, as illustrated in Fig. 7.33. The trend toward lower leakage-current density requirements reverses after the 64-Mbit generation DRAM as a result of memory-cell architecture evolution which significantly reduce the space-charge volume. It can be noticed, assuming a maximum leakage current density of 3.9×10^{-9} A/cm^2, that for a 1 GB DRAM the surface metal concentration must be below 1×10^{12} cm^{-2}.

Embedded flash cells suffer from a higher V_T lowering after 10^4 write/erase cycles for copper contaminated wafers. This is related with the higher density of hot-carrier-stress created interface states in the drain-gate overlap region [217]. On the other hand, a similar degradation is derived for the density of trapped electrons in the tunnel oxide.

CCDs can suffer from a high signal (white defect) by injection of spurious charge or from a dark signal (black pixel) because charge is lost before it reaches the output stage [218]. In other words, both defect-assisted carrier generation (dark current) or recombination can cause a mall-functioning of a CCD. Early studies focused on the impact of metal decoration and gettering on extended defects (SFs,

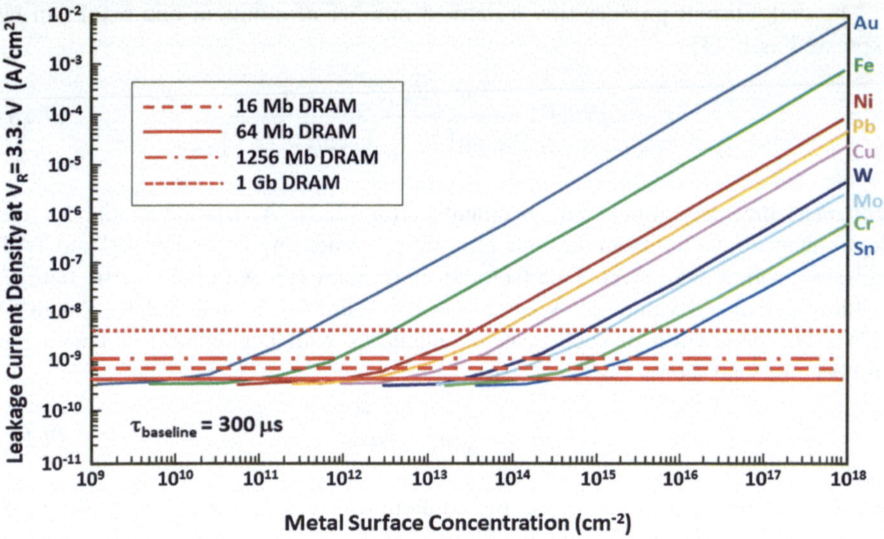

Fig. 7.33 DRAM leakage current density at 3.3 V reverse bias versus metal surface concentration in bulk extrinsic silicon with $N_a = 1 \times 10^{16}$ cm^{-3}. Generation current was calculated based on an 1D SRH model (after Schmid et al. [216]. Copyright 1994 IEEE)

dislocations) affecting the operation of a CCD [218]. In principle, the formation of extended defects can nowadays be avoided so that one should be more concerned about the effects of individual metal atoms. In one case, the impact of Ti impurities on trapping in CCD imagers has been highlighted [219]. Other studies pointed out substitutional gold as the culprit [220, 221]. Given the large sensitive of the dark current of imager circuits to minute amounts of contamination, dedicated analysis methods have been developed. One such technique is the so-called dark current spectroscopy (DCS), originally developed for CCDS [222] and nowadays mainly applied to CMOS image sensors (CIS) [223–226].

DCS investigates the current which is not generated by photons but by deep-level related carrier pair generation. It can detect W and Au contamination effects in silicon down to concentrations of 10^8 at/cm^3, which cannot be reached by any other available tool like DLTS [223, 224]. This high sensitivity is related with the small size of the CIS pixels, which are sensitive to even a single metal atom. DCS measures the distribution of the dark current for a large number of pixels, whereby the histogram becomes quantized depending of the number N of metal atoms which generate the current (in case one dominant defect is present). A first peak in the distribution corresponds with the dark current generated by other processing-induced defects ($N = 0$). The next peaks correspond with $N = 1$, followed by equidistant peaks for $N = 2$, $N - 3$, …, generally reducing in amplitude —the probability for having N Au atoms in a pixel is smaller, following a Poisson distribution—and the FWHM increases [225].

The dark current produced by a defined number of atoms in one pixel can be expressed as [223]

$$I_N = qN\theta(F)\frac{n_i\sqrt{v_{th,n}v_{th,p}}\sqrt{\sigma_n\sigma_p}}{2\cosh\left(\frac{E_T-E_i}{k_BT}+\frac{1}{2}ln\left(\frac{\sigma_n v_{th,n}}{\sigma_p v_{th,p}}\right)\right)} \quad (7.18)$$

with $\theta(F)$ the electric-field enhancement factor [223]. A detailed model for the field-enhancement has been presented in [225], considering Poole-Frenkel and TAT emission from a deep level. This factor strongly increases with the electric field F, yielding calculated values of ~10 at $F = 10^5$ V/cm for W and ~2 for Au in Si [225]. The bandwidth of each peak σ_N increases with temperature and with the number of metal atoms [223]:

$$\sigma_N = \sqrt{\sigma_i^2 + N\sigma_T^2} \quad (7.19)$$

where σ_i is the intrinsic and σ_T is the contaminant atom contribution to the peak dispersion. It is assumed that the distribution of the implanted metal atoms over the pixels follows a Poisson distribution [223–225]. Thus, the probability for a pixel containing N atoms is given by:

$$P_N(N_{av}) = \frac{N_{av}^N}{N!}\exp(-N_{av}) \quad (7.20)$$

with N_{av} the average number of atoms per pixel. A histogram of the number of pixels versus the dark current provides equally spaced peaks which shift with temperature in accordance with the activation energy E_T of the deep level (7.18) [223]. The Nth peak can then be calculated using the function [223–225]:

$$f_N(I) = A_N P_N(N_{av})\exp\left(\frac{(I-I_N)^2}{2\sigma_N^2}\right) \quad (7.21)$$

A_N is a normalization factor. The total histogram can then be written as [223]:

$$F(I) = \sum_{N=0}^{\infty} f_N(I) = A\frac{1}{\sqrt{2\pi}}\sum_{N=0}^{\infty}\frac{P_N(N_{av})}{\sigma_N}\exp\left(\frac{-(I-I_N)^2}{2\sigma_N^2}\right) \quad (7.22)$$

This modeling can be further refined using a Monte Carlo approach to set the number of metal atoms for each pixel [225]. This has been applied to the case of Au_s, yielding a dark current per atom of 39×10^{-18} A at 45 °C and a field of 10^4–10^5 V/cm [225].

A more systematic study of the dark current produced by ion-implanted metals has been reported by Russo et al. [227]. In that case, the dark current histogram has been measured as a function of T and modeled by a diffusion term and a generation term:

$$I_{dark} = I_{diff} + I_{gen} = AT^3 \exp\left(-\frac{E_g}{k_B T}\right) + BT^{3/2} \exp(-\frac{\frac{E_g}{2} + \Delta E}{k_B T}) \tag{7.23}$$

The first term is a measure for the recombination lifetime and is sensitive to deep diffusion of the implanted metal atoms beyond the depletion region of the pixel; the second term is sensitive to carrier generation by implanted metal atoms in the depletion region. After the ion implantation placing the peak in the depletion region of the photodiode a damage anneal at 1020 °C for 7 s has been applied in N_2 [227].

This has resulted in the following observations [227]:

- For ***molybdenum*** (slow diffuser), the number of hot (=high leakage) pixels increases with Mo dose up to 5×10^9 cm^{-2}. No important impact of the implantation energy was observed. $\Delta E \sim 0.24$ eV.
- ***Tungsten*** (slow diffuser) is a very effective generator of dark current, yielding an energy level $\Delta E = 0.19$ eV $\pm$ 0.01 eV. It does not correspond with the E_V + 0.4 eV hole trap of W but with a W-related processing (dry etching?) induced hole trap at E_V + 0.36 eV. Although the W-level is also present, it is believed that the E_V + 0.36 eV level is more effective due to its large σ_p. Also the diffusion current is increased, showing that W has also diffused beyond the depletion region.
- Unexpectedly, ***copper*** shifts the current histogram to larger dark current levels. Both I_{diff} and I_{gen} are larger than in the non-contaminated reference diodes. This suggests trapping of copper in the form of small complexes both in the depletion region and in the silicon bulk. $\Delta E \sim 0$ eV.
- ***Nickel*** has only a limited impact, although slightly higher I_{diff} and I_{gen} are observed. This can be explained by considering the nickel gettering by p$^+$-layer and oxygen precipitates in the Cz bulk wafer. $\Delta E \sim 0.10$ eV.
- A similar limited impact is found for ***iron***, where only I_{gen} is slightly increased. It is concluded that for the particular CIS processing conditions gettering of Fe and Ni is more effective than for Cu. $\Delta E \sim 0.10$ eV.
- ***Chromium***: very similar to the reference diode; can be easily gettered. $\Delta E \sim 0.0$ eV.
- For ***zinc***, I_{gen} is slightly increased. The $\Delta E \sim 0.0$ eV agrees well with the mid-gap position of the zinc level. Zinc can also evaporate completely from the wafer during thermal treatment.
- For the case of ***titanium*** (slow diffuser), I_{gen} is increased while I_{diff} remains unchanged. There appears to be a threshold with respect to the ion implantation energy and dose for the number of hot pixels. $\Delta E \sim 0.25$ eV.
- For ***vanadium***, I_{gen} increases while I_{diff} is unaffected. $\Delta E \sim 0.13$ eV.
- For ***aluminum*** there is no impact; $\Delta E \sim 0$ eV.

The positive impact of gettering has recently been confirmed for Si CMOS image sensors, designed with hydrogen ion-implantation-induced nanocavities and showing reduced dark current for Fe, Cu, Ni and Co contaminated CIS [228]. In addition, the use of a p/p$^+$ epitaxial wafer resulted in segregation gettering of Fe and

Co, while weak gettering was observed for Cu and Ni. A reduction of the white spots has also been achieved by proximity gettering, using implanted carbon complexes formed underneath the contact area of the CMOS image sensors [229].

7.3.3 Impact on Yield

Besides the operation and reliability, metal contamination has a potentially detrimental impact on circuit yield, as already demonstrated in the late seventies-early eighties [230–232]. Dedicated studies on the impact of specific TMs on yield have been published [184, 185, 232] and a scaling law for contamination control developed [233]. It is clear from the latter that the tolerable amount of metals drops significantly upon scaling the device dimensions.

The backside grinding process used for thinning the processed wafers before packaging has an impact on the DRAM yield [234]. The thinning requirements are becoming more stringent for thinner packages and/or the packaging of multiple dies in the same package. The influence on the yield is related to the final roughness of the backside, depending on the size of the polishing particles, and the associated stresses enabling the gettering of metals. However, the stress may lead to a bowing of the wafer resulting in a non-uniform distribution of the electrical parameters as function of the die position on the wafer. Therefore, stress engineering is required for an optimal device yield. The stress can be lowered by using a plasma etching and a dry polishing process. More details on the use of mechanical stress for metal gettering is given in Chap. 8.

References

1. S. Machida, Y. Yamashita, T. Misumi, T. Sugiyama, Effects of trap levels on reverse recovery surge of silicon power diode. Jpn. J. Appl. Phys. **52**, 04CP01/1-6 (2013). https://doi.org/10.7567/jjap.52.04cp01
2. E.H. Snow, A.S. Grove, B.E. Deal, C.T. Sah, Ion transport phenomena in insulating films. J. Appl. Phys. **36**, 1664–1673 (1965). https://doi.org/10.1063/1.1703105
3. E. Yon, W.H. Ko, A.B. Kuper, Sodium distribution in thermal oxide on silicon by radiochemical and MOS analysis. IEEE Trans. Electron Dev. **13**, 276–280 (1966). https://doi.org/10.1109/t-ed.1966.15680
4. J.V. Dalton, J. Drobek, Structure and sodium migration in silicon nitride films. J. Electrochem. Soc. **225**, 865–868 (1968). https://doi.org/10.1149/1.241145
5. P.S.D. Lin, R.B. Marcus, T.T. Sheng, Leakage and breakdown in thin oxide capacitors—correlation with decorated stacking faults. J. Electrochem. Soc. **130**, 1878–1883 (1983). https://doi.org/10.1149/1.2120116
6. D.A. Ramappa, W.B. Henley, Diffusion of iron in silicon dioxide. J. Electrochem. Soc. **146**, 3773–3777 (1999). https://doi.org/10.1149/1.1392548
7. O. Kononchuk, K.G. Korablev, N. Yarykin, G.A. Rozgonyi, Diffusion of iron in the silicon dioxide layer of silicon-on-insulator structures. Appl. Phys. Lett. **73**, 1206–1208 (1998). https://doi.org/10.1063/1.122128

8. K.L. Beaman, A. Agarwal, O. Kononchuk, S. Koveschnikov, I. Bondarenko, G.A. Rozgonyi, Gettering of iron in silicon-on-insulator wafers. Appl. Phys. Lett. **71**, 1107–1109 (1997). https://doi.org/10.1063/1.119741
9. J. Leveneur, M. Langlois, J. Kennedy, J.B. Metson, Excess oxygen limited diffusion and precipitation of iron in amorphous silicon dioxide. J. Appl. Phys. **122**, 135302/1-8 (2017). https://doi.org/10.1063/1.4985819
10. A.R. Smith, R.J. McDonald, H. Manini, D.L. Hurley, E.B. Norman, M.C. Vella, R.W. Odom, Low-background instrumental neutron activation analysis of silicon semiconductor materials. J. Electrochem. Soc. **143**, 339–346 (1996). https://doi.org/10.1149/1.1836433
11. Y. Kamiura, F. Hashimoto, M. Iwami, Observation of iron pileup and reduction of SiO_2 at the Si-SiO_2 interface. Appl. Phys. Lett. **53**, 1711–1713 (1988). https://doi.org/10.1063/1.99802
12. Y. Hoshino, H. Arima, Y. Saito, J. Nakata, Characterization of hot-implanted Fe near the SiO_2/Si interface. Jpn. J. Appl. Phys. **50**, 035601/1-5 (2011). https://doi.org/10.1143/jjap.50.035601
13. J. Wong-Leung, D.J. Eaglesham, J. Sapjeta, D.C. Jacobson, J.M. Poate, J.S. Williams, The precipitation of Fe at the Si-SiO_2 interface. J. Appl. Phys. **83**, 580–584 (1998). https://doi.org/10.1063/1.366643
14. H. Shimizu, H. Hagiwara, Enhanced and retarded SiO_2 growth on thermally oxidized Fe-contaminated n-type Si(001) surfaces. Jpn. J. Appl. Phys. **52**, 041302/1-7 (2013). https://doi.org/10.7567/jjap.52.041302
15. A. De Luca, M. Texier, A. Portavoce, N. Burie, C. Grosjean, S. Morata, F. Michel, Mechanism of β-$FeSi_2$ precipitates growth-and-dissolution and pyramidal defects' formation during oxidation of Fe-contaminated silicon wafers. J. Appl. Phys. **117**, 115302/1-9 (2015). https://doi.org/10.1063/1.4915086
16. R.N. Ghoshtagore, Diffusion of nickel in amorphous silicon dioxide and silicon nitride films. J. Appl. Phys. **40**, 4374–4376 (1969). https://doi.org/10.1063/1657201
17. J. Jablonski, Y. Miyamura, M. Imai, H. Tsuya, Gettering of Cu and Ni impurities in SIMOX wafers. J. Electrochem. Soc. **142**, 2059–2066 (1995). https://doi.org/10.1149/1.2044241
18. K. Honda, T. Nakanishi, A. Ohsawa, N. Toyokura, Catastrophic breakdown in silicon oxides: The effect of Fe impurities at the Si-SiO_2 interface. J. Appl. Phys. **62**, 1960–1963 (1987). https://doi.org/10.1063/1.339534
19. W.B. Henley, L. Jastrzebski, N.F. Haddad, Effects of iron contamination in silicon on thin oxide breakdown and reliability characteristics. J. Non-Cryst. Solids **187**, 134–139 (1995). https://doi.org/10.1016/0022-3093(95)00125-5
20. B.D. Choi, D.K. Schroder, Degradation of ultrathin oxides by iron contamination. Appl. Phys. Lett. **79**, 2645–2647 (2001). https://doi.org/10.1063/1.1410363
21. K. Honda, T. Nakanishi, Influence of Ni impurities at the Si-SiO_2 interface on the metal-oxide-semiconductor characteristics. J. Appl. Phys. **75**, 7394–7399 (1994). https://doi.org/10.1063/1.356654
22. S. Koveshnikov, D. Beauchaine, Z. Radzimski, V. Higgs, Degradation of gate oxide integrity due to Ni and Cu contamination and impurity gettering in epitaxial Si wafers. Electrochem. Soc. Proc. **2**, 387–396 (2000)
23. A.J. Lee, J.-S. Im, H.-B. Kang, S.-W. Lee, Defect formation of nickel-incorporated large-diameter Czochralski-grown silicon and their effect on gate oxide reliability. ECS J. Solid State Sci. Technol. **5**, P3008–P3012 (2016). https://doi.org/10.1149/2.0021604jss
24. A. Ohsawa, K. Honda, N. Toyokura, Metal impurities near the SiO_2-Si interface. J. Electrochem. Soc. **131**, 2964–2969 (1984). https://doi.org/10.1149/1.2115451
25. K. Hiramoto, M. Sano, S. Sadamitsu, N. Fujino, Degradation of gate oxide integrity by metal impurities. Jpn. J. Appl. Phys. **28**, L2109–L2111 (1989). https://doi.org/10.1143/JJAP.28.L2109

26. H. Wendt, H. Cerva, V. Lehmann, W. Pamler, Impact of copper contamination on the quality of silicon oxides. J. Appl. Phys. **65**, 2402–2405 (1989). https://doi.org/10.1063/1.342808
27. A. Correia, D. Ballutaud, J.-L. Maurice, Electron microscopy study of oxidation-induced defects at the silicon-silicon-dioxide interface. Mater. Sci. Eng. B **18**, 269–274 (1993). https://doi.org/10.1016/0921-5107(93)90142-A
28. A. Correia, D. Ballutaud, J.-L. Maurice, Copper precipitation at the silicon-silicon-dioxide interface: role of oxygen. Jpn. J. Appl. Phys. **33**, 1217–1222 (1994). https://doi.org/10.1143/JJAP.33.1217
29. B.O. Kolbesen, H. Cerva, Defects due to metal silicide precipitation in microelectronic device manufacturing: the unlovely face of transition metal silicides. Phys. Status Solidi B **222**, 303–317 (2000). https://doi.org/10.1002/1521-3951(200011)222<301::AID-PSSB303>3.0.CO;2-H
30. J.D. McBrayer, R.M. Swanson, T.W. Sigmon, Diffusion of metals in silicon dioxide. J. Electrochem. Soc. **133**, 1242–1246 (1986). https://doi.org/10.1149/1.2108827
31. J.-Y. Kwon, K.-S. Kim, Y.-C. Joo, K.-B. Kim, Simulation of copper diffusion profile in SiO_2 during bias temperature stress (BTS) test. Jpn. J. Appl. Phys. **41**, L99–L101 (2002). https://doi.org/10.1143/JJAP.41.L99
32. P. Bai, G.R. Yang, T.M. Lu, Intrinsic Cu gettering at a thermally grown SiO_2/Si interface. J. Appl. Phys. **88**, 3313–3316 (1990). https://doi.org/10.1063/1.346383
33. L. Zhong, F. Shimura, Substitutional diffusion of transition metal impurities in silicon. Jpn. J. Appl. Phys. **32**, L1113–L1116 (1993). https://doi.org/10.1143/JJAP.32.L1113
34. K. Hozawa, T. Itoga, S. Isomae, J. Yugami, M. Ohkura, Copper distribution behavior near a SiO_2/Si interface by low-temperature (<400 °C) annealing and its influence on electrical characteristics of MOS capacitors, in *VLSI Technology* Digest of *Technical Papers*, pp. 24–25 (2002). https://doi.org/10.1143/jjap.41.5887
35. K. Hozawa, S. Isomae, J. Yugani, Copper distribution near a SiO_2/Si interface under low-temperature annealing. Jpn. J. Appl. Phys. **41**, 5887–5893 (2002). https://doi.org/10.1143/JJAP.41.5887
36. A. Kohn, E. Lipp, M. Eizenberg, Y. Shacham-Diamand, Copper-related degradation of SiO_2 in metal-oxide-semiconductor capacitors subjected to bias thermal stress: leakage of the minority charge carriers in the inversion layer. Appl. Phys. Lett. **85**, 627–629 (2004). https://doi.org/10.1063/1.1773925
37. A.M. Hoff, S. Aravamudhan, A. Isti, E.I. Oborina, Degradation of oxide properties caused by low-level metallic contamination. J. Electrochem. Soc. **154**, H977–H982 (2007). https://doi.org/10.1149/1.2779959
38. J.M.E. Harper, A. Charai, L. Stolt, F.M. d'Heurle, P.M. Fryer, Room-temperature oxidation of silicon catalyzed by Cu_3Si. Appl. Phys. Lett. **56**, 2519–2521 (1990). https://doi.org/10.1063/1.103260
39. M. Setton, J. Van der Spiegel, B. Rothman, Copper silicide formation by rapid thermal processing and induced room temperature Si oxide growth. Appl. Phys. Lett. **57**, 357–359 (1990). https://doi.org/10.1063/1.104105
40. D. Gräf, M. Grundner, L. Mühlhoff, M. Dellith, Influence of Cu on native oxide growth of Si. J. Appl. Phys. **69**, 7620–7626 (1991). https://doi.org/10.1063/1.347531
41. D. Gupta, K. Vieregge, K.V. Srikrishnan, Copper diffusion in amorphous thin films of 4% phosphorus-silicate glass and hydrogenated silicon nitride. Appl. Phys. Lett. **61**, 2178–2180 (1992). https://doi.org/10.1063/1.108287
42. H. Miyazaki, H. Kojima, A. Hiraiwa, Y. Homma, K. Murakami, Phosphosilicate glass passivation for ULSI Cu metallization. J. Electrochem. Soc. **139**, 3264–3267 (1992). https://doi.org/10.1149/1.2069064
43. M. Vogt, M. Kachel, M. Plötner, K. Drescher, Dielectric barriers for Cu metallization systems. Microelectron. Eng. **37**(38), 181–187 (1997). https://doi.org/10.1016/S6167-9317(97)00110-X

44. H. Miyazaki, H. Kojima, K. Hinode, Passivation effect of silicon nitride against copper diffusion. J. Appl. Phys. **81**, 7746–7750 (1997). https://doi.org/10.1063/1.365380
45. A.L.S. Loke, C. Ryu, C.P. Yue, J.S.H. Cho, S.S. Wong, Kinetics of copper drift in PECVD dielectrics. IEEE Electron Dev. Lett. **17**, 549–551 (1996). https://doi.org/10.1109/55.545766
46. A.L.S. Loke, J.T. Wetzel, P.H. Townsend, T. Tanabe, R.N. Vrtis, M.P. Zussman, D. Kumar, C. Ryu, S.S. Wong, Kinetics of copper drift in low-κ polymer interlevel dielectrics. IEEE Trans. Electron Dev. **46**, 2178–2187 (1999). https://doi.org/10.1109/16.796294
47. P. Motte, J. Torres, J. Palleau, F. Tardif, H. Bernard, Study of Cu contamination during copper integration for subquarter micron technology. Solid State Electron. **43**, 1015–1018 (1999). https://doi.org/10.1016/S0038-1101(99)00017-9
48. H. Cui, I.B. Bhat, S.P. Murarka, H. Lu, W.-J. Hsia, W. Catabay, Copper drift in methyl-doped silicon oxide film. J. Vac. Sci. Technol., B **20**, 1987–1993 (2002). https://doi.org/10.1116/1.1503779
49. K.-S. Kim, Y.-C. Joo, K.-B. Kim, J.-Y. Kwon, Extraction of Cu diffusivities in dielectric materials by numerical calculation and capacitance-voltage measurement. J. Appl. Phys. **100**, 063517/1-6 (2006). https://doi.org/10.1063/1.2353891
50. R. Ali, S.W. King, M. Orlowski, Modeling and simulation of Cu drift in porous low-*k* dielectrics. ECS Trans. **80**(1), 327–337 (2017). https://doi.org/10.1149/08001.0327ecst
51. S. Verhaverbeke, M. Meuris, P.W. Mertens, M.M. Heyns, A. Philipossian, D. Gräf, A. Schnegg, The effect of metallic impurities on the dielectric breakdown of oxides and some new ways of avoiding them, in *Technical Digest of the IEEE International Electron Devices Meeting—IEDM91* (1991), pp. 71–74. https://doi.org/10.1109/iedm.1991.235421
52. R. Hölzl, A. Huber, L. Fabry, K.-J. Range, M. Blietz, Integrity of ultrathin gate oxides with different oxide thickness, substrate wafers and metallic contaminations. Appl. Phys. A **72**, 351–356 (2001). https://doi.org/10.1007/s003390000721
53. D.A. Ramappa, W.B. Henley, Effects of copper contamination in silicon on thin oxide breakdown. J. Electrochem. Soc. **146**, 2258–2260 (1999). https://doi.org/10.1149/1.1391924
54. Y.H. Lin, Y.H. Wu, A. Chin, F.M. Pan, The effect of copper on gate oxide integrity. J. Electrochem. Soc. **147**, 4305–4306 (2000). https://doi.org/10.1149/1.1394059
55. Y.H. Lin, Y.C. Chen, K.T. Chan, F.M. Pan, I.J. Hsieh, A. Chin, The strong degradation of 30 Å gate oxide integrity contaminated by copper. J. Electrochem. Soc. **148**, F73–F76 (2001). https://doi.org/10.1149/1.1357182
56. K. Hozawa, H. Miyazaki, J. Yugami, True influence of wafer-backside copper contamination during the back-end process on device characteristics, in *Technical Digest of the IEEE International Electron Devices Meeting—IEDM02* (2002), pp. 737–740. https://doi.org/10.1109/iedm.2002.1175943
57. K.-S. Lee, E.-H. Kim, Y.-H. Kim, B.-Y. Lee, H.-D. Yoo, Effects of surface roughness and copper contamination on the oxide breakdown of silicon wafer. Jpn. J. Appl. Phys. **41**, 3662–3668 (2002). https://doi.org/10.1143/JJAP.41.3662
58. N. Tokuda, T. Kanda, S. Yamasaki, K. Miki, K. Yamabe, Local electrical characteristics of Cu-contaminated SiO_2 thin films. Electrochem. Soc. Proc. **2002**(20), 311–319 (2002). https://doi.org/10.4028/www.scientific.net/SSP.95-96.641
59. N. Tokuda, T. Kanda, S. Yamasaki, K. Miki, K. Yamabe, Leakage current distribution of Cu-contaminated thin SiO_2. Jpn. J. Appl. Phys. **42**, L160–L162 (2003). https://doi.org/10.1143/JJAP.42.L160
60. N. Tokuda, S. Nishiguchi, S. Yamasaki, K. Miki, K. Yamabe, Leakage current distribution and dielectric breakdown of Cu-contaminated thin SiO_2. J. Electrochem. Soc. **151**, F81–F86 (2004). https://doi.org/10.1149/1.1649984
61. C.S. Liu, L.J. Chen, Catalytic oxidation of (001)Si in the presence of Cu_3Si at room temperature. J. Appl. Phys. **74**, 3611–3613 (1993). https://doi.org/10.1063/1.354499
62. C.S. Liu, L.J. Chen, Room-temperature oxidation of silicon in the presence of Cu_3Si. Thin Solid Films **262**, 187–198 (1995). https://doi.org/10.1016/0040-6090(95)05814-1

63. Y.H. Lin, F.M. Pan, Y.C. Liao, Y.C. Chen, I.J. Hsieh, A. Chin, Cu contamination effect in oxynitride gate dielectrics. J. Electrochem. Soc. **148**, G627–G629 (2001). https://doi.org/10.1149/1.140784
64. K. Matsukawa, H. Yamamoto, Y. Mashiko, A study of metal impurities behavior due to difference in isolation structure on ULSI devices. Jpn. J. Appl. Phys. **41**, 5900–5903 (2002). https://doi.org/10.1143/JJAP.41.5900
65. C.C. Liao, C.F. Cheng, D.S. Yu, A. Chin, The copper contamination effect of Al_2O_3 gate dielectric on Si. J. Electrochem. Soc. **151**, G693–G696 (2004). https://doi.org/10.1149/1.1789391
66. J. Bea, K. Lee, T. Fukushima, T. Tanaka, M. Koyanagi, Evaluation of Cu contamination at backside surface of thinned wafer in 3-D integration by transient-capacitance measurement. IEEE Electron Dev. Lett. **32**, 66–68 (2011). https://doi.org/10.1109/LED.2010.2087004
67. K.-W. Lee, J.-C. Bea, T. Fukushima, T. Tanaka, M. Koyanagi, Electrical evaluation of Cu contamination behavior at the backside surface of a thinned wafer by transient capacitance measurement. Semicond. Sci. Technol. **26**, 025007/1-4 (2011). https://doi.org/10.1088/0268-1242/26/2/025007
68. J. Bea, K. Lee, T. Fukushima, T. Tanaka, M. Koyanagi, Evaluation of Cu diffusion from Cu through-silicon via (TSV) in three-dimensional LSI by transient capacitance measurement. IEEE Electron Dev. Lett. **32**, 940–942 (2011). https://doi.org/10.1109/LED.2011.2141109
69. K.W. Lee, J.-C. Bea, T. Fukushima, T. Tanaka, M. Koyanagi, Cu retardation performance of extrinsic gettering layers in thinned wafers evaluated by transient capacitance measurement. J. Electrochem. Soc. **158**, H795–H799 (2011). https://doi.org/10.1149/1.3597317
70. K. Lee, J. Bea, Y. Ohara, T. Fukushima, T. Tanaka, M. Koyanagi, Impact of Cu diffusion from Cu through-silicon via (TSV) on device reliability in 3-D LSIs evaluated by transient capacitance measurement, in *Proceedings of the IEEE International Reliability Physics Symposium—IRPS12* (2012), pp. 2B.4.1–2B.4.6. https://doi.org/10.1109/irps.2012.6241777
71. K.-W. Lee, J.-C. Bea, Y. Ohara, M. Murugesan, T. Fukushima, T. Tanaka, M. Koyanagi, Impacts of Cu contamination on device reliabilities in 3-D IC integration. IEEE Trans. Dev. Mater. Reliab. **14**, 451–462 (2014). https://doi.org/10.1109/TDMR.2013.2258022
72. L. Lee, H.G. Parks, R.D. Schrimpf, Interpretation of experimentally observed *C-t* responses for copper contaminated devices. Solid State Electron. **39**, 369–375 (1996). https://doi.org/10.1016/0038-1101(95)00133-6
73. Y. Mizushima, H. Kitada, K. Koshikawa, S. Suzuki, T. Nakamura, T. Ohba, Novel through silicon vias leakage current evaluation using infrared-optical beam irradiation. Jpn. J. Appl. Phys. **51**, 05EE03/1-4 (2012). https://doi.org/10.1143/jjap.51.05ee03
74. K.-D. Kim, K.-K. Kim, M.-S. Yoo, Y.-T. Kim, S.-K. Park, S.-J. Hong, C.-H. Park, B.-G. Park, J.-H. Lee, Novel method to characterize the effect from the diffusion of Cu in through silicon via (TSV), in *Technical Digest of the IEEE International Electron Devices Meeting—IEDM15* (2015), pp. 157–160. https://doi.org/10.1109/iedm.2015.7409645
75. J.W. Chan, C.S. Tan, K.C. Lee, X. Cheng, W. Kanert, Observations of copper (Cu) transport in through-silicon vias (TSV) structure by electrical characterization for its reliability evaluation, in *Proceedings of IEEE International Reliability Physics Symposium* (2017), pp. 4A-3.1–4A-3.6. https://doi.org/10.1109/irps.2017.7936303
76. M.S. Angyal, Y. Shacham-Diamand, J.S. Reid, M.A. Nicolet, Performance of tantalum-silicon-nitride diffusion barriers between copper and silicon dioxide. Appl. Phys. Lett. **67**, 2152–2154 (1995). https://doi.org/10.1063/1.114750
77. S.-K. Rha, S.-Y. Lee, W.-J. Lee, Y.-S. Hwang, C.-O. Park, D.W. Kim, Y.-S. Lee, C.-N. Whang, Characterization of TiN barriers against Cu diffusion by capacitance-voltage measurement. J. Vac. Sci. Technol., B **16**, 2019–2026 (1998). https://doi.org/10.1116/1.590123
78. S. Hymes, K.S. Kumar, S.P. Murarka, P.J. Ding, W. Wang, W.A. Lanford, Thermal stability of copper silicide passivation layers in copper-based multilevel interconnects. J. Appl. Phys. **83**, 4507–4512 (1998). https://doi.org/10.1063/1.367235

79. K. Takeda, D. Ryuzaki, T. Mine, K. Hinode, R. Yoneyama, Copper blocking ability of nitrogen-incorporated silicon oxide film. J. Vac. Sci. Technol. B **21**, 1323–1328 (2003). https://doi.org/10.1116/1.1587141
80. S. Rawal, D.P. Norton, K.C. Kim, T.J. Anderson, L. McElwee-White, Ge/HfN_x diffusion barrier for Cu metallization on Si. Appl. Phys. Lett. **89**, 231914/1-3 (2006). https://doi.org/10.1063/1.2400071
81. B.-S. Nguyen, J.-F. Lin, D.-C. Perng, 1-nm-thick graphene tri-layer as the ultimate copper diffusion barrier. Appl. Phys. Lett. **104**, 082105/1-5 (2014). https://doi.org/10.1063/1.4866857
82. C.-L. Lo, K.H. Smithe, R. Mehta, S. Chugh, E. Pop, Z. Chen, Atomically thin diffusion barriers for ultra-scaled Cu interconnects implemented by 2D materials, in *Proceedings of IEEE International Reliability Physics Symposium* (2017), pp. MR-4.1–MR-4.4. https://doi.org/10.1109/irps.2017.7936379
83. R. Gonella, Key reliability issues for copper integration in damascene architecture. Microelectron. Eng. **55**, 245–255 (2001). https://doi.org/10.1016/S0167-9317(00)00454-8
84. J. Noguchi, Dominant factors in TDDB degradation of Cu interconnects. IEEE Trans. Electron Dev. **52**, 1743–1750 (2005). https://doi.org/10.1109/TED.2005.851849
85. W. Wu, X. Duan, J.S. Yuan, A physical model of time-dependent dielectric breakdown in copper metallization, in *Proceedings of IEEE International Reliability Physics Symposium—IRPS03* (2003), pp. 282–286. https://doi.org/10.1109/relphy.2003.1197758
86. K. Takeda, D. Ryuzaki, T. Mine, K. Hinode, R. Yoneyama, Copper-induced dielectric breakdown in silicon oxide deposited by plasma-enhanced chemical vapor deposition using trimethoxysilane. J. Appl. Phys. **94**, 2572–2578 (2003). https://doi.org/10.1063/1.1594812
87. H. Cui, P.A. Burke, Time-dependent dielectric breakdown of hydrogenated silicon carbon nitride thin films under the influence of copper. Appl. Phys. Lett. **84**, 2629–2631 (2004). https://doi.org/10.1063/1.1703839
88. J.R. Lloyd, E. Liniger, T.M. Shaw, Simple model for time-dependent dielectric breakdown in inter- and intralevel low-k dielectrics. J. Appl. Phys. **98**, 084109/1-6 (2005). https://doi.org/10.1063/1.2112171
89. J.R. Lloyd, C.E. Murray, S. Ponoth, S. Cohen, E. Liniger, The effect of Cu diffusion on the TDDB behavior in a low-*k* interlevel dielectrics. Microelectron. Reliab. **46**, 1643–1647 (2006). https://doi.org/10.1016/j.microrel.2006.08.003
90. N. Suzumura, S. Yamamoto, D. Kodama, K. Makabe, J. Komori, E. Murakami, S. Maegawa, K. Kubota, A new TDDB degradation model based on Cu ion drift in Cu interconnect dielectrics, in *Proceedings of IEEE International Reliability Physics Symposium—IRPS06* (2006), pp. 484–489. https://doi.org/10.1109/relphy.2006.251266
91. F. Chen, O. Bravo, K. Chanda, P. McLaughlin, T. Sullivan, J. Gill, J. Lloyd, R. Kontra, J. Aitken, A comprehensive study of low-*k* SiCOH TDDB phenomena and its reliability lifetime model development, in *Proceedings of IEEE International Reliability Physics Symposium—IRPS06* (2006), pp. 46–53. https://doi.org/10.1109/relphy.2006.251190
92. R.S. Achanta, J.L. Plawsky, W.N. Gill, A time dependent dielectric breakdown model for field accelerated low-*k* breakdown due to copper ions. Appl. Phys. Lett. **91**, 234106/1-3 (2007). https://doi.org/10.1063/1.2823576
93. S.S. Hwang, S.-Y. Jup, Y.-C. Joo, The electric field dependence of Cu migration induced dielectric failure in interlayer dielectric for integrated circuits. J. Appl. Phys. **101**, 074501/1-5 (2007). https://doi.org/10.1063/1.2714668
94. L. Zhao, Z. Tőkei, K. Croes, C.J. Wilson, M. Baklanov, G.P. Beyer, C. Claeys, Direct observation of the $1/E$ dependence of time dependent dielectric breakdown in the presence of copper. Appl. Phys. Lett. **98**, 032107/1-3 (2011). https://doi.org/10.1063/1.3543850
95. F. Chen, M. Shinosky, J. Aitken, C.-C. Yang, D. Edelstein, Invasion percolation model for abnormal time-dependent dielectric breakdown characteristics of low-*k* dielectrics due to massive metallic diffusion. Appl. Phys. Lett. **101**, 242904/1-4 (2012). https://doi.org/10.1063/1.4770318

96. F. Chen, M. Shinosky, J. Aitken, C.-C. Yang, D. Edelstein, Temperature and field interrelation study of low-*k* TDDB for Cu interconnects with and without liner—new insights to the roles of Cu for a competing breakdown process, in *Proceedings IEEE of International Reliability Physics Symposium—IRPS13* (2013), pp. 2F.2.1–2F.2.7. https://doi.org/10.1109/irps.2013.6531967
97. S.-Y. Jung, B.-J. Kim, N.-Y. Lee, B.-M. Kim, S.J. Yeom, N.J. Kwak, Y.-C. Joo, Bias polarity and frequency effects of Cu-induced dielectric breakdown in damascene Cu interconnects. Microelectron. Eng. **89**, 58–61 (2012). https://doi.org/10.1016/j.mee.2011.01.070
98. Y.-L. Cheng, Y.-L. Huang, C.-R. Sun, W.-H. Lee, G.-S. Chen, J.-S. Fang, Effect of Cu drift on dielectric breakdown for porous low dielectric constant film under static and dynamic stress. ECS Trans. **72**(2), 241–252 (2016). https://doi.org/10.1149/07202.0241ecst
99. A. Goetzberger, W. Shockley, Metal precipitates in silicon p-n junctions. J. Appl. Phys. **31**, 1821–1824 (1960). https://doi.org/10.1063/1.1735455
100. J.E. Lawrence, Electrical properties of copper segregates in silicon p-n junctions. J. Electrochem. Soc. **112**, 796–800 (1965). https://doi.org/10.1149/1.242369
101. R. Böhm, H. Klose, Copper in silicon n^+-p junctions. Phys. Status Solidi A **9**, K165–K168 (1972). https://doi.org/10.1002/pssa.221009059
102. H.H. Busta, H.A. Waggener, Precipitation-induced currents and generation-recombination currents in intentionally contaminated silicon p^+n junctions. J. Electrochem. Soc. **124**, 1424–1429 (1977). https://doi.org/10.1149/1.2133667
103. R.W. Hamaker, Z.C. Putney, R.L. Ayers, P.H. Smith, Degradation mechanism for silicon p^+-n junctions under forward bias. Solid State Electron. **24**, 1001–1008 (1981). https://doi.org/10.1016/0038-1101(81)90127-1
104. K.V. Ravi, C.J. Varker, C.E. Volk, Electrically active stacking faults in silicon. J. Electrochem. Soc. **120**, 533–541 (1973). https://doi.org/10.1149/1.2403495
105. M. Miyazaki, M. Sano, S. Sumita, N. Fujino, Influence of metal impurities on leakage current of Si n^+p diode. Jpn. J. Appl. Phys. **30**, L295–L297 (1991). https://doi.org/10.1143/JJAP.30.L295
106. H.G. Parks, R.D. Schrimpf, B. Craigin, R. Jones, P. Resnick, Quantifying the impact of homogeneous metal contamination using test structure metrology and device modeling. IEEE Trans. Semicond. Manufact. **7**, 249–258 (1994). https://doi.org/10.1109/66.311326
107. B. Vermeire, L. Lee, H.G. Parks, The effect of copper contamination on field overlap edges and perimeter junction leakage current. IEEE Trans. Semicond. Manufact. **11**, 232–238 (1998). https://doi.org/10.1109/66.670169
108. A.A. Istratov, E.R. Weber, Physics of copper in silicon. J. Electrochem. Soc. **149**, G21–G30 (2002). https://doi.org/10.1149/1.1421348
109. C.C. Wang, H.-H. Lin, M.-C. Chen, Thermal stability of Cu/NiSi-contacted p^+n shallow junction. Jpn. J. Appl. Phys. **43**, 5997–6000 (2004). https://doi.org/10.1143/JJAP.43.5997
110. A.A. Istratov, C. Flink, E.R. Weber, Impact of the unique physical properties of copper in silicon on characterization of copper diffusion barriers. Phys. Status Solidi B **222**, 261–277 (2000). https://doi.org/10.1002/1521-3951(200011)222:1<261:aid-pssb261>3.0.co;2-5
111. E. Kolawa, P.J. Pokela, J.S. Reid, J.S. Chen, R.P. Ruiz, M.A. Nicolet, Sputtered Ta-Si-N diffusion barriers in Cu metallizations for Si. IEEE Electron Dev. Lett. **12**, 321–323 (1991). https://doi.org/10.1109/55.82074
112. E. Kolawa, J.S. Chen, J.S. Reid, P.J. Pokela, M.A. Nicolet, Tantalum-based diffusion barriers in Si/Cu VLSI metallizations. J. Appl. Phys. **70**, 1369–1373 (1991). https://doi.org/10.1063/1.349594
113. T.-S. Chang, W.-C. Wang, L.-P. Wang, J.-C. Hwang, F.-S. Huang, Thermal stability study of $TiN/TiSi_2$ diffusion barrier between Cu and n^+Si. J. Appl. Phys. **75**, 7847–7851 (1994). https://doi.org/10.1063/1.356568
114. K.-M. Chang, T.-H. Yeh, I.-C. Deng, C.-W. Shih, Amorphous like chemical vapor deposited tungsten diffusion barrier for copper metallization and effects of nitrogen addition. J. Appl. Phys. **82**, 1469–1475 (1997). https://doi.org/10.1063/1.365925

115. J. Baumann, Ch. Kaufmann, M. Rennau, T. Werner, T. Gessner, Investigation of copper metallization induced failure of diode structures with and without a barrier layer. Microelectron. Eng. **33**, 283–291 (1997). https://doi.org/10.1016/S0167-9317(96)00056-1
116. M.T. Wang, Y.C. Lin, M.C. Chen, Barrier properties of very thin Ta and TaN layers against copper diffusion. J. Electrochem. Soc. **145**, 2538–2545 (1998). https://doi.org/10.1149/1.1838675
117. S. Nakao, M. Numata, T. Ohmi, Thin and low-resistivity tantalum nitride diffusion barrier and giant-grain copper interconnects for advanced ULSI metallization. Jpn. J. Appl. Phys. **38**, 2401–2405 (1999). https://doi.org/10.1143/JJAP.38.2401
118. J.-C. Chuang, S.-L. Tun, M.-C. Chen, Rapid thermal annealed Cr barrier against Cu diffusion. J. Electrochem. Soc. **146**, 2643–2647 (1999). https://doi.org/10.1149/1.1391985
119. M.T. Wang, L.J. Chen, M.C. Chen, Barrier capabilities of selective chemical vapor deposited W films and WSiN/WSi$_x$/W stacked layers against copper diffusion. J. Electrochem. Soc. **146**, 728–734 (1999). https://doi.org/10.1149/1.1391671
120. P.-T. Liu, T.-C. Chang, J.C. Hu, Y.L. Yang, S.M. Sze, Reliability of multistacked chemical vapor deposited Ti/TiN structure as the diffusion barrier in ultralarge scale integrated metallization. J. Electrochem. Soc. **147**, 368–372 (2000). https://doi.org/10.1149/1.1393202
121. G. Van den bosch, S. Demuynck, Z. Tőkei, G. Beyer, M. Van Hove, G. Groeseneken, Impact of copper contacts on CMOS front-end yield and reliability, in *Technical Digest IEEE International Electron Devices Meeting—IEDM06* (2006), pp. 1–4. https://doi.org/10.1109/iedm.2006.346967
122. Y. Yang, R. Labie, O. Richard, H. Bender, C. Zhao, B. Verlinden, I. De Wolf, The impact of back-side Cu contamination on 3D stacking architecture. Electrochem. Solid State Lett. **13**, H39–H41 (2010). https://doi.org/10.1149/1.3269603
123. H.W. Yeon, S.-Y. Jung, J. Lim, J. Pyun, H. Kim, D. Baek, Y.-C. Joo, Cu contamination of the nMOSFET in a 3-D integrated circuit under thermal and electrical stress. Electrochem. Solid State Lett. **15**, H157–H160 (2012). https://doi.org/10.1149/2.018205esl
124. Y.-J. Chang, C.-T. Ko, T.-H. Yu, C.-H. Chiang, K.N. Chen, Backside-process-induced junction leakage and process improvement of Cu TSV based on Cu/Sn and BCB hybrid bonding. IEEE Electron Dev. Lett. **34**, 435–437 (2013). https://doi.org/10.1109/LED.2013.2238213
125. E. Simoen, P. Clauws, J. Broeckx, J. Vennik, M. Van Sande, L. De Laet, Correlation between DLTS-measurements and the performance of high purity germanium detectors. IEEE Trans. Nucl. Sci. **NS-29**, 789–792 (1982). https://doi.org/10.1109/tns.1982.4335960
126. E. Simoen, P. Clauws, G. Huylebroeck, J. Vennik, Correlation between deep-level parameters and energy resolution of p-type high purity Ge γ-detectors. Nucl. Instrum. Meth. Phys. Res. A **251**, 519–526 (1986). https://doi.org/10.1016/0168-9002(86)90646-7
127. L.K.J. Vandamme, E.P. Vandamme, J.J. Dobbelsteen, Impact of silicon substrate, iron contamination and perimeter on saturation current and noise in n^+p diodes. Solid State Electron. **41**, 901–908 (1997). https://doi.org/10.1016/S0038-1101(97)00023-3
128. S.K. Ghandhi, K.E. Mortensen, J.N. Park, Impact ionization devices. IEEE Trans. Electron Dev. **13**, 515–519 (1966). https://doi.org/10.1109/t-ed.1966.15723
129. A.A. Lebedev, A.T. Mamadalimov, N.A. Sultanov, Principal parameters of diodes made of nickel-doped silicon. Sov. Phys. Semicond. **5**, 1990–1993 (1972)
130. N. Holonyak Jr., S.F. Bevacqua, Oscillations in semiconductors due to deep levels. Appl. Phys. Lett. **2**, 71–73 (1963). https://doi.org/10.1063/1.1753780
131. S.K. Ghandhi, K.E. Mortenson, J.N. Park, Impact ionization in cobalt-doped silicon. Proc. IEEE, 635 (1965). https://doi.org/10.1109/proc.1965.3949
132. J.S. Moore, C.M. Penchina, N. Holonyak Jr., M.D. Sirkis, T. Yamada, Electrical oscillations in silicon compensated with deep levels. J. Appl. Phys. **37**, 2009–2013 (1966). https://doi.org/10.1063/1.1708661
133. J. Amano, K. Nauka, M.P. Scott, J.E. Turner, R. Tsai, Junction leakage in titanium self-aligned silicide devices. Appl. Phys. Lett. **49**, 737–739 (1986). https://doi.org/10.1063/1.97584

134. M. Delfino, A.E. Morgan, E.K. Broadbent, P. Maillot, D.K. Sadana, Effect of post-silicidation annealing on $TiSi_2/p^+$-n Si junctions. J. Appl. Phys. **62**, 1882–1886 (1987). https://doi.org/10.1063/1.339573
135. J. Lin, S. Banerjee, J. Lee, C. Teng, Soft breakdown in titanium-silicided shallow source/drain junctions. IEEE Electron Dev. Lett. **11**, 191–193 (1990). https://doi.org/10.1109/55.55246
136. M. Ada-Hanifi, A. Chantre, D. Levy, J.P. Gonchond, Ph. Delpech, A. Nouailhat, Leakage mechanisms of titanium silicided n^+/p junctions fabricated using rapid thermal processing. Appl. Phys. Lett. **58**, 1280–1282 (1991). https://doi.org/10.1063/1.104336
137. R. Liu, D.S. Williams, W.T. Lynch, A study of the leakage mechanisms of silicided n^+/p junctions. J. Appl. Phys. **63**, 1990–1999 (1988). https://doi.org/10.1063/1.341099
138. H.C. Cheng, M.H. Juang, C.T. Lin, L.M. Huang, A silicidation-induced process consideration for forming scaled-down silicided junctions. IEEE Electron Dev. Lett. **15**, 342–344 (1994). https://doi.org/10.1109/311128
139. K. Goto, A. Fushida, J. Watanabe, T. Sukegawa, K. Kawamura, T. Yamazaki, T. Sugii, Leakage mechanism and optimized conditions for Co salicide process for deep-submicron CMOS devices, in *Technical Digest IEEE International Electron Devices Meeting* (1995), pp. 449–452. https://doi.org/10.1109/iedm.1995.499235
140. K. Goto, A. Fushida, J. Watanabe, T. Sukegawa, Y. Tada, T. Nakamura, T. Yamazaki, T. Sugii, A new leakage mechanism of Co salicide and optimized process conditions. IEEE Trans. Electron Dev. **46**, 117–124 (1999). https://doi.org/10.1109/16.737449
141. E. Simoen, G. Eneman, M. Bargallo Gonzalez, D. Kobayashi, A. Luque Rodríguez, J.-A. Jiménez Tejada, C. Claeys, High doping density/high electric field, stress and heterojunction effects on the characteristics of CMOS compatible p-n junctions. J. Electrochem. Soc. **158**, R27–R36 (2011). https://doi.org/10.1149/1.3555103
142. H.-D. Lee, M.-S. Bae, H.-H. Ji, K.-M. Lee, S.-H. Park, M.-J. Jang, J.-H. Lee, K.-S. Yoon, J.-H. Choi, G.-S. Park, K.-K. Kang, Y.-J. Park, Characterization of the Co-silicide penetration depth into the junction area for 0.15 and sub-0.15 micron CMOS technology. Jpn. J. Appl. Phys. **41**, 2445–2449 (2002). https://doi.org/10.1143/JJAP.41.2445
143. M. Tsuchiaki, C. Hongo, A. Takashima, K. Ohuchi, Intrinsic junction leakage generated by cobalt in-diffusion during $CoSi_2$ formation. Jpn. J. Appl. Phys. **41**, 2437–2444 (2002). https://doi.org/10.1143/JJAP.41.2437
144. M. Tsuchiaki, A. Murakoshi, C. Hongo, Systematic investigation of leakage suppression by pre-silicide implantation for $CoSi_2$ formation on shallow n^+/p Si diodes. Jpn. J. Appl. Phys. **42**, 1847–1854 (2003). https://doi.org/10.1143/JJAP.42.1847
145. W.D. Chen, Y.D. Cui, C.C. Hsu, J. Tao, Interaction of Co with Si and SiO_2 during rapid thermal annealing. J. Appl. Phys. **69**, 7612–7619 (1991). https://doi.org/10.1063/1.347530
146. C.-C. Wang, C.-J. Lin, M.-C. Chen, Formation of NiSi-silicided p^+n shallow junctions using implant-through-silicide and low-temperature furnace annealing. J. Electrochem. Soc. **150**, G557–G562 (2003). https://doi.org/10.1149/1.1599851
147. M. Tsuchiaki, K. Ohuchi, C. Hongo, Junction leakage generation by NiSi thermal instability characterized using damage-free n^+/p silicon diodes. Jpn. J. Appl. Phys. **43**, 5166–5173 (2004). https://doi.org/10.1143/JJAP.43.5166
148. M. Tsuchiaki, A. Nishiyama, Substrate orientation dependence of NiSi junction leakage induced by anisotropic Ni migration in crystal Si. Jpn. J. Appl. Phys. **46**, 1830–1840 (2007). https://doi.org/10.1143/JJAP.46.1830
149. D.Z. Chi, D. Mangelinck, A.S. Zuruzi, A.S.W. Wong, S.K. Lahiri, Nickel silicide as a contact material for submicron CMOS devices. J. Electron. Mater. **30**, 1483–1488 (2001). https://doi.org/10.1007/s11664-001-0162-4
150. D.Z. Chi, D. Mangelinck, J.Y. Dai, S.K. Lahiri, K.L. Pey, C.S. Ho, Nickel-platinum alloy monosilicidation-induced defects in n-type silicon. Appl. Phys. Lett. **76**, 3385–3387 (2000). https://doi.org/10.1063/1.126741

151. D.Z. Chi, D. Mangelinck, S.K. Lahiri, P.S. Lee, K.L. Pey, Comparative study of current-voltage characteristics of Ni and Ni(Pt)-alloy silicided p^+/n diodes. Appl. Phys. Lett. **78**, 3256–3258 (2001). https://doi.org/10.1063/1.1374496
152. T. Yamaguchi, K. Kashihara, T. Okudaira, T. Tsutsumi, K. Maekawa, T. Kosugi, N. Murata, J. Tsuchimoto, K. Shiga, K. Asai, M. Yoneda, Suppression of anomalous gate edge leakage current by control of Ni silicidation region using Si ion implantation technique, in *Proceedings of IEEE International Electron Devices Meeting IEDM06* (2006), pp. 855–858. https://doi.org/10.1109/iedm.2006.346916
153. T. Yamaguchi, K. Kashihara, S. Kudo, K. Hayashi, N. Hashikawa, T. Okudaira, T. Tsutsumi, K. Maekawa, H. Oda, K. Asai, M. Kojima, A novel low leakage-current Ni SALICIDE process in nMOSFETs on Si(110) substrate, in *Proceedings of IEEE International Electron Devices Meeting IEDM07* (2007), pp. 136–142. https://doi.org/10.1109/iedm.2007.4418884
154. S. Kudo, Y. Hirose, T. Hashikawa, K. Kashihara, K. Maekawa, K. Asai, N. Murata, K. Asayama, E. Murakami, Analysis of Ni silicide abnormal growth mechanism using advanced TEM techniques, in *Proceedings of IEEE International Reliability Physics Symposium* (2008), pp. 580–583. https://doi.org/10.1109/relphy.2008.4558948
155. B. Imbert, R. Pantel, S. Zoll, M. Gregoire, R. Beneyton, S. del Medico, O. Thomas, Nickel silicide encroachment formation and characterization. Microelectron. Eng. **87**, 245–248 (2010). https://doi.org/10.1016/j.mee.2009.06.003
156. J.-G. Chang, C.-B. Wu, X.-L. Ji, H.-W. Ma, F. Yan, Y. Shi, R. Zhang, The leakage current improvement of a Ni-silicided SiGe/Si junction using a Si cap layer and the PAI technique. Chin. Phys. Lett. **29**, 058501/1-4 (2012). https://doi.org/10.1088/0256-307x/29/5/058501
157. Y.-J. Chen, B.-Y. Tsui, H.-J. Chou, C.-I. Li, G.-P. Lin, S.-Y. Hu, Reduced junction leakage by hot phosphorus ion implantation of NiGe-contacted germanium n^+/p shallow junction. IEEE Electron Dev. Lett. **38**, 1192–1195 (2017). https://doi.org/10.1109/LED.2017.2726086
158. R.H. Hopkins, R.G. Seidensticker, J.R. Davis, P. Rai-Choudhury, P.D. Blais, J.R. McCormick, Crystal growth considerations in the use of "solar grade" silicon. J. Cryst. Growth **41**, 493–498 (1977). https://doi.org/10.1016/0022-0248(77)90236-4
159. J.R. Davis Jr., A. Rohatgi, R.H. Hopkins, P.D. Blais, P. Rai-Choudhury, J.R. McCormick, H.C. Mollenkopf, Impurities in silicon solar cells. IEEE Trans. Electron Dev. **27**, 677–687 (1980). https://doi.org/10.1109/t-ed.1980.19922
160. R.H. Hopkins, A. Rohatgi, Impurity effects in silicon for high efficiency solar cells. J. Cryst. Growth **75**, 67–79 (1986). https://doi.org/10.1016/0022-0248(86)90226-5
161. S. Pizzini, L. Bigoni, M. Beghi, C. Chemelli, On the effect of impurities on the photovoltaic behavior of solar grade silicon II. Influence of titanium, vanadium, chromium, iron, and zirconium on photovoltaic behavior of polycrystalline solar cells. J. Electrochem. Soc. **133**, 2363–2373 (1986). https://doi.org/10.1149/1.2108409
162. J. Libal, S. Novaglia, M. Acciari, S. Binetti, R. Petres, J. Arumughan, R. Kopecek, A. Prokopenko, Effect of compensation and of metallic impurities on the electrical properties of Cz-grown solar grade silicon. J. Appl. Phys. **104**, 104507/1-8 (2008). https://doi.org/10.1063/1.3021300
163. A.R. Peaker, V.P. Markevich, B. Hamilton, G. Parada, A. Dudas, A. Pap, E. Don, B. Lim, J. Schmidt, L. Yu, Y. Yoon, G. Rozgonyi, Recombination via point defects and their complexes in solar silicon. Phys. Status Solidi A **209**, 1884–1893 (2012). https://doi.org/10.1002/pssa.201200216
164. F. Schindler, B. Michl, J. Schön, W. Kwapil, W. Warta, M.C. Schubert, Solar cell efficiency losses due to impurities from the crucible in multicrystalline silicon. IEEE J. Photovoltaics **4**, 122–129 (2014). https://doi.org/10.1109/JPHOTOV.2013.2284060
165. D. Macdonald, A. Cuevas, Reduced fill factors in multicrystalline silicon solar cells due to injection-level dependent bulk recombination lifetimes. Progr. Photovolt. Res. Appl. **8**, 363–375 (2000). https://doi.org/10.1002/1099-159x(200007/08)8:4<363:aid-pip328>3.0.co;2-y

166. R. Schindler, W. Warta, Improvements and limits of the open circuit voltage of mc-Si solar cells. Phys. Status Solidi B **222**, 389–404 (2000). https://doi.org/10.1002/1521-3951(200011)222.1<389:aid-pssb389>3.0.co;2-z
167. W. Kwapil, P. Gundel, M.C. Schubert, F.D. Heinz, W. Warta, E.R. Weber, A. Goetzberger, G. Martinez-Criado, Observation of metal precipitates at prebreakdown sites in multicrystalline silicon solar cells. Appl. Phys. Lett. **95**, 232113/1-3 (2009). https://doi.org/10.1063/1.327682
168. W. Kwapil, M. Kasemann, P. Gundel, M.C. Schubert, W. Warta, P. Bronsveld, G. Coletti, Diode breakdown related to recombination active defects in block-cast multicrystalline silicon solar cells. J. Appl. Phys. **106**, 063530: 1/7 (2009). https://doi.org/10.1063/1.3224908
169. O. Breitenstein, J. Bauer, K. Bothe, W. Kwapil, D. Lausch, U. Rau, J. Schmidt, M. Schneemann, M.C. Schubert, J.-M. Wagner, W. Warta, Understanding junction breakdown in multicrystalline solar cells. J. Appl. Phys. **109**, 071101/1-10 (2011). https://doi.org/10.1063/1.3562200
170. T. Buonassisi, O.F. Vyvenko, A.A. Istratov, E.R. Weber, G. Hahn, D. Sontag, J. P. Rakotoniaina, O. Breitenstein, J. Isenberg, R. Schindler, Observation of transition metals at shunt locations in multicrystalline silicon solar cells. J. Appl. Phys. **95**, 1556–1561 (2004). https://doi.org/10.1063/1.1636252
171. A. Rohatgi, J.R. Davis, R.H. Hopkins, P. Rai-Choudhury, P.G. McMullin, J.R. McCormick, Effect of titanium, copper and iron on silicon solar cells. Solid-State Electron **23**, 415–422 (1980). https://doi.org/10.1016/0038-1101(80)90076-3
172. M.D. Dlamini, Electrical/electronic effects of titanium and iron impurities in EFG and FZ solar cell silicon: SPV/EBIC analysis. Sol. Energ. Mater Sol. Cells **43**, 353–361 (1996). https://doi.org/10.1016/0927-0248(96)00019-0
173. D. Macdonald, H. Mäckel, A. Cuevas, Effect of gettered iron on recombination in diffused regions of crystalline silicon wafers. Appl. Phys. Lett. **88**, 092105/1-3 (2006). https://doi.org/10.1063/1.2181199
174. S. Dubois, O. Palais, P.J. Ribeyron, N. Enjalbert, M. Pasquinelli, S. Martinuzzi, Effect of intentional bulk contamination with iron on multicrystalline silicon solar cell properties. J. Appl. Phys. **102**, 083525/1-7 (2007). https://doi.org/10.1063/1.2799057
175. S. Dubois, O. Palais, M. Pasquinelli, S. Martinuzzi, C. Jaussaud, N. Rondel, Influence of iron contamination on the performances of single-crystalline silicon solar cells: computed and experimental results. J. Appl. Phys. **100**, 024510/1-7 (2006). https://doi.org/10.1063/1.2218593
176. D.P. Fenning, J. Hofstetter, M.I. Bertoni, S. Hudelson, M. Rinio, J.F. Lelièvre, B. Lai, C. del Cañizo, T. Buonassisi, Iron distribution in silicon after solar cell processing: synchrotron analysis and predictive modeling. Appl. Phys. Lett. **98**, 162103/1-3 (2011). https://doi.org/10.1063/1.3575583
177. T. Mchedlidze, L. Scheffler, J. Weber, M. Herms, J. Neusel, V. Osinniy, C. Möller, K. Lauer, Local detection of deep carrier traps in the pn-junction of silicon solar cells. Appl. Phys. Lett. **103**, 013901/1-3 (2013). https://doi.org/10.1063/1.4807142
178. T. Mchedlidze, C. Möller, K. Lauer, J. Weber, Evolution of iron-containing defects during processing of Si solar cells. J. Appl. Phys. **116**, 245701/1-6 (2014). https://doi.org/10.1063/1.4905027
179. A.M. Salama, Copper precipitation effects in silicon used in solar cells. Electrochem. Soc. Proc. **78**(3), 334–345 (1978). https://doi.org/10.1149/1.2128964
180. X. Li, D. Yang, X. Yu, D. Que, Copper precipitates in multicrystalline silicon for solar cells. ECS Trans. **27**(1), 1135–1140 (2010). https://doi.org/10.1149/1.3360762
181. T. Turmagambetov, S. Dubois, J.-P. Garandet, B. Martel, N. Enjalbert, J. Veirman, E. Pihan, Influence of copper contamination on the illuminated forward and dark reverse current-voltage characteristics of multicrystalline p-type silicon solar cells. Phys. Status Solidi C **11**, 1697–1702 (2014). https://doi.org/10.1002/pssc.201400036

182. J. Bartsch, A. Mondon, K. Bayer, C. Schetter, M. Hörteis, S.W. Glunz, Quick determination of copper-metallization long-term impact on silicon solar cells. J. Electrochem. Soc. **157**, H942–H946 (2010). https://doi.org/10.1149/1.3466984
183. S. Flynn, A. Lennon, Copper penetration in laser-doped selective-emitter silicon solar cells with plated nickel barrier layers. Sol. Energ. Mater Sol. Cells **130**, 309–316 (2014). https://doi.org/10.1016/j.solmat.2014.07.026
184. A. Kraft, C. Wolf, J. Bartsch, M. Glatthaar, Characterization of copper diffusion in silicon solar cells. Energ. Proc. **67**, 93–100 (2015). https://doi.org/10.1016/j.egypro.2015.03.292
185. A. Kraft, C. Wolf, J. Bartsch, M. Glatthaar, S. Glunz, Long term stability of copper front side contacts for crystalline silicon solar cells. Sol. Energ. Mater Sol. Cells **136**, 25–31 (2015). https://doi.org/10.1016/j.solmat.2014.12.024
186. A.M. Salama, L.J. Cheng, The effects of titanium impurities in n^+/p silicon solar cells. J. Electrochem. Soc. **127**, 1164–1167 (1980). https://doi.org/10.1149/1.2129839
187. J.T. Borenstein, J.I. Hanoka, B.R. Bathey, J.P. Kalejs, S. Mil'shtein, Influence of ion-implanted titanium on the performance of edge-defined, film-fed grown silicon solar cells. Appl. Phys. Lett. **62**, 1615–1616 (1993). https://doi.org/10.1063/1.108603
188. S. Dubois, O. Palais, M. Pasquinelli, S. Martinuzzi, C. Jaussaud, Influence of substitutional metallic impurities on the performances of p-type crystalline silicon solar cells: the case of gold. J. Appl. Phys. **100**, 123502/1-8 (2006). https://doi.org/10.1063/1.2400801
189. A. Rohatgi, R.H. Hopkins, J.R. Davis, R.B. Campbell, H.C. Mollenkopf, The impact of molybdenum on silicon and silicon solar cell performance. Solid State Electron. **23**, 1185–1190 (1980). https://doi.org/10.1016/0038-1101(80)90032-5
190. J. Bauer, V. Naumann, S. Großer, C. Hagendorf, M. Schütze, O. Breitenstein, On the mechanism of potential-induced degradation in crystalline silicon solar cells. Phys. Status Solidi RRL **6**, 331–333 (2012). https://doi.org/10.1002/pssr.201206276
191. V. Naumann, D. Lausch, A. Graff, M. Werner, S. Swatek, J. Bauer, A. Hähnel, O. Breitenstein, S. Großer, J. Bagdahn, C. Hagendorf, The role of stacking faults for the formation of shunts during potential-induced degradation of crystalline Si solar cells. Phys. Status Solidi RRL **7**, 315–318 (2012). https://doi.org/10.1002/pssr.201307090
192. J. Hofstetter, J.F. Lelièvre, C. del Cañizo, A. Luque, Acceptable contamination levels in solar grade silicon: from feedstock to solar cell. Mater. Sci. Eng. B **159–160**, 299–304 (2009). https://doi.org/10.1016/j.mseb.2008.05.021
193. G. Coletti, P.C. Bronsveld, G. Hahn, W. Warta, D. Macdonald, B. Ceccaroli, K. Wambach, N. Le Quang, J.M. Fernandez, Impact of metal contamination in silicon solar cells. Adv. Funct. Mater. **21**, 879–890 (2011). https://doi.org/10.1002/adfm.201000849
194. N. Tōyama, Electrical properties of Au-Cu-doped Si Schottky diode. J. Appl. Phys. **54**, 932–936 (1983). https://doi.org/10.1063/1.332017
195. N. Tōyama, Stress effect in Au-Si Schottky diode doped with Cu. J. Appl. Phys. **55**, 4398–4403 (1994). https://doi.org/10.1063/1.333010
196. N. Tōyama, Capacitance decrease due to stress in a Cu-doped, n-type Si Schottky diode. Appl. Phys. Lett. **46**, 892–894 (1985). https://doi.org/10.1063/1.95878
197. A. Bindal, R. Wachnik, W. Ma, Observation of recombination center-assisted tunneling current in Al(Cu)-penetrated PtSi Schottky barrier diodes. J. Appl. Phys. **68**, 6259–6262 (1990). https://doi.org/10.1063/1.346867
198. C. Ahrens, G. Friese, R. Ferretti, B. Schwierzi, W. Hasse, Electrical characterization of Ti/$TiSi_2$ and WN/$TiSi_2$ Cu-diffusion barriers using Schottky diodes. Microelectron. Eng. **33**, 301–307 (1997). https://doi.org/10.1016/S0167-9317(96)00058-5
199. C. Ahrens, R. Ferretti, G. Friese, J.O. Weidner, Thermal stress effects on capacitance and current characteristics of Cu/Si and Cu/TiN/Si Schottky-diodes. Microelectron. Eng. **37**(38), 211–219 (1997). https://doi.org/10.1016/S0167-9317(97)99114-7
200. M. Stavrev, D. Fischer, C. Wenzel, T. Heiser, Study of Ta(N, O) diffusion barrier stability: analytical and electrical characterization of low level Cu contamination in Si. Microelectron. Eng. **37**(38), 245–251 (1997). https://doi.org/10.1016/S0167-9317(97)00118-4

201. E. Simoen, K. Opsomer, C. Claeys, K. Maex, C. Detavernier, R.L. Van Meirhaeghe, S. Forment, P. Clauws, Deep level transient spectroscopy study of nickel-germanide Schottky barriers on n-type germanium. Appl. Phys. Lett. **88**, 183506/1-3 (2006). https://doi.org/10.1063/1.2199615
202. E. Simoen, K. Opsomer, C. Claeys, K. Maex, C. Detavernier, R.L. Van Meirhaeghe, P. Clauws, Point-defect generation in Ni-, Pd-, and Pt-germanide Schottky barriers on n-type germanium. J. Electrochem. Soc. **154**, H857–H861 (2007). https://doi.org/10.1149/2759832
203. E. Simoen, K. Opsomer, C. Claeys, K. Maex, C. Detavernier, R.L. Van Meirhaeghe, P. Clauws, Study of metal-related deep-level defects in germanide Schottky barriers on n-type germanium. J. Appl. Phys. **104**, 023705/1-7 (2008). https://doi.org/10.1063/1.2956708
204. T. Gravier, F. Braud, J. Torres, J. Palleau, A. Chantre, J. Kitsch, Copper contamination effects in 0.5 μm BiCMOS technology. Microelectron. Eng. **33**, 211–216 (1997). https://doi.org/10.1016/S0167-9317(96)00047-0
205. E.P. Burte, W. Aderhold, The impact of iron, copper, and calcium contamination of silicon surfaces on the yield of a MOS DRAM test process. Solid State Electron. **41**, 1021–1025 (1997). https://doi.org/10.1016/S0038-1101(97)00016-6
206. K.C. Tee, K. Prasad, C.S. Lee, H. Gong, C.L. Chu, L. Chan, A.K. See, Effects of deliberate copper contamination from the plating solution on the electrical characteristics of MOSFETs. IEEE Trans. Semicond. Manufact. **14**, 170–172 (2001). https://doi.org/10.1109/66.920729
207. M. Inohara, H. Sakurai, T. Yamaguchi, H. Tomita, T. Iijima, H. Oyamatsu, T. Nakayama, H. Yoshimura, Y. Toyoshima, Copper contamination induced degradation of MOSFET characteristics and reliability, in *Digest 2000 Symposium on VLTS Technology Technical Papers* (2000), pp. 26–27. https://doi.org/10.1109/vlsit.2000.852755
208. Y.-J. Kim, K.-K. Choi, O. Kim, Effects of localized contamination with copper in MOSFET. IEEE Electron Dev. Lett. **23**, 479–481 (2002). https://doi.org/10.1109/LED.2002.801332
209. S.Q. Gu, L. Duong, J. Elmer, S. Prasad, Impact of wafer backside Cu contamination to 0.18 μm node devices. Electrochem. Soc. Proc. **2003**(06), 447–454 (2003)
210. K. Isobe, M. Kanda, T. Murakami, K. Totsuka, M. Moriuchi, T. Miyoshi, S. Yamada, T. Noguchi, Cu contamination induced degradation mechanism of embedded flash cells, in *Technical Digest IEDM* (2001), pp. 715–718. https://doi.org/10.1109/iedm.2001.979612
211. T. Kauerauf, A. Branka, K. Croes, A. Redolfi, Y. Civale, C. Torregiani, G. Groeseneken, E. Beyne, Effect of TSV presence on FEOL yield and reliability, in *Proceedings Reliability Physics Symposium—IRPS13* (2013), pp. 5C.6.1–5C6.4. https://doi.org/10.1109/irps.2013.6532035
212. Y. Dai, Deep-level impurity analysis for p-n junctions of a bipolar transistor from low-frequency g-r noise measurements. Solid State Electron. **32**, 439–443 (1989). https://doi.org/10.1016/S0038-1101(89)90025-7
213. K. Lee, T. Tani, H. Naganuma, Y. Ohara, T. Fukushima, T. Tanaka, M. Koyanagi, Impact of Cu contamination on memory retention characteristics in thinned DRAM chip for 3-D integration. IEEE Electron Dev. Lett. **33**, 1297–1299 (2012). https://doi.org/10.1109/LED.2012.2202631
214. K. Lee, S. Tanikawa, H. Naganuma, J. Bea, M. Murugesan, T. Fukushima, T. Tanaka, M. Koyanagi, Impacts of Cu contamination in 3D integration process on memory retention characteristics in thinned DRAM chip, in *Proceedings of IEEE International Reliability Physics Symposium—IRPS14* (2014), pp. 3E.4.1–3E4.6. https://doi.org/10.1109/irps.2014.6860634
215. K.-W. Lee, S. Tanikawa, M. Murugesan, H. Naganuma, J.-C. Bea, T. Fukushima, T. Tanaka, M. Koyanagi, Impacts of 3-D integration processes on memory retention characteristics in thinned DRAM chip for high-reliable 3-D DRAM. IEEE Trans. Electron Dev. **51**, 379–385 (2014). https://doi.org/10.1109/TED.2013.2295244
216. J.R. Schmid, H.G. Parks, R. Craigin, R.D. Schrimpf, Estimating the effect of contamination-induced leakage current in view of DRAM architectural trends, in

Proceedings of IEEE/SEMI Advanced Semiconductor Manufacturing Conference (1994), pp. 241–250. https://doi.org/10.1109/asmc.1994.588261tre
217. K. Isobe, M. Kanda, T. Murakami, K. Totsuka, M. Moriuchi, T. Miyoshi, S. Yamada, T. Noguchi, Cu contamination induced degradation mechanism of embedded flash cells, in *Technical Digest IEEE International Electron Devices Meeting—IEDM01* (2001), pp. 715–718. https://doi.org/10.1109/iedm.2001.979612
218. L. Jastrzebski, R. Soydan, H. Elabd, W. Henry, E. Savoye, The effect of heavy metal contamination on defects in CCD imagers. J. Electrochem. Soc. **137**, 242–249 (1990). https://doi.org/10.1149/1.2086375
219. N. Krause, H. Soltau, D. Hauff, J. Kemmer, D. Stötter, L. Strüder, J. Weber, Metal contamination analysis of the epitaxial starting material for scientific CCDs. Nucl. Instr. Meth. A **439**, 228–238 (2000). https://doi.org/10.1016/S0168-9002(99)00917-
220. W.J. Toren, J. Bisschop, A CCD delay line to determine low concentrations of bulk traps in silicon. Microelectron. Eng. **19**, 623–626 (1992). https://doi.org/10.1016/0167-9317(92)90509-P
221. W.J. Torn, J. Bisschop, Metal contamination characterization in CCD image sensors, in *IEEE International Electron Devices Meeting IEDM95* (1995), pp. 163–166. https://doi.org/10.1109/iedm.1995.497205
222. C. Tivarus, W.C. McColgin, Dark current spectroscopy of irradiated CCD image sensors. IEEE Trans. Nucl. Sci. **55**, 1719–1724 (2008). https://doi.org/10.1109/TNS.2008.919263
223. F. Domengie, J.L. Regolini, D. Bauza, P. Morin, Impact on device performance and monitoring of a low dose tungsten contamination by dark current spectroscopy, in *Proceedings of IEEE International Reliability Physics Symposium—IRPS10* (2010), pp. 259–264. https://doi.org/10.1109/irps.2010.5488821
224. F. Domengie, J.L. Regolini, D. Bauza, Study of metal contamination in CMOS image sensors by dark-current and deep-level transient spectroscopy. J. Electron. Mater. **39**, 625–629 (2010). https://doi.org/10.1007/s11664-010-1212-6
225. F. Domengie, P. Morin, D. Bauza, Modeling the dark current histogram induced by gold contamination in complementary-metal-oxide-semiconductor image sensors. J. Appl. Phys. **118**, 024501/1-10 (2015). https://doi.org/10.1063/1.4922969
226. J.-P. Carrère, S. Place, J.-P. Oddou, D. Benoit, F. Roy, CMOS image sensors: process impact on dark current, in *Proceedings of IEEE International Reliability Physics Symposium IRPS14* (2014), pp. 3C.1.1–3C.1.6. https://doi.org/10.1109/irps.2014.6860620
227. F. Russo, G. Nardone, M.L. Polignano, A. D'Ercole, F. Pennella, M. Di Felice, A. Del Monte, A. Matarozzo, G. Moccia, G. Polsinelli, A. D'Angelo, M. Liverani, F. Irrera, Dark current spectroscopy of transition metals in CMOS image sensors. ECS J. Solid State Sci. Technol. **6**, P217–P226 (2017). https://doi.org/10.1149/2.0101705jss
228. I.-H. Kim, J.-S. Park, T.-H. Shim, J.-G. Park, Si CMOS image-sensors designed with hydrogen-ion implantation induced nanocavities for enhancing output voltage sensing margin via proximity gettering. IEEE Trans. Electron Dev. **64**, 2345–2349 (2017). https://doi.org/10.1109/TED.2017.2677948
229. T. Yamaguchi, T. Yamashita, T. Kamino, Y. Goto, T. Kuroi, M. Matsuura, White spots reduction by ultimate proximity metal gettering at carbon complexes formed underneath contact area in CMOS image sensors, in *Symposium on VLSI Technology Digest of Technical Papers* (2016). https://doi.org/10.1109/vlsit.2016.7573447
230. J.M. Dishman, S.E. Haszko, R.B. Marcus, S.P. Murarka, T.T. Sheng, Electrically active stacking faults in CMOS integrated circuits. J. Appl. Phys. **50**, 2689–2696 (1979). https://doi.org/10.1063/1.326228
231. P.J. Ward, A survey of iron contamination in silicon substrates and its impact on circuit yield. J. Electrochem. Soc. **129**, 2573–2576 (1982). https://doi.org/10.1149/1.212612
232. W. Bergholz, D. Gilles, Impact of research on defects in silicon on the microelectronic industry. Phys. Status Solidi B **222**, 5–23 (2000). https://doi.org/10.1002/1521-3951(200011)222:5-23AID-PSSB>3.0.CO;2-L

233. A. Hiraiwa, T. Itoga, Scaling law in ULSI contamination control. IEEE Trans. Semicond. Manufact. **7**, 60–67 (1994). https://doi.org/10.1109/66.286832
234. K.K. Kuo, S. Emoto, DRAM technology memory electrical yield improvement by backgrinding induced backside damage, in *Proceedings of Electronic Materials and Packaging—EMAP Conference* (2006), pp. 1–6. https://doi.org/10.1109/emap.2006.4430675
235. D.R. Collins, D.K. Schroder, C.T. Sah, Gold diffusivities in SiO_2 and Si using the MOS structure. Appl. Phys. Lett. **8**, 323–325 (1966). https://doi.org/10.1063/1.1754459
236. W.E. Beadle, J.C.C. Tsai, R.D. Plummer, in *Silicon Processing Quick Reference Manual* (Wiley, New York, 1985), p. 6
237. J. Askill, *Tracer Diffusion Data for Metals, Alloys, and Simple Oxides* (IFI/Plenum, New York, 1970), p. 7

Chapter 8
Gettering and Passivation of Metals in Silicon and Germanium

Even traces of transition metals can be very detrimental for the electrical performance of devices and circuits, as outlined in Chap. 7. Since the early days of IC processing, techniques have been investigated to eliminate or to reduce the concentration of metals or at least to minimize their impact on the electrical device performance. During the last half century a large amount of different gettering techniques have been proposed and investigated, what resulted in an exponential growth of the number of published papers leading even to the editing of dedicated books on the subject. The aim of this chapter is surely not to review all the different approaches that have been reported in the literature, but rather to give a type of synopsis and to focus on the understanding and the new insights that have been obtained since the beginning of the century.

First some general aspects about gettering strategies are discussed, giving a schematic overview of different gettering approaches, before outlining the main gettering mechanisms. At the same time the nomenclature used in the gettering world is explained. Of key importance is the difference between segregation gettering and relaxation gettering. Subsequently the different backside gettering techniques are reviewed, including backside damage, thin layer depositions and ion implantation. New insights related to the phosphorus diffusion gettering will be outlined.

In the gettering world the concept of intrinsic gettering has been playing a dominant role and is still a hot topic. This is discussed in Sect. 8.3. The use of large diameter wafers and the lowering of the thermal budgets in advanced IC fabrication necessitate to review the physical mechanisms for the formation of a denuded zone or the use of nitrogen doping during the crystal growth in order to enhance oxygen precipitation. Instead of controlling the oxygen concentration profile by a high temperature out-diffusion step, attention is now given to control the distribution of point defects, with emphasis on vacancies.

C. Claeys and E. Simoen, *Metal Impurities in Silicon- and Germanium-Based Technologies*, Springer Series in Materials Science 270,
https://doi.org/10.1007/978-3-319-93925-4_8

The lowering of the thermal budget for advanced CMOS processing and the growing importance of TMs with a low diffusivity require the use of frontside gettering techniques, bringing the getter sites closer to the area contaminated with metals. In this part mainly attention will be given to the use of buried layers and ion implantation techniques, where for the latter the focus will be on nanocavities and near-surface proximity gettering.

The two following paragraphs are dealing with gettering in Silicon-on-Insulator technologies and photovoltaics, respectively. Especially for photovoltaics, the essential issues are the difficulty of gettering in polycrystalline silicon due to the presence of grain boundaries and intra-grain extended defects and the fact that the gettering technique is sometimes an inherent part of the processing technology. Typical examples are the use of a heavily doped backside layer operating as a back-surface field, conducting Al layers and porous silicon buried layers as a reflective region.

The last paragraph gives some reflections on ab initio calculations and continuum modeling for gettering, a topic that is discussed in more detail in Chap. 9.

8.1 Gettering Strategies

Gettering of metals is aimed at either the removal of metallic contaminants out of the wafer or attracting them towards non-active regions in the substrate so that there is no impact on the electrical performance of the devices. One of the first papers on gettering of Cu in Ge crystals, based on the creation of lattice defects by roughening the wafer surface, was already published in 1957 [1]. To optimize the performance of Si high-voltage rectifiers, Drake and Ellington [2] refer to 'chemical gettering' based on the deposition of a glass layer and 'physical gettering' using e.g. lattice damage.

Originally, the gettering approaches were classified according to the location in the wafer where the gettering action took place, i.e., backside of the wafer, internal in the bulk of the wafer or more near the front side of the wafer. Depending on whether the gettering action was based on an intrinsic property of the material (concentration of point defects like self-interstitials and vacancies, oxygen content), or the creation by external means of defects and/or stress fields in localized areas, one could differentiate between internal or external gettering. An example of internal gettering, which will later be discussed in detail and often referred to as 'intrinsic gettering' is the use of oxide precipitates in Cz-Si as gettering sites. By controlling the oxygen profile before oxygen precipitation starts, one can limit the precipitation zone to a region in the middle of the wafer. Examples of external gettering are based on the creation of defective layers using e.g. heavily doped layers leading to a dislocation network (e.g. phosphorus gettering), deposition of layers with a different lattice parameter or a high intrinsic stress (polysilicon layers, silicon nitride layer, etc.), high dose implantations resulting in extended defects with an associated stress field, annealing in an ambient with a gettering or passivation effect (chlorine oxidations, hydrogen passivation), deposition of metal layers, etc. A schematic illustration of different possible gettering approaches is given in Fig. 8.1.

Within the scope of this chapter it is surely not the intention to give an in-depth review of all the different gettering strategies, so that reference is made to some books and review papers [3–7]. However, some of the techniques represented in Fig. 8.1 will be further addressed with as main focus to concentrate on the different aspects which are key for an efficient gettering of the metallic contaminants and to report on recent new insights that have been obtained in understanding the gettering mechanisms.

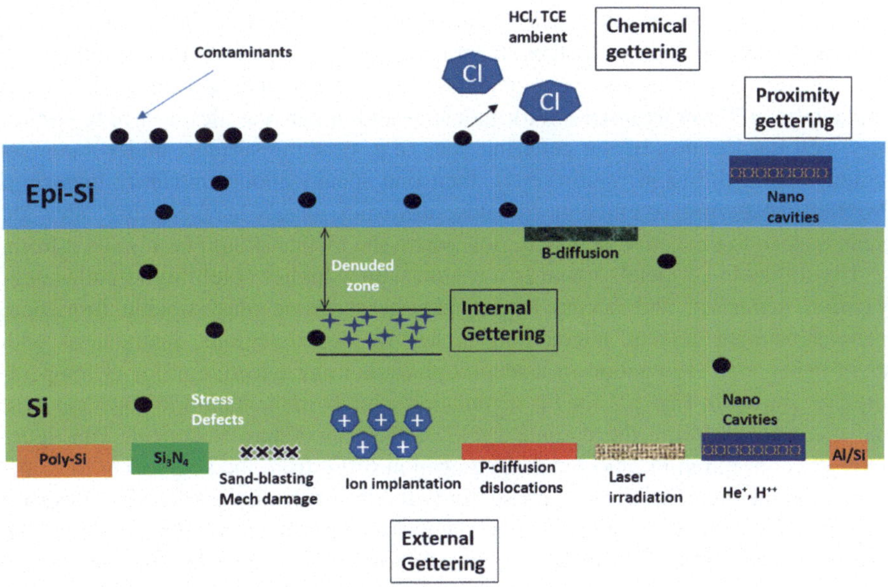

Fig. 8.1 Schematic overview of different gettering strategies

8.1.1 Metal Gettering Mechanisms

For an efficient gettering action of the metals the following steps are important (1) if the metal is already present in a particular phase, the metal-complex has to dissociate, (2) diffusion of the metal to the gettering sites, and (3) trapping of the metals at the gettering sites into stable complexes (e.g. silicides) in order to avoid dissociation during further processing and/or device operation. The nucleation of the gettering sinks can be an internal or external process as shown in Fig. 8.1. The formation of stable complexes at the getter sites can occur by interactions of the metals with e.g. oxide precipitates (internal gettering), doping atoms (ion pairing), self-interstitials or vacancies (silicide formation), trapping at defects, locally enhanced solubility due to the dopants-related Fermi-level shift or the formation of a second phase.

The nomenclature used for the different gettering mechanisms is not always straightforward as different mechanisms can be active in parallel. A good overview is given by Myers et al. [8]. As already mentioned and further discussed in Sect. 8.5 ***internal gettering*** is based on the interaction with oxide precipitates in the bulk of the wafer. Interactions between metal impurities and point-defect gradients associated with dopants like P and B are called ***dopant-diffusion gettering***. Interactions between metals and defects created near or on the surfaces of the wafer or in the bulk using additional process steps (layer deposition, ion implantation, sandblasting etc.) are referred to as the very general group of ***external gettering***. If an ion implantation or laser treatment to form defects and/or nanocavities is used to control very well the location of the gettering sites (e.g. near the surface, under a contact region, near a diffusion region, etc.), then one speaks about ***proximity gettering***. Proximity gettering is efficient for slow diffusing metals as the getter sites are created close to the active device region where the metals should be removed from.

The diffusion of metals towards a region with a higher solubility is called ***segregation gettering***, and occurs by interactions at getter sites, silicide formation, interactions with defects, interactions in heavily doped regions, interactions with strain fields, etc. Segregation in a second phase is more effective at lower temperatures, as shown in Fig. 8.2 for Fe segregation after B implantation to form cavities, and takes place during the cooling cycle of the temperature treatment. The figure gives the concentration ratio for the segregation of Fe from the solution in Si to the solution in the precipitated B–Si phase, for two different B implant doses. In this case the difference in boron dose is not sufficient to see an impact of the Fermi-level on the ionization rate and the equilibrium concentration of the impurities. In general, the enhanced solubility in heavily doped regions is explained by a combined effect of the Fermi-level shift and the ion pairing between the metal and the doping atoms. For segregation gettering in p/p^+ epitaxial wafers, theoretical models have been developed for metals such as Fe [9, 10], Cu [11–13] and Ni [14]. In the latter case the Ni concentration should not be too high and sufficiently low temperatures must be used to achieve optimal gettering, although there are also reports observing either no or only a marginal Ni segregation gettering [11, 15]. As will be discussed later, very effective gettering of Ni is achieved with other techniques [16].

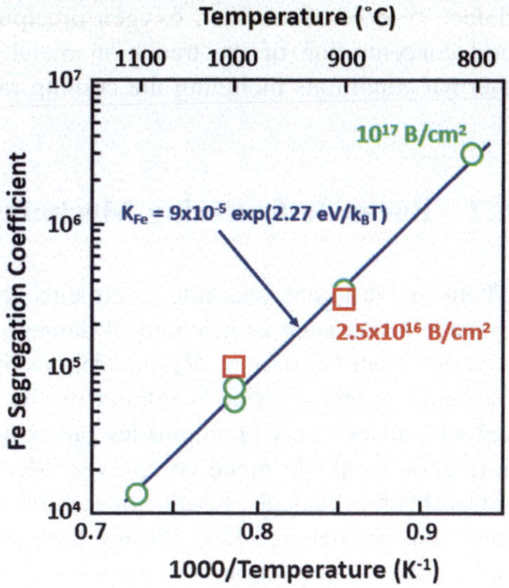

Fig. 8.2 Temperature dependence of the Fe segregation coefficient during gettering by BSi precipitates. Reproduced from Myers et al. [154] by permission of ECS-The Electrochemical Society

Segregation is an equilibrium process taking place at the anneal temperature. During cooling down of the samples, another non-equilibrium process is active causing a diffusion of the dissolved metal atoms toward gettering sites. This heterogeneous precipitation process, with preferential precipitation at structural defects, is called ***relaxation gettering*** and will be influenced by the presence of point defects impacting the metal diffusion process. The cooling rate will thus determine the dominance of either precipitation or relaxation activities. A typical example of relaxation gettering is the internal gettering based on oxide precipitates in Cz silicon. While for segregation gettering the metal solubility is important, for relaxation gettering the density and type of precipitation sites are dominant. The precipitation of supersaturated metals at heterogeneous sites is obeying Ham's law, implying that the process is diffusion-limited. For spherical precipitates this leads to the following expression

$$\frac{dC_M}{dt} \cong [C_{M-diss} - C_M^*]4\pi N_{prec} r_{prec} D_M \tag{8.1}$$

with C_M the total metal concentration, $C_{M\text{-}diss}$ the dissolved metal concentration, C_M* the metal solubility concentration, N_{prec} the density of precipitation sites, r_{prec} the precipitate radius and D_M the metal diffusivity.

Relaxation gettering involves a process where the impurity concentration in excess of the solubility relaxes to the equilibrium concentration. The driving force is not only the free energy gain associated with supersaturation of the impurities, but also the energy release associated with the structural defects. The gettering efficiency depends on a variety of parameters such as the type of defect (extended defect, type of dislocation, oxygen precipitate, etc.) acting as getter site, the type and concentration of the transition metal, the thermal history of the wafer, the thermal conditions including the cooling rate, etc. [17].

8.2 Backside Gettering Mechanisms

There are different backside treatments that can be used to create near-surface damage layers such as mechanical damage, deposition of a glass layer, deposition of a thin layer (e.g. Al, Si_3N_4, amorphous Si, poly-Si, etc.) or ion implantation. The backside gettering action is relying on the interaction of metals with microcracks, defects, dislocations or impurities either acting as getter sites or resulting in the formation of stable metal complexes. Mechanical backside damage can be introduced by mechanical grinding or sandblasting [18], impact sound stressing [19], laser-induced damage [20, 21] and high dose implants [22, 23] leading to a high defect density or even an amorphous region as might be the case when using high dose ion implantation. The laser and ion implantation techniques enable a tight control of the amount of damage and of the location of the damaged region. The mechanical damaging techniques will not be further addressed in this part. Ion implantation gettering has gained a lot of interest due to the possibilities to use it for proximity gettering or to create nanocavities.

8.2.1 Glass Layer Gettering

The milestone paper by Goetzberger and Shockley [24] on the detrimental influence of Au and Cu on the device performance, also outlined for the first time the potential of using a phosphorous (P_2O_5) or boron doped (B_2O_3) glass layer to getter impurities after a high temperature anneal step. Other glasses are less effective [25]. The gettering action is based on the metal solubility in the doped layer due to the effect of both the Fermi-level shift and the ion pairing between the metal and the doping atoms (e.g., M_iP_s). As will be further discussed, new insights in this gettering approach pointed out that gettering can be related to the interaction of the metal with doping-vacancy clusters such as P_nV for phosphorus gettering.

To explain the physics behind this type of gettering mechanism, the ***phosphorus diffusion gettering (PDG)*** is used as an example. The gettering action is based on the enhanced metal solubility in the doped layer due to the effect of both the Fermi-level shift to the conduction band due to the P in-diffusion and the ion pairing between the metal and the doping atoms, e.g., M_sP. The P diffusion also plays an important role as this process has been associated with the injection of self-interstitials, due to the dissociation of the PI pairs, enhancing the gettering effect [26]. If the phosphorus diffusion exceeds that of self-diffusion, non-equilibrium concentrations of point defects are injected into the bulk, while the surface concentration remains at equilibrium. Therefore, the ratio $\frac{C_I}{C_I^*}$, quantifying the non-equilibrium state, should be taken into account. SiP precipitates are during the isothermal heat treatment formed at the interface, enhancing the process with anneal time [27].

Recent investigations based on ab initio calculations pointed out that the above mentioned PDG mechanism has to be reviewed in order to take into account the important role played by the vacancies when high phosphorus concentrations are used for gettering [28]. DFT simulations indicate that P_4V clusters, i.e., a vacancy surrounded by 4 nearest-neighbor phosphorus atoms as illustrated in Fig. 8.3, are energetically stable neutral complexes (formation energy −1.68 eV) and, depending on the temperature, dominant for P concentrations $>10^{20}$ cm^{-3}.

The gettering action is relying on the formation of M-P_4V complexes, whereby the metal atom is taking the vacancy position in the complex forming MP_4. The calculated binding energy of the metal with the P_4V complex is for Fe, Cu, Cr, Mo, Ni, Ti and W equal to −3.2 eV, −4.03 eV, −3.67 eV, −2.57 eV, −3.60 eV, −3.21 eV and −2.90 eV, respectively [28]. The formation energy of M-P_nV complexes is higher for $n < 3$ and thus leading to less stable configurations. The direct binding between M_i and P_s is also energetically unfavorable.

The non-equilibrium concentration of point defects has according to Eq. (2.14) a direct impact on the radius of the metal precipitates, so that the injection of interstitials or the absorption of vacancies will also facilitate the metal complex dissociation. This is in agreement with the observation that the diffusion profile for Au in Si during PDG cannot be reproduced by a model assuming only out-diffusion of Au to an infinite sink at the surface. An additional source of silicon interstitials is needed [29].

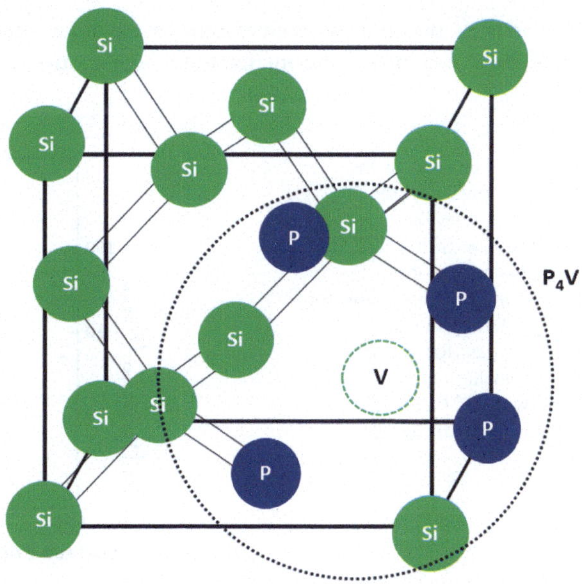

Fig. 8.3 Schematic illustration of a P_4V cluster with 4 nearest neighbor phosphorus atoms around a Si vacancy. After Chen et al. [28]

The getter efficiency depends on the temperature and on the phosphorus concentration as illustrated in Fig. 8.4 for Fe [30]. As can be seen, the getter efficiency increases for higher P-concentrations and lower temperatures. A similar trend has been reported before for the PDG of Au in Si [31, 32]. Although a partial gettering can occur during the phosphorus pre-deposition step, because of the injection of self-interstitials, the full gettering effect is only achieved after the final temperature step [32]. More recent investigations point out that a further increase of the Fe gettering efficiency, up to 99.9%, can be achieved by lowering the temperature to 650 °C [33]. As for such low temperatures the phosphorus diffusion is limited, the increased gettering efficiency is related to the increased segregation coefficient at lower temperatures [34]. The efficient gettering using PDG has also been studied for Cu and Au [35–37].

The mentioned P_4V model for gettering can explain the above mentioned observations keeping in mind that both the P_4V and the MP_4 formation is retarded at higher temperatures, and the phosphorus in-diffusion increases at higher temperatures [28]. Therefore, there is serious doubt about the importance of the silicon self-interstitial injection to the actual metal gettering.

For high phosphorus concentrations a dislocation network may be formed, which can act in parallel to the phosphorus diffusion as metal getter sites [38]. The dislocations are generated by the strain induced in the silicon lattice by the phosphorus atoms. The lattice contraction coefficient β_c for substitutional impurities with covalent radius r_{ext} introduced in the Si lattice (radius r_{Si}) with N_{Si} lattice sites is given by Lawrence [39]

$$\beta_c = \frac{1}{N_{\mathrm{Si}}}\left[1 - \left(\frac{r_{ext}}{r_{\mathrm{Si}}}\right)^3\right] \tag{8.2}$$

For high phosphorus concentrations the total stress is higher than the silicon yield stress leading to the nucleation of dislocations.

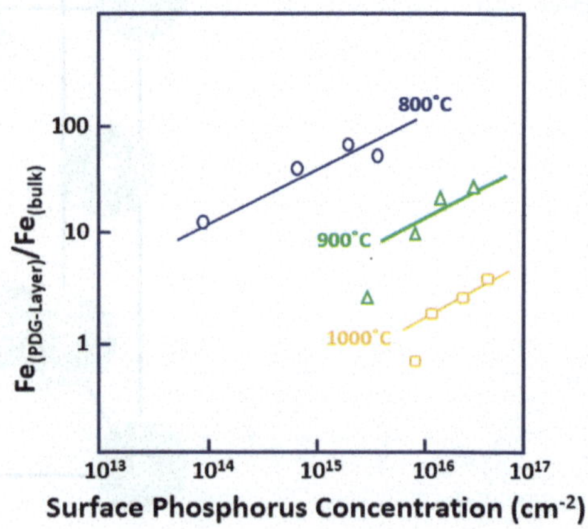

Fig. 8.4 The ratio of the Fe-concentration in the PDG-layer over the concentration in the bulk of the wafer versus the phosphorus concentration, for three different anneal temperatures. Reproduced from Nadahara et al. [30] by permission of ECS-The Electrochemical Society

The PDG condition resulting in a dense interface dislocation network is also effective for reducing oxidation-induced stacking faults (OISFs) by the reaction of the dislocations with the OISF nucleation sites [40]. Also some of the getter techniques reviewed in Sect. 8.2.2 are effective to getter OISF's. As the getter step is performed at the beginning of the processing, before the first oxidation, this approached is referred to as pre-oxidation gettering OISFs (POGO).

Gettering can also be achieved using a highly-doped boron layer [36], while this is not the case for arsenic and antimony [4]. However, boron diffusion gettering (BDG) is much less effective for Fe gettering than phosphorus gettering. An enhanced boron gettering efficiency can be achieved by increasing the boron concentration (increased N_2 flow through the BBr_3 source) and implementing a post-oxidation low temperature anneal step [33]. The improvement by the low temperature step was associated with the formation of Si–B boron precipitates [41]. However, by optimal design of the gettering recipe, good Fe gettering efficiency can also be achieved without the low temperature treatment [33]. In this case the gettering action is attributed to the near surface boron-rich layer which effectively getters Fe.

As will be further discussed in Sect. 8.4 phosphorus and boron can also be used for frontside gettering.

8.2.2 *Thin Layer Gettering*

Damage at the backside of the wafer can be created by a layer with high intrinsic stress, which once above the silicon yield stress will result in the nucleation of dislocations that can be transformed into a dislocation network. Beside the glass layer approach, this can also be achieved by the deposition of thin layers such as Al–Si, Si_3N_4, poly-silicon seal, epitaxial layer doped with Ge, etc.

8.2.2.1 Aluminum Films

A thin Al film on the backside of the wafer is very effective for the gettering of Cu, while this is less the case for Ag and Au [42]. Cobalt can also efficiently be gettered [43]. The getter effect of an Al film deposited on the backside of a Si wafer followed by an anneal step, called ***aluminum gettering (AlG)***, is based on the fact that for temperatures above the eutectic temperature of the binary Al–Si system (577 °C) a liquid Al–Si phase is formed in which the metal solubility is enhanced compared to the solubility in the solid Si so that metal impurities are attracted to the film. The key parameter is the segregation coefficient of the metal k_M given by [44]

$$k_M = \frac{M_{\mathrm{AlSi}}}{M_{\mathrm{Si}}} \propto \exp^{-\left[\frac{E^M_{\mathrm{AlSi}} - E^M_{\mathrm{Si}}}{k_B T}\right]} \tag{8.3}$$

with M_{AlSi} and M_{Si} the metal concentration in the AlSi liquid and Si after reaching equilibrium during the AlG treatment and E^M_{AlSi} and E^M_{Si} the activation energy of the metal in AlSi and Si, respectively. For Fe the segregation coefficient is in the range 10^5–10^7 [45]. The strong and stable gettering of Cu is also the reason why the standard use of Al-4%Cu wiring in integrated circuits is not degrading the electrical performance of the devices.

In contrast to PDG, for AlG the formation of the getter layer is very fast so that fast diffusers like Co can be efficiently gettered. To enable the gettering of interstitial metals M_i present at the frontside of the wafer by a backside getter layer, the diffusion length of the interstitial TM must be equal to or larger than the wafer thickness [46]. Both the anneal temperature and the wafer thickness have an impact on the getter effect. AlG is essentially a low temperature process, where PDG is done at much higher temperatures. For substitutional metals the creation of excess self-interstitials, enabling the kick-out of the metal from a substitutional into an interstitial position, is playing a dominant role in the process. In the case of mc-Si the impact of metal interactions with extended defects and metal silicide precipitates at grain boundaries has to be taken into account. Metals trapped by dislocations at the frontside of the wafers can also be gettered by the AlG technique [47].

A somewhat different approach for Fe gettering is using a negatively charged aluminum-doped dielectric on the backside of the wafer [48]. This is achieved by depositing an Al-doped SiO_2-based spin-on-glass (SOG) with a negative charge density of -1×10^{12} cm^{-2}, which is able to reduce the Fe concentration below the detection limit of 1×10^{10} cm^{-3}. During the film curing negatively charged Al–O complexes are most likely formed. It is assumed that the getter effect is due to the electrostatic attraction between the negatively charged SOG and the positively charged Fe_i [49].

8.2.2.2 Silicon Nitride and Polysilicon Layers

Gettering has also been studied by designing a stress field at the back interface using e.g. a highly stressed Si_3N_4 layer [40]. Although this POGO technique is not very effective for reducing or eliminating the bulk stacking faults, it was observed that the leakage current of diodes improves by gettering the metallic contamination [49]. A minimum Si_3N_4 layer thickness is required for the reduction of the microdefect density observed after reduced pressure epitaxial growth [50]. For too thick Si_3N_4 layers wafer bowing may occur which can generate slip dislocations during further thermal processing. A Si_3N_4 layer can also be used to protect the backside damage caused by e.g. sandblasting or grinding [51].

The metal removal is based on the segregation gettering effect which is operating over a wide temperature range (250–900 °C) [51]. Below 700 °C the Fe getttering is diffusion-limited and for higher temperatures the process becomes reaction-limited. Originally, it was believed that the Fe gettering in PECVD Si_3N_4 films was not caused by the film itself but by the hydrogen passivation effect leading to the formation of FeH pairs [52, 53]. However, recent experimental work on annealing Fe gettered wafers after removal of the Si_3N_4 layer clearly indicates that the Fe is gettered in the layer and doesn't reappear during the annealing [51].

Polysilicon films deposited on the backside of the wafer can getter metallic contaminants (***polysilicon backside gettering—PBG***). In this case the getter sites are grain boundaries, the size of which is controlled by the deposition temperature, with a high thermal stability during the getter anneal step as the grains will not easily realign with the substrate. A lower deposition temperature results in a smaller grain size and therefore a higher getter efficiency [54].

According to Istratov et al. [55] the gettering of Fe is driven by a combined segregation and relaxation mechanism. This model has in recent years been confirmed by other researchers [56]. The impurities segregate at the grain boundaries due to the free energy difference between the perfect silicon and the distorted lattice. The segregation process depends on the grain size.

The segregation and precipitation component for Fe gettering by PBG has been studied experimentally and by simulation. Figure 8.5a shows experimental and simulation results for an initial Fe concentration of 1.6×10^{13} cm^{-3} and a gettering at 650 and 750 °C, respectively. It is clear that iron precipitation occurs at 650 °C. The simulated curves for 750 °C are falling together. The simulated relation between segregation and precipitation is illustrated in Fig. 8.5b for a getter anneal at 750 °C for 5 h. For all simulations the precipitation site density and radius are kept the same (i.e., same diffusion-limited rate in Ham's model). The impact of the segregation coefficient can clearly be seen, i.e., a large segregation coefficient increases the critical contamination level as the supersaturation is much more reduced. A rather high Fe concentration is needed before the precipitation process starts. Figure 8.5 is giving a trend, depending on the experimental conditions, and enables to develop a model for polysilicon backside gettering.

The getter efficiency depends on the anneal temperature and a lower temperature leads to a higher efficiency [54]. The effectiveness of PBG has also been demonstrated for frontside contamination, but one has to take in account that it is a diffusion limed process [57]. The anneal temperature should be low but high enough to allow sufficient diffusion of the metals at the contaminated frontside location to reach the getter site.

The presence of a polysilicon backside layer also has an impact on the oxygen precipitation process in the bulk of Cz wafers [58, 59]. The observed enhanced oxygen precipitation is important for the efficiency of internal gettering as will be discussed in Sect. 8.3

As a final remark it can be mentioned that the use of a Ge backside layer has been evaluated for gettering Au in Si [60]. The Au gettering is caused by the difference in activation coefficient of Au in Si and Au in Ge. Although the Ge layer is also effective for gettering stacking faults, it has not been further investigated.

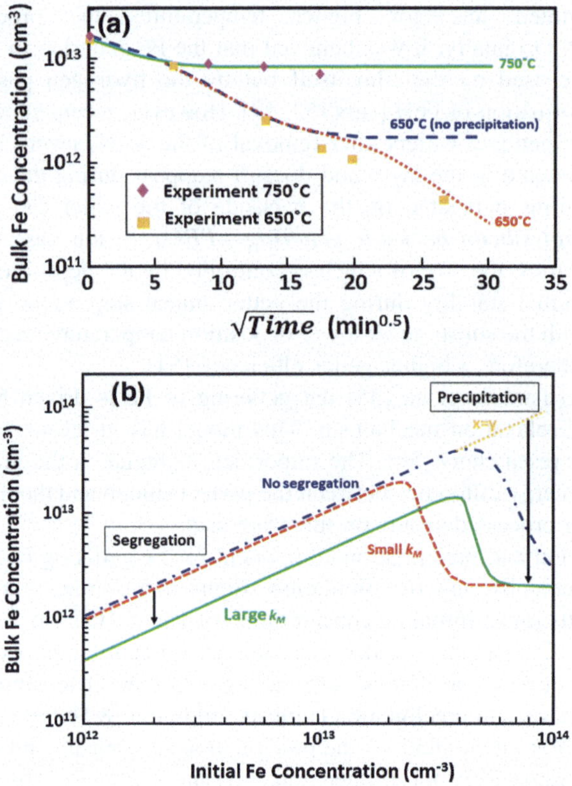

Fig. 8.5 **a** Comparison of experimental and simulated results for 650 and 750 °C (with and without precipitation). Simulations at 750 °C fall together. **b** Bulk Fe concentration after 5 h anneal at 750 °C. A large, small and $k_M = 1$ is taken. The curve $x = y$ is used as a reference. Reproduced from Haarahiltunen et al. [55]. Copyright © 2011 WILEY-VCH Verlag GmbH & Co

8.2.3 Ion Implantation Gettering

Early experiments on implantation gettering have been performed using Si^+ ions in order to create damage without influencing the doping concentration. The getter efficiency depends on the creation of an amorphous layer, which is controlled by

- The implanted species. Heavier ions produce more severe damage [61];
- The used implantation dose and energy [62];
- The substrate temperature during implantation. Without cooling the substrate the temperature can raise up to 500–600 °C [63], resulting in an annealing effect;
- The substrate orientation. The substrate damage is higher for ⟨111⟩ Si than for ⟨100⟩ Si [64].

The metals Fe, Co, Ni, Cu an Au have been studied and it was noticed that higher gettering levels are obtained for faster diffusing TMs like Cu and Ni [22]. The electrical device impact was investigated using a silicon photodiode array and the comparison between P^+, As^+ and Ar^+ implantations pointed out that there exists a critical amorphous region energy corresponding with a critical dose ϕ_c, given by [65]

$$E_c = \phi_c E f_E \tag{8.3}$$

with E_c the critical energy, E the energy per ion, and f_E the fraction of the energy available for lattice disorder. Based on this model a 3×10^{15} cm^{-2} phosphorus implant at 50 keV would be as effective as a 1×10^{15} cm^{-2} arsenic implant at 150 keV [64]. A comparison made for Au gettering using Ar, O, P, Si, As and B implantations indicated that the most efficient gettering was achieved for Ar (10^{16} cm^{-2} at 200 keV), while further ranking is according to O > P > Si > Ar > B [66]. The information given here is only intended as a trend as much is depending on the technological parameters (dose, energy, anneal conditions, etc.) used in the experiments. There is an important difference for implants below or above the critical dose for amorphization since in the latter case solid phase epitaxial regrowth during the thermal treatment can anneal out most of the defects except the end-of-range defects.

Much work on implantation gettering has for a long time been focusing on Ar-implantation gettering [67–73]. The damaged layer can accommodate large amounts of metal. The amount of Au atoms in the damaged layer increases with annealing temperature. However, there is a weak bonding between the Au atoms and the implantation induced defects so that at higher temperature some Au can be released from the traps and redistributed in the bulk [72]. The getter effect is more efficient at lower temperatures but then a longer anneal time is needed to allow the Au toms to diffuse to the getter sites.

Although not further pursued also the concept of using backside Ge implantation to improve the device leakage current has been reported [74].

8.3 Intrinsic Gettering Mechanisms

Czochralski silicon contains a high amount of oxygen (10^{17}–10^{18} cm^{-3}) which during heat treatments precipitates into SiO_x precipitates of different shape (plate-like, octahedral, spherical etc.) depending on the anneal temperature and ambient (see [75–77] and references therein). The amorphous plate-like precipitates, as shown in Fig. 8.6, are square-shaped with $\langle 110 \rangle$ sides and (100) habit plane and can have, depending on the stress around it, punched-out dislocations (SiO_x-PDC). These precipitates are effective sinks for metal impurities like Cu [78]. The first paper on using oxygen precipitates for intrinsic gettering was published in 1976 by Tan et al. [79].

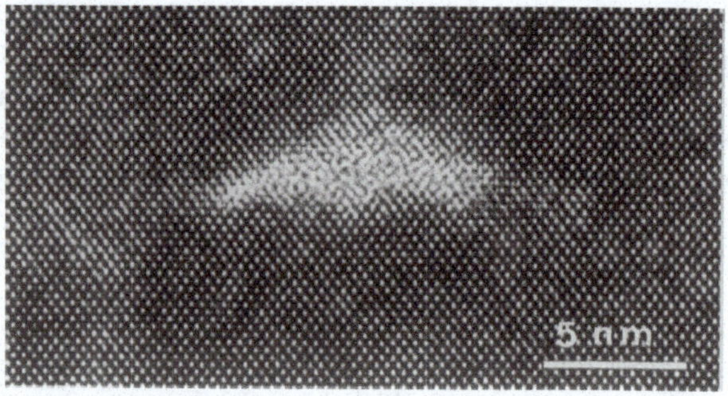

Fig. 8.6 High-voltage TEM image of a plate-like SiO_x precipitate formed after 15 h anneal at 750 °C

Intrinsic gettering is based on the formation of an oxygen precipitate free region near the wafer surfaces and a defective region containing oxygen precipitates in the middle of the wafer as shown in Fig. 8.7. This is achieved by a three step thermal cycle consisting of

- A high temperature step for out-diffusion of oxygen from the surface regions. These near surface regions, the width of which depends on the duration and temperature of the thermal anneal, are defect-free at the end of the IG process and referred to as denuded zone (DZ);
- A low temperature step to enable the nucleation of the oxygen precipitates;
- A high temperature step for the growth of the precipitates.

The bulk defects, sometimes called bulk micro defects (BMDs) will getter the metallic impurities. The precipitation process depends on the initial oxygen concentration in the wafer and the anneal conditions (temperature, time, type of

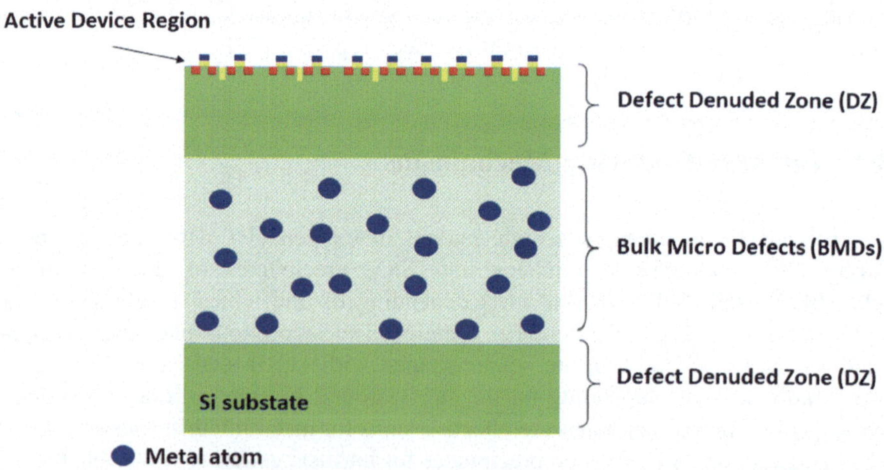

Fig. 8.7 Schematic illustration of the wafer cross-section at the end of the intrinsic gettering process

annealing such as furnace or RTA, cooling cycle, etc). The effectiveness of the IG process will be controlled by the size and the density of the precipitates and the type of the impurity that should be gettered. Several of these parameters are further discussed.

The size and the density of the precipitates are both important for efficient gettering. It has been reported for Fe that for a plate-like precipitate a minimum size of 200 nm is needed, while for a polyhedral precipitate a size of 100 nm is already sufficient [54]. In addition, one also has to take into account that a too high stress field may lead to the nucleation of dislocations. Therefore, there is a minimum precipitate size X_1 above which gettering will occur and there is a maximum size X_2 above which slip will occur under thermal stress. According to Sueoka et al., one can differentiate between four regions as shown in Fig. 8.8a, i.e., (1) optimal region for gettering without slip, (2) region with no slip but also no-gettering, (3) region with gettering but also slip, (4) region with no-gettering but slip [80]. The getter effect is also dependent on the type of metal that has to be gettered. Figure 8.8b gives the same representation for Ni as reported by different authors [80]. To estimate the oxygen radius for the different experimental conditions, Ham's model for the diffusion-limited growth of spherical precipitates was used. The no-gettering in Fig. 8.8 refers to a not complete gettering of all the metal impurities as determined by the optical detection of shallow etch pits and haze on the wafer surface (Chap. 4). A moderate density of small oxygen precipitates getters fast diffusing TMs efficiently. For modern IC processing one has to take into account that the thermal budgets during wafer processing are much reduced so that it is becoming more difficult to achieve intrinsic gettering during standard CMOS processing.

Another solution is to use nitrogen-doped Cz material (NCz) as nitrogen enhances the nucleation of oxygen precipitates in the bulk [81], or to implement at wafer level a treatment leading to a so-called Magic Denuded Zone (MDZ) whereby the microdefect density in the bulk is controlled and a precipitate-free denuded zone is formed [82]. There is no oxygen out-diffusion step at the beginning of the IG cycle, but a RTA step determining the vacancy profile (with a low vacancy concentration near the surface) and the oxygen precipitation process is further controlled by the vacancy rather than oxygen concentration profile [83]. For vacancy concentrations below 5×10^{11} cm^{-3} no oxygen precipitation occurs so that a denuded zone is created near the surface.

These conclusions are in agreement with the observation that in crystals grown in a vacancy-rich regime, the oxide precipitate density is higher than in crystals grown in an interstitial-rich regime [84].

It should also be mentioned that oxygen precipitation can be enhanced by doping the Cz crystals with 10^{16}–10^{19} cm^{-3} Ge [85, 86]. The Ge doping also reduces the thermal donor formation and has a beneficial impact on the mechanical wafer strength.

Proper programming the vacancy distribution forces the wafer to behave in an ideal way so that the oxygen precipitation is independent of the oxygen content and the crystal-history of the wafer. Vacancies in the form of O_2V complexes with a

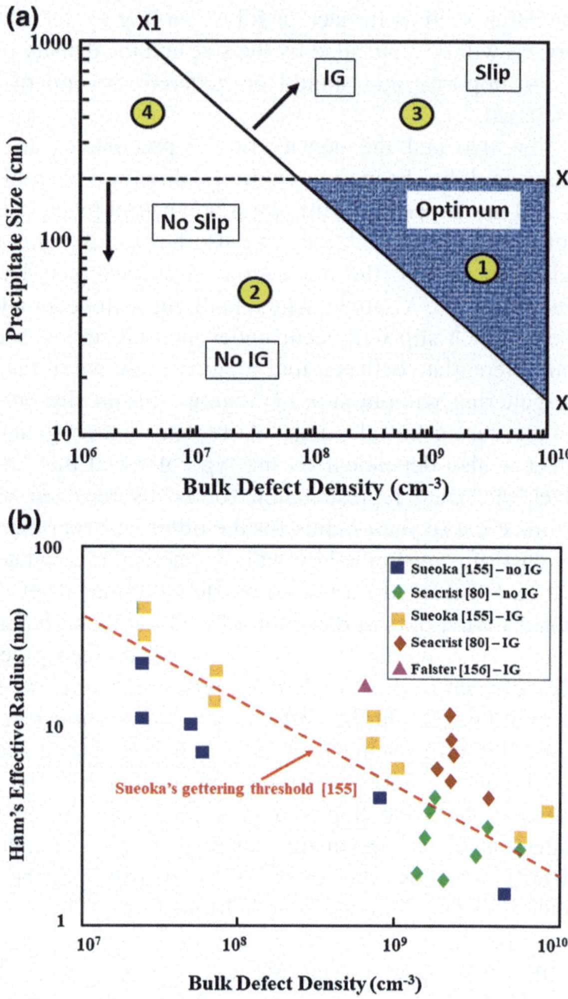

Fig. 8.8 Schematic illustration of the different regions (precipitate density and size) determining where intrinsic gettering is active, with or without slip. Reproduced Sueoka et al. [155] by permission of ECS-The Electrochemical Society. **b** Results for Ni gettering reported in the literature. Reproduced from Seacrist et al. [80] by permission of ECS-The Electrochemical Society

concentration above 1×10^{12} cm^{-3} strongly enhance the oxygen precipitation due to a reduction of the incubation time [83]. The profiles of the vacancies and the O_2V defects are controlled by the thermal treatment as illustrated in Fig. 8.9 [87]. There is a DZ formed near the surface and the O_2V and V concentration in the bulk decrease for higher temperatures. The free V and the O_2V profiles have the same shape and a RTA step is used to tailor the vacancy concentration. Above the critical vacancy concentration oxygen precipitation occurs.

The getter efficiency depends on the metal as illustrated in Fig. 8.10 for Ni and Cu. The threshold value for gettering is 8×10^{-3} and 2×10^{-3} $4\pi r^2 D_{BMD}\, d_{wafer}$ (D_{BMD} = BMD density, d_{wafer}= wafer thickness) for Cu and Ni, respectively [88]. The difference may be related to the reaction-limited precipitation. More recent investigations report a value of 4.55×10^{-3} $4\pi r^2 D_{BMD}\, d_{wafer}$ for both Cu and Ni

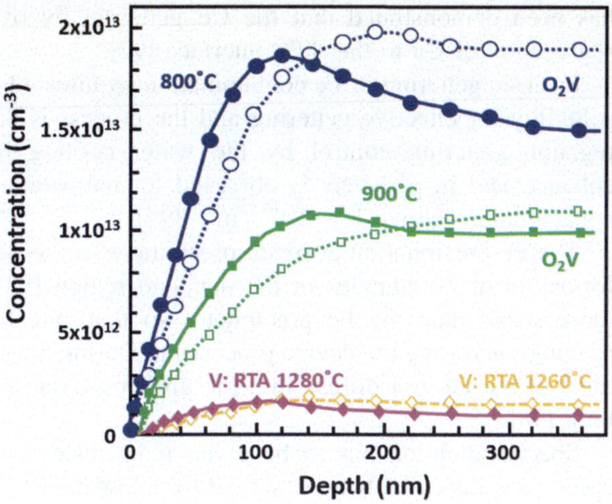

Fig. 8.9 O_2V concentrations at 800 °C (circles) and 900 °C (squares) and V concentrations (triangles) for two RTA simulations: T = 1280 °C (filled symbols) and T = 1260 °C (open symbols). The cooling rate for the simulation is 70 °C/s (after Frewen and Sinno [87]). Copyright 2006 AIP Publishing

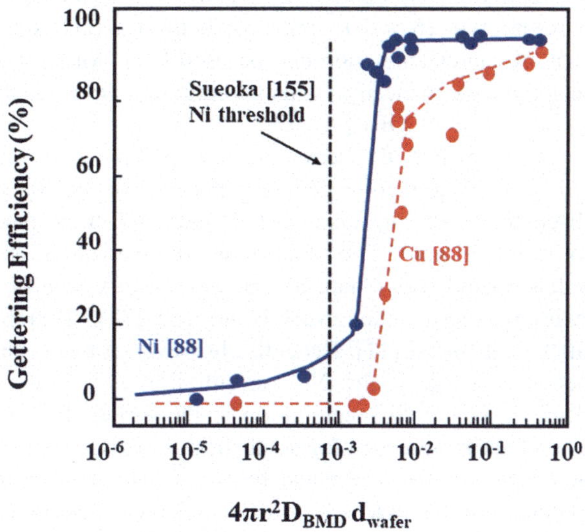

Fig. 8.10 Gettering efficiency of oxygen precipitates as function of the total inner surface of the bulk microdefect for Cu and Ni, respectively. The threshold value reported by Sueoka et al. [155] is also indicated. Data from Hoelzl et al. [88] by permission of ECS—The Electrochemical Society

[89]. By implementing a pre-nucleation RTA anneal at 1250 °C, the threshold limit shifts to lower values, indicating a more efficient gettering by a vacancy controlled oxygen precipitation process. This step also enables to lower the anneal temperature and to shorten the anneal time for the same getter efficiency. A higher vacancy concentration increases the threshold limit thereby reducing the getter efficiency, which may be due to strain release of the oxygen precipitates by the interaction with vacancies [90].

Cu can be gettered at oxide precipitates without punched out dislocations [80, 90]. For an efficient gettering the BMD density should be $>2 \times 10^9$ cm^{-3} [91]. It has been demonstrated that the Cu gettering by oxide precipitates is based on segregation of Cu to the SiO_x interface [92].

Intrinsic gettering of Fe contamination requires a Fe supersaturation above solid solubility for effective gettering and the process is based on reaction-limited segregation gettering control by the wafer cooling rate [93]. The Fe solubility enhancement in p^+-layers is observed for temperatures below 800 °C and boron concentrations above 1×10^{18} cm^{-3} [94].

The Fe precipitation at oxide precipitates in the defect region is faster than the formation of Fe silicides in the denuded region [95]. However, the silicides are more stable than the Fe precipitates so that care has to be taken to avoid Fe re-emission during the device processing. During an anneal in an oxidizing ambient most of the Fe will diffuse into the growing oxide rather than diffusing in the Si bulk [96].

Special attention has to be given if intrinsic gettering is applied to 300 mm epitaxial wafers which are designed for a low density of crystal originated particles (COPs) and surface local light scatterers (LLS), working with reduced oxygen concentrations (so called 'perfect silicon'). In the case of p/p^- epi with low oxygen concentration (6×10^{17} cm^{-3} or below) a RTP step before the low temperature (750 °C) nucleation step can be used [94]. Another solution is to work with NCz material for enhanced oxygen precipitation what has the additional benefit to reduce the size of the COPs [96, 97].

In p/p^+ wafers the heavy boron doped p^+ layer will getter elements like Fe and Cu due to segregation gettering because of the higher solubility in the layer. As discussed before, in this case FeB pairs will be formed. Also Cu^+ can pair with B^- to form CuB pairs. The situation is different for Ni^0 as its donor level lies below the valence band (see Chap. 6) and only a very small enhancement of the solubility (within experimental error) is noticed [98]. Therefore, segregation gettering is inefficient for Ni [11]. Recently, however, it has been reported that Ni segregation becomes active if the Ni concentration is below 1×10^{13} cm^{-2} resulting in a decrease of the Ni precipitation in the surface region of the epitaxial wafers [14]. The effect is observed for a substrate resistivity below 10 mΩcm (1×10^{19} cm^{-3}) and increases for lower resistivity, in agreement with another report studying Ni gettering in 10 mΩcm layers [99]. For Cu and Fe the minimum doping concentrations are 1×10^{17} and 5×10^{18} cm^{-3}, respectively [99].

For n/n^+ epitaxial wafers one has to take into account the reduced oxygen precipitation in heavily As doped silicon. The reduction is becoming important for doping concentrations above 1.6×10^{19} cm^{-3} [100]. The amount of oxygen precipitation is depending on the nucleation temperature and the implementation of an RTA pre-anneal step [101]. Good results have been reported for an RTA step at 1250 °C in Ar and a nucleation step at 650 °C. Higher nucleation anneal temperatures and lower RTA temperatures lower the oxygen precipitation. The enhanced precipitation is related to the formation of As–V–O complexes during the cooling cycle of the RTA step, which act as nucleation centers for heterogeneous precipitation of oxygen [101]. Recent defect simulations confirmed that the reduction of the vacancy concentration leads to a lower concentration of COPs and micro bulk defects [100].

Similar results have been observed for heavily antimony doped silicon. The formation of AsV or SbV precursors for oxygen precipitation is favored over the formation of VO or VO_2 complexes as the DFT calculations indicate that the binding energy of the dopants with vacancies is much larger than that of vacancies with oxygen [102]. An even more enhanced precipitation is observed for an N_2 RTA pre-anneal, whereby not only vacancies are generated but as well N–V–O and N–O complexes are formed, also acting as precursors for heterogeneous oxygen precipitation [103]. The RTA pre-anneal in N_2 is also more effective for Cu gettering after the IG process than the Ar RTA anneal step.

8.4 Frontside Gettering Techniques

For slow diffusing metals backside gettering is not an option so that much attention has been given to frontside gettering approaches. For several years it was popular to use an annealing in a chlorine containing ambient, often referred to as ***HCl-gettering*** [using HCl-gas, trichloroethylene (TCE) or trichloroethane (C33)]. The main process reason to use a HCl-containing gas ambient was to improve the gate oxide quality [104]. Later on the getter capability was studied. This gas ambient not only eliminated the OISFs but also gettered metals by transforming them into volatile metal chloride products. An HCl-gas ambient has also been used to purify metallurgical grade silicon material for photovoltaic applications [105]. Good results have been achieved for an annealing at 650 °C, as at higher temperatures there is a risk to etch the silicon surface. For standard IC fabrication chlorine containing ambients are no longer used due to the green environment regulations in different regions in the world.

Backside gettering techniques are not applicable when double-sided polished wafers are used or when the backside of the wafer is becoming an active part of the device. Due to the restricted thermal budgets used in advanced IC device fabrication, the metal contaminants can only diffuse over a rather short distance so that it has become essential to switch over to frontside gettering approaches. The limited thermal budgets may also pose some restrictions on the use of intrinsic gettering. Some of the approaches schematically illustrated in Fig. 8.1 will be further discussed.

8.4.1 Buried Epitaxial or Porous Si Layer

A buried epitaxial SiGe layer will, depending on the Ge concentration in the film, result in misfit dislocations at the Si-SiGe interface. This gettering approach has first been proposed by Salih et al. [106]. On top of the SiGe layer a defect-free epitaxial

Si layer is grown in which the device fabrication is done. The misfit dislocations are efficient sites to getter metallic impurities. For 0.1–1% Ge the defect density is in the range 10^6–10^8 cm^{-2} and the misfits are about 10 μm apart from each other. Dislocation interactions leading to threading dislocations have to be avoided. Both Au and Cu gettering by the strain field around the dislocations has been demonstrated. Misfit dislocations in 2% SiGe epitaxial layers were used for Ni gettering [107], and low concentrations of Cu and Fe, respectively [108].

Porous Si can be used as a buried layer for gettering purposes. Epitaxial solar cells using a porous Si layer on a p^+-substrate and a high quality epitaxial layer on top for the active layer to generate the photovoltaic power have been studied for their efficiency to getter TMs [109]. The embedded porous Si layer reduces the metal concentration (Fe, Ni and Cu) in the top layer by three to four orders of magnitude. Ab initio simulations pointed out that at least for Fe and Cu the binding energy between the metals and the voids is the driving force for the getter action. The getter efficiency depends on the porosity of the film and the used cooling rate. Smaller pores have a higher gettering efficiency [110] as will be further discussed in Sect. 8.4.2.1.

8.4.2 *Ion Implantation*

The use of backside damage created by ion implantation has already been discussed above. However, ion implantation is a powerful tool to introduce a well-known amount of lattice damage (clusters of point defects) at a certain depth in the silicon (depending on the implant energy) and in localized areas, so that the location of getter sites can be carefully controlled. Performing the implantation at the frontside of the wafer enables to bring the gettering site closer to the active device areas so that even slow diffusers can be removed from the device area. Over the years a large number of different implantation approaches have been studied, using various as-implanted species and only a few cases will be illustrated and discussed here.

A strong gettering of Au and Cu, with a good thermal stability, has been reported for a carbon implantation with a dose in the range 10^{15}–10^{16} cm^{-2} at 100 keV [111]. The gettering action is not due to the carbon enhanced oxygen precipitation in Cz Si but directly to the carbon-related implantation defects, and increases linear with the implant dose. A similar linear effect has been observed for Au gettering in FZ Si by 1 MeV C implants and an anneal at 850 °C for 2 h [112]. The getter effect starts from a C dose of 6×10^{15} cm^{-2}, and the efficiency can be increased from 55 to 82% by an additional 2 h anneal at 900 °C. The gettering remains stable up to 1050 °C. For longer anneal times a saturation in the gettered concentration occurs. This has also been observed for Au gettering by 1 MeV Ar^+ implantations [113]. The increase in getter efficiency by the additional high temperature anneal is due to the stress associated with the increased β-SiC precipitation and nanovoids associated with these precipitates acting as gettering sinks [114].

Experiments with 10 MeV carbon implants indicated that Fe getting starts from a carbon dose of 10^{14} cm^{-2} and complete gettering is obtained for a dose of 10^{16} cm^{-2} [115]. These authors did not observe a similar effect for Si implants leading them to the conclusion that the damage alone is not causing the getter effect, but most likely the carbon and the possible interaction with oxygen. This is in contradiction with the observation that a 10^{15} cm^{-2} Si implant at 2.3 MeV can effectively getter Fe in both FZ and Cz Si [116]. The getter action occurs at the defects around the projected range R_p during the cooling cycle via a relaxation-induced mechanism. However, there is also a segregation-induced getter action at $R_p/2$ where a local solubility enhancement occurs [117]. At higher temperature the latter getter action diminishes. The two gettering regions are shown in Fig. 8.11 for Cu and Ni gettering, respectively [118]. The behavior of the two metals is different. After 900 °C anneal most of the Cu is localized near R_p with a small fraction near $R_p/2$, while Ni exhibits the strongest trapping at $R_p/2$ after 600 °C anneal. For Cu the total amount trapped is the same at both temperatures, while there is a difference of 10% for Ni. For both metals the trapping at $R_p/2$ increases for a lower anneal temperature. The defects in that region are nanovoids and the gettering takes place at the void/Si matrix boundary via the formation of small clusters coherent with the Si lattice, i.e., trapping of Fe_i at nano-sized open-volume defects [119].

Cu gettering at stacking faults has been observed after heavy dose Ge implantation and solid phase epitaxial regrowth [120].

As discussed before, Ar implantation at the backside of the wafer is a well-known gettering technique. A high dose Ar implantation (>10^{13} cm^{-2}) can also be used for Cu gettering in thinned wafers for 3D integration [121]. In Sect. 7.1.2.3 it was demonstrated that Cu contamination on the backside of thinned wafers, caused by the backside grinding technique used to thin the wafers, diffuses in the silicon and can have a degrading impact on the integrity of the gate oxide (see Fig. 7.8). Besides impacting the breakdown voltage there is also an influence on the lifetime, which depends on the used backside treatment such as e.g. dry polishing, a

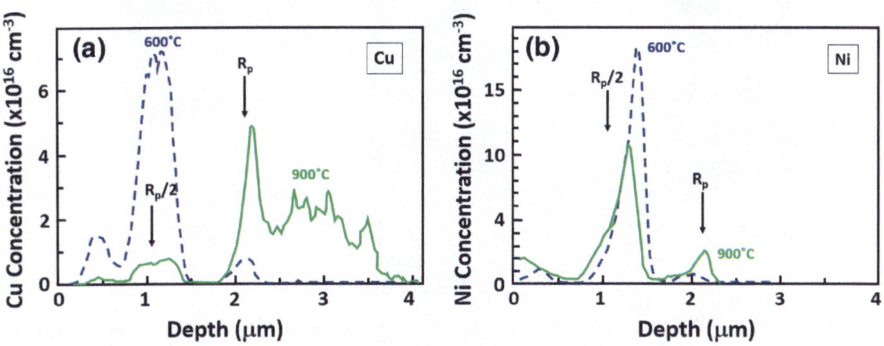

Fig. 8.11 SIMS profiles for Cu (**a**) and Ni (**b**) gettering in epitaxial layers after 2.3 MeV Si implantation and a two-step annealing (900 °C for 1 h + 600 °C or 900 °C for 20 min) After Brown et al. [118]. Copyright 1998 AIP Publishing

CMP treatment of the backside or an ultra-poligrind stress-relief treatment as was shown in Fig. 7.9 [122]. The backside treatment influences the Cu in-diffusion. However, an Ar implantation in the thinned wafer enables to getter the Cu and avoids Cu to diffuse to the frontside of the wafer. Another approach to avoid Cu in-diffusion is the use of a barrier layer such as SiB or SiCN [121, 123]. Cu contamination can also be caused by the Cu filling used in the 3D Through-Silicon-Vias (TSV) process, discussed in Chap. 3.

8.4.2.1 Nano Cavities

In the previous section it was already mentioned that nanocavities have an efficient gettering action. The work has been triggered by the observation that MeV He implantation in Si results in the formation of small voids with the shape of a tetrakaidekahedron with {100}/{111] facets [123].

Void formation has also been studied in germanium. After H implantation small (001) and {111} platelets and {113} defects are nucleated at room temperature,

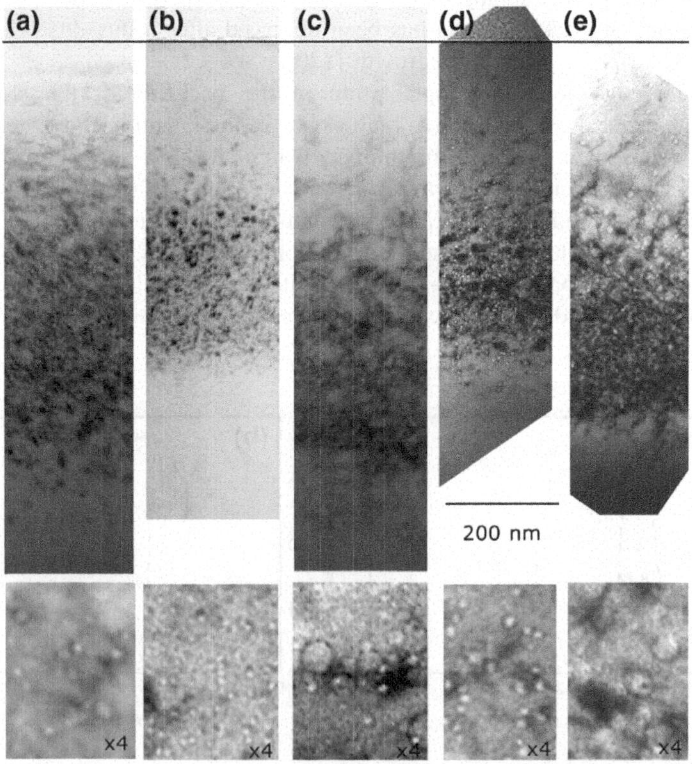

Fig. 8.12 Microstructure of Ge and Si samples implanted with He at 50 keV, 5×10^{16} cm^{-2} and different temperature. **a** Ge: RT, **b** Si: RT, **c** Ge: 200 °C, **d** Si: 200 °C, **e** Si: 400 °C

while for He implantations a continuous cavity layer is formed [124]. Differences between He-implanted voids in Si and Ge are for different implant temperatures illustrated in Fig. 8.12. The size of the cavities in Ge is 3 nm for room temperature implantation and increases 1.5× if the implantation is done at 200 °C. Comparing Ge with Si, the implantation temperature for Si has to be 200 °C higher than for Ge to obtain the same cavity size.

The gettering of TMs by He implantation has been reported by Myers et al. first for Cu and Ni [125] and later also for Ni, Co and Fe [126]. The metal trapping occurs by chemisorption of the metals at cavities after the metal diffusion from their original location. For Cu the binding energy is 0.5 eV larger than in the silicide phase, whereas for Fe and Co the silicide has a greater binding energy although there is still trapping at the cavities. The latter is due to the fact that the initial impurity concentrations are below the threshold for silicide precipitation so that substantial gettering can occur by chemisorption resulting in a reduction of the solubility [127].

For high dose He implantations, bubbles will be formed if the local He concentration is larger than 3.5×10^{20} cm^{-3} [128]. During annealing (e.g. 800 °C for 300 s) He will evaporate so that voids are formed which coalesce with each other to form larger voids. The average void density is [128]

$$D_{void} = 2.3 \times 10^{-10} \exp^{-(0.51/k_B T)} \tag{8.4}$$

These voids are effective for gettering Pt and Au by interacting with the unsaturated Si bonds on the surface of the void [129]. At higher temperatures the bonds are broken and the metal atoms are released but they can be trapped again by another segregation anneal.

H implantations allow, depending on the contamination level, to getter Pt and Ag [129]. Complete gettering by 3×10^{16} cm^{-2} H at 70 keV and 850 °C anneal for 1 h, is achieved up to a dose of 1×10^{14} cm^{-2} and 3×10^{14} cm^{-2} for Pt and Ag, respectively, as shown in Fig. 8.13. For a higher metal implant dose the efficiency

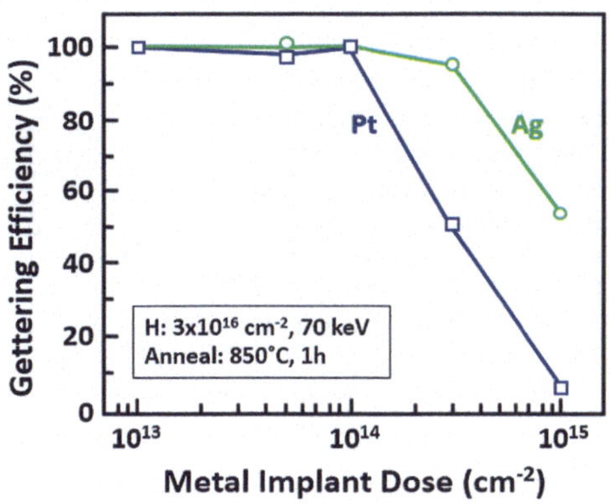

Fig. 8.13 Gettering efficiency of a hydrogen implant (3×10^{16} cm^{-2} at 70 keV and 850 °C anneal for 1 h) for Pt and Ag in Si Reproduced from Kinomura et al. [129] with permission of Elsevier

strongly decreases due to the decreased solubility, triggering silicide formation in the case of Pt.

Instead of using a standard H implantation one can also use H^+ or proton irradiation to getter impurities. This has been demonstrated for Pt in silicon [130]. Instead of creating a damaged layer for gettering, one can rely on the observation that the Pt profile follows the implant damage profile, an approach that can be used to perform lifetime engineering. This technique works also for other atoms such as O, F and Cl [131].

In the case of Cu the getter efficiency is higher at cavities than at dislocations [132]. Gettering of Cu at cavity walls up to 1 monolayer saturation coverage has been reported for Cu concentrations typically observed during processing [133, 134]. The binding energy is 2.2 eV and on the average 6.5 atoms are trapped per μm^2. The binding energy of Cu in precipitated Cu silicides is only 1.7 eV so that the former is a more stable gettering mechanism. The number of gettered Cu atoms will be limited by the number of available chemisorption sites [134].

Frontside MeV Ar implantation with a high dose can be used to form a buried implanted layer underneath the active device area in order to getter metallic impurities like Ni [16] and Au and Pt [135]. Whether or not an amorphous layer is formed by the implantation has no impact on the gettering capability, which is controlled by the anneal temperature and time [135]. The total amount of gettered metal increases with both anneal temperature and anneal time, although for longer anneal times a saturation effect occurs as already discussed before.

8.4.2.2 Near-Surface Proximity Gettering

For slow diffusing transition metals (e.g. Mo and W) the getter sites should be located close to the active device area in order to enable gettering of these metals.

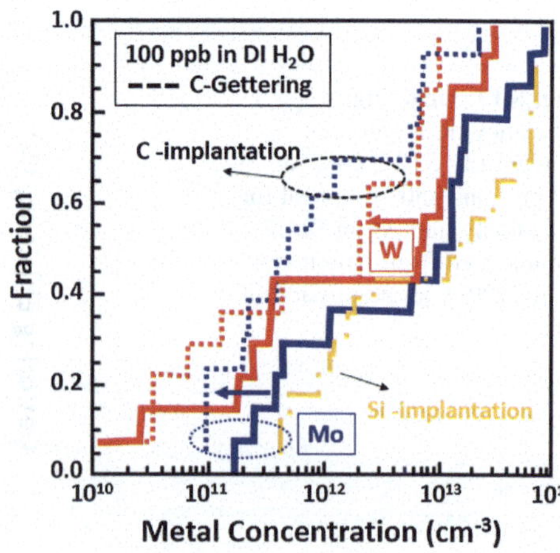

Fig. 8.14 Cumulative distributions of Mo and W concentrations (100 ppb solution in DI H_2O) in wafers without (solid lines) and with an implantation gettering (dotted lines). For Mo (blue curves) the impact of a C- or a Si-implantation is shown, while for W (red curves) only the impact of a C-gettering is given (data from Russo et al. [136])

This approach of 'close by' gettering by sites within e.g. a diffusion length is called ***promixity gettering***.

The proximity gettering capability of a C-implantation (10^{15} cm^{-2}, 5.2 keV) is illustrated in Fig. 8.14 for Mo and W contaminated Si wafers (deionized water solution with 100 ppb Mo or W and subsequently RTA annealed at 1035 °C for 10 s to recrystallize the implanted region and to nucleate getter sites), respectively [136]. For Mo both the impact of a C- (10^{15} cm^{-2}, 5.2 keV) and a Si-implantation (7×10^{14} cm^{-2}, 10 keV) is shown, while for W only the C-implantation is represented. It can clearly be observed that (1) for the same concentration of metal in the solution, there is a higher W than Mo concentration in the Si (comparison of the solid lines), (2) in the case of C-gettering the effect is much more pronounced for Mo than for W, and (3) for Si-implantation no gettering effect is achieved.

The observed lack of gettering after a Si implantation is very important as the used Si implantation condition generates extended defects in the Si, so that the observation of no-gettering after a Si implantation indicates that extended defects don't play a role in the gettering process. The carbon complexes in the defective layer collect the metal contaminants in their strain field, and once metals are gettered, their large binding energies inhibit their release [137]. Proximity gettering is based on the difference in segregation coefficient between the getter sites and the surrounding Si lattice. A crucial parameter is the anneal temperature, which should be high enough to obtain a good diffusivity of the metal but on the other hand sufficiently low as the segregation effect decreases at higher temperatures.

The gettering efficiency also depends on the concentration in the solution and is not occurring below a critical level [136]. The fact that the gettering is enhanced by a high supersaturation has been reported before for Fe gettering by oxygen precipitates, i.e., the getter efficiency is showing an S-curve for the Fe concentration dependence with an increasing precipitated Fe fraction for increasing Fe concentrations and for high Fe concentrations a saturation [138]. The start of the saturation is temperature dependent and shifts to higher Fe concentrations for higher temperatures.

The getter sites should be located close to the device area. Yamaguchi et al. have proposed to perform the C-gettering underneath the contact area in order to trap not only Fe and Ni but also slow diffusing metals such as W during the CMOS image sensor processing [139]. Their ultimate proximity gettering in the shallow surface active area is based on an optimization of not only the C-implantation (defective and strained layer) but also the RTA step, the post-treatment and depends on the design layout of the C-implantation mask (middle, one side only or on both sides of the contact area). The good gettering results for Fe/Ni and W shown in Fig. 8.15 strongly eliminate the white spots in their CIS devices.

Improved performance of high-resolution CIS arrays, fabricated on p/p^{++} epitaxial wafers, has also been obtained by using proximity relaxation gettering via hydrogen-ion induced nanocavities in the substrate underneath the depletion region of the photodiodes [140]. Although such epitaxial wafers enable the control of Fe and Co contamination based on segregation gettering, this is not the case for Cu and Ni. The hydrogen implantation (6×10^{16} cm^{-2} at 26 keV) was done before the

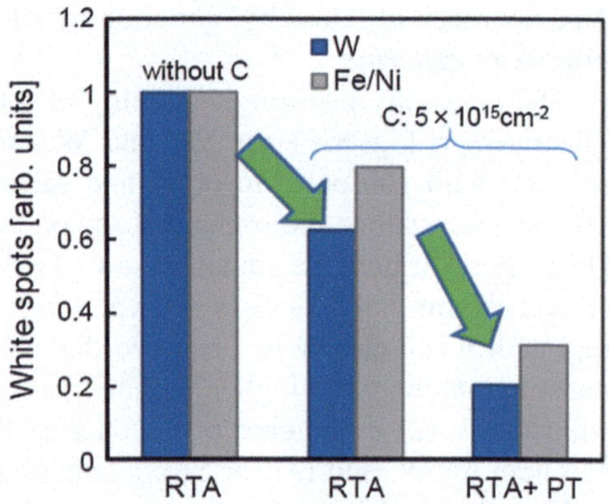

Fig. 8.15 White spots reduction by the optimization of the C-implantation and post treatment (PT) of the proximity gettering process Reproduced from Yamaguchi et al. [139]. Copyright 2016 IEEE

growth of the epitaxial layer. The gettering action occurs by the formation of $FeSi_2$, Cu_3Si, $NiSi_2$ and $CoSi_2$ silicides, respectively, near the surface of the nanocavities due to the interaction with self-interstitials and vacancies, created in order to release the hydrostatic strain in the nanocavity, using a typical relaxation gettering mechanism.

8.5 Gettering in SOI Material

One has to differentiate between the gettering of TMs during the manufacturing of the SOI structure itself and the gettering of TMs during the device processing of SOI wafers. An example of the first case is the fabrication of e.g. the SOI structure by using separation by ion implantation of oxygen (SIMOX). For the second case one can use e.g. smart cut wafers where a high quality buried SiO_2 layer is present at the start of the IC processing and the incoming wafers are inherently free of metallic contamination. For SOI wafers, the buried oxide can be a barrier for the diffusion of TMs from the substrate to the active silicon film on the frontside. However, the situation is becoming different for state-of-the-art SOI wafers for advanced CMOS processing where the buried SiO_2 layer is only 10–20 nm thick so that it is no longer a buffer for metallic contamination picked up in the substrate during the device manufacturing.

8.5.1 *SIMOX SOI Material*

To fabricate the buried oxide in a standard Si wafer, a high dose oxygen implantation (about 10^{17} cm^{-2} O) is performed combined with a high temperature (1200–1300 °C) step to form a continuous SiO_2 layer and to anneal out the defects. It is obvious that his fabrication cycle is very sensitive to metallic contamination. Early

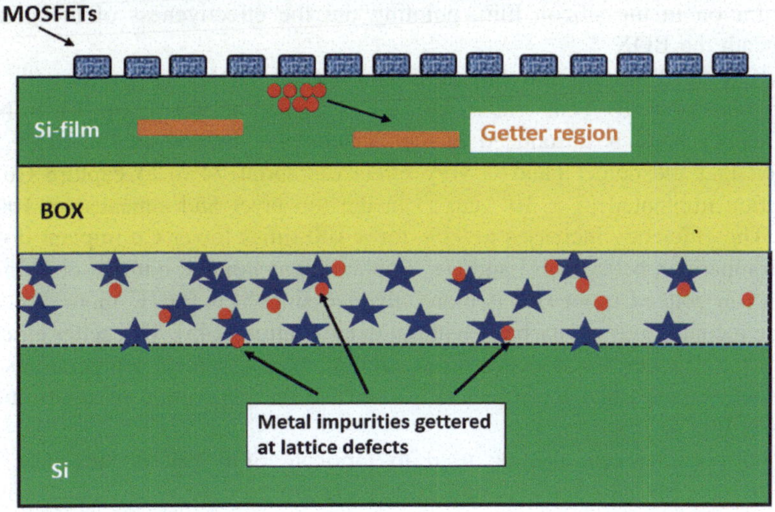

Fig. 8.16 Sketch (not at scale) of a SOI structure and the areas where gettering can occur. Different types of metals can be simultaneous contaminating the substrate and/or the Si film

experiments pointed out that Cu and Cr contamination in the top layer behaves differently [141]: Cr tends to segregate at the Si/SiO_2 front surface so that only a small fraction is diffusing to the damage area, while Cu dominantly segregates at the damage in the buried oxide region. A part of the Cu can diffuse through the buried SiO_2 layer (BOX) and segregate below the buried oxide in the silicon substrate. The gettering can happen in different regions as illustrated in Fig. 8.16, i.e., in the defect region underneath the BOX layer or at getter sites in the top Si-film outside the active device area. Without going into depth about the SIMOX fabrication and processing details, a few examples will be given of metal gettering in SIMOX structures.

He or C implantations have been used as proximity gettering for Fe and Cu [142]. The metal will preferentially segregate near these implants instead of at the buried interface. The C implant effectively getters Cu and Fe, while the He implantation-induced voids preferentially getter Cu indicating the better gettering efficiency of the C implantation. To achieve the same getter effect the C dose (1×10^{16} cm^{-2}) is also one order of magnitude lower than the He implant dose (1×10^{17} cm^{-2}).

The gettering of Ni and Cu was studied for an SOI structure with a 90 nm Si top film and a 370 nm BOX thickness [143]. Both Ni and Cu were gettered in a thin layer underneath the BOX, controlled by the amount and type of defects, due to heterogeneous impurity precipitation at lattice defects. In the temperature range 600–950 °C, Cu and Ni diffuse through the BOX into the silicon film. Metal contamination from either the front or the backside leads to the same final metal concentration in the silicon film, pointing out the effectiveness of the gettering underneath the BOX.

SIMOX (110 nm Si film thickness and 80 nm BOX) structures with a H^+ implantation (4×10^{16} cm^{-2} at 70 keV and annealed at 500 °C for 4 h in N_2) so that a cavity band is introduced in the Si substrate underneath the BOX layer, revealed that the defect band is very effective (about 74%) to capture Cu contamination implanted (5×10^{15} cm^{-2}) in the top layer and annealed at 1000 °C [144]. The efficiency increases to 92% for a 100 times lower Cu implant dose.

A comparison between H^+ and He^+ implantations pointed out that both implant species can getter Cu but He^+ is more effective since it is for H^+ more difficult to produce a continuous cavity band without delamination [145]. The getter efficiency increases for lower Cu concentrations and increasing anneal temperatures. The delaminating properties are beneficially used for the fabrication of smart cut SOI wafers [146].

Wafer bonding can also be used to fabricate SOI wafers. However, these structure have no gettering capabilities as there are no defects associated with the Si/SiO_2 interface. This can be overcome by using a thin polysilicon layer (0.6 μm) between the buried oxide and the device layer [147]. A high Fe gettering efficiency is obtained by a combination of segregation and precipitation. For low Fe concentrations segregation is dominating, while for high Fe concentrations the polysilicon layer thickness (few tens of a micron) must be increased to maintain the segregation effect.

8.5.2 Ultra-thin Body and BOX (UTBB)

Advanced SOI technologies for logic and memory applications are using not only very thin BOX (10–20 nm) but also very thin Si films (10–20 nm), called fully depleted ultra-thin BOX and body (FD-UTBB). For these technologies the only gettering options are relying on either gettering in the silicon substrate under the BOX or backside gettering. One could also rely on the segregation gettering by the highly doped elevated source/drain contact regions but this is only efficient for some metals, as has been shown before.

As already mentioned in the previous section, it is possible to use proximity gettering, creating a getter layer under the thin BOX.

8.6 Gettering Processes for Photovoltaics

For photovoltaic devices many of the gettering approaches mentioned in this chapter are applicable. The main differences compared to IC silicon are (i) the low cost PV feedstock may have a much higher metallic contamination level, (ii) it is more difficult to getter TMs in polycrystalline Si as the grain boundaries are effective getter sinks and the contamination levels are higher, (iii) some of the gettering layers (e.g., high-doped back surface layer, thin film Al layer on the backside, buried porous Si layer, etc.) can be an active part of the cell technology having a direct impact on the cell performance, (iv) in some cases the active device region is the total volume of the Si substrate, and (v) for many solar cells the cost is an important issue so that any additional processing steps should be avoided or their number at least minimized.

A thin Al film on the backside is an effective getter technique (see Sect. 8.2.2.1) but Al is a p-type dopant so that in p-type Si Al deposition forms a p^+-layer, which can be used as a back surface field to reflect electrons and to reduce carrier recombination. Since the early days of IC processing, it is also well known that Al interacts with atomic hydrogen and therefore impacts the defect passivation properties.

For PV applications the carbon concentration in the wafer has an influence on the gettering properties of a frontside phosphorus diffusion and backside Al deposition co-gettering of polysilicon material [148]. There is no evidence for the individual impact of the absolute oxygen or carbon concentration but the excess carbon or oxygen seems to play an dominant role. A high carbon content reduces the getter efficiency.

Several reports claim that the combination of PDG and AlG gettering is more efficient than each on its own, i.e., there is a synergistic effect (see [149] and reference therein). A main challenge for PV remains to achieve homogeneous bulk properties, which can be disturbed by variation in oxygen concentration, carbon concentration and/or a non-uniform dislocation density.

8.7 Modeling Gettering Processes

Due to the different technological process conditions involved, more and more attention is given to the modeling of gettering processes in order to determine the optimum gettering sequence taking into account cost considerations. Although much effort has been concentrated on Cu and Fe, the model should preferentially be valid for different metals. A good starting point are ab initio calculations based on Density Functional Theory to determine the binding energy of metals to getter sites in order to reveal the most stable defect configuration based on minimum energy considerations. This topic is addressed in Chap. 9. When analyzing the PDG approach in Sect. 8.2.1, it was already mentioned that for phosphorus diffusion gettering the P_4V clusters are the dominant gettering sites for high phosphorus concentrations. Ab initio calculations have also been performed for other gettering approaches such as C implantation gettering [150] and gettering by a porous Si layer [106].

The following step is to use continuum models to link the redistribution and trapping of the metals to the device performance. The modeling of the redistribution of metals is based on the drift and diffusion equations. Gettering then occurs by the metal trapping and is described by the associated reactions as e.g. the different reactions describing the interaction between substitutional phosphorus atoms and vacancies resulting in M + $P_4V \rightarrow MP_4V$ for PDG [28]. This enables to determine the reaction rate and the binding energy between the metal and the getter site. Subsequently, the defect distribution has to be coupled to device parameters such as e.g. carrier lifetime or leakage current. Carrier lifetime determination can e.g. be based on the standard Shockley-Read-Hall recombination model. It is essential to calibrate the model parameters based on experimental data.

This combined approach of ab initio calculations and continuum modeling has been used for different types of getter sites such as e.g. P_4V clusters [28, 151], {311} defects [152] and oxygen precipitates [153]. More details are discussed in Sect. 9.3.1.

References

1. M. Kikuchi, S. Iizima, A simpler method for removing coppr from germanium. J. Phys. Soc. Jpn. **12**, 824 (1957). https://10.1143/JPSJ.12.824
2. C.F. Drake, K.L. Ellington, Cause and prevention of high reverse currents in large-area high-voltage diffused-silicon rectifiers. Electr. Commun. **38**, 341–353 (1963)
3. C.F. Cerofolini, M.L. Polignano, A comparison of gettering techniques for very large scale integration. J. Appl. Phys. **55**, 579–876 (1983). https://doi.org/10.1063/1.333066
4. M.L. Polignano, C.F. Cerofolini, H. Bender, C. Claeys, Gettering mechanisms in silicon. J. Appl. Phys. **64**, 869–876 (1988). https://doi.org/10.1063/1.341939
5. S.A. McHugo, H. Hieslmair, Gettering in silicon, in *Encyclopedia of Electrical and Electronics Engineering* (Wiley, New York, 1999). https://doi.org/10.1002/047134608x.w3209
6. M.B. Shabani, T. Yamashita, E. Morita, Study of gettering mechanisms in silicon: Competitive gettering between phosphorus diffusion gettering and other gettering sites. Solid State Phenom. **131–133**, 399–404 (2008). https://doi.org/10.4028/www.scientific.net/SSP.131-133.399
7. M. Seibt, V. Kveder, Gettering processes and the role of extended defects, in *Advanced Silicon Materials for Photovoltaic Applications* (Chap. 4, Wiley, New York, 2012). https://doi.org/10.1002/9781118312193.ch4
8. A.M. Myers, M. Seibt, W. Schröter, Mechanisms of transition-metal gettering in silicon. J. Appl. Phys. **88**, 3795–3819 (2000). https://doi.org/10.1063/1.128927
9. S. Koveshnikov, O. Kononchuk, K. Beaman, G.A. Rozgonyi, F. Gonzalez, Gettering of Fe, Cu and Ni in MeV implanted epitaxial silicon. Electrochem. Soc. Proc. **98**(13), 339–352 (1998)
10. A.L. Smith, K. Wada, L.C. Kimerling, Modeling of transition metal redistribution to enable wafer design for gettering. J. Electochem. Soc. **147**, 1154–1160 (2000). https://doi.org/10.1149/1.139332
11. R. Hoelzl, D. Huber, K.J. Range, L. Fabry, J. Hage, R. Wahlich, Gettering of copper and nickel in p/p^+ epitaxial wafers. J. Electrochem. Soc. **147**, 2704–2710 (2000). https://doi.org/10.1149/1.1393593
12. R. Hoelzl, K.J. Range, L. Fabry, Modeling of Cu gettering in p- an n-type silicon and in poly-silicon. Appl. Phys. A **73**, 137–142 (2002). https://doi.org/10.1007/s003390100846

13. K. Nakamura, H. Iga, J. Tomioka, Analysis of the segregation phenomena of copper in p/p$^+$ epitaxial silicon wafers. ECS Trans. **3**(4), 255–265 (2006). https://doi.org/10.1149/1.2355761
14. K. Torigoe, T. Ono, K. Nakamura, Competitive interaction between segregation gettering and surface precipitation of nickel in p/p$^+$ silicon epitaxial wafers. ECS J. Solid State Sci. Technol. **4**, Q110–Q114 (2015). https://doi.org/10.1149/2.0201509js
15. R. Hoelzl, L. Fabry, K.J. Range, The linkage between macroscopic gettering mechanisms and electronic configuration of $3d$-elements in p/p$^-$ silicon epitaxial wafers. Appl. Phys. A **74**, 35–39 (2002). https://doi.org/10.1007/s003390100866
16. R. Hoelzl, L. Fabry, K.J. Range, Ni reactions with surfaces: dependence of gettering efficiencies for Ni on crystal-growth conditions, backside gettering techniques, oxygen precipitates and thermal treatments. Appl. Phys. A **74**, 711–718 (2002). https://doi.org/10.1007/s003390100946
17. K. Sumino, Basic aspects of impurity gettering. Microelectron. Eng. **66**, 268–280 (2003). https://doi.org/10.1016/S0167-9317(02)00918-8
18. E.J. Mets, Poisoning and gettering effects in silicon junctions. J. Electrochem. Soc. **112**, 420–425 (1965). https://doi.org/10.1149/1.2423560
19. G.H. Schwuttke, K. Yang, H. Kappert, Lifetime control in silicon by impact sound stressing. Phys. Status Solidi A **42**, 553–564 (1977). https://doi.org/10.1002/pssa.2210420218
20. C.W. Pearce, V.J. Zaleckas, A new approach to lattice damage gettering. J. Electrochem. Soc. **126**, 1436–1437 (1979). https://doi.org/10.1149/1.2129298
21. K.H. Yang, G.H. Schwuttke, Minority carrier lifetime improvement in silicon through laser damage gettering. Phys. Status Solidi A **58**, 127–134 (1980). https://doi.org/10.1002/pssa.2210580115
22. T.M. Buck, K.A. Pickar, J.M. Poate, C.M. Shieh, Gettering rates for various fast-diffusing impurities at ion-damaged layers on silicon. Appl. Phys. Lett. **21**, 485–487 (1972). https://doi.org/10.1063/1.1654228
23. A.G. Cullis, T.E. Seidel, R.L. Meek, Comparative study of annealed neon- argon-, and krypton-ion implantation damage in silicon. J. Appl. Phys. **49**, 5188–5198 (1978). https://doi.org/10.1063/1.324414
24. A. Goetzberger, W. Shockley, Metal precipitates in silicon p-n junctions. J. Appl. Phys. **31**, 1821–1824 (1960). https://doi.org/10.1063/1.1735455
25. S.W. Ing, R.E. Morrisson, L.L. Alt, R.W. Aldrich, Gettering of metallic impurities from planar silicon diodes. J. Electrochem. Soc. **110**, 533–537 (1963). https://doi.org/10.1149/1.2425808
26. W. Schröter, A. Döller, A. Zozime, V. Kveder, M. Seibt, E. Spiecker, Phosphorus diffusion gettering of metallic impurities in silicon: mechanisms beyond segregation. Solid-State Phenom. **95**, 527–538 (2004). http://www.scientific.net/SSP.95-96.527
27. A. Armigliato, D. Nobili, M. Servidori, S. Solmi, SiP precipitation within the doped silicon lattice, concomitant with phosphorus predeposition. J. Appl. Phys. **47**, 5489–5491 (1976). https://doi.org/10.1063/1.322549
28. R. Chen, B. Trzynadlowski, S.C. Dunham, Phosphorus vacancy cluster model for phosphorus diffusion gettering of metals in Si. J. Appl. Phys. **115**, 054906/1–6 (2014). https://doi.org/10.1063/1.4864377
29. G.B. Bronner, J.D. Plummer, Gettering of gold in silicon: a tool for understanding the properties of silicon interstitials. J. Appl. Phys. **61**, 5286 (1987). https://doi.org/10.1063/1.338264
30. S. Nadahara, H. Tsunoda, M. Shiozaki, M. Watanabe, K. Yamabe, Low temperature phosphorus diffusion gettering of iron in silicon. Electrochem. Soc. Proc. **91**(9), 667–674 (1991)
31. M.L. Joshi, S. Dash, Distribution and precipitation of gold in phosphorus-diffused silicon. J. Appl. Phys. **37**, 2453–2457 (1966). https://doi.org/10.1063/1.1708836

32. L. Baldi, G.F. Cerofolini, G. Ferla, G. Frigerio, Gold solubility in silicon and gettering by phosphorus. Phys. Status Solidi A **48**, 523–532 (1978). https://doi.org/10.1002/pssa.2210480232
33. S.P. Phang, D. Macdonald, Direct comparison of boron, phosphorus, and aluminum gettering of iron in crystalline silicon. J. Appl. Phys. **109**, 073521/1–6 (2011). https://doi.org/10.1063/1.3569890
34. A. Haarahiltunen, H. Savin, M. Yli-Koski, H. Tavitie, J. Sinkkonen, Modeling phosphorus diffusion gettering of iron in single crystal silicon. J. Appl. Phys. **105**, 023510/1–4 (2009). https://doi.org/10.1063/1.3068337
35. R.L. Meek, T.E. Seidel, A.G. Cullis, Diffusion gettering of Au and Cu in silicon. J. Electrochem. Soc. **122**, 786–796 (1975). https://doi.org/10.1149/1.2134324
36. O. Paz, E. Heran, E. Fayo, POCl and boron gettering of LSI silicon devices: similarities and differences. J. Electrochem. Soc. **126**, 1754–1761 (1979). https://doi.org/10.1149/1.2128791
37. B. Hartiti, M. HageAli, J.C. Muller, P. Siffert, Rutherford backscattering analysis of phosphorus gettering of Au and Cu. Appl. Phys. Lett. **62**, 3476–3478 (1993). https://doi.org/10.1063/1.109025
38. J.E. Lawrence, Metallographic analysis of gettered silicon. Trans. Met. Soc. AIME **242**, 484–489 (1968)
39. J.E. Lawrence, Diffusion-induced stress and lattice disorders in silicon. J. Electrochem. Soc. **113**, 819–824 (1965). https://doi.org/10.1149/1.2424127
40. G.A. Rozgonyi, P.M. Petroff, M.H. Read, Elimination of oxidation-induced stacking faults by pre-oxidation gettering of silicon wafers. J. Electrochem. Soc. **122**, 1725–1729 (1975). https://doi.org/10.1149/1.2134118
41. H. Talvitie, V. Vahanissi, A. Haarahiltunen, M. Yli-Koski, H. Savin, Phosphorus and boron diffusion gettering of iron in monocrystalline silicon. J. Appl. Phys. **109**, 093505-1–093505-5 (2011). https://doi.org/10.1063/1.35820864
42. R.D. Thompson, K.N. Tu, Low temperature gettering of Cu, Ag and Au across a Si wafer by Al. Appl. Phys. Lett. **41**, 440–442 (1982). https://doi.org/10.1063/1.93565
43. M. Apel, I. Hanke, R. Schindler, W. Schröter, Aluminum gettering in silicon. J. Appl. Phys. **76**, 4432–4433 (1994). https://doi.org/10.1063/1.357339
44. C.S. Chen, D.K. Schroder, Kinetics of gettering in silicon. J. Appl. Phys. **71**, 5858–5864 (1992). https://doi.org/10.1063/1.350482
45. H. Hieslmair, S.M. McHugo, E.R. Weber, Aluminum backside segregation, in *Proceedings of Photovoltaic Specialists Conference*(1996), pp. 441–444. https://doi.org/10.1109/pvsc.1996.564038
46. M. Seibt, A. Sattler, C. Rudolf, O. Voss, V. Kveder, W. Schröter, Gettering in silicon photovoltaics: Current status and future perspectives. Phys. Status Solidi A **203**, 696–713 (2004). https://doi.org/10.1002/pssa.200664516
47. S. Martinuzzi, I. Perichaud, J.J. Simon, External gettering by aluminum-silicon alloying observed from carrier recombination at dislocations in float zone silicon wafers. Appl. Phys. Lett. **70**, 2744–2746 (1997). https://doi.org/10.1063/1.119009
48. A. Das, A. Rohatgi, Simultaneous iron getttering and passivation of p-type monocrystalline silicon using a negatively charged aluminum-doped dielectric. Appl. Phys. Lett. **101** 252105/1–4 (2012). https://doi.org/10.1063/1.4771979
49. G.A. Rozgonyi, R.A. Kushner, The elimination of oxidation-induced stacking faults by pre-oxidation gettering of silicon wafers III. Defect etch pit correlation with p-n junction leakage. J. Electrochem. Soc. **123**, 570–576 (1976). https://doi.org/10.1149/1.2132879
50. K. Tanno, F. Shimura, T. Kawamura, Microdefect elimination in reduced pressure epitaxy on silicon wafer by back damage-Si_3N_4 film technique. J. Electrochem. Soc. **128**, 395–399 (1981). https://doi.org/10.1149/1.2127429
51. A.Y. Liu, V.P. Markevich, A.R. Peaker, J.D. Murphy, D. Macdonald, Gettering of interstitial iron in silicon by plasma-enhanced chemical vapor deposited silicon nitride films. J. Appl. Phys. **120**, 193103/1–10 (2016). https://doi.org/10.1063/1.4967914

52. K. Mc Lean, C. Morrow, D. Macdonald Activation energy for the hydrogenation of iron in p-type crystalline silicon wafers, in *Proceedings of IEEE 4th World Conference Photovoltaic Energy Conversion* (2006), pp. 1122–1125. https://doi.org/10.1109/wcpec.2006.279358
53. A.Y. Liu, D. Macdonald, Hydrogen passivation of interstitial iron in boron-doped multi-crystalline silicon during annealing. J. Appl. Phys. **116**, 194902/1–10 (2014). https://doi.org/10.1063/1.4901831
54. S. Ogushi, S. Sadamitsu, K. Marsden, Y. Koike, M. Sana, Gettering characteristics of heavy metal impurities in silicon wafers with polysilicon back seal and internal gettering. Jpn. J. Appl. Phys. **36**, 6601–6606 (1997). https://doi.org/10.1143/JJAP.36.6601
55. A.A. Istratov, W. Huber, E.R. Weber, Experimental evidence for the presence of segregation and relaxation gettering of iron in polycrystalline silicon layers on silicon. Appl. Phys. Lett. **85**, 4472–4474 (2004). https://doi.org/10.1063/1.1819512
56. A. Haarahiltunen, M. Yli-Koski, H. Talvitie, V. Vahanissi, J. Lindroos, H. Savin, Gettering of iron in CZ-silicon by polysilicon layer. Phys. Status Solidi C **8**, 751–754 (2011). https://doi.org/10.1002/pssc.201000194
57. R. Hoelzl, K.J. Range, L. Fabry, Comparison of different gettering techniques for Cu-p^+ versus polysilicon and oxygen precipitates. Appl. Phys. A **75**, 591–595 (2001). https://doi.org/10.1007/s003390101045
58. W. Dyson, L. Hellwig, J. Moody, J. Rossi, Gettering of p^+ (100) Si substrates for epitaxial growth. Electrochem. Soc. Proc. **83**(9), 246–255 (1983)
59. C.C.D. Wong, S. Hahn, F.A. Ponce, Z.U. Rek, Backside gettering and its interactions with intrinsic gettering. Proc. Mat. Res. Soc. **71**, 33–38 (1986)
60. T.A. Baginski, J.R. Monkowski, Germanium backside gettering of gold in silicon. J. Electrochem. Soc. **133**, 142–147 (1986). https://doi.org/10.1149/1.2108510
61. J.R. Dennis, E.B. Hale, Amorphization of silicon by ion implantation: Homogeneous or heterogeneous nucleation? Rad. Effects **30**, 219–225 (1976). https://doi.org/10.1080/00337577608240825
62. F.F. Morehead, B.L. Crowder, R.S. Title, Formation of amorphous silicon by ion bombardment as a function of ion, temperature and dose. J. Appl. Phys. **43**, 1112–1116 (1972). https://doi.org/10.1063/1.1661223
63. P.D. Parry, Target heating during ion implantation. J. Vac. Sci. Technol. **13**, 622–629 (1976). https://doi.org/10.1116/1.569046
64. T.W. Sigmon, L. Csepregi, J.W. Mayer, Ion implantation gettering of gold. J. Electrochem. Soc. **123**, 1116–1117 (1976). https://doi.org/10.1149/1.2133007
65. C.M. Hsieh, J.R. Mathews, H.D. Seidel, K.A. Pickar, C.M. Drum, Ion-implantation-damage gettering effect in silicon photodiode array camera target. Appl. Phys. Lett. **22**, 238–240 (1973). https://doi.org/10.1063/1.1654624
66. T.E. Seidel, R.L. Meek, A.G. Cullis, Ion damage and phosphorus diffusion gettering of Au in Si. Inst. Phys. Conf. Ser. **23**, 494–503 (1975). https://doi.org/10.1063/1.321664
67. K. Murase, H. Harada, Argon implantation gettering for a “through-oxide” arsenic implanted layer. J. Appl. Phys. **48**, 4404–4406 (1977). https://doi.org/10.1063/1.323398
68. B. Golja, A.G. Nassibian, Ar implant damage gettering of generation impurities in silicon employing voltage ramping and nitrogen backscattering. IEE J. Solid-State Electron Dev. **3**, 127–132 (1979). https://doi.org/10.1049/ij-ssed:19790027
69. A.G. Nassibian, B. Golja, Investigation of Ar ion implant gettering of gold in silicon by MOS and Rutherford backscattering techniques. IEE Proc I - Solid State Electron Dev. **127**, 29–36 (1980). https://doi.org/10.1049/ip-i-1.1980.0006
70. B. Golja, A.G. Nassibian, Annealing of Si-SiO_2 interface states using Ar-ion-implant damage-gettering. Solid-State Electron. **23**, 1249–1254 (1980). https://doi.org/10.1016/0038-1101(80)90120-3
71. D. Lecrosnier, J. Paugam, G. Pelous, F. Ricou, M. Salvi, Gold gettering in silicon by phosphorus diffusion and argon implantation: Mechanisms and limitations. J. Appl. Phys. **52**, 5090–5097 (1981). https://doi.org/10.1063/1.329407

72. D. Jaworska, J. Sielanko, E. Tarnowska, Au gettering by Ne and Ar implantation in silicon. Appl. Phys. A **35**, 119–124 (1983). https://doi.org/10.1007/BF00620641
73. S.S. Gong, D.K. Schroder, Implantation gettering in silicon. Solid-State Electron. **30**, 209–211 (1987). https://doi.org/10.1016/0038-110(87)90151
74. T.A. Baginski, Back-side germanium ion implantation gettering of silicon. J. Electrochem. Soc. **135**, 1842–1843 (1988). https://doi.org/10.1149/1.2133007
75. H. Bender, Investigation of the oxygen-related lattice defects in Czochralski silicon by means of electron microscope techniques. Phys. Status Solidi A **86**, 245–261 (1984). https://doi.org/10.1002/pssa.2210860126
76. C. Claeys, J. Vanhellemont, Recent progress in the understanding of crystallographic defects in silicon. J. Cryst. Growth **126**, 41–62 (1993). https://doi.org/10.1016/0022-0248(93)90225-L
77. A. Borghesi, B. Pivac, A. Sassella, A. Stella, Oxygen precipitation in silicon. J. Appl. Phys. **77**, 4169–4244 (1995). https://doi.org/10.1063/1.359479
78. W.K. Tice, T.Y. Tan, Nucleation of CuSi precipitate colonies in oxygen-rich silicon. Appl. Phys. Lett. **28**, 564–565 (1976). https://doi.org/10.1063/1.88825
79. T.Y. Tan, E.E. Gardner, W.K. Tice, Intrinsic gettering by oxide precipitate induced dislocation in Czochralski Si. Appl. Phys. Lett. **30**, 175–176 (1976). https://doi.org/10.1063/1.89340
80. M. Seacrist, M. Stinson, J. Libbert, R. Standley, J. Binns, Determination of minimum oxygen precipitate growth conditions for gettering of copper and nickel. Electrochem. Soc. Proc. **2**, 638–646 (2002)
81. D. Yang, R. Fan, Y. Shen, D. Tian, L. Li, D. Que, Intrinsic gettering in nitrogen doped Czochralski crystal silicon. Electrochem. Soc. Proc. **17**, 357–361 (2000)
82. M.J. Binns, S. Bertolini, R.D. Wise, D.J. Myers, T.A. McKenna, Effective intrinsic gettering for 200 mm and 300 mm p/p^- wafers in a low thermal budget 0.13 μm advanced CMOS logic process. Electrochem. Soc. Proc. **2002**(2): 647–657 (2002)
83. R. Falster, V.V. Voronkov, F. Quast, On the properties of the intrinsic point defects in silicon: a perspective from crystal growth and wafer processing. Phys. Status Solidi B **222**, 219–244 (2000). https://doi.org/10.1002/1521-3951(200011)222:1<219:AID-PSSB219>3.0.CO;2-U
84. J.G. Park, K. Kurita, G.S. Lee, S. Sa, H. Furuya, Dependence of crystal nature on the gettering efficiency of iron and nickel in a Czochralski silicon wafer. Microelectron. Eng. **66**, 247–257 (2003). https://doi.org/10.1016/S0167-9317)02)00916-4
85. J. Chen, D. Yang, H. Li, X. Ma, D. Que, Germanium-doped Czochralski silicon: oxygen precipitates and their annealing behavior. Mater. Sci. Semicond. Eng. **9**, 600–605 (2006). https://doi.org/10.1016/j.mssp.2006.08.009
86. J. Vanhellemont, J. Chen, J. Lauwaert, H. Vrielinck, W. Xu, D. Yang, J.M. Rafi, H. Ohyama, E. Simoen, Germanium doping for improved silicon substrates and devices. J. Cryst. Growth **317**, 8–15 (2010). https://doi.org/10.1016/j.jcrysgro.2010.11.024
87. T.A. Frewen, T. Sinno, Vacancy self-trapping during rapid thermal annealing of silicon wafers. Appl. Phys. Lett. **89**, 191903/1–3 (2006). https://doi.org/10.1063/1.2385069
88. R. Hoelzl, M. Blietz, L. Fabry, R. Schmolke, Gettering efficiencies and their dependence on material parameters and thermal processes: how can this be modeled? Electrochem. Soc. Proc. **2**, 608–625 (2002)
89. D. Kot, G. Kissinger, W. Hackl, A. Sattler, A. von Ammon, The role of vacancies and oxygen for setting up an efficient getter for Cu and Ni in silicon wafers. ECS Trans. **16**(6), 207–218 (2008). https://doi.org/10.1149/1.2980304
90. R.J. Falster, G.R. Fisher, G. Ferrero, Gettering thresholds for transition metals by oxygen-related defects in silicon. Appl. Phys. Lett. **59**, 809–811 (1991). https://doi.org/10.1063/1.105350
91. R. Hoelzl, L. Fabry, K.J. Range, Gettering efficiencies for Cu and Ni as function of the size and the density of oxygen precipitates in p/p^- silicon epitaxial wafers. Appl. Phys. A **73**, 137–142 (2001). https://doi.org/10.1007/s003390100846

92. G. Kissinger, D. Kot, M. Klingsporn, M.A. Schubert, A. Sattler, T. Müller, Investigation of the copper gettering mechanism of oxide precipitates in silicon. ECS J. Solid State Sci. Technol. **4**, N124–N129 (2015). https://doi.org/10.1149/2.0151509jss
93. M. Aoki, A. Hara, A. Ohsawa, Intrinsic gettering of iron impurities in silicon wafers. Jpn. J. Appl. Phys. **30**, 3580–3583 (1991). https://doi.org/10.1143/JJAP.30.3580
94. M.B. Shabani, Y. Shiina, S. Shimanuki, F. Kirscht, Iron solubility in boron-doped silicon and Fe gettering mechanism in p/p^+ epitaxial wafers. Solid State Phenom. **82–84**, 331–340 (2002). https://doi.org/10.4028/www.scientific.net/SSP.82-84.331
95. M. Aoki, A. Hara, Distribution of Fe in intrinsic gettered silicon wafer after annealing at supersaturation temperature. Jpn. J. Appl. Phys. **35**, L1231–L1233 (1996). https://doi.org/10.1143/JJAP.35.L1231
96. Y.H. Kim, K.S. Lee, H.Y. Chung, D.H. Hwang, H.S. Kim, H.Y. Cho, R.Y. Lee, Internal gettering of Fe, Ni and Cu in silicon wafer. J. Kor. Phys. Soc. **39**, S348–S351 (2001)
97. D. Gräff, U. Lambert, R. Schmolke, R. Wahlich, W. Siebert, E. Daub, W. von Ammon, 200 mm epi p/p^- wafer: is there sufficient gettering. Electrochem. Soc. Proc. **17**, 319–330 (2000)
98. R. Schmolke, M. Blietz, R. Hoelzl, D. Menzel, H. Bender, Bulk micro defects of p/p^- epitaxial wafers with nitrogen doped substrates and their gettering behavior. Electrochem. Soc. Proc. **2**, 658–669 (2002)
99. M.B. Shabani, Y. Shiina, Y. Shimanuki, Tuning oxygen concentration at low and high temperature IG process and boron concentration in epitaxial wafer for gettering of metal impurities. Solid State Phenom. **95–96**, 539–546 (2004). https://doi.org/10.4028/www.scientific.net/SSP.95-96.53
100. M. Porrini, S. Haringer, A. Giannattasio, Reduced oxygen precipitation in heavily arsenic-doped Cz-silicon crystals. Phys. Status Solidi C **13**, 766–769 (2016). https://doi.org/10.1002/pssc.201600045
101. B. Wang, X. Zhang, Z. Ma, D. Yang, Effects of high temperature rapid thermal processing on oxygen in heavily arsenic-doped Czochralski silicon. J. Cryst. Growth **318**, 183–186 (2011). https://doi.org/10.1016/j.crygro.2010.11.016
102. X. Zhang, C. Gao, M. Fu, X. Ma, J. Vanhellemont, D. Yang, Impact of rapid thermal processing on oxygen precipitation in heavily arsenic and antimony doped Czochralski silicon. J. Appl. Phys. **113**, 163510/1–7 (2013). https://doi.org/10.1063/1.4803061
103. P. Dong, X. Liang, D. Tian, J. Zhao, C. Cao, X. Ma, D. Yang, Enhanced internal gettering in n/n^+ epitaxial wafer: Coaction of nitrogen impurity and vacancy on oxygen precipitation in substrate. J. Mater. Sci.: Mater. Electron. **25**, 3486–3491 (2014). https://doi.org/10.1007/s10854-014-2043-7
104. C. Claeys, R.F. De Keersmaecker, G.J. Declerck, Technology and kinetics of SiO_2 growth, in *The Si-SiO_2 System,* ed. by P. Balk (Chap 2). Elsevier, New York (1988)
105. J. Hampel, P. Ehrenreich, N. Wiehl, J.V. Kratz, S. Reber, HCl gas gettering of low-cost silicon. Phys. Status Solidi **4**, 767–770 (2013). https://doi.org/10.1002/pssa.201200885
106. A.S. Salih, H.J. Kim, R.F. Davis, G.A. Rozgonyi, Extrinsic gettering via the controlled introduction of misfit dislocations. Appl. Phys. Lett. **46**, 419–421 (1985). https://doi.org/10.1063/1.95598
107. D.M. Lee, J.B. Posthill, F. Shimura, G.A. Rozgonyi, Impurity gettering by misfit dislocations in Si(2%Ge) epitaxy: Nickel. Appl. Phys. Lett. **53**, 370372 (1988). https://doi.org/10.1063/1.99897
108. D.M. Lee, G.A. Rozgonyi, Low temperature gettering of trace iron and copper by misfit dislocations in Si/Si(Ge) epitaxy. Appl. Phys. Lett. **50**, 350–352 (1994). https://doi.org/10.1063/1.112370
109. H.S. Radhakrishnan, C. Ahn, J. Van Hoeymissen, F. Dross, N. Cowern, K. Van Nieuwenhuysen, I. Gordon, R. Mertens, J. Poortmans, Gettering of transition metals by porous silicon in epitaxial silicon solar cells. Phys. Status Solidi A **209**, 1866–1871 (2012). https://doi.org/10.1002/pssa.201200232

110. F. Schiettekatte, C. Wintgens, S. Roorda, Influence of curvature on impurity gettering by nanocavities in Si. Appl. Phys. Lett. **74**, 1857–1859 (1999). https://doi.org/10.1063/1.123692
111. H. Wong-Leung, N.W. Cheung, P.K. Chu, Gettering of gold and copper with implanted carbon in silicon. Appl. Phys. Lett. **52**, 889–891 (1988). https://doi.org/10.1063/1.99263
112. S. Mohapatra, B. Joseph, D.P. Mahapatra, P. Chakraborty, High efficiency gettering of Au in Si(111) by MeV C implantation. Nucl. Instr. Methods Phys. Res. B **217**, 578–582 (2004). https://doi.org/10.1016/j.nimb.2003.12.005
113. P. Rorh, J.J. Grob, P. Siffert, Gold and platinum accumulation on buried defects in silicon. Nucl. Instr. Methods Phys. Res. B **80**, 640–643 (1993). https://doi.org/10.1016/0168-583X(93)96199-M
114. S. Mohapatra, B. Joseph, B. Satpati, D.P. Mahapatra, Gettering of implanted Au in MeV C implanted Si. Appl. Phys. A **82**, 297–304 (2006). https://doi.org/10.1007/s00339-005-3297-y
115. W. Skorupa, R. Kögler, K. Schmalz, P. Gaworzewski, G. Morgensten, H. Syhre, Iron gettering and doping due to MeV carbon implantation. Nucl. Instr. Methods Phys. Res. B **74**, 70–74 (1993). https://doi.org/10.1016/0168-583X(93)95016-X
116. O. Kononchuk, R.A. Brown, Z. Radzimski, G.A. Rozgonyi, F. Gonzalez, Gettering of Fe to below 10^{10} cm^{-3} in MeV self-implanted Czochralski and float zone silicon. Appl. Phys. Lett. **69**, 4203–4205 (1996). https://doi.org/10.1063/1.116986
117. S.V. Koveshnikov, G.A. Rozgonyi, Mechanism of iron gettering in MeV Si ion implanted epitaxial silicon. J. Appl. Phys. **84**, 3078–3084 (1998). https://doi.org/10.1063/1.368462
118. R.A. Brown, O. Konunchuk, G.A. Rozgonyi, S. Koveshnikov, A. Knigths, P.J. Simpson, F. Gonzalez, Impurity gettering to secondary defects created by MeV ion implantation in silicon. J. Appl. Phys. **84**, 2459–2465 (1998). https://doi.org/10.1063/1.368438
119. A. Kvit, R.A. Yankov, G. Duscher, G.A. Rozgonyi, J.M. Glasko, Formation of nanoscale voids and related metallic impurity gettering in high-energy ion-implanted and annealed epitaxial silicon. Appl. Phys. Lett. **83**, 1367–1369 (2003). https://doi.org/10.1063/1.1601678
120. C.J. Peter, J.W. Corbett, C. Deng, Z. Atzmon, The gettering of copper by keV implantation of germanium into silicon. J. Appl. Phys. **78**, 3012–3014 (1995). https://doi.org/10.1063/1.360050
121. K. Hozawa, K. Takeda, K. Torii, Impact of the backside Cu contamination in the 3D integration process. VLSI Techn. Digest 172–173 (2009).
122. K.W. Lee, J.-C. Bea, T. Fukushima, T. Tanaka, M. Koyanagi, Cu retardation performance of extrinsic gettering layers in thinned wafers evaluated by transient capacitance measurement. J. Electrochem. Soc. **158**, H795–H799 (2011). https://doi.org/10.1149/1.3597317
123. D.J. Eaglesham, A.E. White, L.C. Feldman, N. Moriya, D.C. Jacobson, Equilibrium shape of Si. Phys. Rev. Lett. **70**, 1643–1647 (1993). https://doi.org/10.1103/PhysRevLett.70.1643
124. M.L. David, J.F. Barbot, S. Rousselet, F. Pailloux, D. Babonneau, M.F. Beaufort, L. Pizzagalli, M. Drouet, E. Simoen, C. Claeys, Extended defects created by light ion implantation in Ge. ECS Trans. **16**(6), 163–175 (2008). https://doi.org/10.1149/1.2989301
125. S.M. Myers, D.M. Follstaedt, D.M. Bishop, J.W. Medernach, Gettering of metal impurities by cavities in silicon. Electrochem. Soc. Proc. **94**(10), 808–819 (1994)
126. S.M. Myers, D.M. Follstaedt, G.A. Petersen, Ch. Seager, H.J. Stein, W.R. Wampler, Chemical and electrical properties of cavities in silicon and germanium. Nucl. Instr. Methods Phys. Res. B **106**, 379–385 (1995). https://doi.org/10.1016/0168-583X(96)80033-4
127. G.A. Petersen, S.M. Myers, D.M. Follstaedt, Gettering of transition metals by cavities in silicon formed by helium ion implantation. Nucl. Instr. Methods Phys. Res. B **127**, 301–306 (1997). https://doi.org/10.1016/s0168-583x(96)00944-5
128. V. Raineri, P.G. Fallica, G. Percolla, A. Battaglia, M. Barbagallo, S.U. Campisano, Gettering of metals by voids in silicon. J. Appl. Phys. **78**, 3727–3735 (1995). https://doi.org/10.1063/1.359953
129. A. Kinomura, J.S. Williams, J. Wong-Leung, M. Petravic, Gettering of platinum and silver to cavities formed by hydrogen implantation in silicon. Nucl. Instr. Methods Phys. Res. B **127**, 297–300 (1997). https://doi.org/10.1016/S0168-583X(96)00943-3

130. D.C. Schmidt, B.G. Svensson, N. Keskitalo, S. Godey, E. Ntsoenzok, J.F. Barbot, C. Blanchard, Proximity gettering of platinum in proton irradiated silicon. J. Appl. Phys. **84**, 4214–4218 (1998). https://doi.org/10.1063/1.368695
131. B. Holm, K. Bonde Nielsen, Spatial confinement and saturation of substitutional platinum by diffusion into ion-beam damaged silicon. J. Appl. Phys. **78**, 5970–5974 (1995). https://doi.org/10.1063/1.360600
132. B. Stritzker, M. Petravic, J. Wong-Leung, J.S. Williams, Selectivity of nanocavities and dislocations for gettering Cu and Fe in silicon. Appl. Phys. Lett. **78**, 2682–2684 (2001). https://doi.org/10.1063/1.1363689
133. S.M. Myers, D.M. Follstaedt, Interaction of copper with cavities in silicon. J. Appl. Phys. **79**, 1337–1550 (1996). https://doi.org/10.1063/1.361031
134. D.M. Follstaedt, S.M. Myers, G.A. Petersen, J.W. Medernach, Cavity formation and impurity gettering in He-implanted Si. J. Electron. Mater. **25**, 151–164 (1996). https://doi.org/10.1007/BF02666190
135. A. Grob, P. Rohr, G. Mariani, J. Sevely, J.J. Grob, Kinetic of impurity gettering on buried defects creation by MeV argon implantation. Nucl. Instr. Methods Phys. Res. B **112**, 169–172 (1996). https://doi.org/10.1016/0168-583X(95)01014-9
136. F. Russo, G. Moccia, G. Nardone, R. Alfonsetti, G. Polsinessli, A. D'Angelo, A. Patacchiola, M. Liverani, P. Pianezza, T. Lippa, M. Carlini, M.L. Polignano, I. Mica, E. Cazzini, M. Ceresoli, D. Codegoni, Proximity gettering of slow diffuser contamination in CMOS image sensors. Solid-State Electron. **91**, 91–99 (2014). https://doi.org/10.1016/j.sse.2013.10.011
137. S. Shirasawa, K. Sueoka, T. Yamaguchi, K Maekawa (2015) Useful database of effective gettering sites for metal impurities in Si wafers with first principles calculations. ECS J. Solid State Sci. Technol. **4**, P351-P355. https://doi.org/10.1149/2.0051509js
138. A. Haarahiltunen, H. Väinölä, O. Anttila, M. Yli-Koski, J. Sinkkonen, Experimental and theoretical study of heterogeneous iron precipitation in silicon. J. Appl. Phys. **101**, 043507/1–6 (2007). https://doi.org/10.1063/1.2472271
139. T. Yamaguchi, T. Yamashita, T. Kamino, Y. Goto, T. Kuroi, M. Masazumi, White spots reduction by ultimate proximity metal gettering at carbon complexes formed underneath contact area in CMOS image sensors. VLSI Techn Techn Dig 978/1–2 (2016). https://doi.org/10.1109/vlsit.2016.7573447
140. I.H. Kim, J.S. Park, T.A. Shim, J.G. Park, Si CMOS image-sensors designed with hydrogen-ion implantation induced nanocavities for enhancing output voltage sensing margin via proximity gettering. IEEE Trans. Electron Dev. **64**, 2345–2349 (2017). https://doi.org/10.1109/TED.2017.2677948
141. T.I. Kamins, S.Y. Chiang, Heavy metal gettering in silicon-on-insulator structures formed by oxygen implantation into silicon. J. Appl. Phys. **58**, 1559–2563 (1985). https://doi.org/10.1063/1.335910
142. W. Skorupa, N. Hatzopoulos, R.A. Yankov, A.B. Danilin, Proximity gettering of transition metals by implanted oxygen structures. Appl. Phys. Lett. **67**, 1992–2994 (1995). https://doi.org/10.1063/1.114929
143. J. Jablonski, Y. Miyamura, H. Tsuya, Gettering of Cu and Ni impurities in SIMOX wafers. J. Electrochem. Soc. **142**, 2059–2066 (1995). https://doi.org/10.1149/1.2044241
144. M. Zhang, C. Lin, P.L.F. Hemment, K. Gutjahr, U. Gosele, Study of gettering to cavities in separation by implantation of oxygen substrates. Appl. Phys. Lett. **72**, 830–832 (1998). https://doi.org/10.1063/1.120907
145. M. Zhang, C. Lin, X. Duo, Z. Lin, Z. Zhou, Comparison of Cu gettering to H^+ and He^+ implantation-induced cavities in separation-by-implantation-of-oxygen wafers. J. Appl. Phys. **85**, 94–98 (1999). https://doi.org/10.1063/1.369426
146. M. Bruel, Separation of silicon wafers by the smart cut method. Mater. Res. Innovat. **3**, 9–13 (1999). https://doi.org/10.1007/s1001900501

147. M. Yli-Koski, A. Haarahiltunen, J. Hintsala, H. Savin, Iron segregation in silicon-on-insulator wafer with polysilicon interlayer. Phys. Status Solidi A **209**, 724–726 (2012). https://doi.org/10.1002/pssa.201127718
148. K. Mahfoud, M. Loghmarti, J.C. Muller, P. Siffert, Influence of carbon and oxygen on phosphorus and aluminium co-gettering in silicon solar cells. Mater. Sci. Eng. B **36**, 63–67 (1996). https://doi.org/10.1016/0921-5107(95)01284-2
149. I. Périchaud, Gettering of impurities in solar silicon. Solar Energy Mater. Solar Cells **72**, 315–326 (2002). https://doi.org/10.1016/S0927-0248(01)00179-9
150. Y. Jin, S.T. Dunham, Modeling of carbon clustering and associated metal gettering. ECS Trans. **64**(11), 211–218 (2014). https://doi.org/10.1149/06411.0211ecst
151. A. Yazdani, R. Chen, S.T. Dunham, Coupled modeling of the competitive gettering of transition metals and impact on performance of lifetime sensitive devices. J. Appl. Phys. **121**, 095702/1–7 (2017). https://doi.org/10.1063/1.4976525
152. A.H. Gencer, S.T. Dunham, A combined model for {311} defect and dislocation loop evolution: Analytical formulation of kinetic precipitation model. J. Appl. Phys. **91**, 2883–2889 (2002). https://doi.org/10.1063/1.1446223
153. B.C. Trzynadlowski, S.T. Dunham, A reduced moment-based model for oxygen precipitation in silicon. J. Appl. Phys. **114**, 243508/1–10 (2013). https://doi.org/10.1063/1.4849435
154. S.M. Myers, G.A. Petersen, D.M. Follstaedt, C.H. Seager, T.J. Headley, J.R. Michael, W. Deweerd, G. Koops, J. Verheyen, H. Pattyn, Segregation gettering by implantation-formed cavities and B-Si precipitates in silicon. Electrochem. Soc. Proc. **98**(1), 1150–1161 (1998)
155. K. Sueoka, M. Akatsuka, T. Ono, E. Asayama, Y. Koike, N. Adachi, S. Sadamitsu, H. Katahama (2000) Oxygen precipitation behavior and its optimum condition for internal gettering and mechanical strength in epitaxial and polished silicon wafers. Electrochem. Soc. Proc. **17**, 164–179 (2000)
156. R. Falster, W. Bergholz, The gettering of transition metals by oxygen-related defects in silicon. J. Electrochem. Soc. **137**, 1548–1559 (1990). https://doi.org/10.1149/1.2086709

Chapter 9
Modeling of Metal Properties in Si, $Si_{1-x}Ge_x$ and Ge

This Chapter outlines the modeling approaches which have been developed since the late fifties till now to understand the properties of metals in covalent semiconductors. These theories should enable to explain the different configurations of metal atoms in the lattice and in particular, the lowest-energy state, which is important for describing the solid solubility. At the same time, this should give an idea about the possible migration paths of the impurities, which is crucial for their diffusion behavior. Perhaps the most challenging aspect is to describe the electrical properties of the metals, including the energy levels, spin and charge states. The original models were derived from empirical trends established by EPR [1, 2]. In the late seventies, first-principles ab initio studies of metal atoms in silicon took off and have developed since then in a power tool, with Density Functional Theory (DFT) calculations as today's workhorse for defect theorists. The greatest challenge in the understanding of the behavior of metals is for sure the wide variety of their properties, with, for example, diffusivities ranging over more than five decades for the series of 3*d* TMs.

This Chapter is built up as follows. First, a brief review of the evolution of ab initio modeling will be given. In a second paragraph, the properties of individual metal atoms in Si, Ge and $Si_{1-x}Ge_x$ will be described. As will be seen, for most metals in Si, the *T*-site with T_d symmetry is the lowest-energy configuration, while it is the substitutional site for Ge. Next, focus will be on the modeling of the diffusion of metal atoms in covalent semiconductors. For silicon, most metals diffuse from one *T*-site to the next through the hexagonal interstitial site, along a $\langle 111 \rangle$ direction. In a fourth section, the interaction of metal atoms with other impurities and dopants will be discussed from a modeling perspective. These reactions are important from a viewpoint of gettering and passivation of metal contaminants. In a final section, the formation of metal pairs and small clusters will be investigated. In addition, the interactions of metal atoms with radiation and extended defects are discussed. An approach for the understanding of gettering phenomena will also be presented.

C. Claeys and E. Simoen, *Metal Impurities in Silicon- and Germanium-Based Technologies*, Springer Series in Materials Science 270,
https://doi.org/10.1007/978-3-319-93925-4_9

9.1 Modeling Approaches

First-principles modeling of TMs in silicon (and germanium) is more difficult than for lighter impurities, like, H, O or C. This is related to the partially-filled 3*d* shells, which allow for a range of spin multiplicities. The associated wave functions are strongly localized and, therefore, give rise to deep levels. Calculation methods based on plane-waves require a huge basis set. In addition, TMs are notorious for their complex chemical behavior, resulting in all sorts of coordination. This is related to the linear combination of their *s*, *p* and *d* orbitals which enables a wide variety of hybridizations. Depending on the configuration of a TM atom in a group IV semiconductor, the bonding character with its silicon neighbors can range from highly ionic with little wave function overlap to largely covalent, with marked transfer of valence electrons to the silicon neighbors. The chemistry of TMs in a covalent matrix can also lead to substantial lattice relaxations or strong interactions of TM with defects or impurity atoms (H, O, …). These problems can be best approached with those first-principles methods which allow the minimization of total energies of the system with respect to the nuclear coordinates, i.e., ab initio Hartree-Fock (HF) cluster calculations or Density Functional (DF) theory, using periodic supercells [3]. In summary, theoretical studies of TMs in Si must enable the description of a range of chemical environments, which implies that geometries must be systematically optimized.

In this section, a brief overview of the different calculation methods will be given, mainly focusing on 3*d* (and 5*d*) TMs, with most of the work performed on silicon for obvious reasons. In addition, it will be pointed out how in DFT modeling important parameters, like the formation energy E_f or the binding energy E_b can be calculated. Estimation of the electrical properties of metal impurities is more complex and is usually based on the marker method, which will be briefly outlined.

9.1.1 EPR-Based Models

The first theory of 3*d* TMs in covalent semiconductors was based on the pioneering EPR work of Ludwig and Woodbury [1], giving rise to the scheme of Fig. 9.1 for interstitial and substitutional TMs in silicon. The following model has been developed: the atomic 4*s* shells will—after incorporation of a metal atom in a tetrahedral-interstitial site (*T*-site) or a lattice site (both with T_d symmetry)—be promoted to the 3*d*-like orbitals: a 3*d* ion with *n* valence electrons is in a d^n configuration if it is T_d interstitial and in a d^{n-4} state if it is T_d substitutional. In the latter case, four of the *n* valence electrons are supposed to be involved in substituting the four Si electrons removed from the VB, together with the removed Si atom.

By incorporation of a TM ion at one of the two T_d sites in silicon, the atomic 3*d* orbital is split into a doublet *e* and a triplet t_2 state by the tetrahedral crystal field. For interstitial ions, the t_2 states are energetically lower than the *e* doublet states,

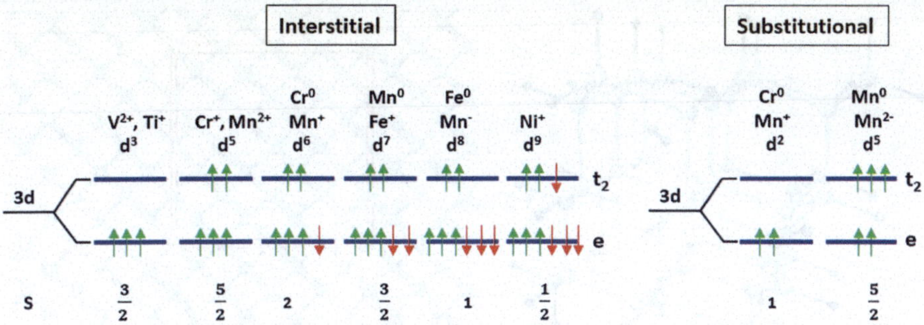

Fig. 9.1 3*d* transition-metal ions observed by EPR in silicon at the interstitial T_d site and the substitutional T_d site. For each EPR-identified 3*d* ion the model of Ludwig and Woodbury [1] is consistent with the known experimental data

while the opposite holds for substitutional ions. A further assumption in the Ludwig-Woodbury model is that for both substitutional and interstitial TM ions in the ground-state configuration, the t_2 and *e* states are populated according to Hund's rule, i.e., the ground states have high spins as in the atom where no crystal field is present. Hund's coupling implies that mutually aligned spins occupy preferentially empty single-particle states on the same site, thereby maximizing the total spin. If N_e is the number of *d* electrons occupying the t_2 and *e* states, the ground state of a TM ion has a total electron spin $S = N_e/2$ if $N_e \leq 5$ and $S = (10 - N_e)/2$ otherwise.

Substitutional heavy TM atoms (e.g., Ag, Au, Cu) can be described by the Watkins vacancy model [2]: fully occupied metallic *d* states lie deep in the valence band, whereas metallic *s* states resonate with the CB, donating their electrons to the vacancy triplet t_2 acceptor state. Thus, the electronic structure of a substitutional metal $M_s^{q'}$ in charge state q' is similar to $V^{q'-1}$, and like the vacancy, metals with partially filled t_2 states are sensitive to Jahn–Teller distortions. However, these turn out to be small, with associated energies in the range of 50 meV, so that the impact on the calculated energy levels is small.

9.1.2 *First-Principles Calculations*

There has been an enormous evolution in theory over the past sixty years, first of all related to the spectacular increase in computing power and, secondly, related to the progress in the methods. Today, the host crystal is much better described than in the past, with larger clusters or supercells—Fig. 9.2 shows a typical example of a silicon cluster (left) or a 64 atom supercell (right) [3]. The basis sets of the impurity atoms are more complete and the calculated deep levels and vibrational spectra are more close to the experimental data. The accuracy of the obtained electrical properties improves all the time and starts to have predictive value, compared with initial studies, which were rather qualitative in nature and at best could reproduce the expected or observed trends.

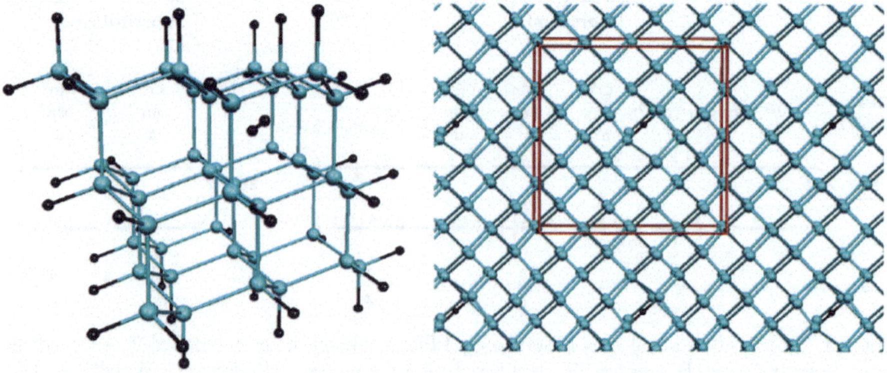

Fig. 9.2 (Left) A cluster of 44 Si atoms containing an interstitial H_2 molecule at the T site. The surface of the cluster is terminated with 42 H atoms. (Right) A 64 Si-atoms periodic supercell is delimited by a box. The cell contains a bond-centered hydrogen atom H_B. Reproduced from Estreicher [3] with permission from Elsevier

Initial numerical calculations of TMs in silicon considered the X_α scattered-wave cluster method [4–7] or the quasiband crystal-field Green's function method [8–13]. Semi-empirical calculations, based on an empirical valence-force potential have also been carried out [14]. The metal atom was assumed to be at an undistorted *T*-, *H*- or substitutional site with T_d symmetry. Using spin-unrestricted self-consistent linear muffin-tin-orbital (LMTO) Green-function calculations for the lowest energy interstitial and substitutional T_d sites in silicon, the Ludwig-Woodbury model for 3*d* TMs was challenged and its validity range explored [11, 12]. It was concluded that the early interstitial TMs (Ti_i^0, V_i^+, Ti_i^-, V_i^0 and V_i^-) and the late substitutional TMs (Fe_S^0, Co_S^+, Ni_S^{2+}, Co_S^0, Ni_S^+, Co_S^-, Ni_S^0 and Cu_S^+) break Hund's rule and, thus, have a low-spin ground-state, which is in conflict with the high-spin state model of Fig. 9.1. This flip from a low- to a high-spin state could also explain the experimental tendency of a characteristic jump of the single-donor (+/0) level between V and Cr [12].

The first total-energy calculations with systematic geometry optimization was related to copper-defects in silicon [15]. These were based on HF theory with hydrogen-saturated clusters containing up to 44 host atoms. A cluster is a fraction of the crystal which consists of a small number of host atoms surrounding the defect under investigation. The surface lattice atoms have dangling bonds which must be passivated, usually with H atoms [3, 15]. The periodicity of the lattice is lost, the band structure is missing and the energy eigenvalues of a defect cannot be placed within the 'gap'. The cluster and the defect it contains are described in real space and, therefore, the basis set consists of localized functions, such as Gaussians or linear combinations of atomic orbitals (LCAOs). The finite size of the cluster artificially confines the wave functions. This affects mainly shallow impurities as the charge tends to distribute itself on the surface of the cluster. However, even small clusters provide a description of the defect that mimics its nearby surrounding and the local chemistry is well replicated.

A supercell is a large unit cell which is repeated periodically in all directions of space. The calculations are done in a single cell and there is no surface to worry about in this case, but the defect is periodic. The interactions between defects in neighboring cells depends on the defect, its charge and dipole moment, the relaxation and distortion around it [3]. A Si_{64} supercell with a bond-centered (BC) hydrogen is shown on the right of Fig. 9.2.

In HF calculations, the total electronic wave function is a Slater determinant of molecular orbitals, which themselves are LCAOs. The HF Hamiltonian describes the electron exchange exactly but does not include electron correlation. The difficult part in HF is the calculation of the four-center integrals because it scales as $N^4/8$, the fourth power of the total number of orbitals. A HF calculation in which all these integrals are calculated is called 'ab initio'. Such calculations predict reliable geometries and a lot of chemical detail is provided, including: orbital overlap, Mulliken charges, spin populations, etc. However, since the total energy differences are often larger than the true ones, the calculated vibrational frequencies can be substantially off and the band gaps are unreasonably large [3].

In density functional theory, the total energy is written in terms of the total electron density and is variationally minimized relative to it. The ground-state density is calculated from non-interacting effective-particle wave functions obeying Schrödinger-like equations with an effective potential which contains all the interactions. The unknown exchange-correlation part is approximated; several schemes exist based on local-density approximation (LDA), the generalized-gradient approximation (GGA) and so on. The atomic core regions are replaced by ab initio pseudopotentials and the valence states are expanded into single-particle wave functions. The type of basis set (plane waves or atomic-like orbitals) affects the choice of the pseudopotential. Central quantities are the total electronic energy (plus the nuclear-nuclear repulsion term) and the ground-state (spin or charge) density [16]. In principle these calculations are not to be considered semi-empirical, since none of the parameters has been adjusted to experimental data. The most important shortcoming of DFT calculations based on LDA or GGA for electron exchange and correlation is the failure to reproduce well the band gap of semiconductors and insulators [17]. In the case of silicon, the limiting value for a large supercell is about half the true value, while Ge is predicted to behave like a metal ($E_g = 0$).

From a first-principles modeling perspective, the treatment of on-site correlations due to the electrons on the TMs implies a departure from the standard local or semi-local approximation to account for electron-electron interactions. One way to include on-site correlation effects is by means of the DFT + U approach. This is a semi-empirical method initially introduced as a Hubbard-corrected local density approximation (LDA + U) [18]. Including this U correction increases the band gap between the occupied and empty states in the defective supercells.

DF calculations in supercells are often carried out in conjunction with molecular-dynamics (MD) simulations, which is useful for example to track the diffusion path of an atom from one equilibrium lowest-energy site to another one. The Born-Oppenheimer approximation is used to separate electrons from the nuclei. For a given configuration of the nuclei at a time t, the DF electronic energy is calculated. The Hellmann–Feynman theorem is next employed to provide the force of each nucleus at time $t + \Delta t$. The nuclei are moved to their new positions and the electronic problem is solved again. The temperature of the cell is related to the kinetic energy of the nuclei and the electrons remain in their ground state (zero electron temperature). A scheme to implement DFT calculations at $T > 0$ K, including the free electron energy has been proposed, so that the thermodynamics of defect complexes can be studied [16].

9.1.3 Calculation of Parameters in DFT

An important parameter of a defect is its formation energy E_f, since it will determine its equilibrium concentration in the lattice. Metals can acquire both interstitial (i) and substitutional (s) lattice positions, so that the ratio of the i-fraction with respect to the s-fraction will be proportional with $exp\left(\frac{E_f^s - E_f^i}{k_B T}\right)$. The formation energy for an interstitial (substitutional) metal atom in a covalent semiconductor (sc) can be derived from [19]:

$$E_f = E(\text{sc with TM dopant}) - E(\text{bulk sc}) - \mu_M + n\mu_{sc} \tag{9.1}$$

with $n = 1$ for substitutional doping and $n = 0$ for interstitial placement. Further, μ_{sc} is the chemical potential of the covalent semiconductor (e.g., the total energy per atom of bulk Si or Ge) and μ_M is the chemical potential of the metal. It should be emphasized that the choice of the reservoir that provides the species to form the impurity is usually a delicate matter, as different types of phases can be considered [19].

In the case of a SiGe alloy, it has been proposed to use the following for a neutral interstitial metal atom at a T-site [20]:

$$E_f^i(v, x) = E_i(v, x) - E_{bulk}(x) - \mu_M(x) \tag{9.2}$$

Here, E_i (v, x) is the total energy of the special quasi-random structure with the interstitial metal atom, i.e., with Ge content x and v the number of nearest neighbor (NN) Si atoms ($v = 1$ to 4) and $\mu_M(x)$ is the metal chemical potential in the $Si_{1-x}Ge_x$ alloy determined as $\mu_M(x) = (1 - x)\mu_{MSi} + x\mu_{MGe} - \mu_{Si(1-x)Gex}$. In the case of a neutral metal atom replacing an X lattice atom (X = Si or Ge), (9.2) becomes [20]:

$$E_f^s(X, v, x) = E_s(X, v, x) - E_{bulk}(x) + \mu_X(x) - \mu_M(x) \tag{9.3}$$

E_s(X, v, x) is the total energy of the special quasi-random structure with the M atom in place of X, which has chemical potential $\mu(x)$ in the Si_{1-x} Ge_x alloy.

In the case of a defect with charge state q_n, the formation energy becomes (see [21] and references therein):

$$E_f(q_n) = E_d(q_n) + q_n\mu_e - \sum_j n_j\mu_j \tag{9.4}$$

where $E_d(q_n)$ is the total energy of a supercell containing a defect composed of n_j atoms of type j with chemical potential μ_j and where μ_e is the electron chemical potential with respect to the top of the valence band of the pure material.

As shown in previous Chaps. 5 and 6, metal ions can become attracted by ionized dopant atoms, forming nearest neighbor pairs. This is particularly the case for TM-acceptors. At the same time, the high reactivity of TMs enables the formation of pairs with other impurities as well, e.g., H, O, donors, … or with native point defects (vacancies and self-interstitials). An important parameter for the stability of a pair defect is its binding energy E_b, which can be calculated from DFT according to [22]:

$$E_b = \{E_{tot}(G_s) + E_{tot}(M)\} - \{E_{tot}(G_sM) + E_{tot}(Si, perfect)\} \tag{9.5}$$

with $E_{tot}(G_s)$ the total energy of the supercell with the capture site (or gettering site) G_s separately (G_s = D, A, V, I, H, O, C, …), $E_{tot}(M)$ the total energy with one metal atom in its tetrahedral (or equilibrium) site, $E_{tot}(G_sM)$ the total energy of the supercell with the G_sM pair in some configuration and E_{tot}(Si, perfect) the total energy of the silicon supercell.

The DFT calculation of the electronic properties of metal atoms in covalent semiconductors is challenging, as there are many possible sources of errors and inaccuracies. The fact that the calculated band gap of silicon and germanium is much smaller than the experimental value already raises concerns about the validity of the energy levels of localized defects. However, a clever technique has been proposed, called the marker method [23, 24], where the result of an unknown impurity level is compared with an experimental result for a well-known impurity (called the standard) with similar characteristics. In this way, one aims at the cancellation of a number of errors, so that the obtained result is accurate within 0.1–0.2 eV, depending on the choice of the marker and of the position of the level. In the case of a deep donor center one can compare the ionization energies of defects with respect to the VB calculated in the same sized clusters and reduce thereby the systematic shift which occurs by the overlap of the defect wave function with the cluster surface [23]. The calculated ionization energy of the defect E_{Id} is compared with that of a standard defect E_{Is}. The position of the donor level is then given by:

$$E_{d0}(+/0) = E_s(+/0) + E_{Id} - E_{Is} \tag{9.6}$$

where the donor level of the standard or marker $E_s(+/0)$ is taken from the experiment. The ionization energy can be calculated from the configuration energy per electron E_{Id} (q', q''), which is simply the difference in the total energy divided by the difference in charge between the two states q' and q'', so that [21]:

$$E_{Id} = \frac{E_{d0}(q'') - E_{d0}(q')}{q' - q''} \tag{9.7}$$

For acceptor levels one can utilize the electron affinities in the same way, with respect to the bottom of the CB. It is possible to use a similar marker method in the case of DFT supercell calculations [21]. In principle, one can also use the silicon supercell as the marker, whereby in the case of donors E_V is the reference level and for acceptors it is E_C. This is a truly ab initio calculation, while if an experimental reference level is used the calculation has to be considered semi-empirical.

It should, finally, be remarked that several open access software packages are available for DFT calculations, like, e.g., SIESTA [25] or VASP [26].

9.2 Configurations of Individual Metal Atoms

In this part, the results of HF and DFT modeling of the properties of individual metal atoms in Si and Ge (and SiGe) will be summarized, with particular emphasis on the 3*d* TMs in silicon. Parameters of interest are first of all the stable positions (i or s), defined by the formation energy, which will determine the solubility to a large extent. In addition, the interaction of metals with intrinsic point defects (Is and Vs) are considered here. Next, the electrical properties (spin state, energy levels in the gap) will be summarized and compared with experimental data. The diffusion properties (migration energy and energy barrier) will be treated in the next section (Sect. 9.3), while the properties of the pairs (Sect. 9.4) and higher-order clusters (Sect. 9.5) are described separately. First, some general trends will be outlined, while next, results for some individual TMs will be given.

*9.2.1 Trends in the Properties of 3*d *TMs in Si and Ge*

Mainly from EPR experiments, some clear trends have been established with respect to the properties of 3*d* TMs in silicon [27]. They are as follows:

(i) Isolated 3*d* TMs in Si are stable at the *T*-site, with the highest spin compatible with the charge of the impurity. In Ge, they prefer substitutional incorporation.
(ii) The high-temperature solubility is smallest for the lightest and largest elements of the series. The TM impurities, introduced at high temperature, are always far above their solid solubility limit at room temperature. As a result, the fast-diffusing TMs (Fe, Cu, Co, Ni) tend to find traps, form clusters, precipitate at surfaces, at GBs and other extended defects, etc.
(iii) The electrical activity of isolated 3*d* TMs decreases from the lightest to the heaviest element. Ti_i has three levels (double donor, donor and single

acceptor) in the band gap; Fe_i has only one donor level and isolated Ni_i has no known band gap state.

(iv) The 3*d* TMs strongly interact with radiation or implantation damage, resulting in new deep level states.

Several modeling efforts have been reported in order to validate theoretically some of these trends in Si [18, 19, 27–29] and Ge [27, 30, 31]. Figure 9.3a summarizes the different metals that have been studied, while Fig. 9.3b gives the lowest energy configuration for the metals in the neutral charge state in Si and Ge [28], showing that in Ge, the substitutional site is the stable state, while in Si, the 3*d* TMs and the 5th row (Y–Tc) and 6th row (Hf–Re) are most stable at the interstitial *T*-site. Exception is Ni, which is slightly more stable in the hexagonal (*H*-site).The other

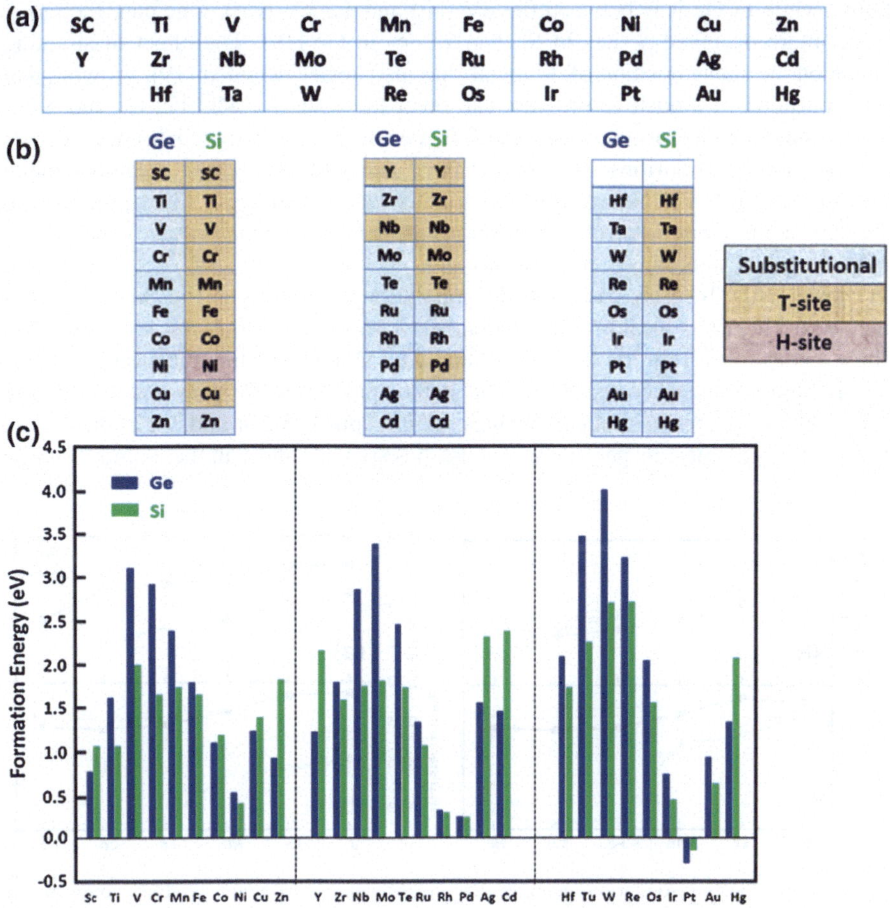

Fig. 9.3 **a** Periodic table of the studied transition metals, **b** lowest-energy configuration for TMs in Ge and Si and **c** calculated formation energy in Ge and Si at the lowest energetic configuration of uncharged individual TMs in Ge and Si. After Yamato et al. [28]

metals are stable at the substitutional site. The corresponding formation energies are represented in Fig. 9.3c, showing maximum values for V, Mo and W both in Si and Ge. For the calculation of the corresponding solubility's, the formation entropies are required, which is difficult to achieve by DFT calculations [28]. In many cases, it is in first instance assumed that the S_f values are constant throughout the series of metals.

In Fig. 9.4, the calculated NN distance between a 3*d* TM atom and Si (left) and Ge (right) for different possible sites (*T*, *H* = hexagonal interstitial site with D_{3d} symmetry and substitutional site) is represented [19]. It is evident that the *T*-site configuration in Si always leads to a dilation of the lattice, i.e., to longer Si–Si bond length than in the perfect crystal. The substitutional 3*d* TMs in Ge are characterized by a shorter bond length than in the perfect crystal, i.e., the lattice is locally contracted. It is also observed from Fig. 9.4 that the difference in E_f between the *T*-site and the substitutional site is in the range of 0.3 eV (Fe and Co) and 0.9 eV (vanadium), while in Ge it is between 0.7 eV (Ni) and 1.3 eV (Co). One can, therefore, expect more occupied *S*-sites in Si than *T*-sites in Ge [19]. The effect of straining silicon on the stable position of 3*d* metals has also been calculated [19], showing that the application of a positive pressure can favor the *S*-site for Mn, Fe, Co and Ni.

DFT calculations of the properties of Ti, Fe and Ni in silicon (Si_{216}) have shown that the isolated impurities are stable at the *T*-site with only small breathing-mode relaxations of the Si NNs, resulting in slightly larger distances [27]. In the neutral charge state, the *atomic* orbital populations are $3d^24s^2$ (Ti), $3d^64s^2$ (Fe) and $3d^84s^2$ (Ni). At the *T*-site, these populations are: $3d^{2.8}4s^{0.7}$ (Ti_i); $3d^{6.5}4s^{0.7}$ (Fe_i) and $3d^{8.5}4s^{0.7}$ (Ni_i). In each case, the 4*s* population decreases as the 3*d* population increases. The TM_is weakly but covalently overlap with each Si 1st and 2nd NNs. Thus, the usual assumption that an isolated 3*d* TM_i behaves like the atomic species is only approximately correct [27]. The stable spin states for each possible charge state match the experimental EPR values: Ti_i has spin 1, 3/2, 1 and 3/2 in the 2+ , +, 0 and − charge states, respectively [27]. Fe_i has spin 3/2 and 1 in the + and 0 charge

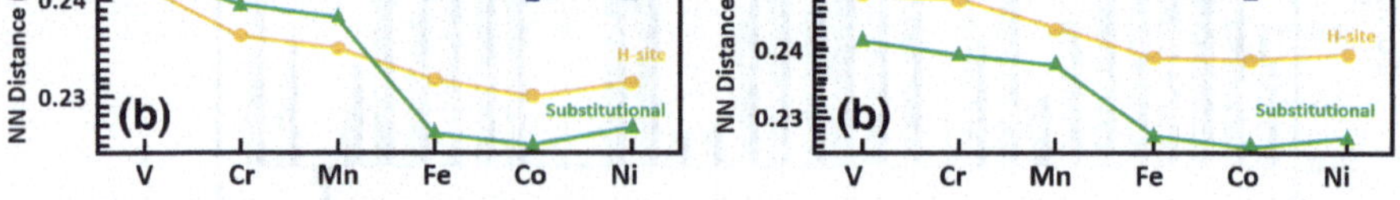

Fig. 9.4 (left) **a** The formation energies for the doping of Si with 3*d* TM atoms at the substitutional site, the interstitial *T*- and *H*-site for the different elements. **b** The nearest neighbor distances between a TM atom and Si for the different doping densities. (Right) The same, but for the case of a Ge matrix. Reproduced from Zhang et al. [19]. Copyright (2008) by The American Physical Society

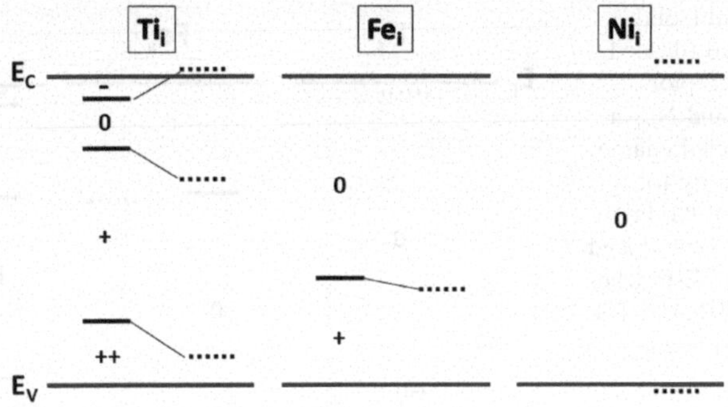

Fig. 9.5 Measured (solid lines) and calculated (dashed lines) donor and acceptor levels for Ti_i, Fe_i and Ni_i in silicon. The D and A levels of Ni_i are found to be in the valence and conduction band, respectively. The charge state of the impurity for various positions of the Fermi level are indicated. Reproduced from Backlund and Estreicher [27]. Copyright (2010) by The American Physical Society

states, while Ni_i is always in the 0 spin state with 0 charge. The calculated energy levels for Ti_i, Fe_i and Ni_i are compared with the measured data in Fig. 9.5, showing reasonable to good agreement and in line with the experimental observation that the electrical activity reduces for higher TMs in silicon.

The interaction of interstitial TMs and pre-existing vacancies, created for example by ion implantation or a thermal treatment, has also been investigated by DFT, in order to characterize the substitutional TMs. It has, furthermore, been assumed that the Fermi level is at mid-gap, representative of a non-equilibrium process with a large free carrier density present. The interacting species are then in the 0 state (Fe_i and Ni_i), except for Ti_i which is in the + charge state. Closer to equilibrium conditions, with E_F closer to the VB in p-type silicon, it is unlikely that Ti_i^+ or Fe_i^+ will interact with the 2+ charged vacancies, so that due to Coulomb repulsion, no interaction is possible in that case. The initial configuration has a TM_i at the 3rd NN T_d-site to the vacancy in the spin and charge states consistent with E_F near mid gap: ${}^{3/2}Ti^+$, ${}^{1}Fe^0$ and ${}^{0}Ni^0$. The final configuration has the metal atom at the substitutional site in the same spin and charge state. The TM migrates along the trigonal axis $T_3 \rightarrow$ hexagonal $\rightarrow T_1 \rightarrow$ V. Once the atom is at the substitutional site, the electronic spin flips to the lowest-energy state (${}^{1/2}Ti^+$, ${}^{0}Fe^0$ and ${}^{0}Ni^0$) with an additional gain in energy in the case of Ti_s and Fe_s [27]. The TMs are on site with very little distortion or breathing mode relaxation. The NN distance between the TMs and the four Si NNs is higher for Ti and slightly lower for Fe and Ni than the unrelaxed calculated Si–Si distance [27] (see also Fig. 9.4).

The corresponding energy gain for the reaction $TM_i + V \rightarrow TM_s$ is on the average 2.05 eV (Ti^+); 3.0 eV (Fe) and 2.64 eV (Ni). In other words, the binding energy of a TM with a vacancy is on the order of 3.0 eV, which is smaller than the formation energy of a vacancy in silicon (on the order of 3.9 eV). This implies that no substitutional TMs will form spontaneously (see also Fig. 9.4) unless a vacancy is supplied by some energetic process [27]. The calculated D and A levels for TMs

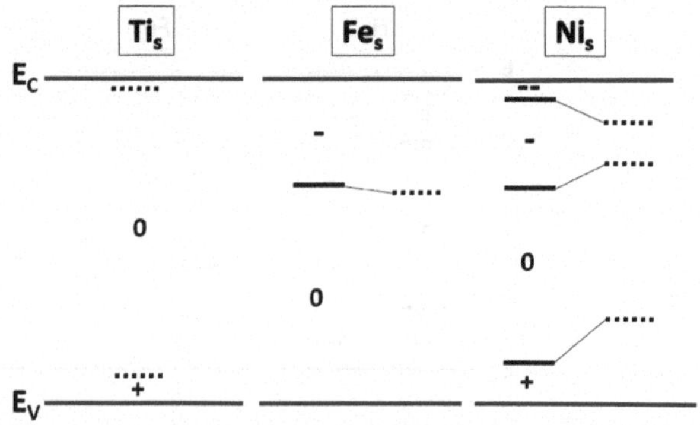

Fig. 9.6 Measured (solid line) and calculated (dashed line) donor and acceptor levels of Ti_s, Fe_s and Ni_s in silicon. The predicted charge state of each impurity for various positions of the Fermi level is indicated. Reproduced from Backlund and Estreicher [27]. Copyright (2010) by The American Physical Society

are represented in Fig. 9.6, showing just the opposite trend than in Fig. 9.5: the electrical activity, measured by the number of gap levels, increases for heavier metal atoms. It suggests that the interaction with a V not only passivates the electrical activity but can also activate it, depending on the metal species. For example, Ti_i with few 3*d* electrons is a major trap for minority carriers in p–Si but is almost electrically inactive as its odd electrons pair up with those in the silicon dangling bonds to form Ti–Si bonds. Depending on the way the interactions of the *d* shell of the Ti impurity are taken care of, this results either in a shallow D and A level (Fig. 9.6) [27] or no levels at all [32]. Fe_i (a deep donor) is a strong recombination center in p–Si, but Fe_s has a deep acceptor level instead and is in the 0 charge state in p–Si, rendering it nearly harmless [27]. It has been shown that the exact position of this acceptor state depends on the calculation specifics, like the type of pseudopotentials and basis functions used and can vary between $E_C - 0.41$ eV to $E_C - 0.29$ eV [27, 33]. It also depends on the choice of the marker level [33]. If the electron affinities of $Cu_s(0/-)$ and $Cu_s(-/2-)$ are employed, the acceptor is positioned at $E_C - 0.34$ eV, not too far from the position of an experimental level for Fe_s [33]. In any case, the electronic structure of Fe_s can be well described by the Watkins vacancy model [2] and results from the resonance between t_2 levels from the iron 3*d* manifold and the t_2 levels from the Si vacancy. Thus, a vacancy can passivate some of the most feared 3*d* TMs in silicon. On the other hand, Ni^0 becomes an active recombination center with several band gap levels when in a substitutional site.

The stable spin states are ½, 0, and ½ for the +, 0, and − charge state of Ti_s. Fe_s has spin 0 and ½ in the 0 and − charge states, Finally, Ni_s occurs in the spin ½, 1, 3/2, and 1 in the +, 0, − and 2− charge states, respectively. Ti and Fe are partially passivated by the vacancy because their odd electrons pair up with those in the silicon dangling bonds of the vacancy, resulting in covalent overlap (energy gain) and low spin. On the other hand, Ni must open up its (initially closed) 3*d* shell (energy cost) in order to achieve overlap with its silicon NNs (energy gain). The open shell gives rise to electrical activity and high spin. This most likely also explains why Ni precipitates are highly electrically active (see Chap. 6). The orbital populations are $3d^{2.4}4s^{1.0}$ (Ti_s); $3d^{6.5}4s^{1.0}$ (Fe_s) and $3d^{8.4}4s^{1.0}$ (Ni_s), showing a much higher covalent overlap than for the interstitial configuration [27]. This covalent overlap between TMs and the four Si NNs amounts to about 0.31 for Ti_s, 0.34 for Fe_s and 0.21 for Ni_s. This shows that the electrons promoted from the 3*d* to the 4*sp* shell hybridize with the silicon dangling bonds, which is responsible for the gain in energy in the reaction $TM_i + V \rightarrow TM_s$ [27].

Calculated energy levels for the early 3*d* and 4*d* TMs in silicon are compared with experimental data from DLTS mainly in Fig. 9.7 [18]. Overall, a good agreement between both has been observed. Moreover, comparing DFT and DFT + *U* calculations reveals in most cases a difference not exceeding 0.1 eV. The effect of the on-site electron correlations are more pronounced for the 3*d* metals than for 4*d* (in particular for Ti_i and V_i). It indicates that Coulomb effects are more important for these cases [18]. In general DFT + *U* shifts the energy levels towards higher energies [18]. This stems from the accumulated correlation energy as more electrons are added to the TM *d* shell. This is especially noticeable for the acceptors and single donor level of the 3*d* row of metals.

DFT calculations concerning the interaction between M_s and a V or self-interstitial in *germanium*, with M = Au, Ag or Cu have been performed by Carvalho et al. [31]. It was found that the Cu_s–V pair can be considered as a perturbed V, with an additional electron donated by the *s* orbital of Cu. Cu_s remains

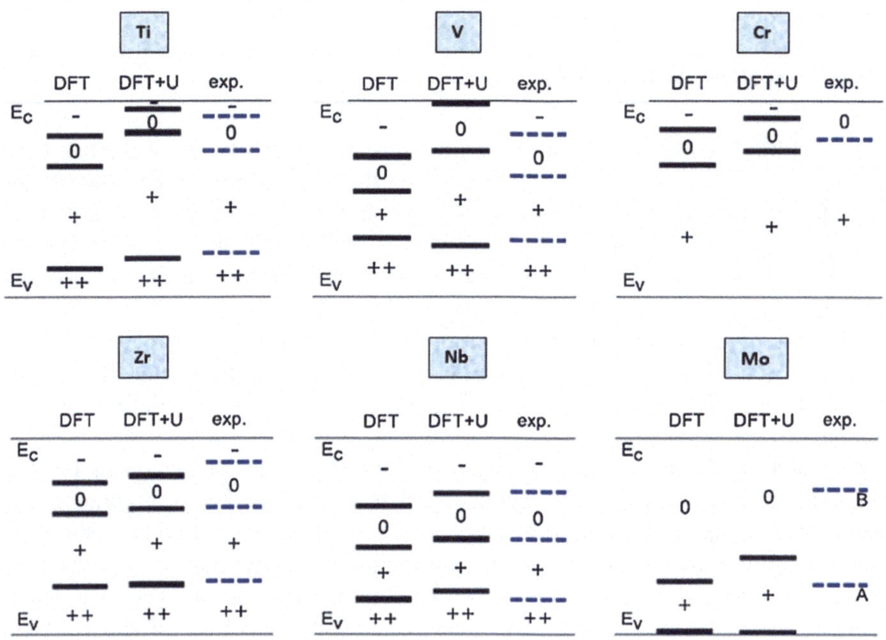

Fig. 9.7 Graphical representation of the electrical levels of the six early 3*d* and 4*d* TMs in silicon from DFT and DFT + *U* calculations. The results were obtained by taking as marker the bulk silicon crystal. The experimental levels are taken from [95] (Ti_i); [96] (V_i); [97, 98] (Cr_i); [99] (Zr_i); [100] (Nb_i) and [66] (Mo_i). after Marinopoulos et al. [18]

near the perfect substitutional site and is aligned with the V along the ⟨111⟩ direction. $\{Cu_s, V\}^-$ has a completely filled doublet arising from the e_u orbital of the undistorted V_2^{2-} and, therefore, is not sensitive to the Jahn–Teller distortion [31]. A -/2- double acceptor level at $E_C - 0.28$ eV is predicted from theory, while the single acceptor overlaps with the VB. The bigger Au_s and Ag_s, on the contrary, prefer to sit in the middle of the V_2 in a so-called split-vacancy configuration. Double acceptor levels in the vicinity of the CB have been derived [31].

Likewise, it has been shown that the neutral interstitial metals (TM_s + I → TM_i) in *germanium* reside in the *T*-site [31]. The split-interstitial structure $\{Au_s, I\}^0$ is barely stable and highly unstable for the lighter metals. It, therefore, switches immediately to the interstitial site upon capture of a hole by the substitutional metal atom [31]. In the *T*-site configuration, there is very little relaxation of the surrounding atoms, with an outward movement below 0.015 nm. In the positive charge state a different situation is found: Cu_i^+ and Au_i^+ prefer the *H*-site, while Ag_i^+ is still more stable in the *T*-site [31]. A possible explanation lies in the fact that the gain in the polarization energy is higher in the *H*- than in the *T*-site for Au_i, resulting in an attractive contribution to the total energy which is higher than the repulsive term, due to the interaction between the electron orbitals of the ion and its neighboring Ge atoms. The opposite may hold for Ag_i^+, which has a lower ionic radius.

9.2.2 *Iron in Si and SiGe*

The solubility of iron in silicon in equilibrium with the β-$FeSi_2$ phase has been calculated from first-principles [33]. Assuming that the entropy of formation is the same for both interstitial and substitutional Fe and with the same number of possible tetrahedral or substitutional sites, i.e., 5×10^{22} cm^{-3}, one can conclude that the Fe_s/Fe_i ratio is about 0.01 in the 900–1200 °C temperature range. This follows from the calculated formation enthalpies of 2.73 eV (Fe_i) and 3.23 eV (Fe_s), in equilibrium with β-$FeSi_2$ [33]. The former value is close to the experimental enthalpy of 2.87 eV, derived from EPR [33]. It also confirms that during the measurements, β-$FeSi_2$ was probably the metal-rich phase in equilibrium with silicon.

Detailed DFT calculations have been performed for the interaction of Fe_i and radiation-induced vacancies in silicon. Ultimately, a stable Fe_s configuration can result, when a barrier between 0.45 and 0.75 eV is surpassed to hop from the $\{Fe_i, V\}$ configuration (nearest neighbor), depending on the charge state and spin and the amount of lattice relaxation allowed [34]. The binding energies of $^1\{Fe_i, V\}^0$ and $^{1/2}\{Fe_i, V\}^+$ relative to $^1Fe_i^0$ and $^{3/2}Fe_i^+$ plus an isolated V are 1.62 eV and 1.50 eV, respectively. Hereby is the interstitial Fe atom located 0.023, 0.018 and 0.020 nm away from ideal T_d-sites in the −, 0, and + charge states. A donor level at $E_V + 0.35$ eV and a deep acceptor at $E_C - 0.71$ eV has been predicted by theory [34].

The isomer shift and quadrupole splitting for the different Fe-related species (Fe_i, Fe_s, …) in Si and Ge, resulting from Mössbauer experiments have been calculated by DFT as well [33]. These parameters probe the shape of the electron density at nuclear sites, depending both on core and valence states and correspond with the electric monopolar and the quadrupolar interaction of a recoilless emission and absorption of γ-radiation from and by the nuclei of bound atoms, respectively. Agreement with experimental data is within a 10% error bar [33] and also in line with earlier calculations [35].

It has been found experimentally that the ionization enthalpy of Fe-related levels in silicon reduces with increasing Ge concentration for alloys up to 8% (Sect. 5.1.2.1). This shift is much faster than expected from the shrinkage of the band gap with Ge content. This issue has also been addressed by DFT [36], showing that Ge does not like to sit as nearest neighbor of Fe_i; the close Fe_i–Ge pairs are not bound and it costs about 0.13 eV to replace a silicon NN by a Ge atom. Even a 2nd nearest neighbor Ge position is unfavorable. This reduces the solubility and diffusivity of iron in SiGe [36]. The presence of a first NN Ge atom increases the ionization energy of Fe_i by 40 meV and for two Ge NNs, by 75 meV. This might explain the satellite peaks found in Laplace DLTS in Fe-doped SiGe (Sect. 5.1.2.1). The long-range alloying effect has also been evaluated, showing a reduction of the ionization energy for Fe_i by about $0.94x$ in good agreement with the experimental data [36]. Similar conclusions were derived for the levels of the FeB and FeAl pairs.

9.2.3 *Copper in Si*

Copper was the first 3*d* TM which has been modeled by DFT [3, 15], given its relevance for the microelectronics industry. Especially the introduction of copper interconnect metallization (instead of Al) has triggered this interest. The equilibrium site for Cu_i^+ is the interstitial *T*-site with T_d symmetry, whereby the nearest neighbor silicon atoms move outward on the order of 0.005 nm. Some controversy exists regarding the electronic configuration and the charge transfer from Cu_i^+ to its silicon NNs: while it was originally shown that there was a significant charge transfer from the 3*d* to the 4*s* orbital of at least one electron [3], this was later corrected in a more accurate analysis, using a larger supercell [37]. It was shown there that copper donates its 4*s* electron to the surroundings, as expected from the experiments, while the 3*d* shell remains almost unaffected by the silicon environment. The resulting Cu_i^+ ion is effectively screened by the lattice—whereby the silicon cage surrounding the metal impurity is almost charge neutral—making the 4*s* state very diffuse [37]. This result was obtained using a consistent set of orbitals for the silicon atoms in the cage surrounding copper. The calculated formation energy of 1.67 eV is very close to the activation energy of the solubility of copper in silicon of 1.7 eV [3]. A corresponding donor level of Cu_i at $E_C - 0.09$ eV has been calculated [38].

The interaction of Cu_i with native point defects (V, I) has been calculated by Estreicher and co-workers [3, 38–41]. Cu_i + V (pre-existing) leads to the formation of Cu_s with a gain in energy, according to the most recent calculations [38], of 3.13 eV in the neutral charge state, which is smaller than E_f of a vacancy in silicon (~3.9 eV). This leads overall to a net energy balance of 0.72 eV in favor of the formation of Cu_i, resulting in a Boltzmann factor of $\exp(-0.72/k_BT)$ and a substitutional fraction in equilibrium conditions of about 0.02% at 700 °C [38]. Thus, like for the other 3*d* TMs, Cu_s will not spontaneously form if no vacancies are created by some energetic process. Most of the gain in energy comes from the formation of four Cu–Si covalent bonds (each ~0.8 eV strength). Virtually no lattice relaxation occurs around Cu_s: the silicon atoms move away from copper by less than 0.01 nm [38–40]. The degree of bonding for each Cu–Si bond is 0.7, a value much larger than for Cu_i^+ but smaller than for an ideal 2-electron bond. The calculated ionization levels for Cu_s are a donor (+/0) at E_V+ 0.19 eV; a single acceptor at E_V+ 0.29 eV (0/−) and a double acceptor at E_C − 0.36 eV (-/2-) [38]. These are in reasonable agreement with experimental data from DLTS analysis (Chap. 5).

If the self-interstitial is in a split-interstitial configuration, the strain around this defect opens up the possibility for Cu_i to react with I. Constant-temperature MD calculations show that the {Cu_i, I} complex forms and that it resembles the {Cu_s, I_2} complex, as copper attempts to grab the vacancy site. The energy gain of this reaction amounts to 1.38 eV [3]. In a way, the interaction with the strain around a split interstitial is representative for the strain which can be found around extended defects (i.e., oxygen precipitates) and may give a clue about possible gettering of copper.

9.2.4 *Cobalt in Si*

DFT calculations in a 216 atoms silicon supercell confirm the *T*-site as lowest-energy state for Co_i in silicon [42]. Co_i is at a slightly relaxed *T*-site in neutral charge state with spin 3/2 ($^{3/2}Co_i^0$) and it does not interact much with its silicon surroundings: 98% of the spin density is in the Co orbitals, 4.7% is distributed on the 1st NNs and −2.8% on the 2nd NNs [42]. The Co–Si distance is slightly larger than for the perfect Si_{216} supercell. The calculated donor level is just below the top of the VB, while the acceptor is far above the CB minimum, so that Co_i is always in the 0 charge state in silicon.

Given the neutral charge state of Co_i, it can interact with a pre-existing silicon vacancy for all Fermi level positions. The reaction $^{3/2}Co_i^0 + {}^0V^0$ results in $^{1/2}Co_s^0$ with a total energy gain of 3.17 eV, including a spin flip from 3/2 → 1/2 (gain 0.45 eV) [42]. This again implies that isolated Co should remain at the lowest-energy *T*-site, unless vacancies are provided by some energetic process. The binding energy of 3.17 eV (rather similar to the one for Cu [38]) corresponds with the four Co–Si covalent bonds, The Co_s–Si distance is 0.2238 nm (0 state) and 0.2245 nm ($^1Co_s^-$ state), which is slightly smaller than the Si–Si distance in the impurity free supercell. An acceptor level at E_C − 0.44 eV has been calculated, with no donor level [42].

The interaction of Co_i and I is accompanied by an energy gain of 0.96 eV [42]. The {I, Co_i} defect has an acceptor level at $E_C - 0.17$ eV, with a spin 0 in negative charge state.

9.2.5 *Vanadium in Si*

According to DFT, interstitial vanadium has a spin of 3/2, 1, 1/2 and 0 in the 2 +, +, 0 and − charge state, respectively [43], whereby the V_i–Si distances are slightly larger than the defect-free Si–Si value in all charge states. Gap levels for V_i have been calculated at E_V+ 0.40 eV (2 +/+), E_V+ 0.65 eV = E_C–0.52 eV (+/0) and E_C–0.26 eV (0/−).

In p-type silicon, there will be a strong long-range Coulomb repulsion between positively charged vacancies and V_i^{2+} so that interactions are highly unlikely. Pair formation is only possible for a range of Fermi levels around mid-gap where one or both species is neutral. This means that the interaction between a neutral vacancy (spin 0) and V_i in the +, 0 or— charge state is possible, resulting in substitutional V_s. The total gain in energy, including a spin flip from 1 to 0 in the + state amounts to 2.19 + 0.13 = 2.32 eV. In the neutral charge, it is 3.02 eV and in the negative state, it becomes 3.72 eV (including a spin flip from 0 to 1) [43]. In all cases, the energy gain is smaller than the formation energy of a silicon vacancy, explaining the preference for the interstitial site. The resulting Si–V_s bond strength ranges from 0.59 to 0.9 eV, showing that adding electrons to vanadium leads to the formation of stronger covalent bonds. An acceptor level at E_C–0.94 eV (0/−) has been calculated. This means that vanadium is in the neutral state in p-type Si and negatively charged in intrinsic and n-Si [43]. The V_s–Si distances are slightly longer than the ideal Si–Si bonds.

Interstitial vanadium is unlikely to interact directly with an isolated self-interstitial. However, due to the strain field surrounding the interstitial in its lowest energy state, the reaction between the two yields a bonding energy of 0.73 eV, with no band gap level [43]. The {I, V_i} pair remains, therefore, always in the positive state. The energy release upon bonding of V_i to an interstitial gives an estimate of the energetics of V_i at strained regions of the silicon crystal, which is relevant for gettering purposes.

9.2.6 *Manganese in Si and Ge*

The observation of ferromagnetism in Mn-doped semiconductors with a Curie temperature above 300 K has drawn the attention of the theorists, resulting in quite a number of DFT calculations of its stable position and corresponding spin state in

Si [20, 29, 44–46], Ge [29, 45] and $Si_{1-x}Ge_x$ [20]. This should enable to forecast what would be favorable conditions for preparing Mn-doped Si or Ge for spintronics applications. In the case of silicon, DFT in a 250 atoms Si supercell, including Hubbard U for exchange and correlation of the valence electrons has shown that the lowest formation energy occurs for parallel-spin *interstitial* Mn pairs [44]. It was also derived that the stability of the favorable interstitial position is rather independent of the supercell size, i.e., of the Mn doping concentration ($x = 0.005$ and 0.03 have been considered) [45]. However, this is not the case for the density of states and the magnetic moments. Drastic effects have been found of the Mn doping on the magnetic moment, which depend heavily on the strong electron correlation effects [45]. Mn was shown to be a source of p-type doping (holes) which is also important in the ferromagnetic ordering and gives rise to localized magnetic moments of about 3 μ_B/Mn atom [46]. Irrespective of the Mn content, Ge-based systems are very close to half metallicity, whereas the silicon-based structures just miss the half metallic behavior due to the crossing of the Fermi level by the lowest conduction bands [46]. Moreover, the ferromagnetic alignment is favored compared to the antiferromagnetic one, with its stabilization generally increasing with Mn content [46]. The possibility for stabilizing a lattice position of individual Mn in Si—and thus avoiding diffusion-assisted clustering—by alloying (diluting) with Ge in a $Si_{1-x}Ge_x$ alloy has been studied in [20]. It was concluded that the substitutional site of Mn becomes preferred over the more mobile interstitial site for a composition $x \geq 0.16$, in Ge-rich neighborhoods. This may allow for the growth of Si-based diluted magnetic semiconductors, for spintronics [20].

DFT calculations of ferromagnetic ordering have been expanded to other 3*d* TMs, like Fe [29, 45] and Cr [29]. In the former case, it has been shown that the magnetic moment seems to be unaffected by the iron doping concentration [45]. The trend of the magnetic moment for the whole 3*d* series has been calculated for Si [19] and Ge [20], giving rise to the results of Fig. 9.8a [19] and 9.8b [30], respectively. It has been pointed out that this can be understood from a simple level filling scheme, where only the 3*d* levels are involved [19]. This means 5 electrons for vanadium and up to 10 electrons for nickel. The corresponding magnetic moment amounts to 5 μ_B, 4 μ_B and 3 μ_B for vanadium, chromium and manganese [19]. For more than the half-filled case, the moment becomes smaller (Fe, Co and Ni). The use of co-doping (Mn, Co); (Mn, Cr) in Ge has also been calculated by DFT [30]. It turns out that while Mn in Ge has a tendency to cluster, the presence of Co and Cr could favor a more uniform impurity distribution. At the same time, it is found that both Co and Cr are far less efficient than Mn in promoting ferromagnetic alignment and high magnetic moment [30].

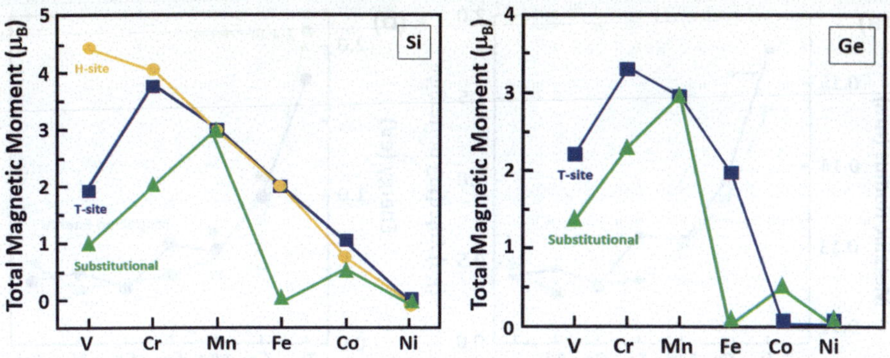

Fig. 9.8 (left) Total magnetic moments for a TM atom doped in Si at a substitutional site, an interstitial site with tetrahedral symmetry and an interstitial site with hexagonal symmetry. Reproduced from Zhang et al. [19]. Copyright (2008) by The American Physical Society. (right) Magnetic moment on the TM impurity, for all the TMs in Ge at the tetrahedral interstitial or the substitutional site. Reproduced from Continenza et al. [30]. Copyright (2008) by The American Physical Society

9.3 Diffusion of Metal Atoms in Si and Ge

There is a huge difference in the diffusion behavior of TMs in group-IV semiconductors. For Si, the diffusivities range over more than five decades (Fig. 2.5) in the 3*d* series of TMs. The activation energy for diffusion is largest for the lightest and smallest for the heaviest elements of the series. It ranges from 1.79 eV for Ti_i, to 0.69 eV for Fe_i, to 0.18 eV for Cu_i. The latter is the fastest diffusing impurity in silicon. In addition, the diffusivity depends on the charge state of the impurity. Over the years, several empirical and ab initio modeling efforts have been spent to explain these trends in the diffusivity, which will be reviewed here. First, semi-empirical modeling approaches will be described, followed by the DFT and MD results.

9.3.1 Elastic Energy Approach

One of the first attempts to understand the origin of the widely varying diffusion behavior of the 3*d* TMs in silicon was based on a consideration of the elastic energy, when an interstitial atom moves from the equilibrium *T*-site to the next one by crossing the potential barrier at the intermediate *H*-site (saddle point for the diffusion) [47]. This is based on the observation in Fig. 9.9 that there is a good correlation between the migration enthalpy H_m and the atomic radii. This suggests that the elastic energy of the moving impurity can account for the migration enthalpy. The opening at the saddle point r_{Hs} and at the equilibrium point r_T from the hard-sphere radius of the atoms is given by $0.45d_0$ (=0.105 nm) and 0.5

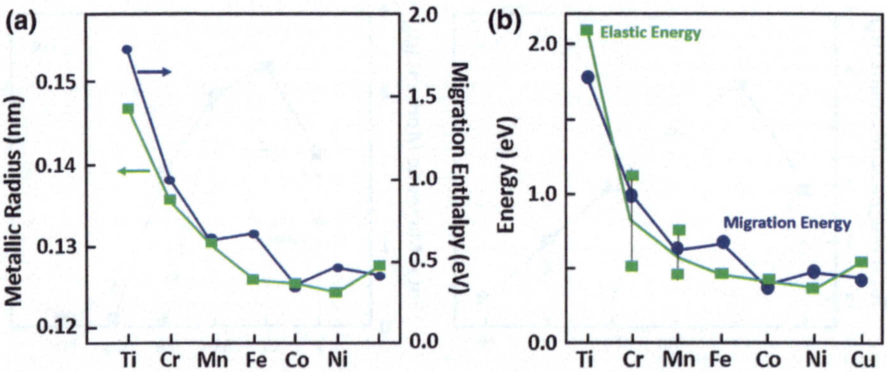

Fig. 9.9 **a** Experimentally determined migration enthalpies and metallic radii for the 3*d* elements (except Sc and V), **b** comparison of the experimental migration enthalpies H_m and the calculated elastic energy difference ΔE_{el} for the 3*d* elements in Si. Reproduced from Utzig [47]. Copyright (1991) by The American Physical Society

d_0 (=0.117 nm), respectively [47], with $d_0/2 = r_{Si}$. This is the maximum hard-sphere radius of an atom which can reside in the *T*- or *H*-site without the need for a dilation of the lattice. The difference between the elastic energies ΔE_{el} between the *T*- and *H*-sites contributes to the migration energy of the diffusing atoms. The latter are given by [47]:

$$E_{el}^{H} = K(r - r_{Hs})^2 z_{Hs}^{NN} \quad \text{if } r \geq r_{Hs} \text{ and 0 otherwise} \tag{9.8a}$$

and:

$$E_{el}^{T} = K(r - r_{T})^2 z_{T}^{NN} \text{if } r \geq r_{T} \text{ and 0 otherwise} \tag{9.8b}$$

where K is the central force constant per atom in Si [K = 3.02 eV/Å^2] and z_H^{NN} or z_T^{NN} is the number of next nearest neighbors at the *H*- or *T*-site. Considering the metallic radii, in general, a dilation is necessary to allow the atoms to fit in the interstitial site, so that $r > r_{Hs}$ and r_T and the difference $E_{el}^H - E_{el}^T > 0$. The atomic radius of the metals depends in Fig. 9.9a on their coordination and becomes smaller for a smaller coordination [47]. As can be seen from Fig. 9.9b, the migration enthalpy and the difference in elastic energy exhibit a nice correlation for the 3*d* TMs in Si [47].

Along the same lines, one can try to predict the pre-exponential factor of the diffusivity [47]. This is given by:

$$D_0 = 1/6 d_0^2 \nu \exp\left(\frac{\Delta S}{k_B}\right) \tag{9.9}$$

with d_0 the jump distance (0.235 nm in Si); ν the attempt frequency (on the order of the phonon frequency of 10^{13} Hz) and S the migration entropy. The latter can be given by [47]:

$$S = \beta \frac{H_m}{T_m} \approx \beta \frac{\Delta E_{el}}{T_m} \quad (9.10)$$

with T_m the melting temperature (1695 K for Si). The factor $\beta = 0.12$ is derived from the temperature dependence of the bulk modulus [47]. Comparing experimental with calculated prefactors (9.9), a reasonable agreement has been obtained [47], although less good as for the activation enthalpy.

Based on this rather simple idea, it should be possible to predict the pressure dependence of the diffusivity. The openings of the H-site will reduce with increasing hydrostatic pressure (more than for the T-site), so that the elastic strains (and ΔH_m) of all 3*d* impurities will increase with pressure [47].

9.3.2 Thermodynamic Approach

The diffusion of TM in *germanium* has been modeled by the so-called $cB\Omega$ model, for the case of Cu [48], Pd [49] and Ag and Au [50]. In this approach, the defect Gibbs energy ΔE_i (with i = defect formation f, self-diffusion activation *act* or migration m) is connected to bulk properties in solids. In the $cB_0\Omega_{vol}$ model, it is assumed that ΔE_i is proportional to the isothermal bulk modulus B_0 and the mean atomic volume Ω_{vol}. This implies that the diffusivity can be expressed as [48–50]:

$$D = f_d d_o^2 \nu \exp\left(-\frac{c^{act} B_0 \Omega_{vol}}{k_B T}\right) \quad (9.11)$$

with f_d the diffusion correlation factor depending on the diffusion mechanism and d_0 the lattice constant (jump distance). According to Fig. 9.10, excellent agreement is found between the experimental data and the $cB_0\Omega_{vol}$ model for the Cu diffusivity in Ge [48], with a c^{act} of 0.0158 eV/Nm. Similar results have been obtained for Pd (c^{act} = 0.0026 eV/Nm) [49] and Au (c^{act}= 0.1332 eV/Nm) and Ag (c^{act}= 0.0394 eV/Nm) [50]. It is assumed that the diffusion of the considered metals in germanium occurs through the Frank–Turnbull dissociative mechanism (Sect. 2.2) assisted by a vacancy.

9.3.3 DFT and MD Calculations

First ab initio HF cluster-based calculations of the diffusivity of 3*d* TMs in silicon were performed in the early nineties [51], showing that the diffusion barrier for Ti_i^+,

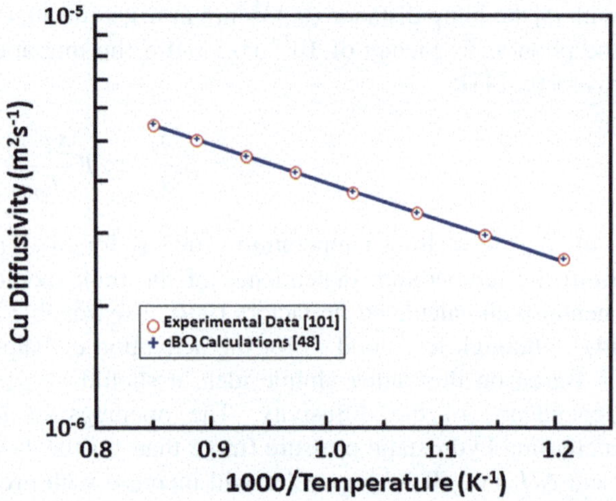

Fig. 9.10 The Arrhenius plot for the Cu diffusion in Ge obtained by experiment [101] and calculated by the $cB_0\Omega_{vol}$ model. Reproduced from Chroneos et al. [48] by permission of Springer

Ti_i^0 and Cu_i^+ along the *T–H–T* path amounted to 3.29 eV, 2.25 eV and 0.24 eV, respectively. More recent calculations confirmed that the diffusion follows a straight *T–T* line with the intermediate *H* point as the saddle point (Fig. 9.11) and corresponding activation energies of 1.79 eV, 1.66 eV, 1.75 eV and 1.66 eV for Ti_i^{2+}, Ti_i^+, Ti_i^0 and Ti_i^-, respectively [27]. In the case of Fe_i, values of 0.69 eV and 0.84 eV were found for Fe_i^+ and Fe_i^0. For Ni_i^0 a value in the range of 0.21 eV has been obtained [27, 52]. Similar small barriers are found for Co_i^0 (0.15 eV) [42]. For V_i the following activation energies have been obtained by DFT: 1.57 eV (2 +); 1.46 eV (+); 1.57 eV (0) and 2.04 eV (−) [43]. Comparing DFT and DFT + *U* calculations of the migration barriers of Ti_i, V_i and Cr_i, a much better fit with the experimental data has been derived from the latter results [18]. A strong charge state dependence has been found of E_d for the early 3*d* and 4*d* TMs in silicon, in particular for V_i, Nb_i and Mo_i [18]. The effect is weaker for Ti_i, Cr_i and Zr_i.

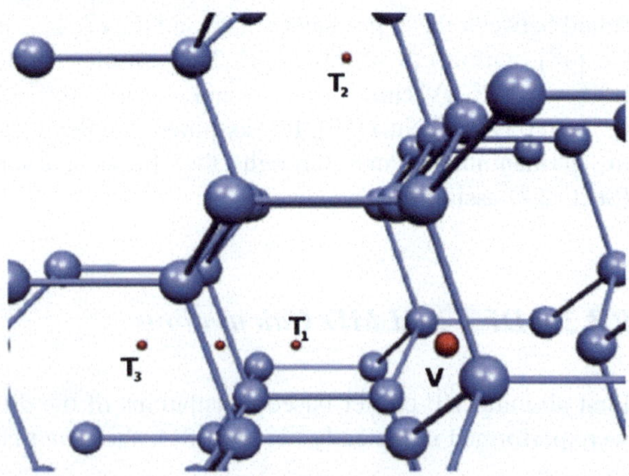

Fig. 9.11 The first-, second-, and third-nearest *T*-sites to the vacancy (red dot) are labeled T_1, T_2, and T_3, respectively. The hexagonal interstitial site is halfway between T_1 and T_3. Reproduced from Backlund and Estreicher [27]. Copyright (2010) by The American Physical Society

The overall general conclusion is that DFT calculations reproduce the trend for lower activation energy going from V to Cu. There seems to be also a tendency for the lowest activation energy for the + charge state, when available. The general trend of reduced migration barrier for increasing atomic number in the series of 3*d* TMs in silicon has been confirmed by a systematic ab initio study [53], although the obtained values and the proposed migration path are not always in good agreement with the DFT values given above.

Less DFT calculations have been performed for the diffusion of metals in germanium. The migration energy has been calculated for Cu_i, Au_i and Ag_i [31], yielding values of 0.16 eV (0.13 eV), 0.24 eV (0.17 eV) and 0.42 eV (0.23 eV), respectively. They correspond with the 0 (+) charge state of the interstitial metal atom.

Ab initio DFT calculations of defects are performed in the adiabatic approximation, yielding activation energies for the diffusion process which are strictly speaking valid for $T = 0$ K [54]. The kinetic energy of the atoms, which vibrate at $T > 0$ K are completely ignored. These effects can be included when performing MD simulations (or dynamic simulations). However, this only works well for fast diffusing atoms [54]: considering a time step of 10 ps between two simulations, this corresponds with a diffusion length of 0.1 nm for a fast diffusivity of 10^{-5} cm^2/s [54]. Slower diffusion can therefore not be distinguished from the displacements caused by lattice vibrations. Cu is fast enough to be treated by first-principles MD calculations [54, 55]. In general, ‘static’ calculations (without lattice vibrations) yield a too high value for the diffusion barrier. The impact of the kinetic energy of the atoms becomes more important for smaller diffusion barriers, like is the case for Cu in Si [54]. One of the consequences of performing MD calculations at $T \neq 0$ K is that the migrating atom is no longer at its lowest energy equilibrium position during the process. On the average, the location of the atom is higher in energy, which tends to reduce the migration barrier. Another effect of the finite T is the fluctuation of the barrier height and the statistical distribution of the migration path [54]. A resulting migration energy of 0.11 eV has been obtained for Cu_i^+ in silicon, which becomes higher as more electrons are added to obtain the 0 or −1 state [55]. It was concluded that the overestimation of the diffusion barrier from 0.18 eV (experimental) to 0.24 eV (static DFT) mostly comes from the thermal energy in the range of 0.1 eV.

It should, finally, be mentioned that the role of SiO_2 or Si_3N_4 layers on top of a silicon wafer, as potential diffusion barrier for TM contamination, has been calculated by DFT [56, 57]. It was concluded that the diffusion barrier of Fe in β-Si_3N_4 is high and that of Cu low, almost the same as those in cristobalite SiO_2. The diffusion barrier of Cr in a CVD Si_3N_4 film appears to be also smaller than in a thermally-grown SiO_2 layer [56].

9.4 Interactions of Metals with Dopants, H, O, C in Si and Ge

It has been shown in Chaps. 5 and 6 that highly mobile interstitial metals are very reactive and can be attracted by shallow dopant atoms to form electrically active pairs. This can be used beneficially in gettering schemes, relying on highly doped epitaxial substrates. Conversely, hydrogen is also very mobile and reactive and can pair with (im)mobile metal atoms. This may result in a passivation of the impurity, removing all levels from the band gap, but at the same time, inactive species can become active by introducing new (or more) levels in the band gap. A third possibility is the shift of existing levels deeper in the band gap, making them more harmful to degrade some electrical performance parameter. Other common impurities in mainly silicon wafers are oxygen (Cz material) and carbon. Again, one may be interested in using the interaction of carbon or oxygen to attract metal contaminants away from the active regions. These interactions have also been investigated by DFT, as these can be helpful in identifying the many deep levels which are often found in metal-diffused Si or Ge.

9.4.1 *Interaction with Dopants*

Originally, the formation of pairs between fast diffusing interstitial metal ions (Cu, Co, Fe, …) and acceptors in silicon has been interpreted in terms of a Coulomb attraction point charge model (see Sect. 2.2.1) where the binding energy is given by the Coulomb energy for the acceptor in a substitutional site and the metal ion in the nearest *T*-site. While this qualitative picture provides a good first order description, it does not explain the impact of the acceptor species in the pair stability and the position of the energy levels. An improved ionic pair model has been derived for the case of Fe_i in silicon [58], which includes the following points:

- A short-range attractive component in the point charge Coulomb interaction and the lattice distortion around the acceptor;
- A repulsive interaction between Fe_i and the acceptors is approximated by a softened Lennard-Jones type of potential;
- The silicon crystal is treated like a dielectric medium.

In addition the possible configurations of Fig. 9.12 have been considered. The driving force for the interaction remains the Coulomb attraction, which is charge-state controlled. The association process is dominated by the long range Fe_i migration. The classic point charge model works only well when the atomic size difference between the acceptor and silicon is negligible. This is the case for Al, while for B and In the elastic energy contribution due to the strain field coming from B–Si and In–Si bond relaxation should be accounted for [58]. In fact, for In

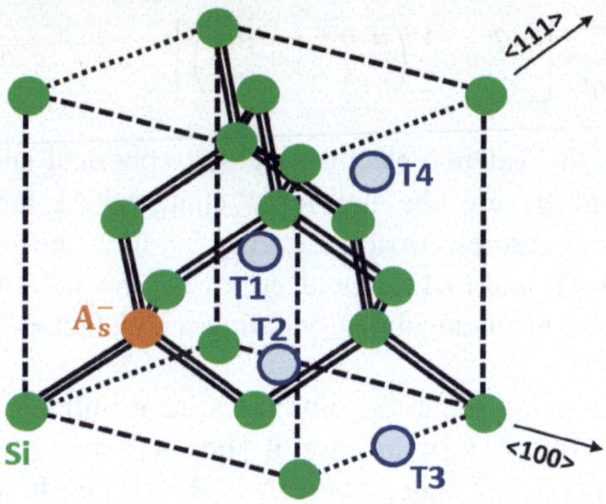

Fig. 9.12 The Fe_iA_s pair configurational symmetry is determined by the Fe_i position relative to the substitutional acceptor A_s position. *Tm* indicates the m-th neighbor tetrahedral interstitial site relative to the substitutional acceptor. The Fe_iA_s pair exhibits $\langle 111 \rangle$—trigonal symmetry if the Fe_i is at the *T1* or *T4* sites. It exhibits $\langle 100 \rangle$—orthorhombic symmetry if the Fe_i is at a *T2* site

the elastic interaction is strong enough to compete with the Coulomb attraction so that In stays in the *T2* site (Fig. 9.12) rather than in *T1*. In other words, the stability position is reversed.

The elastic term in the pair energy has been modeled by a repulsive, softened Lennard-Jones-type potential. It mainly stems from the repulsion due to the electron shell overlap between atoms and the electrostatic attraction between A_s^- and Fe_i^+ or Fe_i^{2+} [58]. The second component in the pair energy, namely, the electrostatic component is composed by the point charge Coulomb attraction and a short-range enhancement in the interaction due to the induced valence electronic cloud polarization. The short range repulsive interaction between Fe_i and the lattice (Si and A_s) is the major source of the elastic energy. The interatomic potential can be represented by:

$$V_{rep}(r) = 4\epsilon(\frac{\theta}{r})^6 \tag{9.12}$$

In (9.12), r is the interatomic distance and θ,ϵ are empirical parameters. To determine the latter, it is assumed that the Fe_i migration enthalpy originates purely from the elastic energy and that the migration path is $T \rightarrow H \rightarrow T$ [47].

The electrostatic interaction is composed of the sum of two contributions, the long-range Coulomb attraction and a short-range polarization term. The latter originates from the fact that the separation between Fe_i and A_s is comparable to the sum of the ionic radii. In that case, the interaction due to the valence electronic cloud polarization becomes important [58]. As a result, the attractive interaction in this Polarization Model can be described by [58]:

$$V_{att}(r) = \frac{1}{4\pi\varepsilon_0\varepsilon_s}\left[\frac{q_1q_2}{r} - \frac{1}{2}\left(\frac{a_2^3q_1^2}{r^4} + \frac{a_1^3q_2^2}{r^4}\right)\right] \quad \text{for } r \gg a_1 + a_2 \tag{9.13}$$

where a_1 and a_2 are the valence electronic cloud spherical surface radii (74% of atom radius), q_1 and q_2 are the net ionic charges. At large separation, V_{att} approaches the point charge electrostatic energy, as the - term becomes negligibly small. Combining (9.12) and (9.13), the deep levels, the stability and metastability of the Fe_iA_s pairs have been calculated, yielding a satisfactory agreement with the experimental data [58].

Ab initio calculations on metal-dopants pairs have initially been performed for Fe [33, 59, 60] and Cu [3, 41]. In the case of $\{Fe_i, A\}^0$ the ½ spin state is higher in energy than the high-spin 3/2 state, where A = B, Al, Ga, In and Tl [60]. For the case of FeB it amounts to about 0.42 eV. Two competing geometries have been found for each $\{Fe_i, A\}$ pair: a trigonal one with Fe very near (or at) one of the four nearest *T*-sites and an orthorhombic configuration with Fe at one of the twelve second-nearest *T*-sites (see also Fig. 9.12). The energy difference ΔE between the two configurations calculated at $T = 0$ K is shown in Fig. 9.13, for the different possible charge states of the pair [60]. In this figure, ΔE is negative if the trigonal configuration is stable and positive the other way around. The ΔE has also been calculated as a function of temperature, considering only vibrational free energies, showing that the predicted stable configurations for the Fe_iA_s pairs match the experimental ones [60].

The binding energies of the pairs is calculated based on (9.5) and is thus obtained by comparing the energies of each pair in its lowest-energy configuration to its dissociation products in different supercells [60]. The resulting E_b amount to 0.65 eV for $\{Fe_i, B\}$, 0.52 eV for $\{Fe_i, Al\}$, 0.42 eV for $\{Fe_i, Ga\}$, 0.44 eV for $\{Fe_i, In\}$ and 0.55 eV for $\{Fe_i, Tl\}$, showing clearly the impact of the acceptor species and in reasonable agreement with experimental data (Sect. 5.1.2.1). The origin of the binding energy is mostly Coulombic. However, the TM impurity does weakly overlap with the acceptor, explaining the differences. Finally, the donor and

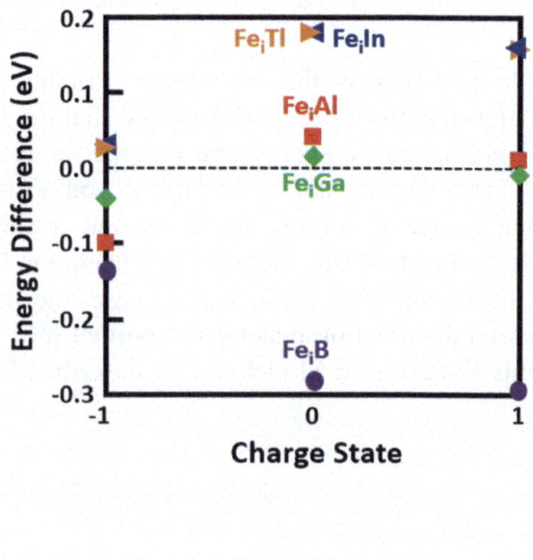

Fig. 9.13 Calculated potential energy difference between the trigonal and orthorhombic configurations of the $\{Fe_i, A_s\}$ pairs in the three possible charge states of the pairs. ΔE is negative if the trigonal configuration is more stable. Reproduced Sanati et al. [60]. Copyright (2008) by The American Physical Society

acceptor levels for the {Fe_i, A} pairs have been calculated, showing good agreement with the available experimental data [60].

In the case of copper in silicon, it has been shown that $Cu_i^+ + B_s^-$ leads to a neutral (passivated) {Cu_iB_s} pair [3, 15, 39], which has Cu at the antibonding (AB) site of the B atom with a weak but covalent Cu–B bond. Cu is displaced from the *T*-site toward B by about 0.02 nm and the B atom is five-fold coordinated (to its four Si NN and to Cu). Each of these bonds have a degree of bonding of ~0.65, implying that all boron's three valence electrons participate. The calculated bonding energies for {Cu_i, B}, {Cu_i, Al} and {Cu_i, Ga} are 0.69 eV, 0.88 eV and 0.92 eV, respectively, confirming the experimental trend, but overestimating the measured data [3]. The reactivation energy has also been calculated by DFT [41] and is much smaller than the dissociation energy with the two components at infinite distance. Values of 3.27 eV for B, 3.46 eV for Al and 3.54 eV for Ga have been obtained. This is consistent with the ease with which shallow acceptors trap Cu_i^+. A thermally dissociated {Cu_i, A} pair will likely have the copper ion not too far from A_s^- unless a strong electric field drives the positive ion to the edge of the space-charge region.

Pairing of a wide range of metal atoms with dopants (A or D) has been systematically studied by Sueoka and co-workers [61–64], aiming to predict and design stable gettering sites in silicon [62–64] and germanium [28, 62]. For silicon, the most stable site for most of the calculated metals near a B atom is the 1st nearest neighbor *T*-site [28]. Some metals in the 6th row elements are stable at the substitutional site next to B. For phosphorus in silicon, metals in group 3 and 7 to 12 are the most stable at the substitutional site next to the P atom, while those in group 4 to 6 are most stable at the 2nd neighbor *T*-site from the P atom [28]. The binding energy of a whole range of metals to B, P and As has been calculated according to (9.5) and is shown in Fig. 9.14 [63]. It is clear that B forms stable pairs with most of the metals studied, while this is less the case for substitutional P and As. In addition, substitutional C is for most metals not a favorable partner to form pairs. When studying the pair formation between B and P, on the one hand and Fe, Ti, Ni and Hf on the other, it has been shown that metals can be trapped in near-neighbor *T*-sites

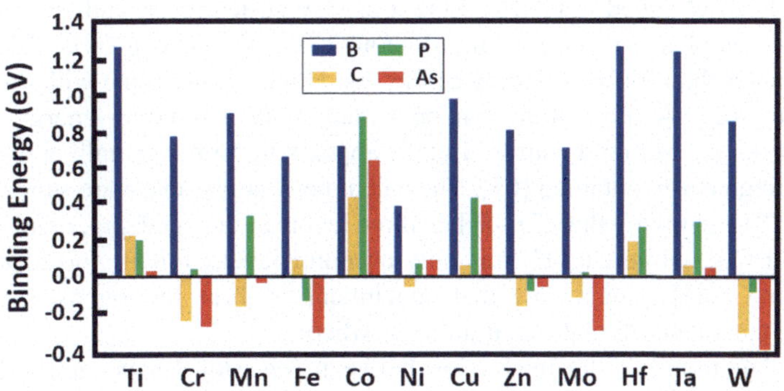

Fig. 9.14 Binding energies of B, C, P, and As with metal atoms in silicon. Reproduced from Shirasawa et al. [63] by permission of ECS-The Electrochemical Society

up to the 8th NN at least, due to long-range Coulomb attraction [64], whereby the trend in bonding energy is not necessarily monotonous: a local maximum is found for one of the 4th NN sites for Fe, Ti and Hf. At the same time, it has been shown that Ni is not effectively bound to neither B or P, because of the neutral charge state [64].

In the case of ***germanium***, it has been shown that for Ga doping, the most stable configuration for the pairs is with the metal at the 1st nearest neighbor *T*-site for the elements in group 3 to 7 (except Re) [28]. On the other hand, for those in group 8 to 12 and Re, the metals are most stable at the substitutional site near the Ga. For Sb doping, all metals are the most stable at the substitutional site next to Sb [28].

9.4.2 Interaction with Hydrogen

Initial ab initio calculations of the interaction between H and 3*d* TMs focused on Ti_i and Cu_i [51], coming to the conclusion that Ti_i^+ forms a bond with interstitial H, while Cu_i^+ does not. In the meantime, calculations have been refined and the passivation of ***substitutional copper*** by hydrogen has been investigated [3, 21, 65]. It was shown that different configurations are possible for Cu_s–H complexes. In all charge states, the H atom is close to a cubic site and directly bonded to the Cu atom [21]. This is located in a [100] direction on a (110) plane between two equivalent Si atoms. The exact location of the H atom depends on the charge state of the defect complex. In the − charge state, the H atom is equidistant from the two nearest neighbor Si atoms and occupies the same (110) plane as them and the Cu atom. This corresponds with a C_{2v} symmetry [21]. In the 0 and + charge state the minimum energy structure has either C_s or C_1 symmetry [21]. In the neutral charge state, there exists a significant bonding energy between interstitial H and substitutional copper. The reaction $Cu_s^0 + H_{BC}^0 \rightarrow \{Cu_s, H\}_{C_s}^0$ yields an energy gain of ~ 1.70 eV (H_{BC}: bond-centered hydrogen in neutral state). This is about 0.75 eV smaller than found in a previous study [65].

For Cu_s–H_2 (copper dihydride) two different structures with nearly equal energies have been obtained [21]. The lowest-energy structure depends on the charge state of the complex. An energy gain of about 1.65 eV has been derived, i.e., about 0.1 eV smaller than for the bonding of the first H atom and in line with the work of West et al. [65]. A third atom can be added, with a bonding energy of about 1.76 eV. Finally, adding a fourth H atom appears to be rather unlikely, given the small binding energy obtained [65]. The calculated energy levels are summarized in Table 9.1 [21], showing that Cu_s–H has three levels in the band gap, quite similar to the ones of Cu_s, while Cu_s–H_2 has only acceptor levels. Finally, no levels where derived for Cu_s–H_3, suggesting that attachment of three hydrogens to Cu_s fully passivates the substitutional metal atom in silicon.

DFT calculations of the interaction between ***interstitial iron*** and hydrogen in silicon show that in p-type silicon, it is unlikely that positive interstitial hydrogen and Fe_i^+ form a bond [60, 67, 68]. However, for intrinsic or n-type silicon,

Table 9.1 Summary of the "best" estimates for the energies of electrical levels in eV for model defects of copper and hydrogen impurities in silicon calculated with respect to pure silicon supercells. After Latham et al. [21]

Defect	+/0	0/−	−/2−
Cu_i	$E_C - 0.03$		
Cu_s	$E_V + 0.17$	$E_C - 0.50$	$E_C - 0.24$
Cu_s–H	$E_V + 0.08$	$E_C - 0.49$	$E_C - 0.39$
Cu_s–H_2	$E_V + 0.10$	$E_C - 0.57$	

interaction between neutral Fe_i^0 and hydrogen becomes possible. In that case, hydrogen can diffuse as H_{BC}^0 or, more slowly, as H_T^- [68]. The resulting {Fe_i,H} pair is characterized by Fe_i moving to the hexagonal interstitial site, with H pointing toward the *T*-site along the trigonal axis with Fe–H = 0.151 nm. The calculated binding energy for the reaction ${}^1Fe_i^0 + {}^{1/2}H_{BC}^0 \rightarrow {}^{1/2}\{Fe_iH\}^0$ is 0.82 eV. This pair has both a donor level at E_V + 0.36 eV and an acceptor at E_C–0.26 eV [68]. In the + and − charge states, the lowest-energy spin state is 1 and the binding energy is about 0.4 eV smaller. The pair between H and ***substitutional iron*** has also been calculated [60, 67, 68], showing that hydrogen binds directly to Fe_s without perturbing one of the four Fe_s–Si bonds. The calculated binding energy is 1.39 eV for the ${}^{½}\{Fe_s, H\}^0$ pair [68] (is EPR active), which is more stable than the {Fe_i, H} pair. This should be consistent with a higher annealing temperature in the range of 225 °C [68]. {Fe_s, H} has no donor level but a calculated acceptor at E_C–0.62 eV. The lowest-energy states are ${}^{½}\{Fe_s, H\}^0$ (Fe_s–H along ⟨100⟩) and ${}^0\{Fe_s, H\}^-$(Fe_s–H along ⟨111⟩). The pair is EPR active in the 0 charge state with spin ½ [68]. It was derived that on the average {Fe_s, H} has tetrahedral symmetry with H rotating (or tunneling) around Fe_s. A second H can be trapped according to the scheme: ${}^{1/2}\{Fe_s,H\}^0 + {}^{1/2}H_{BC}^0 \rightarrow {}^0\{HFe_iH\}^0 + 1.75$ eV. The two hydrogens are found to rotate around Fe_s with activation energy 0.22 eV. An acceptor level at E_C–0.71 eV is expected from the calculations [68].

The interaction of hydrogen and nickel in silicon has been calculated on a number of occasions [52, 69, 70]. It was concluded that Ni_i weakly interacts with bond-centered hydrogen (H_{BC}^+) in p-type and intrinsic silicon [52]. No Ni–H bond forms but Ni_i is 0.24 eV more stable near H_{BC}^+ than far away from it. On the other hand, H_{BC} is trapped at ***substitutional nickel*** [70], with a gain in energy of 1.44 eV. A second hydrogen can be trapped with a binding energy of 1.39 eV. If a third H atom interacts with {Ni_s, H_2}, the Ni atom spontaneously moves away from the substitutional site toward the *T*-site with an energy gain of 0.39 eV in the neutral state. The result is a partially saturated VH_3 vacancy and the electrically inactive Ni_i [70]. It was found that {Ni_s, H} has four levels in the band gap: double and single acceptor levels at E_C–0.22 eV and E_C–0.34 eV and double and single donors at E_V+ 0.06 eV and E_V+ 0.20 eV. {Ni_s, H_2} has only one calculated acceptor level at E_C–0.24 eV [70].

Table 9.2 Calculated electrical levels in eV of Ti_iH_n [71] and Ti_sH_n [70] complexes in silicon along with the binding energies $E_{b,mol}$ for progressive hydrogenation of Ti_i with molecular hydrogen. All values are in eV

Level	Ti_i	Ti_iH	Ti_iH_2	Ti_iH_3	Ti_sH	Ti_sH_2
$E_C - E(-/2-)$					0.10	
$E_C - E(0/-)$	0.10	0.13			0.21	0.20
$E_C - E(+/0)$	0.46	0.48	0.50	0.39		
$E(+/0) - E_V$	0.71	0.69	0.67	0.77	0.07	0.05
$E(2+/+) - E_V$	0.22	0.30	0.36			
$E_{b,mol}$	1.05	0.76	0.40			

The calculated lowest-energy states of Ti_i–H_n ($1 < n < 4$) complexes had the H atoms at antibonding sites forming up to four H–Ti–Si units along ⟨111⟩ directions [71]. The calculated energy levels are given in Table 9.2. It is clear that a gradual hydrogenation of Ti_i leads to a progressive removal of electronic states from the Ti_i ion and to the creation of Ti–H bonding states within the VB. The maximum number of H atoms that can decorate Ti_i depends on the Fermi level. In p-type Si, Ti_iH_4 (fully passivated) may not be stable against release of BC H^+, where only TiH_2^{2+} and TiH_3^+ may actually form [71]. For intrinsic and n-type silicon, the binding energies of the progressive hydrogenation of Ti vary from about 0.6 eV for Ti–H to about 0.2 eV for the fourth Ti–H bond. These are mostly ionic in character, where a *d*-like electron is donated from the Ti_i^+ cation to the H^- hydride anion [71]. In the 0 charge state, H traps at ***substitutional titanium*** [70], with a gain in energy of 1.54 eV. The resulting {Ti_s, H} pair has H bound directly to titanium along the C_{2v} axis [70]. The trapping of a second hydrogen leads to the formation of {Ti_s, H_2} with a binding energy of 1.86 eV. In this case, the two hydrogens bind to silicon atoms in a bonding configuration and the Ti atom moves away from the substitutional site to the *T*-site. A third H atom completes the kick-out process and the result is $Ti_i + VH_3$ with a gain of 2.12 eV [70].

In the case of ***interstitial cobalt***, it was shown that Co_i reacts with hydrogen but the binding energies are small [42]. This results from the near full 3*d* orbital of cobalt. An acceptor level at $E_C - 0.18$ eV has been calculated but is at best marginally stable at room temperature. ***Substitutional cobalt*** is more likely to trap hydrogen [42]. In the neutral charge state Co_s reacts with $^{1/2}H^0_{BC}$ with an energy gain of 1.43 eV. An acceptor level at $E_C - 0.27$ eV is found [42]. The hydrogen atom is directly attached to the metal, rather than bonding to a nearby silicon atom. A second H can be attached (1.67 eV) corresponding with a calculated acceptor at $E_C - 0.42$ eV [42].

In p–Si, interactions between *interstitial vanadium* and H are highly unlikely because of the long-range Coulomb repulsion between $V^{2+\ +}_{2i}$ and H^+_{BC} (the stable state of H in p–Si) [43]. For the same reason, the same applies for heavily doped n-type silicon. In moderately doped n- or intrinsic silicon interactions involving at least one neutral species are possible. The reaction between $^{1/2}V^0_i$ and $^0H^+_{BC}$ results in a trigonal complex with H in a hexagonal site along the ⟨111⟩ axis [43], with a binding energy of 1.15 eV. This is consistent with an expected annealing temperature in the range of 150–200 °C. ${}^{0}H_{T}^{-}$ and ${}^{1}V_{i}^{+}$ produce a similar pair with 0.96 eV energy gain [43]. The {V_i, H} complex is electrically active with a donor level at E_V + 0.56 eV and an acceptor just below the CB. Hydrogen can interact with ***substitutional vanadium*** V_s in p- and intrinsic silicon, In n-type Si, the Coulomb repulsion between ${}^{1}V_{s}^{-}$ and H_{T}^{-} prohibits pair formation [43]. The reactions ${}^{1}V_{s}^{-} + {}^{0}H_{BC}^{+} \rightarrow {}^{1}\{V_sH\}^{0} + 1.10\,\mathrm{eV}$ and ${}^{1/2}V_{s}^{0} + {}^{0}H_{BC}^{+} \rightarrow {}^{1/2}\{V_sH\}^{+} + 1.03\,\mathrm{eV}$ produces the same configuration with a V–H bond length of 0.173 nm. An acceptor level at E_C − 0.70 eV is expected [43]. {V_s, H} can trap a second H with an energy gain of 1.44 eV, a bond length of 0.170 nm and an acceptor at E_C–0.51 eV [43].

Most of the ab initio calculations have been devoted to the interaction of hydrogen with 3*d* TMs, which are the most feared from a perspective of contamination and device performance degradation (Chaps. 3 and 7). Some calculations have also been performed for the noble metals (Ag, Au, Pt and Pd) in silicon, considering the defect structure [72] or the energy levels [23, 73]. Using a marker method, the electrical levels of various combinations of TM-H_n complexes in Si have been derived. The shifts of the levels have been interpreted in terms of a displacement and splitting of the gap t_2 manifold of states. It has been shown that Ag_sH_3 and Au_sH_3 are fully passivated [23].

9.4.3 Interaction with Oxygen and Carbon

Carbon and oxygen are common impurities in mc-Si and Cz-Si wafers and could, therefore, interact with metal atoms. The binding energy between substitutional carbon and interstitial oxygen in silicon and a wide range of metals has been calculated by DFT, based on the total energy difference of (9.5); the results are compared in Fig. 9.15 with hydrogen and substitutional P as traps for metal atoms [22]. Using the B_s–Fe_i binding energy as a gettering criterion (dashed line in Fig. 9.15), it was concluded that C_s and O_i are in general no stable traps for most of the metals.

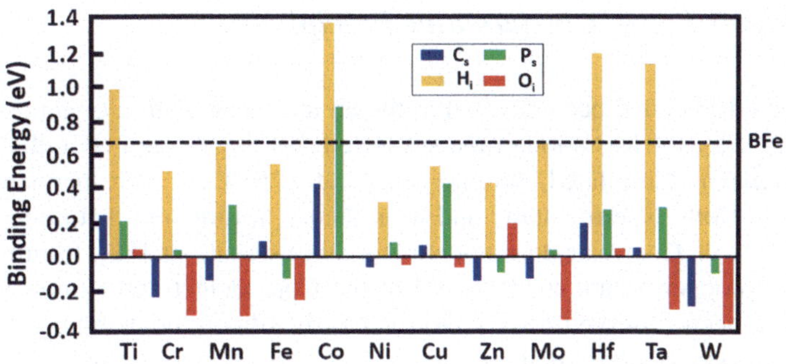

Fig. 9.15 Binding energies E_b of C_s, H_i, P_s and O_i atoms to metals in silicon. Reproduced from Shirasawa et al. [22] with permission of Elsevier

More detailed calculations have been performed for some 3*d* TMs in silicon. In principle $\boldsymbol{Cu_i^+}$ can be attracted by O_i because oxygen locally distorts the crystal, resulting in an interstitial space near O_i larger than away from it [63]. The calculated binding energy is 0.31 eV, with Cu–O at a distance of 0.20 nm and no covalent overlap. Overall, O_i has no driving force to weaken (or break) the very strong Si–O bond to form a much less energetic Cu–O one. The same conclusion is drawn for Fe_i and O_i and C_s [34] and Ni_i with O_i and C_s [52]: no stable bonds are likely to be formed. The case is slightly different for the bonding between Fe_i and C_i [34], although the binding energies are rather small. For the case of ${}^1Fe_i^0 + {}^0C_i^0 \rightarrow {}^1\{Fe_i, C_i\}^0$ the energy gain amounts to 0.52 eV, while it is 0.73 eV for ${}^{3/2}Fe_i^+ + C_i^0 \rightarrow {}^{3/2}\{Fe_i, C_i\}^+$. Donor and acceptor levels of the $\{Fe_i, C_i\}$ pair are at $E_V + 0.67$ eV and $E_C - 0.44$ eV, respectively. The stable spin states correspond with ${}^{3/2}\{Fe_i, C_i\}^-$, ${}^1\{Fe_i, C_i\}^0$ and ${}^{3/2}\{Fe_i, C_i\}^+$. The structure was found nearly identical in the three charge states [34].

9.5 Interactions of Metals with Other Defects, Clustering and Gettering

The retrograde solid solubility of metals in covalent semiconductors creates a large supersaturation upon cooling from high temperatures. For mobile interstitial metal atoms, this forms a tremendous driving force for clustering and precipitation at other defects, which can be point or extended defects. This is the basic idea of many of the relaxation gettering techniques (Chap. 8), which rely on the capture of metal atoms at some defective sites, where they may undergo some chemical binding or transformation, so that stable trapping occurs. In this part, the ab initio calculation of the formation of metal pairs, small clusters and precipitates will be discussed, followed by the interaction with other defects (radiation and extended defects). In a final paragraph, the current fundamental understanding of gettering phenomena from a first-principles perspective will be outlined.

9.5.1 Metal Pairs, Clusters and Precipitates

Quite some efforts have been devoted to the ab initio calculation of copper pairs in silicon [74, 75], as it was a long time suspected that these could be at the origin of the well-known Cu-related luminescence at 1.014 eV zero-phonon energy (Sect. 5.2.2). Both experimentally and from theory, it turns out that a model based on a simple $\{Cu_i, Cu_s\}$ pair does not explain all observations [38]. It turns out that at least four copper atoms are involved in the Cu_{PL} defect and similar PL active defects have been observed for Ag, Au and Li in silicon, containing at least four atoms (also combinations of these metals have been demonstrated). These complexes have been studied in a recent DFT effort, where up to ten different combinations of four copper atoms were considered [38]. It turns out that Cu_sCu_{i3} has the lowest formation energy with about 0.91 eV energy gain per copper atom in the neutral charge state. It has the proper trigonal symmetry, established experimentally and it should reorient swiftly under uniaxial stress since the activation energy for diffusion of Cu_i is very small [38]. The three interstitial copper atoms are in neighboring *T*-sites of Cu_s. A donor level at $E_V + 0.25$ eV was derived, with the acceptor level virtually resonant with the CB. The model can also explain why complexes with only 4 metal atoms and none with 2, 3 or higher form [38]. Initially, Cu_s and Cu_i occur normally in opposite charge states in intrinsic silicon, so that Cu_s can trap Cu_i with a gain in energy of 0.78 eV. This gives rise to a trigonal pair, with the Cu_i in a *T*-site adjacent to Cu_s. This copper pair has ionization levels very close to Cu_s itself, so that Cu_{s1} Cu_{i1}^- is a trap for Cu_i^+. The gain in energy for trapping the second Cu_i is similar (0.76 eV), for the main part coming from Coulomb attraction. The complex $Cu_{s1}Cu_{i2}$ has an acceptor level in the lower band gap, so that it is also negatively charged in intrinsic silicon. Consequently, a third interstitial copper can be trapped, with 0.77 eV binding energy. This is also consistent with the experimental binding enthalpy of 0.84 eV. The resulting $Cu_{s1}Cu_{i3}$ defect now has an acceptor level nearly resonant with the CB, so that it normally exists in the neutral charge state. The absence of Coulombic attraction hinders another Cu_i trapping and stops the further clustering. The model also predicts that no Cu_{PL} should form in p-type silicon with E_F close to the VB, which needs further experimental verification [38]. Finally, it has been shown that Cu_s–Cu_i pairs in germanium give rise to a double acceptor and the only level in the band gap is placed at $E_C - 0.39$ eV [31]. At the same time, it was shown that Au_i–Au_s and Ag_i–Ag_s pairs can give rise to a single donor and a single and double acceptor level.

It has been experimentally established that manganese tends to form dimers and higher-order clusters in covalent semiconductors. This is of high interest for ferromagnetic applications, so that some first-principles calculations of Mn dimers [76] and higher-order clusters [77] have been performed. As usual, for isolated Mn atoms in silicon, *T*-site interstitials are more stable than the substitutional configuration. The difference in formation energy is larger in p-type material and reduces when the Fermi level increases. For $E_F > 0.97$ eV, the double acceptor Mn_s becomes energetically favorable over Mn_i [76], so that in n-type silicon, Mn_i and Mn_s can coexist and pairs between them are in principle possible. A candidate Mn_s^- charge state has been derived, which is non-magnetic and cannot be detected by EPR [76]. The formation of Mn_s–Mn_i pairs releases about 0.6 eV energy in n-type silicon but also Mn_s–Mn_s pairs are found to be stable [76]. The double interstitial pair is always too high in energy to be an energetically stable configuration. The binding energy between two Mn atoms in a pair depends on the Fermi energy but is in the range of 1 eV, so that pair formation is favored, if Mn can diffuse [76].

The formation of Mn dimers has important consequences for the magnetic properties of Mn-doped silicon. It was concluded that Mn dimers prefer ferromagnetic (FM) spin alignment in the ground state, with the only exception of $Mn_sMn_s^{2-/4-}$ and $Mn_i - Mn_i^0$ [76]. This finding has important consequences for the magnetic properties of Mn-doped silicon. Pairing does not automatically quench the magnetic properties related to isolated Mn atoms. For $E_F < 0.6$ eV, Mn_s–Mn_i dimers—as well as isolated Mn_i—are defects with a large magnetic moment, from 2 to 4 Bohr magnetons [76]. For higher E_F, substitutional pairs are the lowest energy configuration, so that formation of dimers transitions to an anti-ferromagnetic alignment. It is derived that only p-type conditions are favorable for the existence of a magnetic activity in Mn-doped Si, while n-type silicon with large concentrations of Mn does not necessarily lead to a net magnetization of the material. In the case of Ge, it was found that anti-ferromagnetic pairing prevails when Mn impurities are brought together to form pairs [76]. Another calculation concluded that ferromagnetic spin configuration is stabilized when two manganese atoms are located along the $\langle 110 \rangle$ direction in Si and Ge hosts, due to the strong ferromagnetic exchange interaction between Mn impurities [77]. It was, furthermore, shown that a ferromagnetic Mn cluster is more stable than the ferromagnetic Mn configuration along the $\langle 110 \rangle$ direction. This implies that Mn impurities tend to gather and create undesired magnetic clusters in a Ge and Si matrix. Co-doping of Mn in germanium with Cr and Co, and the associated pair formation has also been studied in order to find optimum conditions for the growth of FM materials [20, 30].

The possibility of forming a so-called intermediate-band (IB) semiconductor in silicon by Ti implantation has been theoretically investigated in [78]. These IB materials are of potential interest for the fabrication of high-efficiency solar cells and consist of a semiconductor doped with a high density of impurities beyond the Mott transition ($\sim 10^{19}$ cm^{-3}) in order to form an impurity band in the band gap by wavefunction overlap between neighboring dopants, but not overlapping the band edges. The atomic structure of different species has been investigated, including $\{Ti_i, Ti_i\}$ or $\{Ti_i, Ti_s\}$ pairs. It was concluded that the formation of close Ti_i pairs provides an energy gain of 0.36 eV compared with both metal atoms at infinite distance (=twice E_f of a single Ti_i or ~ 3 eV), so that there exists a tendency for clustering [78]. The formation of a $\{Ti_i, Ti_s\}$ pair is 0.54 eV more favorable than a close interstitial pair (~ 2.09 eV). The formation energy is even smaller than for a single substitutional Ti_s ($E_f = 2.16$ eV) However, at a concentration of 5×10^{19} cm^{-3}, the probability of interstitial pair formation will be small and most of the implanted titanium will be in isolated interstitial form [78]. The corresponding density of states for interstitial Ti atoms may well be responsible for the observation of IB behavior in Ti-implanted silicon, corresponding with a donor band close to the CB [78].

Under conditions of high supersaturation, homogeneous nucleation of metal-silicide precipitates can occur (Chaps. 5 and 6). The formation and structure of different types of silicides and the interface with the silicon matrix has been calculated among others for $NiSi_2$ [79, 80], $FeSi_2$ [79] and $CoSi_2$ [79–82].

9.5.2 *Interaction with Implantation and Extended Defects*

Metals also interact with radiation and implantation defects, like the divacancy (V_2), the oxygen-vacancy (VO), the di-oxygen-divacancy (VO_2), etc. This can be exploited in state-of-the-art proximity gettering schemes, where the gettering defects are created close to the active device layer as was discussed in Chap. 8. Given the reducing thermal budgets, the interest in 3D integration of thinned wafers and the growing impact of thin-silicon-film solar cells, proximity gettering will only gain more weight in future process technologies. DFT calculations can be helpful in the understanding of the interaction of metals with implantation-induced defects. A systematic study has been performed, considering simple defects, like VO and VO_2, VH_n complexes, etc. [22, 62]. The binding energy for a series of metals has been calculated and the result for VO and VO_2 is represented in Fig. 9.16 [22]. It turns out that in case C_3H_5 cluster implantations are employed in a proximity gettering scheme, the created VH_n and VO complexes are the best candidate gettering sites [62], given their highest E_b for most of the metals investigated. When considering VO_2, the efficiency is reduced due to the fact that one O_i atom terminates two dangling bonds in one vacancy, reducing the overall binding energy.

The interaction of some metals with radiation defects/clusters has been calculated in more detail. While Cu_i in silicon only weakly binds to O_i, a much stronger interaction with the A-center has been derived by DFT [3, 65]. Constant-temperature MD runs at 300 and 500 K show that Cu_i^+ near V easily displaces O from the vacancy and becomes Cu_s, while O_i bridges an adjacent Si–Si bond. This reaction releases 1.69 eV [65] and occurs without much CuO covalent overlap, i.e., without much chemical interaction between the two elements.

The interaction of copper with internal voids has been investigated by DFT considering the ring-hexa-vacancy as a prototype defect for a void [39–41]. The latter can be formed during H^+ or He^+ ion implantations, followed by a thermal anneal and can be useful in the proximity gettering of metals. The association of a Cu_i^+ with V_6 is an energetically favorable process yielding about 3.36 eV. Copper

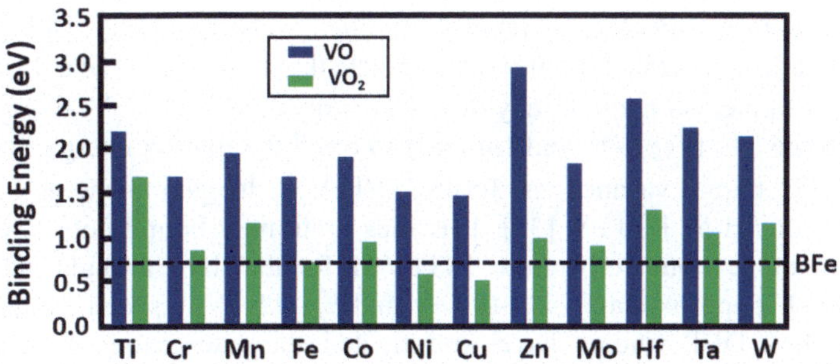

Fig. 9.16 Binding energies of VO and VO_2 complexes to the metals under consideration in silicon. Reproduced from Shirasawa et al. [22] with permission of Elsevier

is found to move away from the center of the void and weakly binds to four silicon atoms and will also become displaced when several H atoms are trapped by V_6 [41]. The configuration for $\{Cu_i^0, V_6\}$ is very similar. A second Cu_i^+ fits nicely inside $\{Cu_i^+, V_6\}$. The energy gained by placing the copper ion inside the void outweighs the energy cost of the Coulomb repulsion of both positively charged ions [41]. The energy release is estimated 0.93 eV. The configuration of both copper atoms is nearly identical. There is no overlap between them, while they are sitting at opposite inner surfaces. The electrical activity of the defect increases with the size of the void and the number of trapped copper atoms. V_6, on the other hand, is by itself found not active by DFT calculations [41]. The numerous band gap levels are probably related with the bonding/antibonding eigenvalues of the weak Cu–Si bonds [41].

The A-center in Si also interacts with Fe_i and Ni_i [27], giving rise to an annealing of it at much lower temperatures than in non-metal-doped material. While normal annealing occurs between 300 and 375 °C, this is lowered to the 200–275 °C window in the presence of Fe, Ni and Cu [27]. The calculated $\{TM_i, O, V\}$ configuration has a TM_i trapped near the most favorable *T*-site adjacent to the A-center. The stability comes from the increased covalent overlap between the TM and the silicon atom displaced by the oxygen from its ideal substitutional site [27]. The TM_i takes advantage of the lattice distortion near the A-center to increase its overlap with more silicon neighbors. A second stable consideration consists of a substitutional TM with an adjacent O_i, where the interstitial TM has grabbed the vacancy from the A-center. Its stability comes from the energy gained from the $TM_i \rightarrow TM_s$ reaction (gain $\sim$3 eV) minus the energy cost involved in displacing oxygen from the A-center configuration to the interstitial one (about 2 eV). In the neutral charge state, this results in calculated gains of 1.31, 1.19 and 1.21 eV for Ti, Fe and Ni, respectively. There is thus a potential barrier separating $\{TM_i, O, V\}$ and $\{TM_s, O_i\}$ configurations but both can coexist. For Ti, the former one is favored, while for Fe and Ni the two configurations are degenerate [27]. Moreover, it has been established that these complexes are electrically active, giving rise to several levels in the silicon band gap [27].

The interaction between Fe_i and the divacancy has also been considered [33, 34]. {VFeV} has a sixfold Fe located halfway between the two vacancies. It are all trigonal structures with D_{3d} symmetry. In the — charge state $^{1/2}\{VFeV\}^-$ and $^{3/2}\{VFeV\}^-$ are degenerated. In the 0 charge state the stable spin state is 1. If isolated $^1Fe_i^0$ traps at a preexisting divacancy than the energy gain is 3.02 eV. If, on the other hand, $^0V^0$ traps at $^1\{Fe_i, V\}^0$ and forms $^1\{VFeV\}^0$ the gain in energy is 2.86 eV. Finally, if $^0Fe_s^0$ traps a vacancy and forms $^1\{VFeV\}^0$, the gain in energy is 1.56 eV [34], later reduced to 1.33 eV [33]. It is thus a strongly bound defect which can survive well above room temperature. {VFeV} is found to have a single and a double acceptor level at approximately E_C–0.7 eV and E_C–0.5 eV, respectively [34].

Only a few DFT studies have investigated the interaction of metals with extended defects [83–86]. A calculation has been performed of the impact of substitutional copper at reconstructed 90° and 30° partial dislocations in silicon

[83]. These are when undecorated electrically inactive. Copper atoms prefer a substitutional site in the faulted region of a dislocation core, with binding energies in the range of 1.5–2.0 eV. The resulting energy levels have also been calculated and it was found that the donor level was upshifted by 0.3 eV and the single acceptor level downshifted by about 0.15 eV in the band gap, compared with Cu_s in bulk Si [83].

The interaction of metallic impurities with GBs has been studied from the perspective of mc-Si for PV applications [84, 85]. It was demonstrated that segregation of TMs is more favored at substitutional sites, owing to the formation of TM-Si bonds that destabilize the GB plane [84]. The calculated segregation energies rank in the order Fe > Co > Ni > Cr at substitutional sites and Cr > Cu > Fe > Ni at interstitial sites. The fact that positive values were obtained indicates that segregation is not readily favored on Σ3 (111) GBs [84]. When the metal impurity is placed at a substitutional site, a new state was observed, indicating enhanced electrical activity of the GB. The electronic properties and the diffusion behavior of Na on stacking faults in silicon have been investigated in order to explain the observation of shunts in both c-Si and mc-silicon solar cells [86]. The origin of the Na contamination is due to the fact that soda lime glass is used as a support for the thin-film cells.

9.5.3 First-Principles Studies of Metal Gettering

First-principles calculations can also be helpful in the atomistic understanding of gettering and may assist in identifying new gettering sites and schemes for metals in semiconductors [28, 61–64, 87]. The evaluation of the suitability as a gettering site is based on the calculation of the binding energy to a particular defect (9.5); another important factor is the dissociation energy, which gives an idea about the stability of the gettering and is composed of E_b and the migration barrier for metal diffusion E_d [87]. In order to evaluate the gettering efficiency in the presence of several different types of gettering sites, a statistical model has been introduced: the so-called Hakoniwa approach [63, 64, 88]. It is based on statistical thermodynamics and DFT and relies on the existence probability p^{i} of the metal atom at the *i*-th *T*-site from the G_s atom, representing a gettering site [64]:

$$p^i = \left\{ N^i \exp\left[-\frac{E_f^i}{k_B T} \right] \right\} / Z \tag{9.14}$$

Here, E_f^i and N^i are the formation energy and the statistical weight (degeneracy number) of the metal atom at the *i*-th *T*-site. *Z* is the microcanonical ensemble partition function given by:

$$Z = \sum_i N^i \exp\left[-\frac{E_f^i}{k_B T}\right] \tag{9.15}$$

The average binding energy to a specific gettering site $\langle E_b \rangle$ is calculated as [64]:

$$\langle E_b \rangle = \sum_i p^i E_b^i \tag{9.16}$$

The gettering efficiency of a specific site is then calculated from [64]:

$$R_{\frac{nearG_s}{perfectSi}} = \frac{C_{nearG_s}^{eq}}{C^{eq}} = \frac{\sum_i N^i \exp\left[-\frac{E_f^i}{k_B T}\right]}{94\exp\left[-\frac{E_f}{k_B T}\right]} \tag{9.17}$$

The number 94 stems from the total number of T-sites in a 64-atom supercell, up to the 8th nearest neighbor shell. The ratio $R_{\frac{nearG_s}{perfectSi}}$ corresponds with the increase in solubility of metal atoms in the G_s doped silicon with concentration of 7.8×10^{20} cm^{-3} (one G_s atom or defect in a Si_{64} supercell), compared with the perfect silicon cell [63].

This procedure has been applied to the case of Cu [88] and Fe, Ni, Ti and Hf gettering [64] by a p/p^+ epitaxial wafer (see also Chaps. 5 and 8). An example is shown in Fig. 9.17 for the case of Fe [64], comparing different doping densities of the p^+-substrate, with a 1×10^{15} cm^{-3} B-doped p-type epitaxial layer on top. One can derive that in case the p^+-substrate has the same doping density than the epi layer, no gettering is obtained (the ratio is then given by the thickness ratio of the epi layer and the wafer which is 10^{-3} in this case) [58]. The gettering efficiency increases (or $R_{epi/sub}$ decreases) for increasing p^+ doping density and for lower process temperatures. In other words, the solubility of Fe in the p^+-substrate becomes much higher than in the epi layer, intended for device fabrication. Ultimately, near room temperature, the ratio becomes equal to the ratio of the B

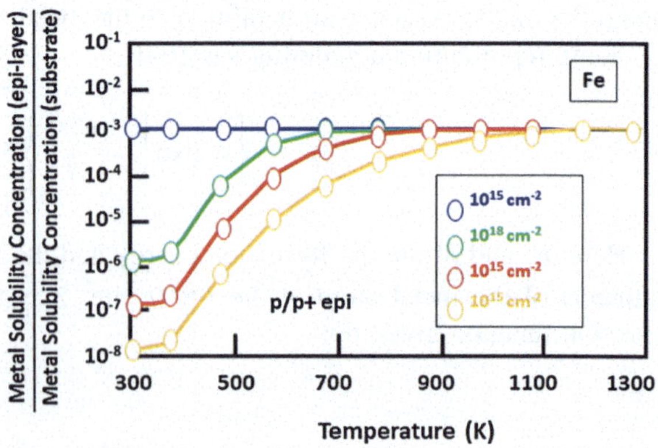

Fig. 9.17 Temperature dependence of the ratio of the equilibrium metal solubility concentration in the epi-layer and in the substrate for Fe in p/p^+ epitaxial silicon wafers. Reproduced from Yamada and Sueoka [64] by permission of ECS-The Electrochemical Society

concentration in the epi layer and the highly doped substrate. It is assumed in these calculations that the Fe contamination level is (much) lower than 1×10^{15} cm^{-3}, so that the gettering of each Fe atom by Fe–B pairing is independent, i.e., no two Fe atoms are trapped at a different *T*-site near the same B atom [64]. It was concluded that this type of gettering works well below 400 K for Fe, below 300 K for Ni (so no efficient gettering by p^+-silicon in agreement with experimental observations) and for Ti and Hf it should be effective below about 700 K [64].

Other gettering mechanisms/sites have also been evaluated by DFT calculations: the defects at the interface between direct silicon bonded (110)/(100) substrates [89], the pores in highly p^+-Si, used for the fabrication of epitaxial thin-film solar cells [90] and the carbon clusters (by ion implantation) [91]. In the latter case, it was concluded that most metals are moderately bound to C/I clusters, while a strong binding was derived to SiC precipitates.

While DFT calculations are mainly used to provide fundamental insight in the properties of TM in silicon and germanium, it can also be incorporated in a more global scheme, where the ab initio approach is combined with continuum modeling and device simulations [92, 93, 94]. In this way, the impact of certain critical process steps on the final device performance can be assessed. This has been done for the case of PDG in silicon solar cells [92, 93, 94].The role of DFT in this is to identify the most stable gettering sites, their formation energy and concentration and how stable the binding of TM to these sites is [92]. As already mentioned in Chap. 8, for PDG it has been shown that P_4V is the dominant cluster, whereby the TM is attracted by the V site and becomes trapped there, with a formation energy between 3 and 4 eV, which is the highest for the considered P_nV clusters ($n = 0$ to 4) [92]. At the same time, no gap states have been derived for the M–P_4V center, indicating that the metal atoms (Cu, Cr, Ni, Fe, Ti, W and Mo) become inactive [92]. A binding energy of 1.52 eV has been calculated for the case of Fe_i [92]. The observed strong and non-linear Fe gettering by PDG near the surface of the highly P-doped layer supports P_4V as the dominant gettering site. This results in a super-linear relationship between the phosphorus concentration and the metal profile in the near surface region [92].

Once the key defect parameters have been defined they can be used as input for the continuum modeling of the trapping kinetics of TM by the gettering sites [94], considering both diffusion and clustering of the doping and the metal impurities. These process simulations, relying on realistic thermal budgets yield concentration profiles on all metal species (single and clustered/trapped atoms) in the device structures, which can then be implemented in the device performance simulation, i.e., to predict the dark current of a solar cell [94]. This requires good knowledge about the electrical parameters of the different species, which are either available from experiments (Chaps. 5 and 6) or from DFT calculations.

References

1. G.W. Ludwig, H.H. Woodbury, Electron spin resonance in semiconductors. Solid State Phys. **13**, 223–304 (1962). https://doi.org/10.1016/S0081-1947(08)60458-0
2. G.D. Watkins, P.M. Williams, Vacancy model for substitutional Ni^-, Pd^-, Pt^-, and Au^0 in silicon. Phys. Rev. B **52**, 16575–16580 (1995). https://doi.org/10.1103/PhysRevB.52.16575
3. S.K. Estreicher, First-principles theory of copper in silicon. Mater. Sci. Semicond. Process. **7**, 101–111 (2004). https://doi.org/10.1016/j.mssp.2004.06.004
4. L.A. Hemstreet, Electronic states of simple-transition-metal impurities in silicon. Phys. Rev. B **15**, 834–839 (1977). https://doi.org/10.1103/PhysRevB.15.834
5. G.G. DeLeo, G.D. Watkins, W.B. Fowler, Theory of interstitial transition-metal impurities in silicon. Phys. Rev. B **23**, 1851–1858 (1981). https://doi.org/10.1103/PhysRevB.23:1851
6. G.G. DeLeo, G.D. Watkins, W.B. Fowler, Many-electron effects for interstitial transition-metal impurities in silicon. Phys. Rev. B **25**, 4962–4971 (1982). https://doi.org/10.1103/PhysRevB.25.4962
7. G.G. DeLeo, G.D. Watkins, W.B. Fowler, Level positions of interstitial transition-metal impurities in silicon. Phys. Rev. B **25**, 4972–4980 (1982). https://doi.org/10.1103/PhysRevB.25:4972
8. A. Zunger, U. Lindefelt, Theory of substitutional and interstitial 3*d* impurities in silicon. Phys. Rev. B **26**, 5989–5992 (1982). https://doi.org/10.1103/PhysRevB.26.5989
9. A. Zunger, U. Lindefelt, Electronic structure of transition-atom impurities in semiconductors: substitutional 3*d* impurities in silicon. Phys. Rev. B **27**, 1191–1227 (1983). https://doi.org/10.1103/PhysRevB.27.1191
10. H. Katayama-Yoshida, A. Zunger, Chemical trends in ground- and excited-state properties of interstitial 3*d* impurities in silicon. Phys. Rev. B **31**, 8317–8320 (1985). https://doi.org/10.1103/PhysRevB.31.8317
11. F. Beeler, O.K. Andersen, M. Scheffler, Theoretical evidence for low-spin ground states of early interstitial and late substitutional 3*d* transition-metal ions in silicon. Phys. Rev. Lett. **55**, 1498–1501 (1985). https://doi.org/10.1103/PhysRevLett.55.1498
12. F. Beeler, O.K. Andersen, M. Scheffler, Electronic and magnetic structure of 3*d*-transition-metal point defects in silicon calculated from first principles. Phys. Rev. B **41**, 1603–1624 (1990). https://doi.org/10.1103/PhysRevB.41.1603
13. L.V.C. Assali, J.R. Leite, A. Fazzio, Theoretical model of the Au-Fe complex in silicon. Phys. Rev. B **32**, 8085–8091 (1985). https://doi.org/10.1103/PhysRevB.32.8085
14. U. Lindefelt, Symmetric lattice distortions around deep-level impurities in semiconductors: vacancy and substitutional Cu in silicon. Phys. Rev. B **28**, 4510–4518 (1983). https://doi.org/10.1103/PhysRevB.28.4510
15. S.K. Estreicher, Copper, lithium, and hydrogen passivation of boron in c-Si. Phys. Rev. B **41**, 5447–5450 (1990). https://doi.org/10.1103/PhysRevB.41.5447
16. S.K. Estreicher, M. Sanati, D. West, F. Ruymgaart, Thermodynamics of impurities in semiconductors. Phys. Rev. B **70**, 125209 (2004). https://doi.org/10.1103/physrevb.70.125209
17. K. Sueoka, E. Kamiyama, P. Śpiewac, J. Vanhellemont, Review—properties of intrinsic point defects in Si and Ge assessed by density functional theory. ECS J. Solid State Sci. Technol. **5**, P3176–P3195 (2016). https://doi.org/10.1149/2.0251604jss
18. A.G. Marinopoulos, P. Santos, J. Coutinho, DFT + *U* study of electrical levels and migration barriers of early 3d and 4d transition metals in silicon. Phys. Rev. B **92**, 075124 (2015). https://doi.org/10.1103/physrevb.92.075124
19. Z.Z. Zhang, B. Partoens, K. Chang, F.M. Peeters, First-principles study of transition metal impurities in Si. Phys. Rev. B **77**, 155201 (2008). https://doi.org/10.1103/physrevb.77.155201

20. A.J.R. da Silva, A. Fazzio, A. Antonelli, Stabilization of substitutional Mn in silicon-based semiconductors. Phys. Rev. B **70**, 193205 (2004). https://doi.org/10.1103/physrevb.70.193205
21. C.D. Latham, M. Alatalo, R.M. Nieminen, R. Jones, S. Öberg, P.R. Briddon, Passivation of copper in silicon by hydrogen. Phys. Rev. B **72**, 235205 (2005). https://doi.org/10.1103/physrevb.72.235205
22. S. Shirasawa, K. Sueoka, T. Yamaguchi, K. Maekawa, Density functional theory calculations for estimation of gettering sites of C, H, intrinsic point defects and related complexes in Si wafers. Mater. Sci. Semicond. Process. **44**, 13–17 (2016). https://doi.org/10.1016/j.mssp.2016.01.001
23. A. Resende, R. Jones, S. Öberg, P.R. Briddon, Calculations of electrical levels of deep centers: application to Au-H and Ag-H defects in silicon. Phys. Rev. Lett. **82**, 2111–2114 (1999). https://doi.org/10.1103/PhysRevLett.82.2111
24. J.P. Goss, M.J. Shaw, P.R. Briddon, Marker-method calculations for electrical levels using Gaussian-orbital basis sets. Topics Appl. Phys. **104**, 69–93 (2007). https://doi.org/10.1007/11690320_4
25. E. Artacho, D. Sánchez-Portal, P. Ordejón, A. García, J.M. Soler, Linear-scaling ab-initio calculations for large and complex systems. Phys. Status Solidi B **215**, 809–817 (1999). https://doi.org/10.1002/(sici)1521-3951(199909)215:1<809::aid-pssb809>3.0.co;2-0
26. G. Kresse, J. Furthmuller, Efficiency of ab-initio total energy calculations for metals and semiconductors using a plane-wave basis set. Comput. Mater. Sci. **6**, 15–50 (1996). https://doi.org/10.1016/0927-0256(96)00008-0
27. D.J. Backlund, S.K. Estreicher, Ti, Fe, and Ni in Si and their interactions with the vacancy and the A-center: a theoretical study. Phys. Rev. B **81**, 235213 (2010). https://doi.org/10.1103/physrevb.81.235213
28. T. Yamato, K. Sueoka, T. Maeta, First-principles analysis on interaction between dopant (Ga, Sb) and contamination metal atoms in Ge crystals. Solid State Phenom. **205–206**, 417–421 (2014). www.scientific.net/SSP.205-206.417
29. H. Weng, J. Dong, First-principles investigation of transition-metal-doped group-IV semiconductors: R_xY_{1-x} (R = Cr, Mn, Fe; Y = Si, Ge). Phys. Rev. B **71**, 035201 (2005). https://doi.org/10.1103/physrevb.71.035201
30. A. Continenza, G. Profeta, S. Picozzi, Transition metal impurities in Ge: chemical trends and codoping studied by electronic structure calculations. Phys. Rev. B **73**, 035212 (2006). https://doi.org/10.1103/physrevb.73.035212
31. A. Carvalho, J. Coutinho, R. Jones, E. Silva, S. Öberg, P.R. Briddon, Density-functional theory study of Au, Ag and Cu defects in germanium. Mater. Sci. Semicond. Process. **11**, 340–343 (2008). https://doi.org/10.1016/j.mssp.2008.10.007
32. V.P. Markevich, S. Leonard, A.R. Peaker, B. Hamilton, A.G. Marinopoulos, J. Coutinho, Titanium in silicon: lattice positions and electronic properties. Appl. Phys. Lett. **104**, 152105 (2014). https://doi.org/10.1063/1.4871702
33. E. Wright, J. Coutinho, S. Öberg, V.J.B. Torres, Mössbauer parameters of Fe-related defects in group-IV semiconductors: first principles calculations. J. Appl. Phys. **119**, 181509 (2016). https://doi.org/10.1063/1.4948243
34. S.K. Estreicher, M. Sanati, N. Gonzalez Szwacki, Iron in silicon: interactions with radiation defects, carbon, and oxygen. Phys. Rev. B **77**, 125214 (2008). https://doi.org/10.1103/physrevb.77.125214
35. Y. Abreu, C.M. Cruz, I. Piñera, A. Leyva, A.E. Cabal, P. Van Espen, N. Van Remortel, Hyperfine electric parameters calculation in Si samples implanted with $^{57}Mn \rightarrow {}^{57}Fe$. Physica B **445**, 1–4 (2014). https://doi.org/10.1016/j.physb.2014.03.028
36. A. Carvalho, J. Coutinho, R. Jones, M. Barroso, J.P. Goss, P.R. Briddon, Density-functional theory study of interstitial iron and its complexes with B and Al in dilute SiGe alloys. Mater. Sci. Semicond. Process. **11**, 332–335 (2008). https://doi.org/10.1016/j.mssp.2008.07.010

37. F.J.H. Ehlers, A.P. Horsfield, D.R. Bowler, Electronic state of interstitial Cu in bulk Si: density functional calculations. Phys. Rev. B **73**, 165207 (2006). https://doi.org/10.1103/physrevb.73.165207
38. A. Carvalho, D.J. Backlund, S.K. Estreicher, Four-copper complexes in Si and the Cu-photoluminescence defect: a first-principles study. Phys. Rev. B **84**(155322), 1–8 (2011). https://doi.org/10.1103/PhysRevB.84.155322
39. S.K. Estreicher, Rich chemistry of copper in crystalline silicon. Phys. Rev. B **60**, 5375–5382 (1999). https://doi.org/10.1103/PhysRevB.60.5375
40. S.K. Estreicher, Copper-related defects in silicon. Physica B **273–274**, 424–428 (1999). https://doi.org/10.1016/S0921-4526(99)00496-2
41. S.K. Estreicher, J.L. Hastings, Cu-related complexes in silicon. Mater. Sci. Eng. B **58**, 155–158 (1999). https://doi.org/10.1016/S0921-5107(98)00289-X
42. T.M. Gibbons, D.J. Backlund, S.K. Estreicher, Cobalt-related defects in silicon. J. Appl. Phys. **121**(045704), 1–6 (2017). https://doi.org/10.1063/1.4975034
43. D.J. Backlund, T.M. Gibbons, S.K. Estreicher, Vanadium interactions in crystalline silicon. Phys. Rev. B **94**, 195210 (2016). https://doi.org/10.1103/physrevb.94.195210
44. B.R. Sahu, S.K. Banerjee, L. Kleinman, Density-functional study of bulk silicon lightly doped with manganese. Phys. Rev. B **77**, 155202 (2008). https://doi.org/10.1103/physrevb.77.155202
45. F. Küwen, R. Leitsmann, F. Bechstedt, Mn and Fe doping of bulk Si: concentration influence on electronic and magnetic properties. Phys. Rev. B **80**, 045203 (2009). https://doi.org/10.1103/physrevb.80.045203
46. A. Stroppa, S. Picozzi, A. Continenza, A.J. Freeman, Electronic structure and ferromagnetism of Mn-doped group-IV semiconductors. Phys. Rev. B **68**, 155203 (2003). https://doi.org/10.1103/physrevb.68.155203
47. J. Utzig, An elastic energy approach to the interstitial diffusion of 3*d* elements in silicon. J. Appl. Phys. **65**, 3868–3871 (1989). https://doi.org/10.1063/1.343349
48. A. Chroneos, Y. Panayiotatos, R.V. Vovk, Copper diffusion in germanium: connecting point defect parameters with bulk properties. J. Mater. Sci.: Mater. Electron. **26**, 2693–2696 (2015). https://doi.org/10.1007/s10854-015-2744-6
49. A. Chroneos, R.V. Vovk, Palladium diffusion in germanium. J. Mater. Sci.: Mater. Electron. **26**, 3787–3789 (2015). https://doi.org/10.1007/s10854-015-2903-9
50. Y. Panayiotatos, R.V. Vovk, A. Chroneos, Gold and silver diffusion in germanium: a thermodynamic approach. J. Mater. Sci.: Mater. Electron. **28**, 1966–1970 (2017). https://doi.org/10.1007/s10854-016-5750-4
51. D.E. Woon, D.S. Marynick, S.K. Estreicher, Titanium and copper in silicon: barriers for diffusion and interactions with hydrogen. Phys. Rev. B **45**, 13383–13389 (1992). https://doi.org/10.1103/PhysRevB.45.13383
52. J. Lindroos, D.P. Fenning, D.J. Backlund, E. Verlage, A. Gorgulla, S.K. Estreicher, H. Savin, T. Buonassisi, Nickel: a very fast diffuser in silicon. J. Appl. Phys. **113**, 204906 (2013). https://doi.org/10.1063/1.4807799
53. Y. Kamon, H. Harima, A. Yanase, H. Katayama-Yoshida, Ultra-fast diffusion mechanism of the late 3*d* transition metal impurities in silicon. Physica B **308–310**, 391–395 (2001). https://doi.org/10.1016/S0921-4526(01)00754-2
54. K. Shirai, T. Michikita, H. Katayama-Yoshida, Molecular dynamics study of fast diffusion of Cu in silicon. Jpn. J. Appl. Phys. **44**, 7760–7764 (2005). https://doi.org/10.1143/JJAP.44.7760
55. K. Matsukawa, K. Shirai, H. Yamaguchi, H. Katayama-Yoshida, Diffusion of transition-metal impurities in silicon. Physica B **401–402**, 151–154 (2007). https://doi.org/10.1016/j.physb.2007.08.134
56. K. Saga, R. Ohno, D. Shibata, S. Kobayashi, K. Sueoka, Behavior of transition metals penetrating silicon substrate through SiO_2 and Si_3N_4 films by arsenic ion implantation and annealing. ECS J. Solid State Sci. Technol. **4**, P131–P136 (2015). https://doi.org/10.1149/2.0061505jss

57. K. Saga, Metallic contamination control in advanced ULSI processing. ECS Trans. **69**(8), 57–68 (2015). https://doi.org/10.1149/06908.0057ecst
58. S. Zhao, L.V.C. Assali, J.F. Justo, G.H. Gilmer, L.C. Kimerling, Iron-acceptor pairs in silicon: structure and formation processes. J. Appl. Phys. **90**, 2744–2754 (2001). https://doi.org/10.1063/1.1389763
59. L.V.C. Assali, J.R. Leite, Electronic properties of the iron-boron impurity pair in silicon. Phys. Rev. B **36**, 1296–1299 (1987). https://doi.org/10.1103/PhysRevB.36.1296
60. M. Sanati, N. Gonzalez Szwacki, S.K. Estreicher, Interstitial Fe in Si and its interactions with hydrogen and shallow dopants. Phys. Rev. B **76**, 125204 (2007). https://doi.org/10.1103/physrevb.76.125204
61. K. Sueoka, S. Ohara, S. Shiba, S. Fukutani, First principles calculation for Cu gettering by dopant or dopant-vacancy complex in silicon crystal. ECS Trans. **2**(2), 261–273 (2006). https://doi.org/10.1149/1.2195664
62. K. Sueoka, Application of DFT calculation for the development of high quality Si and Ge substrates: from ultra large diameter crystal pulling to metal gettering. ECS Trans. **75**(4), 111–127 (2016). https://doi.org/10.1149/07504.0111ecst
63. S. Shirasawa, K. Sueoka, T. Yamaguchi, K. Maekawa, Useful database of effective gettering sites for metal impurities in Si wafers with first principles calculation. ECS J. Solid State Sci. Technol. **4**, P351–P355 (2015). https://doi.org/10.1149/2.0051509jss
64. A. Yamada, K. Sueoka, Density functional theory study on formation energy and diffusion path of metal atom near dopant in Si crystals. ECS J. Solid State Sci. Technol. **6**, P125–P131 (2017). https://doi.org/10.1149/2.0131704jss
65. D. West, S.K. Estreicher, S. Knack, J. Weber, Copper interactions with H, O, and the self-interstitial. Phys. Rev. B **68**, 035210 (2003). https://doi.org/10.1103/physrevb.68.035210
66. A. Rohatgi, R.H. Hopkins, J.R. Davis, R.B. Campbell, The impact of molybdenum on silicon and silicon solar-cell performance. Solid-State Electron. **23**, 1185–1190 (1980). https://doi.org/10.1016/0038-1101(80)90032-5
67. N. Gonzalez Szwacki, S.K. Estreicher, First-principles investigations of Fe-H interactions in silicon. Physica B **401–402**, 171–174 (2007). https://doi.org/10.1016/j.physb.2007.08.139
68. N. Gonzalez Szwacki, M. Sanati, S.K. Estreicher, Two FeH pairs in n-type Si and their implications: a theoretical study. Phys. Rev. B **78**, 113202 (2008). https://doi.org/10.1103/physrevb.78.113202
69. R. Jones, S. Öberg, J. Goss, P.R. Briddon, A. Resende, Theory of nickel and nickel-hydrogen complexes in silicon. Phys. Rev. Lett. **75**, 2734–2737 (1995). https://doi.org/10.1103/PhysRevLett.75.2734
70. D.J. Backlund, S.K. Estreicher, Structural, electrical, and vibrational properties of Ti–H and Ni–H complexes in Si. Phys. Rev. B **82**, 155208 (2010). https://doi.org/10.1103/physrevb.82.155208
71. P. Santos, J. Coutinho, V.J.B. Torres, M.J. Rayson, P.R. Briddon, Hydrogen passivation of titanium impurities in silicon: effect of doping conditions. Appl. Phys. Lett. **105**, 032108 (2014). https://doi.org/10.1063/1.4891575
72. A. Resende, R. Jones, S. Öberg, P.R. Briddon, The structural properties of transition metal hydrogen complexes in silicon. Mater. Sci. Eng. B **58**, 146–148 (1999). https://doi.org/10.1016/S0921-5107(98)00285-2
73. R. Jones, A. Resende, S. Öberg, P.R. Briddon, The electronic properties of transition metal hydrogen complexes in silicon. Mater. Sci. Eng. B **58**, 113–117 (1999). https://doi.org/10.1016/S0921-5107(98)00293-1
74. S.K. Estreicher, D. West, J. Goss, S. Knack, J. Weber, First-principles calculations of pseudolocal vibrational modes: the case of Cu_s and Cu^i pairs in Si. Phys. Rev. Lett **90**, 035504 (2003). https://doi.org/10.1103/physrevlett.90.035504
75. S.K. Estreicher, D. West, M. Sanati, *Cu0: a metastable configuration of the Cus, Cui pair in Si. Phys. Rev. B **72**, 121201R (2005). https://doi.org/10.1103/physrevb.72.121201r

76. F. Bernardini, S. Picozzi, A. Continenza, Energetic stability and magnetic properties of Mn dimers in silicon. Appl. Phys. Lett. **84**, 2289–2291 (2004). https://doi.org/10.1063/1.1688002
77. Y. Miura, M. Shirai, K. Nagao, First-principles design of ferromagnetic nanostructures based on group-IV semiconductors. J. Phys.: Condens. Matter **16**, S5735–S5738 (2004). https://doi.org/10.1088/0953-8984/16/48/038
78. K. Sánchez, I. Aguilera, P. Palacios, P. Wahnón, Assessment through first-principles calculations of an intermediate-band photovoltaic material based on Ti-implanted silicon: interstitial versus substitutional origin. Phys. Rev. B **79**, 165203 (2009). https://doi.org/10.1103/physrevb.79.165203
79. M.G. Wardle, J.P. Goss, P.R. Briddon, R. Jones, Structural and electronic properties of thin fluorite-structure $NiSi_2$, $CoSi_2$ and $FeSi_2$ interfaces and precipitates in Si. Phys. Status Solidi A **202**, 883–888 (2005). https://doi.org/10.1002/pssa.200460508
80. P.J. van den Hoek, W. Ravenek, E.J. Baerends, MSi_2/Si(111) (M = Co, Ni) interface chemical bond. Phys. Rev. Lett. **60**, 1743–1746 (1988). https://doi.org/10.1103/PhysRevLett.60.1743
81. V.A. Borodin, M.-O. Ruault, M.G. Ganchenkova, F. Fortuna, Microscopic mechanisms of cobalt disilicide nucleation in silicon. Solid State Phenom. **108–109**, 133–138 (2005). https://doi.org/10.4028/www.scientific.net/SSP.108-109.133
82. V.A. Borodin, M.G. Ganchenkova, M.-O. Ruault, F. Fortuna, Sensitivity of $CoSi_2$ precipitation in silicon to extra-low dopant concentrations II. First-principles calculations. J. Appl. Phys. **117**, 045704 (2015). https://doi.org/10.1063/1.4906403
83. N. Fujita, R. Jones, S. Öberg, P.R. Briddon, A.T. Blumenau, A theoretical study of copper contaminated dislocations in silicon. Solid State Phenom. **131–133**, 259–264 (2008). https://doi.org/10.4028/www.scientific.net/SSP.131-133.259
84. A. Suvitha, N.S. Venkataramanan, R. Sahara, H. Mizuseki, Y. Kawazoe, First-principles calculations on Σ3 grain boundary transition metal impurities in multicrystalline silicon. Jpn. J. Appl. Phys. **49**, 04DP02 (2010). https://doi.org/10.1143/jjap.49.04dp02
85. Y. Ohno, K. Inoue, K. Kutsukake, M. Deura, T. Ohsawa, I. Yonenaga, H. Yoshida, S. Takeda, R. Taniguchi, H. Otubo, S.R. Nishitani, N. Ebisawa, Y. Shimizu, H. Takamizawa, K. Inoue, Y. Nagai, Nanoscopic mechanism of Cu precipitation at small-angle tilt boundaries in Si. Phys. Rev. B **91**, 235315 (2015). https://doi.org/10.1103/physrevb.91.235315
86. B. Ziebarth, M. Mrovec, C. Elsässer, P. Gumbsch, Potential-induced degradation in solar cells: electronic structure and diffusion mechanism of sodium in stacking faults of silicon. J. Appl. Phys. **116**, 093510 (2014). https://doi.org/10.1063/1.4894007
87. T. Michikita, K. Shirai, H. Katayama-Yoshida, Materials design for Cu gettering by electronic dopants in silicon. Jpn. J. Appl. Phys. **44**, 7904–7906 (2005). https://doi.org/10.1143/JJAP.44.7904
88. E. Kamiyama, K. Sueoka, J. Vanhellemont, A statistical model for the gettering of impurities on an atomistic scale. Phys. Status Solidi C **13**, 746–749 (2016). https://doi.org/10.1002/pssc.201600031
89. H. Kariyazaki, T. Aoki, K. Izunome, K. Sueoka, Molecular simulation on interfacial structure and gettering efficiency of direct silicon bonded (110)/(100) substrates. J. Appl. Phys. **107**, 113509 (2010). https://doi.org/10.10163/1.3407525
90. H.S. Radhakrishnan, C. Ahn, J. Van Hoeymissen, F. Dross, N. Cowern, K. Van Nieuwenhuysen, I. Gordon, R. Mertens, J. Poortmans, Phys. Status Solidi A **209**, 1866–1871 (2012). https://doi.org/10.1002/pssa.201200232
91. Y. Jin, S.T. Dunham, Modeling of carbon clustering and associated metal gettering. ECS Trans **64**(11), 211–218 (2014). https://doi.org/10.1149/06411.0211ecst
92. R. Chen, B. Trzynadlowski, S.T. Dunham, Phosphorus vacancy cluster model for phosphorus diffusion gettering of metals in Si. J. Appl. Phys. **115**, 054906 (2014). https://doi.org/10.1063/1.4864377

93. B. Trzynadlozyki, A. Yazdani, R. Chen, S.T. Dunham, Coupled modeling of evolution of impurity/defect distribution and cell performance. Proc. IEEE Photovolt. Special Conf. 217–220 (2011). https://doi.org/10.1109/pvsc.2012.6317604
94. A. Yazdani, R. Chen, S.T. Dunham, Coupled modeling of the competitive gettering of transition metals and impact on performance of lifetime sensitive devices. J. Appl. Phys. **121**, 095702 (2017). https://doi.org/10.1063/1.4976525
95. D. Mathiot, S. Hocine, Titanium-related deep levels in silicon: a reexamination. J. Appl. Phys. **66**, 5862–5867 (1989). https://doi.org/10.1063/1.343608
96. T. Sadoh, H. Nakashima, T. Tsurushima, Deep levels of vanadium and vanadium-hydrogen complex in silicon. J. Appl. Phys. **72**, 520–524 (1992). https://doi.org/10.1063/1.352353
97. H. Conzelmann, K. Graff, E.R. Weber, Chromium and chromium-boron pairs in silicon. Appl. Phys. A **30**, 169–175 (1983). https://doi.org/10.1007/BF00620536
98. T. Kunio, T. Yamazaki, E. Ohta, M. Sakata, defect levels in chromium-doped silicon. Solid-State Electron. **26**, 155–160 (1983). https://doi.org/10.1016/0038-1101(83)90117-X
99. H. Lemke, Properties of silicon-crystals doped with zirconium or hafnium. Phys. Status Solidi A **122**, 617–630 (1990). https://doi.org/10.1002/pssa.2211220223
100. M.L. Polignano, D. Codegoni, G. Borionetti, F. Bonoli, J. Brivio, S. Greco, A. Marino, P. Monge, I. Patoprsta, V. Privitera, C. Riva, Niobium contamination in silicon. ECS Trans. **33**(11), 133–144 (2010). https://doi.org/10.1149/1.3485688
101. H. Bracht, Copper related diffusion phenomena in germanium and silicon. Mater. Sci. Semicond. Process. **7**, 113–124 (2004). https://doi.org/10.1016/j.mssp.2004.06.001

Index

C. Claeys and E. Simoen, *Metal Impurities in Silicon- and Germanium-Based Technologies*, Springer Series in Materials Science 270,
https://doi.org/10.1007/978-3-319-93925-4

N

O

P

Q

R

S

T

The manufacturer's authorised representative in the EU is Springer Nature Customer Service Centre GmbH, Europaplatz 3, 69115 Heidelberg, Germany. If you have any concerns regarding our products, please contact ProductSafety@springernature.com

Printed and bound by CPI Group (UK) Ltd, Croydon, CR0 4YY
05/10/2026
02228138-0018